W9-CRB-642

ELECTRONIC PROPERTIES OF NOVEL MATERIALS— MOLECULAR NANOSTRUCTURES

Previous Proceedings in the Series of International Kirchberg Winterschools

	Year	Held in	Publisher	ISBN
XIII	1999	Kirchberg, Austria	AIP Conf. Proceedings vol. 486	1-56396-900-9
XII	1998	Kirchberg, Austria	AIP Conf. Proceedings vol. 442	1-56396-808-8
XI	1997	Kirchberg, Austria	World Scientific Publishers	981-02-3261-6
X	1996	Kirchberg, Austria	World Scientific Publishers	981-02-2853-8

Other Related Titles from AIP Conference Proceedings

535 Fundamental Physics of Ferroelectrics 2000: Aspen Center for Physics Winter Workshop
Edited by Ronald E. Cohen, September 2000, 1-56396-959-9

527 Lectures on the Physics of Highly Correlated Electron Systems IV: Fourth Training Course in the Physics of Correlated Electron Systems and High-Tc Superconductors
Edited by Ferdinando Mancini, July 2000, 1-56396-950-5

507 X-Ray Microscopy: Proceedings of the Sixth International Conference
Edited by Werner Meyer-Ilse, Tony Warwick, and David Attwood, March 2000, 1-56396-926-2

483 High Temperature Superconductivity
Edited by Stewart E. Barnes, Joseph Ashkenazi, Joshua L. Cohn, and Fulin Zuo, August 1999, 1-56396-880-0

477 Atomic Physics 16: Sixteenth International Conference on Atomic Physics
Edited by William E. Baylis and Gordon W. F. Drake, May 1999, 1-56396-752-9

466 Quantum 1/f Noise and Other Low Frequency Fluctuations in Electronic Devices: Seventh Symposium
Edited by Peter H. Handel and Alma L. Chung, March 1999, 1-56396-854-1

To learn more about these titles, or the AIP Conference Proceedings Series, please visit the webpage **http://www.aip.org/catalog/aboutconf.html**

ELECTRONIC PROPERTIES OF NOVEL MATERIALS— MOLECULAR NANOSTRUCTURES

XIV International Winterschool/Euroconference

Kirchberg, Tirol, Austria   4–11 March 2000

EDITORS

Hans Kuzmany
Universität Wien, Austria

Jörg Fink
*Institut für Festkörper- und Werkstoff-Forschung
Dresden, Germany*

Michael Mehring
Universität Stuttgart, Germany

Siegmar Roth
*Max-Planck Institut für Festkörperforschung
Stuttgart, Germany*

Melville, New York, 2000
AIP CONFERENCE PROCEEDINGS ■ VOLUME 544

Editors:

Hans Kuzmany
Institut für Materialphysik
Universität Wien
Strudlhofgasse 4
A-1090 Wien
AUSTRIA

E-mail: kuzman@ap.univie.ac.at

Jörg Fink
Institut für Festkörper- und Werkstoff-Forschung
Postfach 270016
D-01171 Dresden
GERMANY

E-mail: j.fink@ifw-dresden.de

Michael Mehring
2. Physikalisches Institut
Universität Stuttgart
Pfaffenwaldring 57
D-70550 Stuttgart
GERMANY

E-mail: sk2@physik.uni-stuttgart.de

Siegmar Roth
Max-Planck-Institut für Festkörperforschung
Heisenbergstr. 1
D-70569 Stuttgart
GERMANY

E-mail: roth@klizix.mpi-stuttgart.mpg.de

L.C. Catalog Card No. 00-109460
ISBN 1-56396-973-4
ISSN 0094-243X
Printed in the United States of America

CONTENTS

I. INTRODUCTION

II. FULLERENES AND FULLERIDES

VII. OTHER MOLECULAR MATERIALS

VIII. APPLICATIONS

PREFACE

The present book contains the proceedings of the 14th International Winterschool on Electronic Properties of Novel Materials in Kirchberg, Tirol, Austria. It was held from 4th March to 11th March 2000 in Hotel Sonnalp. The series of these schools started in 1985. Originally the school was held every second year and was devoted to conducting polymers. After the discovery of high temperature superconductors, the periodicity changed to annual format and the topic alternated between conjugated polymers and superconductors. Since fullerenes are both conjugated compounds and in some cases superconductors, it was tempting to choose fullerenes as topic for the Kirchberg schools. The evident extension of this topic is carbon nanotubes and so the title changed from Fullerenes via Fullerene Derivatives and Fullerene Nanostructures to Molecular Nanostructures. This gradual change enables us to keep a fairly large interdisciplinary scientific community together and to stimulate numerous international cooperations. A compilation of the previous Kirchberg Winterschools will be presented in the table at the end of this preface.

The term Molecular Nanostructures implies the "bottom-up" (synthetic) approach, as opposed to the "top-down" (lithography and etching) techniques in semiconductor technology. As for the physics, we are in a field where solid state physics and molecular physics overlap. This is nicely seen on the example of carbon nanotubes. Their diameter is in the order of a few nanometers, and thus perpendicular to their axis, nanotubes are molecular (different diameters lead to different electronic structures, while along their axis they are extended solids.

The materials treated at the 14th Winterschool are mainly fullerenes, fullerides and carbon nanotubes, but there are also contributions on conducting polymers, on molecular clusters, and on nucleic acids. The topics reach from synthesis and sample characterization over chemical, mechanical, and electronic properties to industrial applications. There was a special discussion session on hydrogen storage on one of the evenings, and a report on that is given at the end of the proceedings.

The meeting could not have taken place without the support of the Bundesministerium für Wissenschaft und Forschung in Wien, the Verein zur Förderung der Winterschulen in Kirchberg, and the Commission of the European Communities Directorate General for Research, as well as from numerous industrial sponsors. Without their contribution, all the enthusiasm and dedication could be wasted and so we express our gratitude to the sponsors and supporters.

Finally, we are indebted, as ever, to the managers of the Hotel Sonnalp, Herrn Gradnitzer and Frau Jurgeit, and to their staff for their continuous support and for their patience with the many special arrangements required during the meeting.

The present volume is a Festschrift devoted to the 60th birthday of Hans Kuzmany. Hans is the spiritus rector of the Kirchberg Winterschools. He had the idea to start the series of these schools, and it is his effort and energy which kept it running. With this Festschrift we, the co-organizers of the Winterschools, want to express our appreciation and our gratitude to Hans.

J. Fink, K. Mehring, S. Roth

ORGANIZER
Institut für Materialphysik
Universität Wien

PATRONAGE
Elisabeth Gehrer
Bundesminister für Wissenschaft und Verkehr

Magnifizenz
Univ. Prof. Dr. Georg Winckler
Rektor der Universität Wien

Herbert Noichel
Bürgermeister of Kirchberg

SUPPORTERS
Bundesministerium für Wissenschaft und Forschung, AT
Verein zur Förderung der Internationalen Winterschulen in Kirchberg, AT
The Commiccion of the European Communities Directorate General for Research

SPONSORS
ALDRICH, P.O. Box 355, Milwaukee, WI 53201, USA
ATOS GmbH, Robert-Bosch-Straße, D-64319 Pfungstadt, Germany
AVENTIS Research & Technologies, Industriel Park, Building G830,
D-65926 Frankfurt am Main 80, Germany
AVL LIST GmbH, Hans List Platz 1, A-8020 Graz, Austria
BRUKER Analytische Meßtechnik GmbH, Wikingerstraße 13,
D-76189 Karlsruhe, Germany
CREDITANSTALT BANKVEREIN, Nußdorferstraße 2, A-1090 Wien, Austria

The financial assistance from the sponsors and from the supporters is greatly acknowledged

Table of Previous Kirchberg Winterschools

Year	Titel	Published by
1985	Electronic Properties of Polymers and Related Compounds	Springer Series in Solid-State Sciences 63
1987	Electronic Properties of Conjugated Polymers	Springer Series in Solid-State Sciences 76
1989	Electronic Properties of Conjugated Polymers III - Basic Models and Applications	Springer Series in Solid-State Sciences 91
1990	Electronic Properties of High-T_c Superconductors and Related Compounds	Springer Series in Solid-State Sciences 99
1991	Electronic Properties of Polymers - Orientation and Dimensionality of Conjugated Systems	Springer Series in Solid-State Sciences 107
1992	Electronic Properties of High-T_c Superconductors	Springer Series in Solid-State Sciences 113
1993	Electronic Properties of Fullerenes	Springer Series in Solid-State Sciences 117
1994	Progress in Fullerene Research	World Scientific Publ. 1994
1995	Physics and Chemistry of Fullerenes and Derivatives	World Scientific Publ. 1995
1996	Fullerenes and Fullerene Nanostructures	World Scientific Publ. 1996
1997	Molecular Nanostructures	World Scientific Publ. 1998
1998	Electronic Properties of Novel Materials – Progress in Molecular Nanostructures	AIP Conference Proceedings 442 (1998)
1999	Electronic Properties of Novel Materials – Science and Technology of Molecular Nanostructures	AIP Conference Proceedings 486 (1999)

I. INTRODUCTION

Fullerenes and Nanotubes - An Introduction

Wolfgang Krätschmer

Max Planck Institut für Kernphysik, D-69251 Heidelberg, P.O.Box 103980, Germany

In 1991 Sumio Iijima examined the soot from a fullerene generator by his transmission electron microscope (TEM). This did not happen by accident since Iijima's interest in fullerenes reached many years back. As an experienced electron microscopist he had already in the early 1980s carefully studied the thin carbon films by which TEM sample holders are usually covered, discovering spots in the films in which the carbon was arranged in concentric, onion-like structures. The innermost shell of such onions had a diameter of about 0.8 nm. When in 1985 C_{60} was discovered [1], Iijima remembered the onion structures he had seen before. His observation pretty much supported the basic idea of Kroto and co-workers on fullerenes, i.e. closed cage carbon molecules. Thus, in the years between 1985 and 1990 when the fullerene concept was still severely disputed, Iijima provided independent and strong evidence for the existence of such species [2]. In his examination of fullerene soot Iijima was probably disappointed so see nothing exciting - just carbon nanoparticles with irregular and chaotic structures. Fortunately, he also examined parts of the graphite electrodes of the generator and at the graphite rod which served as cathode he found surprisingly regular needle-like structures. A closer look revealed that each needle consisted of several concentrically arranged graphene tubes, regularly separated by about 3.4 nm space, the well known distance between graphitic layers. These structures are nowadays called multi-walled carbon nanotubes (MWCNTs) [3]. Electron diffraction performed by TEM on individual needles revealed that the tubules seem to consist of rolled up graphene layers. Each tubule appeared to be rolled up in a different fashion, i.e. exhibited different helicity (see Fig. 1). The tubes, which could extend in length up to micrometers, were closed at the end by half-spherical or faceted caps and thus seem to resemble extremely elongated, large diameter fullerenes. All in all, the MWCNTs exhibited a rather complex structure. The formation of these nanotubes is still not well understood, but apparently the electric field prevailing between the graphite electrodes during arc burning dictates the growth of the tubes into a uniform direction.

Considering the fact that the carbon-sp^2 bond is stronger than the sp^3 bond (as e.g. in diamond), the individual tubules in MWCNTs should have extraordinary mechanical strength along their axis. Furthermore, early electronic band calculations performed on carbon tubes already indicated interesting properties: depending on their structure, i.e.

CP544, *Electronic Properties of Novel Materials—Molecular Nanostructures*, edited by H. Kuzmany, et al.
© 2000 American Institute of Physics 1-56396-973-4/00/$17.00

diameter and helicity, individual tubes should be metallic or semiconducting [4,5]. The sharp tips of conducting nanotubes may thus be used for field-emission sources e.g. in displays [6] or in scanning tunnel or force microscopes. An efficient production of either multi- or single-walled nanotubes thus seemed to be rewarding. The former task was soon accomplished by Thomas Ebbesen, who like Iijima was then working at the Fundamental Research Laboratories of the NEC Corporation near Tokyo. Ebbesen used an DC arc burning in an atmosphere of 500 torr of Helium between graphite electrodes, and found that the growing cathode deposit consisted mainly of aligned

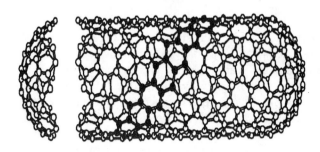

Figure 1. Sketch of the structure of a single carbon nanotube with caps at both ends. The tube has helicity as indicated by the black colored carbon atoms forming a screw-like array of hexagons.

MWCNTs [7]. He also worked out combustion methods to clean the MWCNT samples from carbon nanoparticles which were also produced by the arc-process. Not very much later, the production of single-walled carbon nanotubes (SWCNTs) was accomplished. With the help of catalysts, two research teams reported success at almost at the same time, and thus both papers, that by Iijima and by Donald Bethune and their respective co-workers were published in the same issue of Nature in 1993 [8,9]. Both teams used a fullerene generator but had doped the graphite electrodes with Fe respectively Co (Ref. [8] also added methane to the helium buffer gas). In these early experiments, the yields of SWCNTs were not very high. In a cleaning process tubes had to be freed from particles consisting of catalyst material and of carbon soot. Richard Smalley and co-workers developed an alternative technique. These researchers were adapting the Bethune idea of doping the graphite electrodes with a catalyst but used a laser-furnace rather than a discharge-based fullerene generator. The laser-furnace had been invented 1992 by the Rice group in order to find out how measurable yields of fullerenes could be obtained by the laser-ablation method [10]. The success was remarkable: SWCNTs of almost uniform diameter could be produced in high yield [11]. The tubes were obtained not as individuals but in the form of bundles or ropes, i.e. many individual SWCNTs were tied together. The diameter distribution of the individual tubes was so narrow that the ropes formed a kind of "crystal" which gave rise to regular X-ray diffraction patterns. Originally it was assumed that all these

SWCNTs are monodisperse, i.e. of only one uniform structure, namely of type (10,10) according to the graphene roll-up nomenclature (see, e.g. [5]). Such tubes are also called as of "armchair" type, have no helicity, and should be metallic conductors. More recent data suggest that the SWCNTs produced by the laser-furnace technique are not strictly monodisperse but exhibit a spread in diameters and helicities (see, e.g. [12]). Thus, although the development of production methods for obtaining really monodisperse SWCNTs remains a challenge for the future, it seems that with the present techniques one is already quite close. The production of SWCNTs by the arc process has also progressed. SWCNTs produced by arc are less uniform in structure, though the total amount of tube material obtained is usually much larger.

The development of carbon nanotubes in recent years was so rapid that one almost forgot about the fullerenes and the exciting discoveries which we owe to these species. While in the old days - say before the Nobel-price year 1996 - the nanotubes were regarded as rather elongated fullerenes, nowadays one is inclined to define fullerenes as very short nanotubes! Fortunately, the organisers of this winterschool have kept in mind that both, fullerenes and nanotubes belong intimately together. In fact, both species are usually produced simultaneously – TEM observations of nanotubes filled by fullerenes demonstrate this fact very strikingly (see, e.g. [13]).

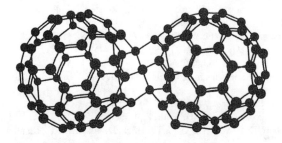

Figure 2. The peanut shaped carbon cluster C_{119} [17] - an intermediate on the way to carbon nanorods?

One may speculate on approaches to produce new nanomaterials which bridge the gap between fullerenes and nanotubes. One possibility: Why not, e.g. by a still to be developed process, cut nanotubes into fullerene-like rods or pieces? Some catalysts may help to achieve this in large scale. The decisive role of catalysts in the production of nanotubes is well known, even though the why and how is not yet understood. A striking example for the effects of catalysts is e.g. given in [14] where the formation of cubic rather than cylindrical carbon structures is reported. Another possible approach starting from the "fullerene direction": Why not use our experience with fullerenes - which can be easily obtained monodispersed (e.g. in case of C_{60} and C_{70}) and which can be subjected to solution-based chemistry - to make something like nanotubes, or at least some kind of rod-like linear structures? The well known linear arrays of C_{60} in Alkali-C_{60} compounds [15] or otherwise polymerised C_{60} (see, e.g. [16]) may already serve as examples. In our work on C_{60} dimers we recently synthesised and completely characterised the peanut shaped carbon cluster C_{119} (see Fig. 2) which consists of two C_{58} fullerene "baskets" connected by three bridging sp^3 carbon atoms [17,18]. The synthesis of even larger monodisperse C_{60} dumbbell-shaped dimers, like e.g. C_{120}, C_{121}, and C_{122} has also been reported [19,20,21]. Are we on the way to make nanotubes-like structures from C_{60}? It almost appears so even though the difficulties in converting C_{60} dimers into short fullerene-tubes probably are formidable. Again, a suitable catalyst may help. Naturally, this all sounds very much like black magic rather than science. As an excuse one may consider that in the field of fullerenes and nanotubes most of the big leaps forward so far came from serendipity discoveries and from trial and error experiments. After all, we are still beginners in a field which in the future may develop into a carbon-based nanoarchitecture.

REFERENCES

1. Kroto, H.W., Heath, J.R., O'Brien, S.C., Smalley, R.E., *Nature* **318**, 162-163 (1985)
2. Iijima, S., *J.Phys. Chem.* **91**, 3466-3467 (1987)
3. Iijima, S., *Nature* **354**, 56-58 (1991)
4. Mintmire, J.W., Dunlap, B.I., White, C.T., *Phys. Rev. Lett.* **68** (5), 631-634 (1992)
5. Dresselhaus, M.S., Dresselhaus, G., Eklund, P.C., „C_{60}-Related Tubules and Spherules" in Science of Fullerenes and Carbon Nanotubes, Academic Press, 756-869 (1996)
6. de Heer, W.A., Chatelain, A., Ugarte, D., *Science* **270**, 1179-1180 (1995)
7. Ebbesen, T. W., *Physics Today* **1996**, 26-32 (1996)
8. Iijima, S., Ichihashi, T., *Nature* **363**, 603-607 (1993)
9. Bethune, D.S., Klang, C.H., deVries m.s., Gorman, G., Savoy R., Vazquez J., Beyers R., *Nature* **363**, 605-607 (1993)
10. Smalley, R.E., *Accounts of Chem. Research* **25**, 98-105 (1992)
11. Thess, A., Lee, R., Nikolaev, P., Dai, H., Petit, P., Robert, J., Xu, Ch., Lee, Y.H., Kim, S.G., Rinzler, A.G., Colbert, D.T., Scuseria, G.E., Tomanek, D., Fischer, J.E., Smalley, R.E., *Science* **273**, 483-487 (1996)
12. Milnera, M., Kürti, J., Hulman, M., Kuzmany, H., *Rev Phys.. Lett.* **84** (6) 1324-1327 (2000)
13. Smith, B.W., Monthioux, M., Luzzi, D.E., *Chem Phys. Lett.* **315**, 31-36 (1999)
14. Saito, Y., Matsumoto, T., *Nature* **392**, 237 (1998)
15. Stephens, P.W., Bortel, G., Faigel G., Tegze, M., Jánossy, A., Pekker, S., Oszlanyi, G., Forró, L., *Nature* **370**, 636-639 (1994)
16. Sundqvist, B., *Advances in Physics* **48** (1), 1-134 (1999)
17. Gromov, A., Ballenweg, S., Giesa S., Lebedkin, S., Hull, W.E., Krätschmer, W., *Chem. Phys. Lett.* **267**, 460-466 (1997)
18. Lebedkin, S., Rietschel H., Adams G.B., Page J.B., Hull W.E., Hennrich F.H., Eisler H.-J., Kappes M.M., Krätschmer W., *J. Chem. Phys.* **110** (24), 11768-11778 (1999)
19. Wang, G.W., Komatsu, K., Murata, Y., Shiro, M., *Nature* **387**, 583-586 (1997)
20. Fabre, T.S., Treleaven, W.D., McCarley, T.D., Newton, C.L., Landry, R.M., Saraiva, M.C., Strongin, R.M., *J. Org. Chem.* **63**, 3522-3523 (1998)
21. Dragoe, N., Tanibayashi S., Nakahara K., Nakao S., Shimotani H., Xiao L., Kitazawa K., Achiba Y., Kikuchi, K., Nojima, K., *Chem Commun.* **1999,** 85-86 (1999)

II. FULLERENES AND FULLERIDES

Resistivity of metallic A_3C_{60} (A= K, Rb): No lower limit to the mean free path?

O. Gunnarsson and J.E. Han

Max-Planck-Institut für Festkörperforschung, D-70506 Stuttgart, Germany

Abstract. We calculate the electrical resistivity due to electron-phonon scattering for a model of A_3C_{60} (A= K, Rb), using an essentially exact quantum Monte-Carlo method. In agreement with experiment, we obtain a very large metallic resistivity at large temperatures T. This illustrates that the apparent mean free path can be much shorter than the separation of the molecules. An interpretation of this result is given.

INTRODUCTION

The alkali-doped Fullerenes A_3C_{60} (A= K, Rb) are metallic systems with exceptionally large resistivities [1]. Typically, the measured resistivities are of the order $\rho \sim 0.5 - 1$ mΩcm at low temperatures and $\rho \sim 2 - 5$ mΩcm at $T \sim 500$ K [2–5]. Assuming a spherical Fermi surface containing the three conduction electrons of A_3C_{60}, we derive a mean free path $l \sim 4 - 10$ Å at low T and $l \sim 1 - 2$ Å at $T \sim 500$ K. This suggests that $l \ll d$ at high T, where $d = 10$ Å is the separation between two C_{60} molecules [6]. In a semi-classical theory it is hard to imagine $l < d$, since an electron could at worst be scattered at every molecule (giving $l \sim d$) [7]. Most metallic systems melt long before $l \sim d$ becomes a possibility, but in th 70's and 80's certain transition metal compounds were found which have $l \sim d$ at high T. For these systems it seems that l never is (much) smaller than d [8,9]. This raises the the important principle question if $l \ll d$ is possible. It also raises the question if really $l \ll d$ for A_3C_{60}. For instance, the system could be in an exotic state where only a small fraction of the conduction electrons contribute to the conductivity and where l is then correspondingly larger. A few other systems, e.g., some high temperature superconductors, apparently also have $l \ll d$.

In this paper we perform an essentially exact calculation of ρ for a model of A_3C_{60}, including the electron-phonon scattering. We show that in this model ρ becomes very large for large T, implying $l \ll d$. This shows that there is no fundamental lower limit on l. On the other hand, in a model calculation of the resistivity due to *electron-electron* scattering, we find saturation. This difference is traced to the difference between bosons and fermions. Although the Boltzmann

CP544, *Electronic Properties of Novel Materials—Molecular Nanostructures*, edited by H. Kuzmany, et al.
© 2000 American Institute of Physics 1-56396-973-4/00/$17.00

equation is usually derived semi-classically, assuming $l \gg d$. for our model there is no qualitative break-down of the Boltzmann equation at large T, where $l \ll d$.

MODEL AND METHOD

The conduction in A_3C_{60} takes place in a partly filled t_{1u} band. The T-dependent part of the resistivity is assumed to be due to scattering against phonons with H_g symmetry. We therefor consider a model with a three-fold degenerate t_{1u} level on each molecule, the hopping between the molecules, a five-fold degenerate H_g phonon on each molecule and the coupling between the electrons and the phonons. The hopping between the molecules takes into account [11] the orientational disorder of the molecules [12]. The one-particle band width is $W = 0.6$ eV and the phonon frequency is ω_{ph}. The electron-phonon coupling is described by the dimensionless parameter λ.

We perform a finite temperature Quantum Monte Carlo (QMC) calculation [14], treating the phonons quantum mechanically. We calculate the current-current correlation function for imaginary times and make a transformation to real frequencies, using a maximum entropy method [15]. The QMC method has no "sign-problem", and the resistivity can be calculated essentially exactly down to quite small T.

RESULTS

Fig. 1 shows the resistivity for a cluster of 48 C_{60} molecules for $\lambda = 0.5$ and $\omega_{ph} = 0.2$ eV. The essentially exact QMC calculation (full line) shows that the resistivity can become very large, corresponding to $l \sim 0.7$ Å at $T = 0.5$ eV. By also considering an unrealisticly large T, we emphasize the lack of a limitation of the type $l > d$.

To interpret these results we use a diagrammatic approach. In the Kubo formalism this requires the calculation of a bubble diagram including vertex corrections. We assume that the vertex corrections can be neglected, and calculate the bubble diagram using the electron Green's function from the QMC calculation. The resulting resistivity (dashed line in Fig. 1) is practically identical to the QMC result, justifying the neglect of vertex corrections for the present model. It was shown by Holstein [16] that in the limit of a broad electronic band, all vertex corrections except ladder diagrams can be neglected and that a Boltzmann equation can be derived. Holstein's proof is not valid for the narrow band considered here, but our calculations show that his arguments are still rather accurate. For our model with a q-independent electron-phonon coupling, even the ladder diagrams can be neglected. Essentially following Holstein we obtain approximately a Boltzmann like conductivity

$$\sigma(T) \sim \int d\omega N(\omega)(-\frac{df(\omega)}{d\omega})\frac{1}{\mathrm{Im}\Sigma(\omega)}|j_k|^2_{\varepsilon_k=\omega}, \qquad (1)$$

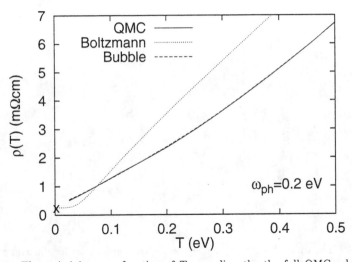

FIGURE 1. The resistivity as a function of T according the the full QMC calculation, the Boltzmann equation (Bloch-Grüneisen) and the bubble diagram with the QMC Green's function. The symbol $\times$ shows the ρ due to the orientational disorder. The figure illustrates that ρ can become extremely large, that the self-consistent bubble calculation is quite accurate and that there is no qualitative break-down of the Boltzmann equation at high T.

where $N(\omega)$ is the density of states, f is the Fermi function, $\Sigma(\omega)$ is the electron self-energy, j_k is the current matrix element for a state with the label k and the energy ε_k. We interpret Im Σ as the inverse of the relaxation time. For a large T, Im Σ becomes comparable to or larger than the one-particle band width and the quasi-particles are ill-defined.

The resistivity thus depends crucially on Σ. To understand its behavior, we consider the lowest order electron self-energy diagram. We use bare Green's functions and for simplicity neglect the orbital degeneracy

$$\Sigma^{(1)}(\mathbf{k}, \omega) = g^2 \sum_{\mathbf{q}} [\frac{n_B(\omega_{ph}) + 1 - f(\varepsilon_{\mathbf{q}})}{\omega - \omega_{ph} - \varepsilon_{\mathbf{q}}} + \frac{n_B(\omega_{ph}) + f(\varepsilon_{\mathbf{q}})}{\omega + \omega_{ph} - \varepsilon_{\mathbf{q}}}], \qquad (2)$$

where

$$n_B(\omega_{ph}) = \frac{1}{e^{\omega_{ph}/T} - 1} \xrightarrow[T \to \infty]{} \frac{T}{\omega_{ph}} \qquad (3)$$

is the Bose occupation number and g is a coupling constant. For large T, n_B becomes large, leading to a large Im Σ, a small σ and a large ρ. The Bose nature of the phonons is therefore crucial for our results.

To further illustrate this, we compare with the resistivity due to the electron-electron scattering. Using the dynamical mean-field theory (DMFT) [17] we have

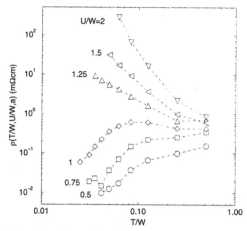

FIGURE 2. The resistivity for the nondegenerate Hubbard model for different values of the Coulomb repulsion U. The figure illustrates how the resistivity saturates for large T in the electron-electron scattering case for our half-filled model.

calculated the resistivity for a nondegenerate Hubbard model with the band width W and a simple cubic lattice with the lattice parameter $a = 1$ Å. We focus on the half-filled case, which is relevant for A_3C_{60}, and we do not consider the case of a doped Mott insulator [18]. Fig. 2 shows $\rho(T)$ for different values of the on-site Coulomb interaction U. For $U < W$ the system is a metal and $\rho(T)$ grows with T, while for $U > W$ it is an insulator and $\rho(T)$ decreases with T. The important observation is that in the metallic case $\rho(T)$ saturates at $\rho \sim 0.4$ mΩcm, which corresponds to $l/a \sim 1/3$. Thus, in contrast to the electron-phonon scattering case, electron-electron scattering does not lead to a l which is very much smaller than a in the metallic state and for this model.

To understand this, we study the electron self-energy Σ to second order in U, since Σ determines ρ in the DMFT. For low T, there is little scattering due to the small phase space available, as controlled by the Fermi functions. As T increases, the available phase space grows and ρ increases. However, for large T, ρ essentially saturates, since the Fermi functions approach a constant value. This is in strong contrast to the Bose occupation numbers (Eq. (3)). The qualitative difference between the two scattering mechanisms for large T can then be traced to the difference between fermions and bosons.

We now address the validity of the Boltzmann equation for the case of the electron-phonon scattering in view of $l \ll d$. We have calculated the resistivity using the Ziman version of the Bloch-Grüneisen solution of the Boltzmann equation [19] and added the resistivity due to the orientational disorder as a T-independent contribution [20]. Using the plasma frequency $\omega_p = 1.2$ eV, we obtain the dotted line in Fig. 1. The Boltzmann result is larger than the QMC result for large T, but there is no qualitative break down of the Boltzmann equation, although $l \ll d$ and the quasi-particle concept is not applicable. The justification for the Boltzmann equation in this limit is not the semi-classical derivation, but the (approximate) derivation from the full quantum mechanical Kubo formulation (Eq. (1)). The proper language is not in terms of a very short mean free path, but in terms of a very broad spectral function (Im Σ large), as discussed above. Essential for this result is that the coupling to intramolecular phonons causes fluctuations in the t_{1u} level positions which are comparable to the t_{1u} band width.

We are grateful to M. Jarrell for making his MaxEnt program available. The work has been supported by the Max-Planck-Forschungspreis.

REFERENCES

1. Gunnarsson, O. Superconductivity of Fullerides. *Rev. Mod. Phys.* **69**, 575 (1997).
2. A.F. Hebard *et al.*, Phys. Rev. B **48**, 9945 (1993).
3. L. Degiorgi *et al.*, Phys. Rev. Lett. **69**, 2987 (1992).
4. J.G. Hou *et al.*, Solid State Commun. **86**, 643 (1993).
5. J.G. Hou *et al.*, Solid State Commun. **93**, 973 (1995).
6. The experimental data have substantial uncertainties, but this is unlikely to influence the qualitative discussion of l. Different experimental methods (direct and optical) for different types of samples (thin films and doped single crystals) all suggest that $l \ll d$ for large T.
7. It has been suggested that the electrons are scattered inside the molecules. This would imply scattering from the t_{1u} band to bands that are at least 0.5-1 eV higher. This should play a very small role for experimental temperatures. For higher T this scattering plays a role, but it then *reduces* the resistivity by providing an additional channel for conduction.
8. Z. Fisk and G.W. Webb, Phys. Rev. Lett. **36**, 1084 (1976).
9. P.B. Allen, in *Superconductivity in d- and f-Band Metals* Eds. H. Suhl and M.B. Maple, Academic (New York,1980) p. 291.
10. W.A. Vareka and A. Zettl, Phys. Rev. Lett. **72**, 4121 (1994).
11. O. Gunnarsson *et al.*, Phys. Rev. Lett. **67**, 3002 (1991); S. Satpathy *et al.*, Phys. Rev. B **46**, 1773 (1992); I.I. Mazin *et al.*, Phys. Rev. Lett. **26**, 4142 (1993).
12. P.W. Stephens *et al.*, Nature **351**, 632 (1991).
13. O. Gunnarsson, Phys. Rev. B **51**, 3493 (1995).
14. R. Blankenbecler, D.J. Scalapino, and R.L. Sugar, Phys. Rev. D **24**, 2278 (1981); D.J. Scalapino and R.L. Sugar, Phys. Rev. B **24**, 4295 (1981).
15. M. Jarrell and J.E. Gubernatis, Phys. Rep. **269**, 133 (1996).
16. T. Holstein, Ann. Phys. **29**, 410 (1964).
17. A. Georges *et al.*, Rev. Mod. Phys. **68**, 13 (1996).
18. G. Pålsson and G. Kotliar, Thermoelectric response near the density driven Mott transition. *Phys. Rev. Lett.* **80**, 4775 (1998)).
19. G. Grimvall, *The electron-phonon interaction in metals*, North-Holland (Amsterdam, 1981).
20. M.P. Gelfand and J.P. Lu, Phys. Rev. B **46**, 4367 (1992); *ibid* 47, 4149 (1993).

ELECTRONIC STRUCTURE OF BODY-CENTERED LATTICE FULLERIDES

SUSUMU SAITO and KOICHIRO UMEMOTO

Department of Physics, Tokyo Institute of Technology
2-12-1 Oh-okayama, Meguro-ku, Tokyo 152–8551, JAPAN

We study the electronic structure of body-centered lattice fullerides in the framework of the density-functional theory. The body-centered orthorhombic Cs_4C_{60}, which is an isostructural material of the superconducting Ba_4C_{60}, is found to have two holes in the t_{1u} conduction band. Therefore the Cs_4C_{60} possesses the electron-hole symmetric electronic structure of the Ba_4C_{60}. Its t_{1u} band is found to be narrow and the Fermi-level density of states is large. We also design the body-centered cubic $Ba_3Br_3C_{60}$, which is found to have a half-filled t_{1u} conduction band.

1. Introduction

Although the body-centered cubic (bcc) K_6C_{60} and Cs_6C_{60} were the first fullerides of which geometry was determined [1], so far they have attracted less interests than the face-centered cubic (fcc) A_3C_{60} (A=K, Rb, or combination of Na, K, Rb, and Cs) because of the discovery of high-transition temperature superconductivity in fcc fullerides [2,3]. Accordingly, the electronic structure of A_3C_{60} materials has been studied in detail [4,5]. Their t_{1u} conduction band which can accommodate up to six electrons is known to be half filled by three electrons without noticiable modification of the band dispersion from that of the pristine fcc C_{60} upon an inclusion of alkali intercalants, indicating rather simple charge-transfer nature from A to C_{60} [6]. This charge-transfer picture had been widely assumed also for bcc A_6C_{60} fullerides. However, our recent comparative study of the bcc pristine C_{60} and several bcc fullerides, K_6C_{60} as well as $A_3Ba_3C_{60}$ (A=K, Rb), has revealed interesting differences between bcc and fcc fullerides as for the A-C_{60} interaction nature [7]. In bcc fullerides studied, a definite widening of the lower conduction bands upon an inclusion of alkali atoms into the pristine C_{60} lattice is observed. This, together with a considerable valence-electron distribution around A sites, clearly indicates the presence of the hybridization between C_{60} and alkali electronic states.

CP544, *Electronic Properties of Novel Materials—Molecular Nanostructures*, edited by H. Kuzmany, et al.

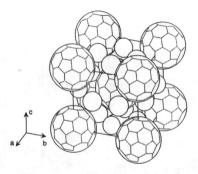

Figure 1: Body-centered orthorhombic fullerides, Cs_4C_{60} and Ba_4C_{60} ($a < b < c$). They possess a perfect orientational order of C_{60}.

The origin of the this interesting difference of the A-C_{60} interaction nature in fcc and bcc fullerides can be attributed to a non-spherical geometry of the C_{60} cluster. The lowest t_{1u} and the second-lowest t_{1g} conduction bands of C_{60}-based solids originate from corresponding lower unoccupied states of the C_{60} cluster, which are known to be spatially distributed mostly on pentagon areas [8]. Alkali intercalants in fcc fullerides do not contact pentagons and therefore their electronic states interact little with C_{60} states, while intercalants at distorted tetrahedral interstitial sites in bcc fullerides are surrounded by two pentagons and their electronic states can hybridize with C_{60} states.

Considering the above origin, characteristic electronic properties of the bcc fullerides are expected to be also present in other body-centered lattice fullerides having a similar alkali-pentagon contact. Therefore the body-centered orthorhombic (bco) Cs_4C_{60}, which is an isostructural material of the superconducting Ba_4C_{60} (Fig. 1) [9,10], may show interesting hybridization behavior and its electronic structure and conducting properties are worth studying in detail. Although K_4C_{60} and Rb_4C_{60} take the body-centered tetragonal structure with so-called merohedral disorder in their C_{60} orientation, conducting properties of all A_4C_{60} fullerides including bco Cs_4C_{60} has been discussed altogether. Also for this reason, Cs_4C_{60} should be studied further.

Moreover, the bco Cs_4C_{60} is a very interesting material from the viewpoint of the possible electron-hole symmetric character of its conduction band with respect to that of Ba_4C_{60}. From the density-functional study of the band structure of the bcc K_6C_{60} as well as that of the pristine bcc C_{60} lattice, it is known that their t_{1u} and t_{1g} bands show dispersions which are almost mirror symmetric to each other [11]. This should be due to their opposite transformation properties under a space-inversion operation. This symmetric character of t_{1u} and t_{1g} bands should hold, at least to some extent, also in bco fullerides which possess the inversion symmetry. Counting the number of valence electrons in Cs_4C_{60} and Ba_4C_{60}, these two fullerides may be regarded as electron-hole symmetry systems, with two holes in the t_{1u} band of Cs_4C_{60} and two

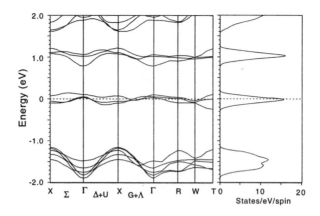

Figure 2: Electronic band structure (left panel) and the density of states (right pannel) of the bco Cs_4C_{60}. Energy is measured from the Fermi level.

electrons in the t_{1g} band of Ba_4C_{60}.

Finally, it should be mentioned that there is a strong relationshiop between Cs_4C_{60} and Cs_3C_{60} in their geometries. While the fcc Cs_3C_{60} has not been produced, the presence of bco and A15 Cs_3C_{60} phases has been revealed [12]. Interestingly, the bco Cs_3C_{60} is reported to have the same crystalline lattice as the bco Cs_4C_{60} with a fractional occupation of 0.75 at each Cs site. Moreover, superconductivity under pressure with the transition temperature of as high as 40 K in Cs_3C_{60} was reported although the superconducting phase has not been identified yet [13]. Hence, the structural similarity between bco Cs_4C_{60} and Cs_3C_{60} again gives high interest for conducting properties of Cs_4C_{60}.

Here we report the electronic structure of this bco Cs_4C_{60} studied in the framework of the density-functional theory with the local-density approximation (LDA). Norm-conserving pseudopotentials of the Troullier-Martins type [14] is adopted with a plane-wave basis set [15]. The cutoff energy used is 50 Ryd. In addition, using the same procedure, we design a candidate for a new class of bcc fullerides, $Ba_3Br_3C_{60}$. Its half-filled t_{1u} conduction band is found to be narrower than the t_{1g} conduction bands of superconducting body-centered-lattice fullerides, Ba_4C_{60} and $A_3Ba_3C_{60}$ [7,16]. Consequently, $Ba_3Br_3C_{60}$ has a large Fermi-level density of states.

2. Results and Discussion

The electronic band structures of the bco Cs_4C_{60} together with its density of states (DOS) obtained is shown in Fig. 2. Here, the internal geometry of the material

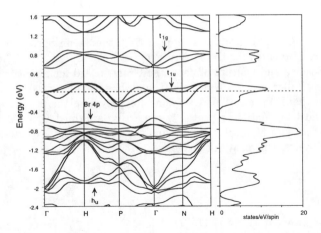

Figure 3: Electronic band structure (left panel) and the density of states (right pannel) of the $Ba_3Br_3C_{60}$. Energy is measured from the Fermi level.

has been optimized uder the experimentally observed lattice constants. Although the mirror symmetric character of the t_{1u} and t_{1g} bands is less evident in this bco fulleride, the t_{1u} conduction band of the Cs_4C_{60} is isolated from other bands and possesses two holes. Therefore, the bco Cs_4C_{60} can be regarded as an electron-hole symmetric system of the isostructural Ba_4C_{60}.

Quantitatively, the t_{1u} conduction band is found to be much narrower than the t_{1g} conduction band of the Ba_4C_{60} studied previously. Still, from the comparative study of the bcc Cs_4C_{60}, the hybridization between Cs and C_{60} states is found to be present [17]. This narrow conduction band, on one hand, gives rise to a large Fermi-level DOS value ($N(E_F)$) for the Cs_4C_{60}, which is found to be three times larger than that for the Ba_4C_{60}, and is a preferable factor for superconductivity. On the other hand, due to the residual electron correlation beyond the LDA used here, the narrow conduction band may prefer non-metallic conducting nature as was discussed in K_4C_{60} etc. [18].

3. $Ba_3Br_3C_{60}$

Most of the superconducting fullerides identified so far are known to have cubic geometries and half-filled conduction bands. The bcc $A_3Ba_3C_{60}$ superconductors have the half-filled t_{1g} band while fcc A_3C_{60} superconductors have the half-filled t_{1u} band. However, a half-filled t_{1u}-band bcc fulleride has not been produced so far. Here we report the electronic structure of the $Ba_3Ba_3C_{60}$ which is designed to belong to this new class of fullerides. The electronic structure obtained is shown in Fig. 3. The material is acutually found to be a candidate for the first half-filled t_{1u}-band bcc fulleride [19].

Acknowledgements – This work was supported by the Japan Society for the Promotion of Science (Research for the Future Program, No. 96P00203), The Ministry of Education, Science and Culture of Japan as Grant-in-Aid for Scientific Research on Priority Area "Fullerenes and Nanotubes", and The Nissan Science Foundation. We would like to thank Professor A. Oshiyama for providing the program used in this work. Numerical calculations reported were partly performed at the Computer Center of the Institute for Molecular Science and the Supercomputer Center, Institute for Solid State Physics, University of Tokyo.

References

1. O. Zhou *et al.*, *Nature* **351** (1991) 462.
2. A. F. Hebard *et al.*, *Nature* **350** (1991) 600.
3. K. Tanigaki *et al.*, *Nature* **352** (1991) 222.
4. S. Saito and A. Oshiyama, *Phys. Rev.* **B 44** (1991) 11536.
5. S. C. Erwin and W. E. Pickett, *Science* **254** (1991) 892.
6. J. L. Martins and N. Troullier, *Phys. Rev.* **B 46** (1992) 1766.
7. K. Umemoto, S. Saito, and A. Oshiyama, *Phys. Rev.* **B 60** (1999) 16186.
8. P. Dahlke, P. F. Henry, and M. J. Rosseinsky, *J. Mat. Chem.* **8** (1998) 1571.
9. C. M. Brown *et al.*, *Phys. Rev. Lett.* **83** (1999) 2258.
10. S. C. Erwin and M. R. Pederson, *Phys. Rev. Lett.* **67** (1991) 1610.
11. Y. Yoshida *et al.*, *Chem. Phys. Lett.* **291** (1998) 31.
12. T. T. M. Palstra *et al.*, *Solid State Commun.* **93** (1995) 327.
13. N. Troullier and J. L. Martins, *Phys. Rev.* **B 43** (1990) 1993.
14. O. Sugino and A. Oshiyama, *Phys. Rev. Lett.* **68** (1992) 1858.
15. K. Umemoto and S. Saito, *Phys. Rev.* **61** (2000) 14204.
16. K. Umemoto and S. Saito, *to be published.*
17. M. Knupfer and J. Fink, *Phys. Rev. Lett.* **79** (1997) 2714.
18. K. Umemoto, *Ph.D Thesis, Tokyo Insitute of Technology* (2000).

Broken-symmetry band structure description of the spectroscopy of K_3C_{60}

L.F.Chibotaru and A. Ceulemans

Department of Chemistry, University of Leuven,

Celestijnenlaan 200F, B-3001 Leuven, Belgium

Abstract. The electronic structure of K_3C_{60} is investigated in a band approach allowing for arbitrary distribution of the charge density among the three LUMO orbitals at each fullerene site (a model analogue of the LDA+U method). The predictions for optical absorption and EELS of these broken-symmetry band structure calculations are compared with available experimental data. We find that such calculations reproduce the three absorption bands in the low energy region of the optical conductivity spectrum and give the correct dc limit for the conductivity. Simulations of the loss function show the right position and width of the metallic plasmon peak. Similar calculations within the conventional band structure approach fail to reproduce the above spectroscopic features.

I INTRODUCTION

Despite intensive investigations during recent years concerning alkali doped fullerides $A_nC_{60}, n = 3, 4, A = K, Rb, Cs$, a coherent theoretical description of their normal state properties is still lacking. No attempts were made to describe key experiments like optical conductivity, electron energy loss spectroscopy and photoemission within one model. The reason is that several complex interactions present in the LUMO band of these compounds should be taken into account simultaneously: interelectron repulsion, Jahn-Teller effect and Hund's rule coupling on each fullerene site, the merohedral disorder of C_{60} molecules and the long range electron interaction [1].

The interplay between these interactions results in quite diverse physical properties, which are not yet fully understood for most of fullerides. Among them the electronic properties of K_3C_{60} seem to be less controversial. For this compound, the measurements of NMR [2], resistivity [3] and magnetic susceptibility [4] show conventional metallic behaviour consistent with the Fermi liquid description. Recently it was also found [5] that despite strong electron correlation the screening in metallic A_3C_{60} is well described within RPA. This provides evidence that the band approach should be a good starting point for the electronic structure of K_3C_{60}.

CP544, *Electronic Properties of Novel Materials—Molecular Nanostructures*, edited by H. Kuzmany, et al.
© 2000 American Institute of Physics 1-56396-973-4/00/$17.00

A strong enough electron correlation shows up in the band structure calculations via broken-symmetry solutions for band spin-orbitals leading to phenomena of spin, charge and orbital ordering. The uniqueness of fullerides as crystals with threefold degenerate conduction band orbitals is that the Jahn-Teller stabilization on each site overcomes the Hund energy, making the interorbital charge disproportionation of the conduction electron density (orbital ordering at fullerene sites) a leading electronic instability in these materials [6]. In the present paper we investigate the manifestation of this instability in the optical conductivity and electron energy loss spectroscopy (EELS) of K_3C_{60}.

II THE THEORY

We start from the LUMO Hamiltonian including all bielectronic and vibronic interactions within the t_{1u} orbitals at each fullerene site and the electron transfer between nearest neighbour sites [7]. The transfer part was parametrized in order to reproduce the LDA results for both ordered and merohedrally disordered fullerenes in the fcc lattice of K_3C_{60}, while the bielectronic and vibronic parameters were taken from experiment. Supposing arbitrary static Jahn-Teller distortions of each C_{60}, their amplitudes can be expressed exactly through the elements of the intrasite one particle density matrix, $D_{mi\sigma}^{mj\sigma}$ ($\mathbf{m}$ denotes sites and i, j are t_{1u} orbital components) [6]. The resulting Fock operator, describing band orbitals with symmetry breaking in the orbital sector, has the following form in the site representation:

$$F_{\mathbf{m}i;\mathbf{n}j} = h_{\mathbf{m}i;\mathbf{n}j} + \frac{5}{6}(U_{\parallel} - 2J)n\delta_{ij}\delta_{\mathbf{m},\mathbf{n}} - U_d\delta_{\mathbf{m},\mathbf{n}}[\Delta N_{mi\sigma}\delta_{ij} + D_{mi\sigma}^{nj\sigma}(1 - \delta_{ij})],$$
$$U_d = U_{\parallel} + 6E_H - 5J, \tag{1}$$

where $U_{\parallel}$ and J are intrasite Coulomb and exchange parameters, $h_{\mathbf{m}i;\mathbf{n}j}$ are transfer amplitudes and E_H is the Jahn-Teller stabilizattion energy of a C_{60}^- anion in the space of H_g coordinates. n is the average population on the t_{1u} shell ($n = 3$ in the present case) and $\Delta N_{\mathbf{m}i\sigma}$ is the deviation of the population of the orbital component i (with spin $\sigma = \alpha$ or β) from the cubic symmetry value ($1/2$).

The merohedral disorder was simulated by a random distribution of two standard orientations of fullerene molecules in an enlarged unit cell containing 256 sites, which proved to be sufficient to reach the asymptotic value of the oscillator strengths of t_{1u} optical transitions [8]. The next-LUMO (t_{1g}) band was described by the tight binding Hamiltonian giving the dispersion and the shift relative to the t_{1u} band in agrement with conventional LDA calculations [9]. To describe the contribution of the t_{1g} bands to the optical conductivity, the intrafullerene $t_{1u} \rightarrow t_{1g}$ transitions have been additionaly taken into account with the dipole matrix element estimated as 1.64 D [8].

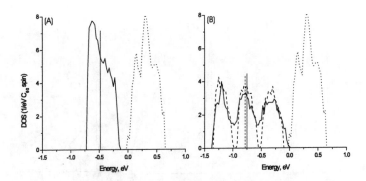

FIGURE 1. Density of states of the the LUMO and next-LUMO bands in K_3C_{60} obtained by conventional (left) and broken-symmetry (right) band structure calculations including merohedral disorder. In the right pannel the t_{1u} band is resolved into three subbands: the fully occupied (lowest), the half filled (middle) and the empty (upper) shown for the case of translational order (solid lines) and full disorder (dashed lines) of the corresponding orbital components on the sites. In both pannels the zero of energy is fixed at the bottom of the t_{1g} band shown by dotted line. Vertical lines correspond to the Fermi level.

III   RESULTS AND DISCUSSION

The resulting density of states (DOS) pictures are shown in Fig.1. Starting from the cubic electron distribution ($\Delta N_{mi\sigma} = 0$) we obtain for the t_{1u} band a structureless DOS profile (left pannel) coinciding with previous results [10]. Nonequal population of the t_{1u} orbital components in the initial electron distribution ($\Delta N_{mi\sigma} \neq 0$) results in the splitting of the cubic band into three subbands (right pannel). Each t_{1u} orbital component on fullerene sites contributes almost exclusively to one of the three subbands. Therefore the relative shift of the subbands is described by the difference of the corresponding terms $U_d \Delta N_{mi\sigma}$ in Eq.(1). The magnitude of these shifts is much exaggerated at the Hartree-Fock level of calculations but becomes less pronounced when the screening of the exchange interaction between band states is taken into account, as it was shown by GW calculations for K_4C_{60} [11]. Accordingly, in the present calculations we consider the shift of the side subbands relative to the middle one as a free parameter (the only one) used to fit the position of the lowest subband in Fig1.b to the broad peak in the photoemission spectrum [12].

Fig. 2 shows the calculated optical conductivity spectra together with the experimental one taken from Ref. [13]. The intra t_{1u} band contribution to the optical conductivity in the conventional band structure calculation (Fig. 2a) looks like a broad Drude peak whereas in the broken-symmetry aproach (Fig. 2b) it contains two additional absorption bands corresponding to transitions between

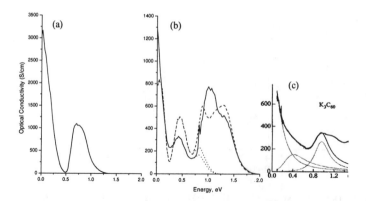

FIGURE 2. Conductivity of K_3C_{60}: conventional band structure calculations (left), broken-symmetry calculations (middle) and experiment [13] (right). Solid and dashed lines in the middle pannel correspond to translational order and full disorder of disproportionated t_{1u} orbitals at fullerene sites. Dotted lines show the intra t_{1u} band contribution to the conductivity in the overlapping region.

neighbour subbands ($\approx$0.4 eV) and between the lower and upper subbands ($\approx$0.8 eV). The transition from the middle t_{1u} subband to t_{1g} is almost superimposed on the highest $t_{1u} \rightarrow t_{1u}$ absorption band while the transition from the lowest t_{1u} subband to t_{1g} gives the third absorption band at $\approx$1.3 eV. These three bands coincide well with experiment (Fig. 2c). Moreover the *dc* limit in Fig. 2b matches the measured value of 1300 S/cm in Ref. [14]. The calculations based on the conventional approach (Fig. 2a) fail to reproduce these features.

The calculated optical conductivity was further used to simulate the electron loss function via the Drude-Lorentz fit of the spectra in Fig.2. We discuss here the metallic plasmon peak and consider for simplicity only the contribution from the t_{1u} band. Applying such a transformation to the experimental conductivity in Fig. 2c gives the position and the width of the plasmon peak which are in good agreement with the EELS data [15]. Clearly the peak which results from conventional band structure calculations is placed too high in energy and is much narrower (Fig. 3a). By contrast broken-symmetry calculations give a plasmon peak closer to the right position and enlarged by shoulders corresponding to transitions between t_{1u} subbands (Fig. 3b). Actually one should expect a significant enlargement of the one-particle DOS bands when going beyond Hartree-Fock approximation [11], which broadens the one-electron transition bands in optical conductivity and EELS. If one enlarges the side peaks in Fig. 3b twice, they will merge with the central one (Fig. 3c) thus explaining the origin of the broad plasmon peak seen in K_3C_{60}. Note that in the broken-symmetry bands picture the metallic plasmon falls in the region of one-electron transitions between t_{1u}

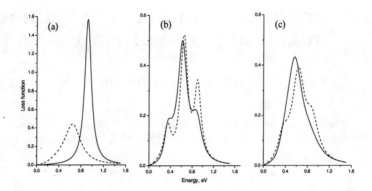

FIGURE 3. The t_{1u} band contribution to the electron loss function for K_3C_{60}. Left pannel: derived from the conventional band structure calculation (solid line) and from the measured conductivity [13] (dashed line). Middle pannel: derived from broken-symmetry band structure calculations (cf. Fig. 2b) for ordered (solid lines) and disordered (dashed lines) cases; Right pannel: the same as in the middle but for twice enlarged widths of the two inter t_{1u} transition bands in Fig. 2b.

subbands (cf. Figs. 1b and 2b) which will give an additional contribution to the plasmon broadening.

REFERENCES

1. O. Gunnarsson, Rev. Mod. Phys. **69**, 575 (1997).
2. C.H. Pennington and V.A. Stenger, Rev. Mod. Phys. **68**, 855 (1996).
3. T.T.M. Palstra *et al.*, Phys. Rev. B **50**, 3462 (1994).
4. J. Robert *et al.*, Phys. Rev. B **57**, 1226 (1998).
5. E. Koch, O. Gunnarsson, and R. Martin, Phys. Rev. Lett. **83**, 620 (1999).
6. L.F. Chibotaru and A. Ceulemans, Phys. Rev. B **53**, 15522 (1996).
7. A. Ceulemans, L.F. Chibotaru and F. Cimpoesu, Phys. Rev. Lett. **78**, 3725 (1997).
8. L.F. Chibotaru and A. Ceulemans, in *Electron-Phonon Dynamics and Jahn-Teller effect*, edited by G. Bevilacqua et al., World scientific, Singapore, 1999, p. 233.
9. S.C. Erwin and W.E. Pickett, Science **254**, 842 (1991).
10. I.I. Mazin et al., Phys. Rev. Lett. **70**, 4142 (1993).
11. L.F. Chibotaru, A. Ceulemans, and S.P. Cojocaru, Phys. Rev. B **59**, R12728 (1999).
12. M. Knupfer et al., Phys. Rev. B **47**, 13944 (1993).
13. Y. Iwasa, T. Kaneyasu, Phys. Rev. B **51**, 3678 (1995).
14. L. Degiorgi *et al*, Phys. Rev. B **49**, 7012 (1994).
15. M. Knupfer, J. Fink, and J.F. Armbruster., Z. Phys. B **101**, 57 (1996).

Molecular motions and T site symmetry in the sc phase of Na$_2$C$_{60}$

V. Brouet[1], H. Alloul[1], T. Saito[1] and L. Forró[2]

[1] *Physique des solides, UMR 8502 CNRS, UPS, 91 405 Orsay (France)*
[2] *Physique des solides semicristallins, IGA, EPFL, 1015 Lausanne (Switzerland)*

Abstract. We present an NMR study of Na$_2$C$_{60}$ with a particular emphasis on the behavior of molecular motions in this compound. We show that the special orientations of the C$_{60}$ molecules in the sc phase promote a displacement of Na ion from the center of the T site. This is quite a unique feature of Na containing compounds and the implication for the electronic structure of these materials is discussed.

I   INTRODUCTION

Na$_2$C$_{60}$ is the only example among alkali fullerides with 2 electrons per C$_{60}$. One could expect its electronic properties to be similar to A$_4$C$_{60}$ (A = K, Rb, Cs) due its symmetric position with respect to half-doping in the triply degenerate t$_{1u}$ band, but detailed experimental studies are still lacking. It is well known that A$_4$C$_{60}$ display an insulating behavior in contradiction with a simple band picture [1]. Yet, as the structutre is not cubic in these compounds (but *bct*), the comparaison with A$_3$C$_{60}$ is not direct and the insulating behavior has sometimes been assigned to the structural deviation [2]. This problem is not encountered with Na$_2$C$_{60}$ which stays cubic for all temperatures, even though it displays a C$_{60}$ orientational transition at 300 K that reduces the symmetry from *fcc* to *sc*.

In this paper, we discuss the informations that one can gain by NMR about structural/electronic modifications at the orientational transition. On one hand, ^{13}C NMR is a useful tool to study the dynamics of molecular motions. On the other hand, quadrupole effects on ^{23}Na can allow to estimate structural distortions with respect to the ideal cubic symmetry. This study is also motivated by the fact that a similar orientational transition takes place in superconductring Na$_2$AC$_{60}$ (contrary to A$_3$C$_{60}$ compounds) and that a possible change of the electronic structure related to the *sc* symmetry has often been invoked to explain the different variation of *Tc* with the lattice constant *a* in Na$_2$AC$_{60}$ compared to A$_3$C$_{60}$.

CP544, *Electronic Properties of Novel Materials—Molecular Nanostructures*, edited by H. Kuzmany, et al.

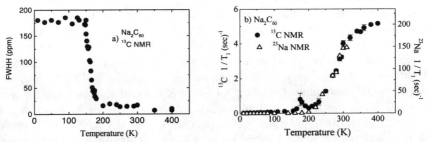

FIGURE 1. a) ^{13}C NMR linewidth as a function of temperature showing the slowing down of molecular motions. b)Comparaison of $1/T_1$ for ^{13}C and ^{23}Na to evidence the molecular motion peak at 180 K.

II MOLECULAR MOTION IN NA$_2$C$_{60}$

NMR has proved to be a very efficient tool to study molecular motions in fullerenes [3,4]. First, the temperature dependence of the ^{13}C spectrum linewidth indicates the temperature where the motions are static on the NMR time scale (≈ 1 ms). When the molecule does not rotate, the different orientations of the p_z orbital on the ^{13}C sites with respect to the applied magnetic field lead to a distribution of shifts which results in a broad powder spectrum characteristic of the anisotropy of the orbitals. On the other hand, molecular rotations at high temperature average the different orientations and the spectrum narrows. This behavior is clearly seen on figure 1a and points out that molecular rotations persist down to about 150 K in Na$_2$C$_{60}$. This is lower than in other fullerides like K$_3$C$_{60}$ ($\simeq 200$ K) or Rb$_3$C$_{60}$ ($\simeq 350$ K), which demonstrates indicate that molecular rotations are less hindered by the small Na ions, as could be expected. At low T, the spectrum is almost as broad as in pure C$_{60}$.

Second, the existence of molecular motions can show up in the relaxation behavior. The fluctuations of the local field associated with the reorientation of one C$_{60}$ ball can be a source of relaxation if they match the NMR Larmor frequency ($\simeq 75$ MHz). This contribution is then expected at a slightly higher temperature than the one of the linewidth broadening and can indeed be detected on figure 1b as a small peak at 180 K which adds to the electronic contribution that will be discussed elsewhere. The fact that this peak is not seen in ^{23}Na NMR $1/T_1$ which follows otherwise the same temperature dependence is a very clear proof of its origin.

Although this behavior is similar to what has been observed in other fullerides, the shape of the peak is quite different. The simplest model for the relaxation is to assume that the rotation is characterized by a thermally activated fluctuation time τ [3] and :

$$\frac{1}{T_1} = (\gamma H_{loc})^2 \frac{2\tau}{1 + \omega_0^2 \tau^2} \quad \text{with} \quad \tau = \tau_0 \exp\left(\frac{Ea}{k_B T}\right)$$

γH_{loc} characterizes the amplitude of the fluctuations of the local field. The linewidth at low T gives an upper value for this prefactor, which is definitely too large to account for the small amplitude of the peak here. We note quite a distribution of these prefactor values in the analysis of different fullerenes compounds which has not been clearly explained so far. A full study should take into account both the particular symmetry of the motion and the real anisotropy to be averaged as in ref. [6]. Nevertheless, we find a temperature dependence for τ which is much steeper than in other fullerides and this reflects that the peak here is unusually sharp. We obtain $\tau_0 = 2.8 \ 10^{-24\pm5}$ sec and $E_a/k_B = 6000$ K $\pm$ 2000 K, compared to $2 \ 10^{-14}$ sec and 2600 K for K_3C_{60}.

The first possibility is that the difference is due to the sc symmetry which very likely affects the dynamics of the molecular rotations. For example, to preserve the orientational order, the rotation around the principal axis could be less isotropic than in fcc phases, resulting in restricted motion as suggested by the very high value of E_a. But the parameters extracted from the relaxation in the sc phase of pure C_{60} are very similar to K_3C_{60} [7].

On the other hand, we have measured another Na_2C_{60} sample with poorer quality where the peak at 180 K was also clearly observed but was indeed broader. Within the same analysis, this would yield rotation parameters closer from the ones of K_3C_{60}. This raises a doubt about the intrinsic character of the values extracted from such fits in the various systems studied so far.

III QUADRUPOLE EFFECTS ON ^{23}NA IN NA$_2$C$_{60}$

^{23}Na spectra can give complementary informations because ^{23}Na is a spin 3/2 and is therefore sensitive to quadrupole effects, i.e. to the presence of electric field gradient (EFG) due to deviation from the cubic symmetry at Na site. Figure 2 shows the evolution of the ^{23}Na spectra in Na_2C_{60} through the orientational transition from fcc to sc phase. The position of the line moves, so that the intrinsic properties of the two phases can be distinguished even when they coexist (290 K - 310 K). In the fcc phase, there are no quadrupole effects, as expected in the cubic environment of one tetrahedral site and the line does not exhibit any "T, T' splitting" as observed in Na_2CsC_{60} [8] for the very same structure indicating that this splitting is a fingerprint of n = 3.

In the sc phase, quadrupole effects are present and only the central transition $(1/2 \rightarrow -1/2)$ is observed as shown by a decrease in the NMR intensity. The splitting observed at 200 K and below is characteristic of second order broadening of the central transition. Figure 2c shows that it can be well fitted assuming a quadrupole frequency $\nu_q = 700 \ kHz$ and a small asymmetry of the EFG tensor $\eta = 0.2$. A very similar behavior was previously observed in the sc phase of Na_2AC_{60} [8] indicating that it is a general feature of sc compounds. Figure 2b shows that ν_q progressively decreases when the temperature is raised from 230 to 310 K.

Why is the symmetry reduced ? As noted by Prassides et al. [9], the environment

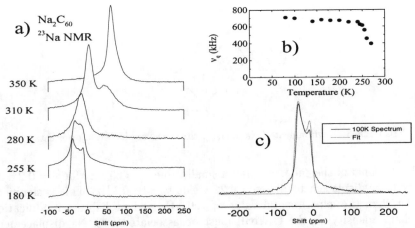

FIGURE 2. a) ^{23}Na spectra on both sides of the sc-fcc tansition. b) Linewidth in the sc phase that shows stabilization of quadrupole effects. c) Fit of the ^{23}Na spectrum at 100 K to a model of second order quadrupole broadening of the central transition.

of the Na T site is quite different in *sc* and *fcc* phases. In the *fcc* phase, the orientations of the four neighbouring C$_{60}$ balls are such that Na faces four hexagonal rings. As represented on figure 3, in the sc phase, it faces only one hexagonal ring and three double bonds. This asymmetric environment could favor a displacement of Na along the cube diagonal (indicated by the arrow) towards the hexagonal ring, which would create an electric field gradient.

It seems clear that molecular motions must be reduced to allow this displacement. More precisely, rotation around one axis could still exist (and it probably freezes only at 180 K as suggested by ^{13}C NMR) but reorientation of the rotation axis must be nearly prohibited. Such a decomposition of the molecular rotation was proposed for K$_3$C$_{60}$ and the peak observed in 1/T$_1$ around 240 K was also assigned to the fast motion around one preferrential axis. Here, we believe that the "slow" reorientation motion is almost frozen when we observe static quadrupole effects, around 250 K.

In Na$_2$CsC$_{60}$, a displacement of Na from the center of T site was observed [9] to position (0.241, 0.241, 0.241). Such a displacement in Na$_2$C$_{60}$ would correspond to an electric field gradient, which can be obtained by calculating the gradient V$_{xx}$, V$_{yy}$, V$_{zz}$ of the electrostatic potential where Na and C$_{60}$ are taken as respectively +e and -2e point charges. The quadrupole effects can be characterized by a quadrupole frequency ν_q and an assymetry η with :

$$\nu_q = \frac{3\,(1-\gamma_\infty)\,eV_{zz}\,Q}{2I\,(2I-1)\,h} \quad \text{and} \quad \eta = \frac{|V_{xx}-V_{yy}|}{V_{zz}}$$

where $(1-\gamma_\infty)$ (the Sternheimer antishielding factor) is taken here to be 5 [10] and V$_{zz}$ is by convention the larger eigenvalue of the EFG tensor, that corresponds here

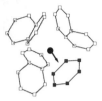

FIGURE 3. Environment of Na in the T site of the sc phase (major orientation).

to the diagonal of the cube. We find a small value $\nu_q = 65$ kHz for x = 0.24. Even a very large displacement like x = 0.22 yields $\nu_q = 260$ kHz. This is significantly smaller than the experimental value of 700 kHz and probably a modification of the charge density on the C_{60} ball must be associated with Na displacement to emphasize this effect.

IV CONCLUSION

Molecular motions in the sc phase of Na_2C_{60} can be described by a combination of a fast rotation around one axis that freezes around 180 K and less frequent jumps of the rotation axis above 250 K. In addition, we have shown that the special symmetry of the sc phase combined with the higher mobility of Na ion allows a displacement of Na from the center of the T site which is likely to modify the charge density on a C_{60} ball. This effect might be important in view of recent ideas that Na cannot be considered as a simple lattice spacer [11] in Na_2CsC_{60}, as a very similar behavior is expected there. The characteristics of the rotation at 180 K are apparently quite different from the ones observed in other fullerides which might also indicate different dynamics related to the sc symmetry and/or the presence of Na.

REFERENCES

1. M. Knupfer and J. Fink, *Phys. Rev. Letters* **79**, 2714 (97) and references therein.
2. J.E. Han, E. Koch and O. Gunnarsson, *Phys. Rev. Letters* **84**, 1276 (00)
3. Y. Yoshinari et al. *Phys. Rev. Letters* **71**, 2413 (93)
4. G. Zimmer et al., *Phys. Rev. B* **52**, 13300 (95)
5. C.H. Pennington and V. A. Stenger, *Rev. of modern Physics* **68**, 855 (96)
6. Y. Yoshinari et al. *Phys. Rev. B* **54**, 6155 (96)
7. R. Tycko et al. *Phys. Rev. Letters* **67**, 1886 (91)
8. T. Saito et al., *Journal of the phys. Soc. of Japan*, 64, 4513 (95)
9. K. Prassides et al. Science, **263**, 950 (94)
10. A. Abragam, *Principles of Nuclear Magnetism*, Oxford Science publications, p. 168
11. N. Cegar et al., submitted to Phys. Rev. Letters

Crystal Structure and
Magnetic Properties of $Eu_xSr_{6-x}C_{60}$

K. Ishii*, H. Ootoshi*, Y. Nishi*, A. Fujiwara*,
H. Suematsu*, and Y. Kubozono[†]

*Department of Physics, The University of Tokyo, Tokyo 113-0033, Japan
†Department of Chemistry, Okayama University, Okayama 700-8530, Japan

Abstract. Crystal structure and magnetic properties of $Eu_xSr_{6-x}C_{60}$ ($x = 1$, 3, 6) were studied. They all have a *bcc* structure which is an isostructure to other M_6C_{60} (M is an alkali or alkaline earth metal). The lattice constants follow the Vegard's law, suggesting a formation of solid solution at $x = 1$ and 3. Magnetic measurements reveals all Eu atoms are in the divalent state with a magnetic moment of $7\mu_B$ ($S = 7/2$), which is consistent with XANES experiments. Ferromagnetic correlation develops at low temperature in all three samples and seems to be caused the π-f interaction through C_{60} rather than the direct exchange interaction between Eu atoms.

INTRODUCTION

C_{60} is known to make compounds with various atoms and molecules. Especially, alkali and alkaline earth fullerides are intensively studied because of the variation of the compounds and superconductivity [1,2]. Rare earth metals can be also good candidates to make compound with C_{60} and superconductivity was reported in some rare earth C_{60} compounds [3,4]. Other interest in rare earth fulleride is magnetism originated from localized $4f$ electrons. As for europium, which has a magnetic moment of $7\mu_B$ in divalent state, some experiments have been reported. A photoemission study of C_{60} evaporated on Eu revealed the formation of fulleride and the charge transfer from Eu to C_{60} [5]. The bulk compounds of Eu_xC_{60} was synthesized by solid-solid reaction at high temperature and magnetic properties were measured [6]. High resolution x-ray diffraction experiments of Eu_xC_{60} confirmed two stable phases of $x = 3$ and 6 and determined that Eu_6C_{60} has a *bcc* structure which is an isostructure to other M_6C_{60} (M is an alkali or alkaline earth metal) [7].

In this paper, we report the crystal structure and magnetic properties of Eu_6C_{60} and its strontium-substituted compounds. Because Sr^{2+} has a similar ionic radius to Eu^{2+} and has no magnetic moments, substitution from Eu to Sr can give us a good information to investigate a magnetic interaction between Eu atoms.

CP544, *Electronic Properties of Novel Materials—Molecular Nanostructures*, edited by H. Kuzmany, et al.
© 2000 American Institute of Physics 1-56396-973-4/00/$17.00

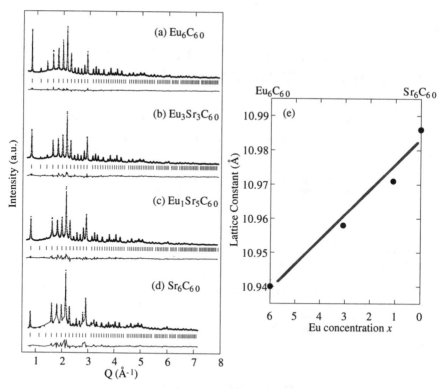

FIGURE 1. (a)-(d) : Powder x-ray diffraction spectra of $Eu_xSr_{6-x}C_{60}$. The cross marks represent observed intensity, and solid lines are the results of Rietveld refinements. The spectrum of Sr_6C_{60} was collected at the wavelength of 0.801 Å and others were at 0.851 Å. (e) : Eu concentration (x) dependence of the lattice constants obtained from Rietveld refinements. The solid line is a guide to eyes.

EXPERIMENTAL

The polycrystalline samples of $Eu_xSr_{6-x}C_{60}$ were synthesized by heat treatment of stoichiometric mixture of metal and C_{60}. Powder x-ray diffraction experiments were carried out with use of synchrotron radiation x-rays at Photon Factory (BL-1B), KEK, Tsukuba. Eu L_{III}-edge XANES was measured in the fluorescence method at BL01B1 of SPring-8, Harima. Magnetic measurements were performed using a SQUID magnetometer.

RESULTS AND DISCUSSION

Figures 1(a)-(d) show the powder x-ray diffraction spectra of $Eu_xSr_{6-x}C_{60}$. The spectra of Eu_6C_{60} and Sr_6C_{60} are consistent with previous works [7,8] and those of

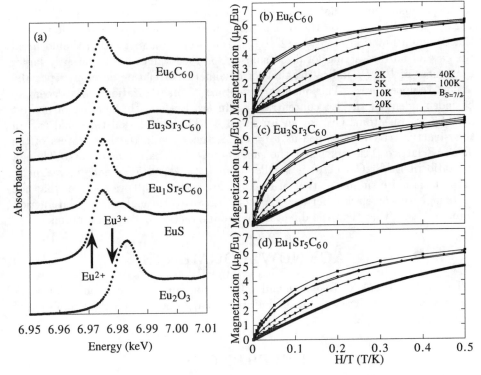

FIGURE 2. (a) : Eu L_{III}-edge XANES spectra. Absorption edge of $Eu_xSr_{6-x}C_{60}$ is close to that of EuS, meaning that all Eu atoms are divalent in $Eu_xSr_{6-x}C_{60}$. (b)-(d) : H/T dependence of magnetization in $Eu_xSr_{6-x}C_{60}$. Ferromagnetic correlation develops at low temperature.

$Eu_3Sr_3C_{60}$ ($x = 3$) and $Eu_1Sr_5C_{60}$ ($x = 1$) are quite similar. The lattice constants obey the Vegard's law; that is, the lattice constant changes linearly with the Eu concentration x. This confirms the formation of solid solution at $x = 1$ and 3. Rietveld refinements were carried out with use of the RIETAN program [9], in which the atomic ratio of Eu and Sr is refined with total number of metal atoms conserved. The weighted pattern R factor (R_{wp}) is 3.81% for Eu_6C_{60}, 6.20% for $Eu_3Sr_3C_{60}$, 6.37% for $Eu_1Sr_5C_{60}$, and 7.03% for Sr_6C_{60}. The refined Eu concentration x and lattice constant are plotted in Fig. 1(e).

Eu L_{III}-edge XANES spectra are shown in Fig 2(a). The spectra of EuS and Eu_2O_3 are also presented as a reference of divalent and trivalent Eu, respectively. The results suggest all Eu atoms are divalent in $Eu_xSr_{6-x}C_{60}$. Recently Eu endohedral C_{60}, Eu@C_{60} was extracted and the valence state of Eu atom was determined to be +2 [10]. Eu atom seems to prefer divalent state in the C_{60} compounds.

Magnetic measurements also confirm the divalent state of Eu. Above 100 K, temperature dependence of magnetization can be understood by Curie-Weiss law and effective Bohr magneton is close to the theoretical value $7.94\mu_B$. Moreover the saturation moment is roughly $7\mu_B$ which is consistent to the divalent state of $S = 7/2$. Figures 2(b)-(d) show the H/T dependence of magnetization. We also present the Brillouin function of $S = 7/2$ $(B_{S=7/2})$ on which magnetization should lie in the case of no interacting localized spins. Observed values are apparently larger than those of $B_{S=7/2}$, indicating that ferromagnetic correlation develops especially at low temperatures. The qualitative feature of magnetization curves seem to be independent of the Eu concentration. In bcc Eu_6C_{60}, the number of nearest neighbor of Eu atoms (N) for an Eu site is 4. The percolation threshold (p_c) in this structure can be assumed as $p_c = 0.428$, the value in the site process of the diamond lattice [11], which has the same number of nearest neighbors. Though Eu ratio in $Eu_1Sr_5C_{60}$ (1/6) is smaller than both values $1/N$ and p_c, magnetic property is quite similar to that of Eu_6C_{60}. From this result we consider that the ferromagnetic correlation between Eu atoms is caused through the π-f interaction through C_{60}, rather than the direct exchange interaction between Eu atoms.

ACKNOWLEDGMENTS

The authors thank T. Takenobu and Y. Iwasa for valuable discussions. This work was supported by "Research for the Future" of Japan Society for the Promotion of Science (JSPS), Japan.

REFERENCES

1. Hebard, A. F., et al., Nature (London), **350**, 600-601 (1991).
2. Kortan, A. R., et al., Nature (London), **355**, 529-532 (1992).
3. Özdaş, E., et al., Nature (London), **375**, 126-129 (1995).
4. Chen, X. H., and Roth, G., Phys. Rev. B, **52**, 15534-15536 (1995).
5. Yoshikawa, H., et al., Chem. Phys. Lett., **239**, 103-106 (1995).
6. Ksari-Habiles, Y., et al., J. Phys. Chem. Solids, **58**, 1771-1778 (1997).
7. Ootoshi, H., et al., in Proceedings of 10th International Symposium on Intercalation Compounds, to be published in Mol. Cryst. and Liq. Cryst..
8. Kortan, A. R., et al., Chem. Phys. Lett. **223** 501-505 (1994).
9. Izumi, F., The Rietveld Method, Oxford University Press, Oxford, 1993, ch. 13.
10. Inoue, T., et al., Chem. Phys. Lett. **316** 381-386 (2000).
11. Stauffer, D., Introduction to Percolation Theory, Taylor & Francis, London, 1985.

Mott-Hubbard Transition and Antiferromagnetism in Ammoniated Alkali Fullerides

T. Takenobu, Y. Iwasa, and T. Mitani

Japan Advanced Institute of Science and Technology, Tatsunokuchi, Ishikawa 923-1292, Japan

Abstract. Intercalation of neutral ammonia molecules into trivalent face-centered-cubic (fcc) fullerides induces an electronic phase transition from cubic superconductivity to an orthorhombic antiferromagnetic insulating state. Monoammoniated alkali fullerides salt $(NH_3)K_{3-x}Rb_xC_{60}$ ($0 = x = 3$), forming an isostructual orthorhombic series followed by a Mott-Hubbard type antiferromagnetic transition. The Néel temperature T_N was found to first increase with the interfullerene spacing and then decreases for $(NH_3)Rb_3C_{60}$, forming a maximum at 76K. This behavior can not be understood by a simple localized electron model and provides direct evidence for the important role of electron correlation effects.

INTRODUCTION

The fullerenes have attracted much interest, not least because of the superconductivity of A_3C_{60} (A = K, Rb). The experimentally estimated Coulomb interaction U between two electrons on the same C_{60} molecule may be a factor 2~3 larger than the width W of the (partly occupied) t_{1u} band, leading to a conjecture that the fullerene compounds are highly correlated electron materials [1]. In fact, the spin density wave (SDW) states found in the polymeric A_1C_{60} and $NaCs_xRb_{1-x}C_{60}$ are examples of electron correlation effects combined with the low dimensionality [2,3]. The intercalation of ammonia into the K_3C_{60} superconductor induces a slight structural distortion from cubic to orthorhombic, forming $(NH_3)K_3C_{60}$, where superconductivity (SC) is replaced by an antiferromagnetic (AF) insulating state [4-6]. The resulting AF state with a Néel temperature (T_N) of 40K is confirmed by several methods [5,7-9]. These investigations on the trivalent compounds ($NaCs_xRb_{1-x}C_{60}$ and $(NH_3)K_3C_{60}$) revealed that slight modification of the fcc structure which affords various superconductors dramatically changes the electronic properties.

In this paper, we present synthesis and EPR experiments of a new series of the $(NH_3)K_3C_{60}$ type compounds, in order to uncover the nature of AF state. We found that T_N systematically changes, and has a maximum as a function of the interfullerene distance. These result are explained by the general phase diagram of the Mott-

CP544, *Electronic Properties of Novel Materials—Molecular Nanostructures*, edited by H. Kuzmany, et al.
© 2000 American Institute of Physics 1-56396-973-4/00/$17.00

Hubbard system, indicating that the intercalation of ammonia into fcc superconducting fullerides results in highly correlated electron systems.

EXPERIMENTAL

While $(NH_3)K_3C_{60}$ is synthesized by exposing the preformed K_3C_{60} to ammonia gas [4], $(NH_3)K_{3-x}Rb_xC_{60}$ $(0 < x = 3)$ were obtained by removing of ammonia from the ammonia rich phase. After dissolving stoichiometric amount of C_{60} and alkali metals into dry liquid ammonia kept at $-65°C$, an unidentified ammonia rich phase was obtained by a slow evaporation of ammonia. Ammonia was further removed by a ten-minute annealing at $100°C$ to obtain high quality sample $(NH_3)K_{3-x}Rb_xC_{60}$. The sample was sealed in a thin glass capillary. 9GHz EPR spectra have been measured from 40K to 300K using a JEOL EPR spectrometer equipped with an APD cryostat.

RESULT AND DISCUSSION

The observed x-ray patterns of $(NH_3)K_{3-x}Rb_xC_{60}$ $(0 < x = 3)$ agreed fairly well with the simulation by the Lazy-Pulverix software (K. Yvon, W. Jeitschko, E. Parthe, unpublished) on the orthorhombic $(NH_3)K_3C_{60}$ type structural model, where a single A-NH_3 pairs occupies every octahedral site. The ESR signals of samples consist of two components. Through a Lorentzian fit, we deduced the integrated intensities and the full width at half maximum linewidths $?H_{1/2}$ for the two components. Since the intensity of the narrow line $(?H_{1/2} = 3.2mT$ at 300K for $(NH_3)KRb_2C_{60})$ displays a Curie-type temperature dependence throughout 40-300K, the narrow line is ascribed to a paramagnetic impurity possibly due to lattice imperfections. The temperature dependence of the integrated, normalized by room temperature value, of the intrinsic peak is summarized in Figure 1. Dramatic anomalies were observed at 40K for $(NH_3)K_3C_{60}$, 67K for $(NH_3)K_2RbC_{60}$, 76K for $(NH_3)KRb_2C_{60}$ and 58K for $(NH_3)Rb_3C_{60}$, receptively. Importantly, the decrease of intensity at 40K for the standard compound $(NH_3)K_3C_{60}$ is caused by the AF transition, which has been confirmed by other measurements [5,7-9]. The rapid broadening of the EPR line across these temperatures, which is a sign of buildup of internal fields due to the magnetic ordering, is also observed for every sample. The drop in intensity does not necessarily mean the decrease of spin susceptibility, but that the signal becomes invisible due to the huge magnetic broadening, as observed in $(NH_3)K_3C_{60}$. Taking into consideration the close correlation of chemical, structural and EPR properties with those of $(NH_3)K_3C_{60}$, we concluded that the newly compounds $(NH_3)K_2RbC_{60}$, $(NH_3)KRb_2C_{60}$ and $(NH_3)Rb_3C_{60}$ are antiferromagnets with $T_N = 67K$, 76K, and 58K, respectively.

Figure 2 shows the plots of T_N against the Volume per C_{60}. The observed change of T_N is totally different from what is expected from a simple localized moment model, since this model predicts that the increase of interfullerene distance causes the

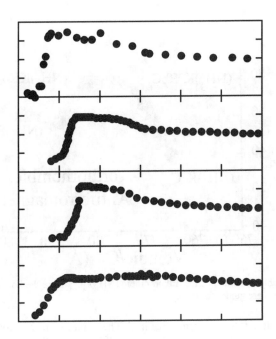

FIGURE 1. Temperature dependence of integrated intensity (normalized by the room temperature value) for $(NH_3)K_{3-x}Rb_xC_{60}$ ($0 = x = 3$).

decrease of the exchange interaction J and thus the reduction of T_N. The present results reminds us a phase diagram of the Mott-Hubbard system, where T_N shows a crossover behavior from the itinerant to localized picture against U/W [10]. T_N increases with U/W in the metallic regime (small U/W), while T_N decreases in the localized regime (large U/W). This feature closely resembles Fig. 2, since the V is a good parameter for $1/W$. In other words, the electronic states of the $(NH_3)A_3C_{60}$ system should be explained not by a simple localized model, but in terms of the Hubbard model.

An important question arises from this result: what is the origin of such a dramatic change of electronic properties on ammoniation? Since $(NH_3)K_3C_{60}$ is expected to be a superconductor according to an empirical correlation between T_c and V, we postulate that this transition is not caused by a simple lattice expansion, but that the symmetry reduction from fcc to orthorhombic crystal plays a crucial role. According to Lu and Gunnarsson, the doped C_{60} is essentially a strongly correlated electron system, and the triple degeneracy of the LUMO of C_{60} is crucially important to maintain the metallic state [11,12]. Ammoniation removes this degeneracy by the structural transformation from fcc to orthorhombic.

In summary, we have synthesized an orthorhombic series of alkali ammonia trivalent fulleride compounds, and showed that AF transition T_N exhibits a systematic

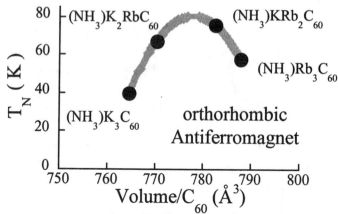

FIGURE 2. T-V phase diagram of the orthorhombic $(NH_3)_4 C_{60}$. Here the horizontal axis V represents the volume/C_{60}. lines are guides for eyes.

change against the interfullerene spacing. The change of T_N, forming a peak as a function of V, is possibly explained in terms of the phase diagram of Hubbard model.

ACKNOWLEDGMENTS

We thank Y. Murakami, H. Shimoda, T. Muro, T. Itou, D. H. Chi, S. Moriyama and M. Miyake for experimental collaborations. This work has been supported by the JSPS "Future Program", RFTF96P00104, and also by the Grant-In-Aid for Scientific Research on the Priority Area "Fullerenes and Nanotubes" by the Ministry of Education, Sport, Science and Culture of Japan.

REFERENCES

1. Gunnarsson, O. *Rev. Modern Phys.* **69**, 575- (1997)
2. Chauvet, O. *et. al.*, *Phys. Rev. Lett.* **72**, 2721-2724 (1994).
3. Arcon, D. *et. al.*, *Phys. Rev. Lett.* **84**, 562-565 (2000).
4. Rosseinsky, M. J. *et. al.*, *Nature (London)* **364**, 425-427 (1993).
5. Iwasa, Y. *et. al.*, *Phys. Rev. B* **53**, R8836-8839 (1996).
6. Allen, K. M. *et. al.*, *J. Mater. Chem.* **6**, 1445- (1996).
7. Simon, F. *et. al.*, *Phys. Rev. B* **61**, R3826-3829 (2000).
8. Tou, H. *et. al.*, *Physica B* **261**, 868- (1999).
9. Prassides, K. *et. al.*, *J. Am. Chem. Soc.* **121**, 11227-11228 (1999).
10. Moriya, T. and Hasegawa, H. *J. Phys. Soc. Jpn.* **48**, 1490- (1980).
11. Lu, J. P. *Phys. Rev. B* **49**, 5687-5687 (1994).
12. Gunnarsson, O. *et. al.*, *Phys. Rev. B* **54**, R11026-11029 (1996).

Structural Studies of $(NH_3)K_3C_{60}$ at Ambient and High Pressure

S. Margadonna,[1] K. Prassides,[1] H. Shimoda[2] and Y. Iwasa[2]

[1]School of Chemistry, Physics and Environmental Studies, University of Sussex,
Brighton BN1 9QJ, U.K.
[2]Japan Advanced Institute of Science and Technology, Tatsunokuchi, Ishikawa, Japan.

Abstract. The structural features of the ammoniated fulleride $(NH_3)K_3C_{60}$ are studied by neutron and synchrotron X–ray powder diffraction at low temperature and high pressure, respectively. Below 150 K the system adopts an orthorhombic structure in space group *Fddd*. Rietveld refinements led to the determination of the exact orientation and geometry of both the C_{60} units (antiferrorotative ordering) and the ammonia molecules in the pseudo octahedral sites (antiferroelectric ordering). We also performed synchrotron X-ray powder diffraction measurements at pressures between ambient and 3.9 GPa. Near ~1 GPa, $(NH_3)K_3C_{60}$ undergoes a structural phase transition from orthorhombic *Fmmm* (isostructural with the ambient pressure high temperature phase) to orthorhombic *Fddd* (isostructural with the ambient pressure low temperature phase).

INTRODUCTION

Alkali fullerides A_3C_{60} (A= alkali metal) exhibit superconductivity with T_c as high as 33 K at ambient pressure ($RbCs_2C_{60}$: *fcc* structure). The well known relationship in which the transition temperature, T_c scales monotonically with the cubic unit cell size, a_0 can be rationalised in terms of increasing density-of-states at the Fermi level, $N(\varepsilon_F)$ with increasing interfullerene separation, resulting from the decrease in the overlap between the molecules that leads to band narrowing. As a consequence, in order to obtain high-T_c fullerides, large interfullerene spacings are needed. The lattice expansion which can be achieved by using alkali metals alone in the interstitial sites of the cubic structure is limited by the maximum size of the alkali metal cations ($r_{max}\leq r(Cs^+)$= 1.69 Å) used for intercalation. Larger lattice constants can be achieved by using as structural "spacers" alkali ions solvated with neutral molecules, such as ammonia; in these cases, the extent of charge transfer is also maintained.

In K_3C_{60}, the introduction of one ammonia molecule in the octahedral site of the *fcc* structure to form $(NH_3)K_3C_{60}$ produces an anisotropic expansion of the fulleride array, inducing a symmetry reduction and suppression of superconductivity at ambient pressure [1]; application of pressure >1 GPa leads to recovery of superconductivity with T_c= 28 K [2]. A structural phase transition has been identified by synchrotron X-ray powder diffraction below 150 K and attributed to the ordering of the orientation of the K^+-NH_3 pairs residing in the interstitial sites [3]. Measurements of the electronic and magnetic properties by ESR and ^{13}C NMR have shown $(NH_3)K_3C_{60}$ to be a narrow-band metal which exhibits a transition to an insulating ground state at 40 K [4,5]. Subsequent μ^+SR experiments characterized the ground state as a long range

CP544, *Electronic Properties of Novel Materials—Molecular Nanostructures*, edited by H. Kuzmany, et al.
© 2000 American Institute of Physics 1-56396-973-4/00/$17.00

ordered antiferromagnet implying that the suppression of superconductivity and the metal-insulator transition are associated with effects of magnetic origin, in analogy with the phenomenology in high-T_c and organic superconductors [6].

By performing high resolution powder neutron diffraction measurements, we could introduce explicitly the NH_3 molecule in the refinement of the structure instead of employing a hypothetical isoelectronic Ne atom. Data were collected at low temperatures on the deuterated analogue of $(NH_3)K_3C_{60}$, as isotopic substitution is necessary for neutron diffraction studies because of the large incoherent scattering cross section of hydrogen. A detailed structural study of $(NH_3)K_3C_{60}$ as a function of pressure is also necessary to detect any changes in symmetry that could be linked to the recovery of superconductivity. We collected synchrotron X-ray powder diffraction data at room temperature at various pressures ranging between ambient and 3.9 GPa.

RESULTS

(a) Structural Results at Ambient Pressure

High resolution neutron diffraction measurements on $(ND_3)K_3C_{60}$ were performed at 10 K ($\lambda= 1.5944$ Å). The presence of a minor K_3C_{60} impurity necessitated the use in the Rietveld analysis of a two-phase model of orthorhombic $(ND_3)K_3C_{60}$ and cubic K_3C_{60}.

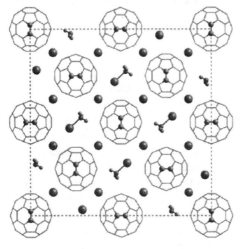

Our starting structural model for $(ND_3)K_3C_{60}$ was based on the orthorhombic structure reported by Ishii *et al.* (space group *Fddd*) [3]. The C_{60} units are located at $(0,0,0)$ and $(\frac{1}{2},\frac{1}{2},\frac{1}{2})$ and are oriented in such a way that the three orthogonal twofold axes are aligned with the unit cell vectors and a 6:6 C-C bond of the molecule at the origin aligned parallel to *a*. The off-centred octahedral K^+ is placed on the $32h$ position $(\frac{1}{4}+\delta_x,\delta_y,0)$,

Fig. 1 Projection of the face-centred orthorhombic structure of $(ND_3)K_3C_{60}$ on the (110) plane. The antiferrorotative order of C_{60} at the $(0,0,0)$ and $(\frac{1}{2},0,0)$ sites is shown.

where $(\frac{1}{4},0,0)$ is the centre of the pseudo-octahedral site and $(\delta_x \approx 0.03, \delta_y \approx 0.03, 0)$ is the displacement vector. In our model we introduced explicitly the NH_3 molecule. The N atom was placed in the $32h$ off-centred position and the fractional coordinates were first refined ($\delta_x \approx -0.03, \delta_y \approx -0.02, 0$; the displacement is in the opposite direction from K^+). The positions of the D atoms were consequently derived by assuming that all D-N-D angles were equal to $109.5°$ and that the three N-D distances are equal to 1.00 Å.

Despite the introduction of the NH_3 molecule the results of the Rietveld refinement in the whole 2θ range were quite poor, especially at high 2θ as the intensities of numerous reflections were not well reproduced but damped to the background. In general, the high 2θ region of the neutron diffraction profiles of fulleride salts is strongly influenced by the orientational state of the C_{60} units, suggesting that the X-ray structural model of $(ND_3)K_3C_{60}$ may be inadequate in describing correctly the C_{60} orientations. For this reason, we performed a series of refinements in an attempt to determine the optimum orientational state of the fulleride units. The agreement factors of the refinement improved significantly by introducing an antiferrorotative ordering of the C_{60} molecules for which the fulleride units located at (0,0,0) and (½,½,½) adopt two different orientations related by a 90° rotation about the c orthorhombic axis and order along the a axis (Fig. 1).

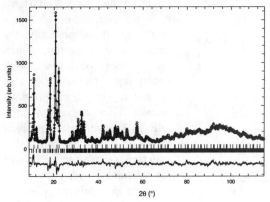

Fig. 2 Final observed (points) and calculated (solid line) neutron (λ= 1.5944 Å) powder diffraction profile for $(ND_3)K_3C_{60}$ at 10 K. The bottom line shows the difference profiles and the ticks mark the position of the Bragg reflections of $(ND_3)K_3C_{60}$ (bottom) and K_3C_{60} (top).

The next step was to determine the exact orientation and geometry of ammonia in the octahedral void together with its coordination environment to the fulleride units. For this reason, we performed a series of refinements keeping fixed the geometry of the C_{60} units and the positions of the K and N atoms, and allowing only the D atoms to rotate around the ND_3 molecular threefold axis. The variation of the agreement factor χ^2 of the Rietveld refinements as a function of the rotation angle, ϕ showed a minimum for a clockwise rotation of 110° from the starting orientation (R_{wp}= 9.39%, R_{exp}= 6.81%). After arriving at the optimum orientation adopted by the ammonia molecule, we checked the ND_3 geometry by varying the D-N-D angle in steps of 1° between 106° and 112°. By following again the evolution of χ^2 as a function of the D-N-D angle, a minimum was found for the value of 108°. A final Rietveld refinement using the optimised orientation and geometry of the ammonia molecule in the pseudo octahedral void was then performed and the results are shown in Fig. 2. The final values of the lattice constants at 10 K are a= 29.929(9) Å, b= 29.787(9) Å, c= 27.136(4) Å (R_{wp}= 8.68%, R_{exp}= 6.81%).

(a) Structural results at high pressure

Synchrotron X-ray powder diffraction profiles of $(NH_3)K_3C_{60}$ were collected at pressures between ambient and 3.9 GPa. Inspection of the diffraction data of $(NH_3)K_3C_{60}$ indicated that the patterns at ambient, 0.4 and 0.8 GPa could be indexed with a face-centred ortho-rhombic cell (space group *Fmmm*) consistent with the high temperature ambient pressure structure [1]. How-ever, at pressures higher that 0.8 GPa, a number of extra peaks starts to appear, growing in intensity with increasing pressure.

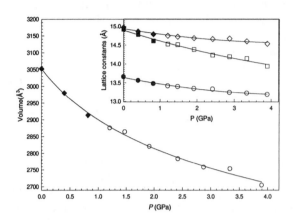

The appearance of forbidden peaks with half-integer Miller indices implies the existence of a super-structure, arising by doubling the unit cell constants of the structure at low pressures. It is precisely the same phase transition, which has been observed with decreasing temperature, as discussed in the previous section. Thus above 0.8 GPa, $(NH_3)K_3C_{60}$ undergoes a phase transition to an orthorhombic structure (space group *Fddd*) isostructural with the low temperature ambient pressure phase. To extract reliable values of the lattice constants, the diffraction profiles at pressures up to 3.9 GPa were analysed by the LeBail method in the 2θ range 2°-19°. The X-ray diffraction profiles up to 0.8 GPa were treated using a two-phase model of a coexisting major $(NH_3)K_3C_{60}$ orthorhombic phase (space group *Fmmm*) and a minor K_3C_{60} cubic phase (space group $Fm\overline{3}m$), while for pressures in excess of 1.2 GPa, two-phase refinements were performed using for the $(NH_3)K_3C_{60}$ phase the orthorhombic *Fddd* structural model employed before for the low temperature neutron data.

Fig. 3 Pressure dependence of the volume of orthorhombic $(NH_3)K_3C_{60}$ ($V(Fmmm)$: diamonds, ($V/8)(Fddd)$: open circles). The line through the data represents a least squares fit to the Murnaghan EOS. Inset: pressure evolution of the orthorhombic lattice constants (a: diamonds, b: squares, c: circles). The data for the high pressure phase are divided by a factor of 2.

Fig. 3 shows the pressure evolution of the volume of the orthorhombic unit cell of $(NH_3)K_3C_{60}$ (scaled to that of the *Fmmm* phase) together with a least squares fit of its ambient temperature equation-of-state (EOS) to the semi-empirical second-order Murnaghan EOS. The extracted value of the bulk modulus is K_0= 14(2) GPa and that of its pressure derivative, K_0'= 13(1). The volume compressibility, κ= 0.073(5) GPa^{-1} is larger than those measured for the merohedrally disordered *fcc* fulleride phases, K_3C_{60} (0.036(3) GPa^{-1}) and Rb_3C_{60} (0.046(3) and 0.058(1) GPa^{-1}), a result presumably reflecting the expansion of the unit cell due to the introduction of the ammonia in the octahedral site and the modified orientational state of the C_{60}^{3-} ions in the orthorhombic structure.

The evolution of the lattice constants with pressure shows that the system displays a strongly anisotropic compressibility along the three directions (Fig 3 inset). The most compressible axis is b, while the c and a axes show almost the same compressibility. While at ambient pressure, the structure of $(NH_3)K_3C_{60}$ can be regarded almost as tetragonal $(a \approx b)$, with increasing pressure the difference in the values of a and b increases leading to a more anisotropic orthorhombic crystal structure. The introduction of the ammonia in the octahedral void of the cubic structure of K_3C_{60} results in an anisotropic expansion of the lattice, which primarily occurs on the $\{hk0\}$ planes. As previously suggested by Zhou et al. [2], this additional expansion in the $\{hk0\}$ planes could lead to electron localisation, inducing the metal-insulator transition and the disappearance of superconductivity. As the application of hydrostatic pressure causes primarily a significant compression of the b axis with consequent rapid contraction of the $\{hk0\}$ planes, this could lead to band broadening, delocalisation of the t_{1u} electrons and the recovery of superconductivity

CONCLUSIONS

Neutron powder diffraction was employed to study the structure of $(ND_3)K_3C_{60}$ which at 10 K is orthorhombic (space group $Fddd$). A series of Rietveld refinements established the exact orientation and geometry of both the C_{60} units and the ammonia molecules in the pseudo octahedral sites. The ND_3 molecule is orientationally ordered along the [110] directions, inclined at an angle of 3.9° to the face diagonal directions. A detailed investigation of the possible orientational states of the fullerides units was performed, leading to orientationally ordered C_{60} units, which adopt two different orientations related by a 90° rotation about the c axis and order along the a axis. The introduction of the ND_3 molecules in the octahedral sites leads to a large expansion of the ab basal plane. The resulting increase in the interball separation could have important consequences for electron localisation and the loss of superconductivity.

We also performed synchrotron X-ray powder diffraction measurements with pressures varying between ambient and 3.9 GPa. $(NH_3)K_3C_{60}$ undergoes near ~1 GPa a structural phase transition from orthorhombic $Fmmm$ to orthorhombic $Fddd$. The evolution of the unit cell volume with pressure showed a larger compressibility than that of the parent K_3C_{60}, consistent with the larger unit cell caused by the introduction of the ammonia molecule. With increasing pressure, orthorhombic $(NH_3)K_3C_{60}$ becomes increasingly anisotropic with an increased size difference between a and b, as a result of the larger compressibility along b. The latter causes a rapid contraction with pressure of the ab planes, and consequently of the interball separation, leading to electron delocalisation and recovery of superconductivity.

REFERENCES

[1] Rosseinsky, M. J. et al. Nature 1993, 364, 425.
[2] Zhou, O et al. Phys. Rev. B 1995, 52, 483.
[3] Ishii, K. et al. Phys. Rev. B 1999, 59, 3956.
[4] Iwasa, Y. et al. Phys. Rev. B 1996, 53, R8836.
[5] Allen, K. M. et al. Mater. Chem. 1996, 6, 1445.
[6] Prassides, K. et al. J. Am. Chem. Soc. 1999, 121, 11227.

First NMR study on a C_{60} ternary "superconductor" compound: $K_6 C_{60} (C_6 H_6)_{1.5}$

A.-S. Grell, B. Burteaux*, A. Hamwi* and F. Masin

Université Libre de Bruxelles, CP 232, Bvd. du Triomphe, 1050 Bruxelles, Belgium
* LMI, UPRESA CNRS 6002, Université B. Pascal de Clermont-Ferrand, Aubière, France

Abstract. ^{13}C and ^1H NMR measurements on $K_6 C_{60} (C_6 H_6)_{1.5}$ as function of temperature are presented. At room temperature, the ^{13}C line shift of the C_{60} molecules of our sample is equal to 218.63 ppm relative to TMS and the line shape is similar to the one obtained for $RbCs_2C_{60}$. The ^{13}C chemical shift does not change between 300 and 10 K. The ^1H line shift of the benzene molecules is equal to - 2 ppm (relative to TMS) at room temperature and varies between -2 and 11 ppm between 10 and 300 K, the most important variation in this shift occurs between 60 and 25 K. The ^{13}C spin-lattice relaxation time (T_1) as function of temperature varies between 17 and 110 s. The ^1H spin-lattice relaxation time shows a sudden change around 60 K.

INTRODUCTION

Nuclear magnetic resonance (NMR) has proven to be a remarkably versatile tool for the elucidation of key properties of superconductors. The alkali-doped solid A_3C_{60} (where A is an alkali metal), which are superconductors with transition temperature among the highest known apart from the high-T_c cuprates [1]. In this paper we present the first NMR study on a possible C_{60} ternary "superconductor" $K_6 C_{60} (C_6 H_6)_{1.5}$ [2-4]. As the superconductivity temperature of this compound is the same as the one obtained for K_3C_{60}, it is possible that some K_3C_{60} clusters are present in our compound explaining why the susceptibility curve [3,4] of the sample indicates that this sample is superconductor at 17.2 K. As NMR acts as a local probe of the nuclei environment, it is interesting:

- to check the ^{13}C chemical shift of the C_{60} molecules
- to detect whether there are other types of compounds in this sample (K_3C_{60}, K_4C_{60})
- to examine the ^{13}C environment of the benzene molecules.

We tried to characterize this sample by MAS; the sample however could not be made to rotate in our stretch rotor in spite of all precautions taken to balance the sample. This is probably due to the high conductivity of the compound. To study the ^{13}C of the benzene in addition to the ^{13}C of the C_{60}, we have also tried some cross-polarization (CP)

CP544, *Electronic Properties of Novel Materials—Molecular Nanostructures*, edited by H. Kuzmany, et al.
© 2000 American Institute of Physics 1-56396-973-4/00/$17.00

measurements but with no result. This is probably due to electro-accoustic effects produced by the motion of small conducting crystals induced by radio frequency pulses in the magnetic field. To overcome this problem, we should dilute the sample in something unreactive. In this paper, only static NMR results will be presented.

EXPERIMENTAL

The supposed orthorhombic crystals [2] of $K_6C_{60}(C_6H_6)_{1.5}$ were first obtained by the synthesis of the basic binary compound K_6C_{60} followed by the immersion of this binary compound in liquid benzene during 4 to 6 days.

The ^{13}C NMR measurements were performed with a Bruker MSL300 spectrometer. All chemical shift are given with respect to the ^{13}C and ^1H resonance frequency in tetramethylsilane (TMS). ^{13}C spin-lattice relaxation times (T_1) were obtained by using a saturation-recovery pulse sequence. The sample (inside a sealed glass tube of 5 mm diameter) was placed in a Bruker (HP LTBB SOL5) low temperature probe and the temperature varied by means of an Oxford CF1200 cryostat. Magic Angle Spinning (MAS and CP MAS) experiments were run with a 7 mm Bruker MAS probe equipped with an elongated "Stretch" stator. For all measurements, standard CP pulse sequence has been used. To record the ^1H and ^{13}C NMR spectra a "one pulse" $\frac{\pi}{2}$ acquisition with phase alternation was used.

RESULTS AND DISCUSSION

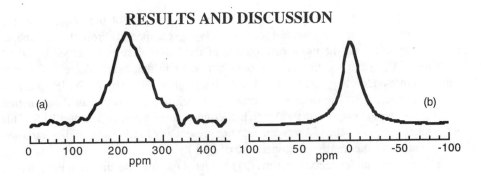

Figure 1: ^{13}C line shape of C_{60} molecules (a) and ^1H line shape of benzene molecule (b) in $K_6C_{60}(C_6H_6)_{1.5}$, at room temperature. The ^{13}C chemical shift is 218 ppm and the proton chemical shift is 2 ppm (relative to TMS).

^{13}C study of C_{60} molecules

The first result obtained from the ^{13}C NMR line of $K_6C_{60}(C_6H_6)_{1.5}$ presented in Fig. 1(a) is the relatively high value of the chemical shift equal to 218.63 ppm (187 ppm

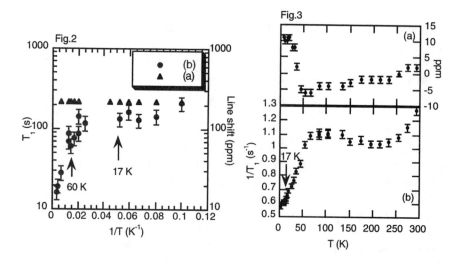

Figure 2: ^{13}C chemical shift (a) (according to TMS) and the spin-lattice relaxation time (b) versus temperature.

Figure 3: ^{1}H chemical shift of benzene molecules (a) and (b) ^{1}H spin-lattice relaxation rate $(1/T_1)$ versus temperature.

for K_3C_{60} [1]) corresponding to 75 ppm away from the shift of pure C_{60}. The line shape is similar to the one observed for the $RbCs_2C_{60}$ compound. From the line shape, it clearly appears that the usual fast rotation of the C_{60} molecules observed by NMR in different C_{60} samples, like K_3C_{60}, does not occur in this case. There is however sufficient motion left to present a complete exchange after 750 μs by 2D-NMR spectrum at room temperature. The surprising result is the evolution of this line shift as function of temperature, see Fig. 2 (curve a) which does not change between 300 and 10 K. This means that the ^{13}C of the C_{60} molecule do not present any local susceptibility variation above and below the transition temperature of 17.2 K.

The ^{13}C spin-lattice relaxation time (T_1) results of C_{60} are also shown in Fig.2 (curve b). T_1 values are obtained by fitting the ^{13}C magnetization recovery curves with a stretched-exponential $(m(t) = m_0(1 - e^{-(t/T_1)^\beta})$ with $\beta = 0.47$)[1]. To fit these data, and considering that we are in presence of a superconductor we use the following equation : $\frac{1}{T_1} = \alpha e^{\frac{\Delta_0}{kT}}$ (with $\Delta_0 = 1.76 \ kT_c$ in the weak-coupling limit). In this case we obtain $\Delta_0 = 0.4 \ kT_c$.

^{1}H study of benzene molecules

The ^{1}H line shape of our sample at room temperature is presented in Fig.1b. The chemical shift of 2 ppm should be compared to that of liquid benzene equal to 7.15 ppm. The evolution of the ^{1}H chemical shift of the benzene molecules as function of temperature is shown in Fig.3a. We observe a big step in the shift values around 60 K. Between 300 and 60 K, there is a slow decrease, from 2 ppm to -5 ppm and then a sudden rise to 11 ppm.

The ^1H spin-lattice relaxation rates ($1/T_1$) versus temperature are given in Fig.3b. T_1 values are obtained, in this case, by fitting the ^1H magnetization recovery curves with a single exponential ($m(t) = m_0(1 - e^{-t/T_1})$). This curve presents three main parts: (i) between 300 and 70 K where a relatively small evolution of T_1 is observed; (ii) between 60 and 17 K the T_1 values increase rapidly and (iii) between 17 and 10 K this increase slows down.

CONCLUSIONS

In this compound, the C_{60} molecules do not rotate isotropically like in K_3C_{60}, as shown by the broad ^{13}C line width. The ^{13}C chemical shift which is 75 ppm larger than the 143 ppm observed for pure C_{60}, indicates that this compound is a conductor. This fact is also confirmed by the failure of MAS and CP NMR measurements. If this sample becomes superconductor, it is really surprising that the ^{13}C chemical shift does not vary at $T_c = 17.2$ K.

The ^1H chemical shift presents a relatively strong evolution below 60 K. The ^1H spin-lattice relaxation time as function of temperature shows the same type of evolution. These results obtained by ^1H NMR measurements are really surprising and raise some questions:

  - How does the charge transfer happen?
  - Where are the conductor strings?

ACKNOWLEDGMENTS

This work was supported by the BNB (Banque Nationale de Belgique) and the "Fonds Defay", A.-S. Grell acknowledges financial support from FRIA (Belgium). Finally, we thank our colleagues W. Stone and G. Gusman for reading the manuscript.

REFERENCES

1. C.H. Pennington, and V. A. Stenger *Rev. of Mod. Phys.* **68**, 855-910 (1996), and all references mentionned in it.
2. B. Burteaux, *PhD. thesis*, Université de Clermont-Ferrand (1997).
3. A. Hamwi, *Images de la Recherche*, Les fullerènes, CNRS (1998) 63-65.
4. B. Burteaux, A. Hamwi, Y. Ksari, G. Chouteau and A. Sulpice, *Mol. Cryst. Liq. Cryst.* **310**, 143-148 (1998).

Jahn-Teller Distortion and Rotational Ordering Transition of C_{60}^- in $[A^+(C_6H_5)_4]_2 \, C_{60}^- B^-$-salts $(A^+ = P^+, As^+; B^- = I^-, Cl^-)$

W. Bietsch, J. Bao, P. Medick, S. Traßl

Lehrstuhl für Experimentalphysik II, Universität Bayreuth, 95440 Bayreuth

Abstract. We used single crystals of $[A^+(C_6H_5)_4]_2C_{60}^- B^-$ as an ideal model system to determine accurately principal values and principal axes of the g-tensor of C_{60} mono radical anion embeded in a crystal field. The g-tensor corresponds to the (static) Jahn-Teller distortion of C_{60}^-. In contrast to the Jahn-Teller distortion of isolated C_{60}^- with D_{5d} or D_{3d} symmetry the crystal field stabilizes the D_{2h} symmetry. Extending DSC, NMR and ESR to temperatures above 300 K we found the rotational ordering transition of C_{60}^- in these salts. Both rotational ordering of C_{60}^- and the transition from static to dynamic Jahn-Teller effect depend on the counterions $A^+(C_6H_5)_4$.

INTRODUCTION

Interesting physical properties like superconductivity or ferromagnetic ordering of fullerides are found so far in C_{60}-based compounds only. The high molecular symmetry of the C_{60} molecule leads in the fullerides for the negatively charged C_{60} ions to a Jahn-Teller (JT) distortion which survives as a molecular property in the solid state due to weakly bound molecules. Our investigations of TDAE-C_{60} [1] motivated us to determine the JT distortion of C_{60}^- and its orientation relative to the C_{60} structure in the solid state (i.e. within a crystal field) by ESR according to numerous ESR studies of the JT effect of transition ions, especially of Cu^{2+} complexes. Historically, the magnetically diluted Cu^{2+} salts provided the first experimental evidence of the JT effect [2]. To monitor the JT distortion of C_{60}^- and its relation to molecular reorientational dynamics we used the air stable $[A^+(C_6H_5)_4]_2 \, C_{60}^- B^-$ fullerene salts ($A^+ = P^+, As^+; B^- = I^-, Cl^-$) which were grown by electrocrystallization techniques [3]. In these salts C_{60}^- radical ions are well separated due to relatively large organic cations $A^+(C_6H_5)_4$ which is a requirement that the JT distortion of C_{60}^- becomes visible in the ESR spectra [4].

RESULTS AND DISCUSSION

Using a home built goniometer (Fig. 1) which allows to rotate a single crystal in the ESR resonator around two perpendicular axes at 34 GHz and at low temperatures we could refine our g-factor measurements [4] and accurately determine the principal

CP544, *Electronic Properties of Novel Materials—Molecular Nanostructures*, edited by H. Kuzmany, et al.

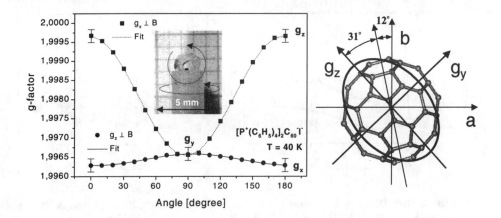

FIGURE 1. left: angular dependence of the g-factor of C_{60}^- at 40 K when the crystal is rotated around two principle axes of the g-tensor. left inset: goniometer with crystal. right: anisotropy of the g-tensor in the a-b plane (view along c axis) relative to the C_{60}^- structure (note: The ellipse represents the anisotropy of the g-tensor only).

values and principle axes of the g-tensor[1] of C_{60}^- at 40 K: g_{xx} = 1.9962, g_{yy} = 1.9965, g_{zz} = 1.9995 (all: ±0.00005). Fig. 1 left shows the angular dependence of the g-factor of C_{60}^- when the crystal is rotated around two principle axes of the g-tensor of C_{60}^-. Note only the angular dependency for one C_{60}^- orientation is plotted in Fig. 1 [4]. The second angular dependency (not shown) is shifted by 90° in agreement with two possible C_{60}^- orientations (merohedral disorder) found by x-ray [3]. Fig. 1 right depicts the anisotropy of the g-tensor in the a-b plane relative to the C_{60} structure and the crystallographic axes. The g-tensor, which is a direct measure of the JT distortion, is elongated along the pentagons of C_{60} pointing towards the phenyl rings of the positive counterions. In contrast to earlier measurements [5,6] we observed within experimental errors a very small but distinct deviation from the axial symmetry (η = 0.14) which can be seen best from the rotation of the crystal around the molecular g_z axis (Fig. 1 left). This means that crystal field has stabilized a static JT distortion with D_{2h} symmetry (i.e. no axial g-tensor) compared to the JT distortion with D_{5d} or D_{3d} symmetry (i.e. axial g-tensor: g_{xx} = g_{yy}) of isolated C_{60}^-. This stabilization of D_{2h} symmetry by crystal fields has been predicted theoretically [7] and is in accordance with symmetry considerations for C_{60}^- (molecular two-fold axis parallel to the four-fold axis c) and tetragonal crystal structure (four-fold axis c) [3]. No definite difference for the g-factor results has been found among systems containing different cations $A^+(C_6H_5)_4$ and halogen anions B^-.

We extended to our knowledge for the first time DSC, NMR and ESR measurements of these salts to the temperature range above 300 K and found the transition where the free rotation of whole C_{60}^- ions begins. As can be seen from the comparison of DSC data (Fig. 2, left) the transition temperature in these systems is shifted to

---

[1] we used the convention: $g_{xx} < g_{yy} < g_{zz}$

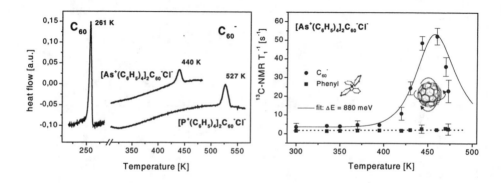

FIGURE 2. left: Differential Scanning Calorimetric data of salts with $P^+(C_6H_5)_4$ and $As^+(C_6H_5)_4$ counterions. For comparison data of pristine C_{60} are plotted, too. right: ^{13}C-NMR spin-lattice relaxation rates, measured by saturation-$90°$-τ-$180°$ echo sequence with ^1H-decoupling.

higher temperatures relative to pristine C_{60} and depends on the positive counterions $A^+(C_6H_5)_4$. This rotational dynamics can be probed by ^{13}C-NMR spin-lattice relaxation measurements (Fig. 2 right) according to A_4C_{60} [8] or ternary AB_2C_{60} [9] compounds and is also found in the ESR linewidth (Fig. 3 left). According to carbon atoms in the phenyl rings and carbon atoms of C_{60}^- the ^{13}C-NMR spin lattice relaxation rate T_1^{-1} consists of two components (Fig. 2 right). While the T_1^{-1} due to carbon atoms in the phenyl rings is almost temperature independent T_1^{-1} due to carbon atoms of C_{60}^- shows a distinct peak separating the regions of slower motion ("ratchet" uniaxial rotations) and extreme motional narrowing (free C_{60}^- rotation). We ascribe the relative large activation energy of about 880 meV to both coulomb interactions between C_{60}^- and counterions [9] and face to face coordination of phenyl rings and C_{60}^- [10]. Similar to the rotational ordering transition at high temperatures the counterions $A^+(C_6H_5)_4$ determine the transition temperature from the static to the dynamic JT distortion at low temperatures [5,4] which is seen in the averaging of the g-factor anisotropy above 140 K (Fig. 3, right). While the ESR linewidth probes the dynamic competition between different JT distortions (distortions of same symmetry along different molecular axes or distortions of different symmetry) with an activation energy of 24 meV (Fig. 3 left) the averaging of the g-anisotropy probes the (uniaxial) molecular reorientation with an activation energy of about 40 meV (Fig. 3 right) which is in agreement with the value of temperature dependence of the ^{13}C-NMR linewidth of C_{60}^-.

CONCLUSIONS

At low temperature $[A^+(C_6H_5)_4]_2C_{60}^-B^-$ salts serve as simple model systems to monitor the JT distortion of C_{60}^- via the g-tensor as the crystal field induces a stabilization of a static JT distortion along a specific molecular axis. In agreement with symmetry considerations for molecular g-tensor of C_{60}^- the crystal field stabilizes D_{2h} symmetry. The largest elongation is not parallel to the c axis showing that the JT-effect is detected as a molecular property. Both the transition from the static to the dynamic JT effect

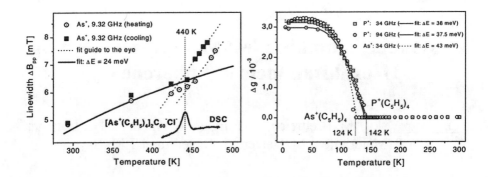

FIGURE 3. left: Temperature dependence of ESR linewidth of $[As^+(C_6H_5)_4]_2C_{60}^-Cl^-$ above 300 K. For comparison DSC data are plotted, too. Note: ΔE was obtained by a fit to data between 4 and 450 K. right: Temperature dependence of the g-factor anisotropy of salts with $P^+(C_6H_5)_4$ and $As^+(C_6H_5)_4$ counterions.

and the rotational ordering transition of C_{60}^- depend on the counterions $A^+(C_6H_5)_4$ indicating that reorientational phenomena are determined by the crystal properties.

ACKNOWLEDGMENTS

We thank I. Bauer and J. Gmeiner for crystal growing, Prof. I. Ovchinnikov and V. Petrashen (Kazan Physical-Technical Institute of Russian Academy of Sciences) for performing ESR measurements at high temperatures, Prof. M. Schwoerer and Prof. E. Rößler for stimulating discussions. Financial support by Sonderforschungsbereich 279 and Fonds der Chemischen Industrie is acknowledged. J. Bao acknowledges financial support by Volkswagen-Stiftung.

REFERENCES

[1] Schilder, A., Bietsch, W., Schwoerer, M., *New Journal of Physics* **1**, 5.1-5.11 (1999).
[2] Bleaney, B., Bowers, K.D., *Proc. Phys. Soc.* (London) **A65**, 667-668 (1952).
[3] Pénicaud, A., Peréz-Benitez, A., Gleason, R., Munoz, V.E., Escudero, R., *J. Am. Chem. Soc.* **115**, 10392-10393 (1993).
[4] Bietsch, W., Bao, J., Schilder, A., Schwoerer, M., "The Jahn-Teller Distortion of C_{60}^- as Determined by ESR" in *Electronic Properties of Novel Materials*, edited by H. Kuzmany et. al. API Conference Proceedings 486, New York, 1999, pp. 3-6.
[5] Gotschy, B., Völkel, G., *Appl. Magn. Reson.* **11**, 229-238 (1996) and references therein.
[6] Kadoma, T., Kato, M., Mogi, K., Aoyagi, M., Kato, T., *Mol. Phys. Reports* **18/19**, 121-126 (1997).
[7] Chancey, C.C., O'Brien, C.M., *The Jahn-Teller Effect in C_{60} and other Icosahedral Complexes*, Princeton, Princeton Universtiy Press, 1997.
[8] Zimmer, G., Mehring, M., Goze, C., Rachdi, F., *Phys. Rev. B* **52**, 13300-13305 (1995).
[9] Thier, K.-F., Goze, C., Mehring, M., Rachdi, F., Yildirim, T., Fischer, J.E., *Phys. Rev. B* **59**, 10536-10540 (1999).
[10] Scudder, M., Dance, I., *J. Chem. Soc., Dalton Trans.*, 3155-3165 (1998).

Globular Amphiphiles: Membrane Forming Hexakisadducts of Fullerene C$_{60}$

O. Vostrowsky,[a] M. Brettreich,[a] S. Burghardt,[a] A. Hirsch,[a] C. Böttcher,[b]
A.P. Maierhofer,[c] M. Hetzer,[c] S. Bayerl[c] and T.M. Bayerl[c]

[a]Institut für Organische Chemie der Universität Erlangen-Nürnberg, D-91054 Erlangen, Germany
[b]Institute for Organic Chemistry, Freie Universität Berlin, D-14195 Berlin, Germany
[c]Institute for Physics EP-5, University of Würzburg, D-97074 Würzburg, Germany

Abstract. The synthesis, membrane and vesicle formation of a new class of artificial amphiphilic lipids is reported. These lipids have a globular architecture, the spherical C$_{60}$ been used as a structure determining tecton. In aqueous solution they form unilamellar spherical vesicles, the properties of which were studied by cryo-TEM recording. Monolayers at the air water interface were investigated at different lateral pressures by a combination of film balance techniques, neutron reflection (NR) and infrared reflection-absorption spectroscopy (IRRAS).

SYNTHESIS OF FULLERENE-BASED AMPHIPHILES

For the synthesis of globular amphiphile **5** the C$_{60}$ monoadduct **1** was reacted by a modified nucleophilic cyclopropanation [1,2] under reversible dimethylanthracene template activation [3] with an excess of didodecyl malonate **2** in the presence of CBr$_4$/DBU. The all-*e*-substituted [1:5]-mixed hexakisadduct *tert*-butyl ester **3** was deprotected by treatment with TFA and the diacid formed condensed with a second generation Newkome-type [4] amide dendron **4** under peptide synthesis conditions. After subsequent deprotection the amphiphilic hexakisadduct **5** with 18 carboxylic end groups was obtained as a yellow powder.[5] Similarly, four pairs of lipophilic aliphatic dodecyl chains were covalently attached at the doubly substituted bisadduct **6** by formation of [2:4]-mixed hexakisadduct **7**. Consequently, the *tert*-butyl protection group was cleaved off and the resulting acid condensed with a first generation amide dendron **8**. The amphiphilic mixed hexakisadduct **9** with 12 carboxylic groups and an octahedral T_h-addition pattern was obtained (Scheme 1).

Because of the balance between hydrophilic and hydrophobic substituents, the compounds **5** and **9** reveal a new type of amphiphiles. They show a globular architecture with addends arranged in an octahedral [1:5]- and [2:4]-addition pattern, the C$_{60}$ fullerene serving as the structure determining core. Molecule dynamic simulations of **5** and **9** (Hyperchem, geometry optimization with force field MM+) revealed two separate hemispheres formed by the two different types of addends, the

CP544, *Electronic Properties of Novel Materials—Molecular Nanostructures*, edited by H. Kuzmany, et al.
© 2000 American Institute of Physics 1-56396-973-4/00/$17.00

densily packed lipophilic and hydrophilic spheres covering completely the C_{60} core. The average dimension along the main axis is about 3.5 nm, comparable with a natural phospholipid. The cross section diameter with 1.4 - 2.3 nm is about two to threefold the cross section of a double chain lipid molecule. The polar as well as the apolar groups can be varied synthetically and thus the amphiphilic properties of the fullero-amphiphiles can be tuned and adjusted.[5]

FORMATION OF UNILAMELLARY VESICLES

The amphiphiles dissolve slowly in water at p_H 7.4 (phosphate buffer) forming a yellowish opalescent colloidal solution. At higher concentrations (< 3 mg/ml), a highly viscous light scattering phase is separated below the remaining clear solution

The globular amphiphile **5** forms vesicles of 100-400 nm diameters in aqueous solution. Freeze fracture electron microscopy and Cryo-TEM recording revealed unilamellar bilayer structures of various shape and of different size ranging from 50-400 nm.

The tendency to form unilamellar structures is favoured by the high density of negative charges at p_H 7.4. With cryo-TEM micrographs of the vesicles a higher resolution is obtained and even the membrane double layer structure (7 nm thickness) can be visualized. Beside the dominant vesicle structures also rod shape aggregates with 20 - 200 nm length and 7 nm cross-section diameter are observed.[6]

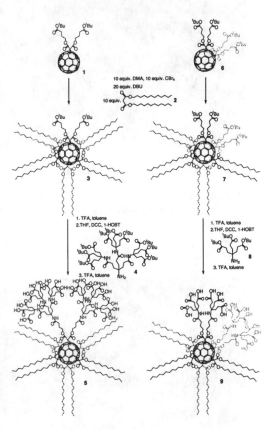

Scheme 1. Synthesis of spherical fullerene based amphiphiles **5** and **9**.[5]

Monolayers of **5** at the air water interface were studied at different lateral pressures by a combination of film balance techniques, neutron reflection (NR) and infrared reflection-absorption spectroscopy (IRRAS). The amphiphil monolayers can be compressed and expanded without significant hysteresis and the alkyl chains remained fluid at all pressures. The thickness of the monolayer at high lateral pressure was 30 Å, similar to that of typical phospholipid monolayers in the condensed state. In contrast, the molecular area was about sixfold higher than that of phospholipids at high pressure. By a titration series, the pK value of the monolayer was determined as 7.5 and pH dependent measurements allowed a variation of the negative headgroup charge by about 18 charges.

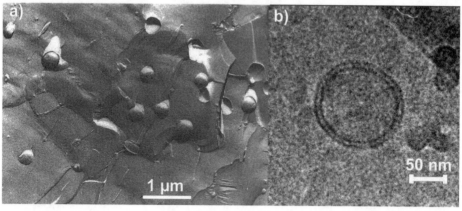

Figure 1. a) Freeze fracture micrographs FFM of various unilamellar vesicle structures (100 - 400 nm) from a solution of 2 mg/ml **1** at p_H 7.4, chilled with liquid N_2. FFM experimental details lit.[6] **b)** Cryo TEM micrographs, projection image of a bilayer membrane vesicle (Ø 80 nm). Membrane thickness ~ 7 nm, hydrophilic headgroups ~ 2 nm and ~ 3 nm for the inner hydrophobic alkyl chain sphere.

Beside the dominant vesicle structures (Fig.2a) also rod shaped cylindrical aggregates with 20-200 nm length and 7 nm cross-section diameter are observed (Fig. 2b).

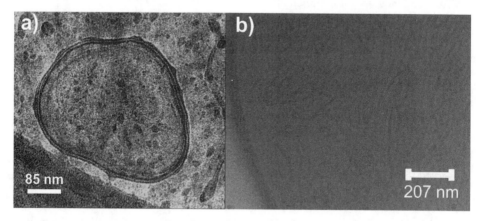

FIGURE 2. a) Cryo-TEM micrograph of a deformed large vesicle (Ø 400 nm), indicating the begin of a multilamellar growth of the membrane. **b)** A population of rod shaped aggregates with the usual double layer profile. Because of the great number of aggregates a circular cross-section can be assumed.

The amphiphile **5** has a very small CMC. Aggregates could be detected down to a concentration of < 3 mg/l. Interactions of **5** with artificial DPPC membranes were investigated by differential scanning calorimetry (Fig. 3a) and deuterium NMR spectroscopy. Monolayers of **5** at the air/water interface were studied at different lateral pressures by film balance technique.[6] The thickness of the 5-monolayer at high lateral pressure was 30 Å, similar to that of phospholipid monolayers in the condensed state. By titration, the pK value of the 5-monolayer was determined (Fig. 3b) and an 18fold charging of the negative carboxylate groups was obtained.[6]

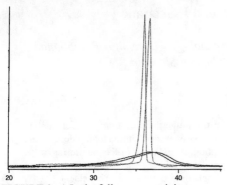

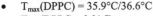

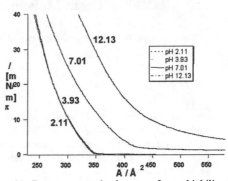

FIGURE 3. a) In the fullerene-containing membrane a broadening of the phase transition and an increase of the phase transition temperature is observed
- T_{max}(DPPC) = 35.9°C/36.6°C
- T_{max}(DPPC + 3 % **5**) = 36.5°C/37.4°C

b) Pressure-area isotherms of amphiphilic **5** monolayers at T = 20.0 °C at different subphase p_H values.

The hydration capacity of the hydrophilic headgroup of **5** is significantly higher than that of phospholipids. The negatively charged monolayer showed a strong coupling of the water soluble protein cytochrome c from the subphase, leading to the formation of a 29 Δ thick protein layer. The protein content of this layer varied drastically with the pH value. The properties of the AF monolayers may be useful in the design of dedicated biomimetic surfaces.

[1] C. Bingel, *Chem. Ber.* **126**, 1957-1959 (1993).
[2] X. Camps, A. Hirsch, *J. Chem. Soc. Perkin Trans. I* **1997**, 1595-1596 (1997).
[3] a) I. Lamparth, C. Maichle-Mössmer, A. Hirsch, *Angew. Chem.* **107**, 1755-1757 (1995); *Angew. Chem. Int. Ed. Engl.* **34**, 1607-1609 (1995); b) I. Lamparth, A. Herzog, A. Hirsch, *Tetrahedron* **52**, 5065-5075 (1996); c) A. Hirsch, *Top. Curr. Chem.* **199**, 1-65 (1999).
[4] G.R. Newkome, R.K. Behera, C. N. Moorefield, G.R. Baker, *J. Org. Chem.* **56**, 7162 (1991); G.R. Newkome, A. Nayak, R.K. Behera, C.N. Moorefield, G.R. Baker, *J. Org. Chem.* **57**, 358 (1992).
[5] M. Brettreich, S. Burghardt, C. Böttcher, T. Bayerl, S. Bayerl, A. Hirsch, *Angew. Chem.* **2000**, in press.
[6] A.P. Maierhofer, M. Brettreich, O. Vostrowsky, A. Hirsch, S. Langridge, T. Bayerl, *Langmuir* **2000**, in press.

Ion Chromatographic Determination of T_h-hexakis-dicarboxymethanofullerene-C_{60}

J. Cerar, M. Pompe, and J. Škerjanc

Faculty of Chemistry and Chemical Technology, University of Ljubljana
Aškerèeva 5, 1000 Ljubljana, Slovenia

Abstract. Separation of T_h-hexakis-dicarboxymethanofullerene-C_{60} from its degradation products has been performed using Dionex ASA4 ion exchange chromatographic column. Previously developed [1] separation method was optimized and used for the monitoring of the degradation processes of T_h-hexakis-dicarboxymethanofullerene-C_{60}. A gradient elution method was used for the baseline separation of fullerene derivatives. The eluent (KOH) concentration was increased from 0.4 M up to 1M. T_h-hexakis-dicarboxymethanofullerene-C_{60} and its degradation products were detected by UV absorption at 330 nm. Stability of T_h-hexakis-dicarboxymethanofullerene-C_{60} in aqueous solutions has been investigated at different conditions. It was found that T_h-hexakis-dicarboxymethanofullerene-C_{60} is stable in basic aqueous solutions while in acidic and neutral solutions degradation was observed. Elevated temperature accelerates degradation.

INTRODUCTION

An explanation of some experimental results [2] has evoked need for a new analytical method that could separate T_h-hexakis-dicarboxymethanofullerene-C_{60} (fullerenehexamalonic acid), T_h-$C_{66}(COOH)_{12}$, from its degradation products, seen in

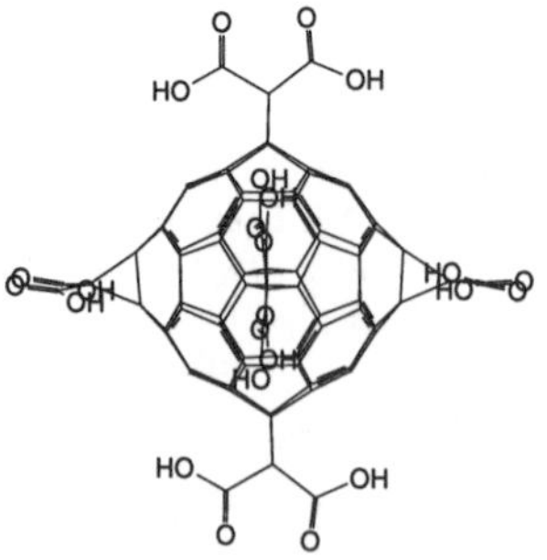

FIGURE 1. T_h-hexakis-dicarboxymethanofullerene-C_{60}

54

NMR spectra. The acid and particularly its sodium salt are practically insoluble in the commonly used organic solvents. Hydrophobicity of fullerene core is screened by hydrophilic character of six ionized malonic groups attached to the fullerene sphere. Strong ionic character of this molecule allowed us to separate T_h-$C_{66}(COOH)_{12}$ from its degradation products by using ion chromatography.

EXPERIMENTAL

HPLC instrumentation we used consisted of a gradient pump (Merck-Hitachi, L-6200A), an autosampler (Merck-Hitachi, AS-2000A), an UV-VIS detector (Merck-Hitachi, L-4250) and corresponding software (Merck-Hitachi, HPLC manager, version 2). Eluent rate was set to 2 ml/min and detection was obtained by UV absorption at 330 nm. All chromatographic runs were carried out at room temperature.

Separation started with the eluent concentration of 0.4 M KOH for 3 minutes. Then the concentration of eluent was linearly increased to 0.65 KOH in 12 minutes and further to 1 M in additional 3 minutes and remained the same for 12 minutes. The total separation time was 30 minutes.

NMR spectra were performed at room temperature on Bruker Avance DPX 300 and Varian Unity Inova 600 MHz instruments using D_2O as solvent and acetone-d_6 as internal standard.

RESULTS AND DISCUSSION

The baseline separation of fullerene derivatives was achieved using KOH as an eluent and gradient elution program. Typical chromatogram can be seen on Fig. 2.

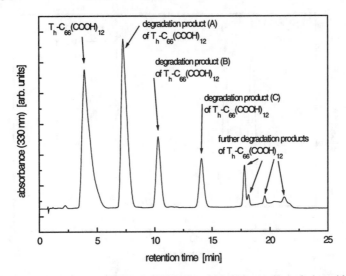

FIGURE 2. Ionic chromatogram of T_h-$C_{66}(COOH)_{12}$ obtained by gradient elution with KOH.

The main disadvantage of previously used method [1] was permanent retention of some fullerene derivatives on the separation column. A simple experiment was performed in order to ensure that all compounds are quantitatively eluted from the separation column. Two subsequent chromatographic runs were made. During the first one the separation column was removed from the separation system. The unseparated sample was flushed directly to the detector and the integral of the absorbance at 330 nm was determined. This integral represents the sum of the absorbances of the individual components in the sample. During the second run the sample was injected on the separation column and the integrals of the absorbances of the individual chromatographic peaks were calculated.

The sum of integrals, belonging to eluted peaks, was equal to integral obtained in the run without separation column. Therefore it was postulated that all components that are present in our sample are eluted from the separation column within single chromatographic run.

Characteristic identification of T_h-$C_{66}(COOH)_{12}$ peak was obtained by ^{13}C NMR. UV-VIS spectra of all degradation products presented on Fig. 2 have the same features as the parent T_h-$C_{66}(COOH)_{12}$ compound and could not be distinguished one from another on the diode-array detector. This fact implies that during degradation the T_h addition pattern on the fullerene core remains untouched. ^{13}C NMR spectrum of degradation product (A) revealed distorted molecule symmetry, seen as multiplet in sp^2 region.

Stability of T_h-$C_{66}(COOH)_{12}$ in aqueous solutions was examined at room and elevated temperature at various pH values. While compound is stable at least 24 hours at room temperature, elevated temperature accelerates its decomposition in water and hydrochloric acid. See Fig. 3.

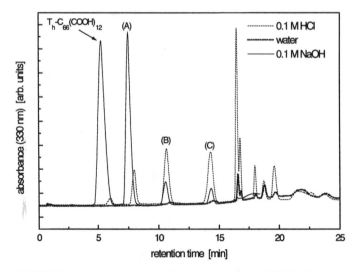

FIGURE 3. Ionic chromatogram of sample, exposed to 80°C for 5 hours.

During degradation absorbance at 330 nm was decreased for few percents. Degradation products weren't characteristically identified yet but we suspect that partial decarboxylation of malonic groups occurs. This is to some extent supported by ^{13}C NMR spectra, UV-VIS spectra, literature data [3] and conductometric titration curve of degraded samples.

CONCLUSIONS

The presented method is as far as we know the first application of ion chromatography for separation of fullerene species. In the particular case this procedure enables analytical separation of T_h-$C_{66}(COOH)_{12}$ from its degradation products - an achievement that has not been be reached so far with other methods. Slight degradation dependence of absorption coefficient of T_h-$C_{66}(COOH)_{12}$ related compounds in UV-VIS region enables semiquantitative determination of T_h-$C_{66}(COOH)_{12}$.

Stability of T_h-$C_{66}(COOH)_{12}$ was tested using ion chromatographic method. It was found that T_h-$C_{66}(COOH)_{12}$ degrades slowly in acidic and neutral aqueous solution at room temperature. Increased temperature accelerates degradation of T_h-$C_{66}(COOH)_{12}$ in aqueous solutions.

ACKNOWLEDGMENT

We are grateful to dr. Janez Plavec from Slovenian NMR Center at the National Institute of Chemistry, Ljubljana, and dr. Janez Cerkovnik from the Faculty of Chemistry and Chemical Technology, Ljubljana, for performing NMR spectra.

REFERENCES

[1] Cerar, J., Pompe, M., Cerkovnik, J., and Škerjanc, J., *in preparation*
[2] Cerar, J., Cerkovnik, J., and Škerjanc, J., *J. Phys. Chem. B* **102**, 7377-7381 (1998).
[3] Lamparth, I., Schick, G., and Hirsch, A., *Liebigs Ann.* , 253 (1997).

Small Reorganization Energies in Fullerenes

D.M. Guldi[a], K.-D. Asmus[b], A. Hirsch[c], and M. Prato[d]

[a]University of Notre Dame, Radiation Laboratory
Notre Dame, IN 46556, USA
[b]University of Notre Dame
Radiation Laboratory and Department of Chemistry & Biochemistry
Notre Dame, IN 46556, USA
[c]Institut für Organische Chemie der Universität Erlangen-Nürnberg, Henkestrasse 42
D-91054 Erlangen, Germany
[d]Dipartimento di Scienze Farmaceutiche, Università di Trieste, Piazzale Europa
34127 Trieste, Italy

Abstract. This article highlights the advantages of employing fullerenes as a viable electron accepting building block in novel donor acceptor systems. The selected examples characterize the small reorganization energies of fullerenes in electron transfer reactions and illustrate the continuing interest and potential of fullerenes as multifunctional electron storage moieties in well-ordered multicomponent composites. Furthermore, the importance of C_{60} as a promising model system for unraveling fundamental aspects of ET theory is emphasized.

INTRODUCTION

In the photosynthetic reaction center (PRC), a variety of short-range electron transfer (ET) and energy transfer (ENT) events occur between well-arranged organic pigments and other cofactors. Thereby, charges are separated with remarkable efficiency to yield a spatially and electronically well-isolated radical pair and thus eliminate the energy wasting back electron transfer (BET). The arrangement of the donor-acceptor couples in the PRC is accomplished *via* non-covalent incorporation into a well-defined protein matrix.[1]

Owing to the importance and complexity of natural photosynthesis, the study thereof necessitates suitable simpler models. The ultimate goal is to design and assemble synthetic systems, which can efficiently convert solar energy into useful chemical energy. An important approach to PRC modeling has been the covalent linking of a photoexcitable chromophore with an electron acceptor or an electron donor. It is important to note that in these artificial systems the organizing motif is the covalent linkage between the redox active moieties.[2]

In the following contribution some noteworthy features are summarized concerning C_{60} as a new, three-dimensional electron acceptor unit in artificial reaction centers.

RESULTS AND DISCUSION
Small reorganization energies

The choice of a 3-dimensional electron acceptor, such as C_{60}, arises from a series of well-established reasons. One of the most fascinating phenomena in the field of

CP544, *Electronic Properties of Novel Materials—Molecular Nanostructures*, edited by H. Kuzmany, et al.
© 2000 American Institute of Physics 1-56396-973-4/00/$17.00

fullerene chemistry is the small reorganization energy associated with almost all their reactions, especially in photoinduced electron transfer (PET).[3] This is an important requisite for the directional control and also the efficiency of ET reactions, as illustrated by the well-organized special pair (*e.g.*, bacterial chlorophyll and ubiquinone) in the photosynthetic reaction center. The total reorganization energy (λ) is the sum of a solvent-independent term λ_i and the solvent reorganization energy λ_s. The λ_i contribution stems from the nuclear configurations, associated with the transformation of the molecule, for instance, in a photochemical reaction from an initial to a final state. It is notable that the rigid structure of the fullerene core leads to small Raman shifts under reductive conditions, and small Stokes shifts in excitation experiments.[4] A reasonable interpretation for these observations is the structural similarity between C_{60} in the ground, reduced and also excited states.

It is also believed that the solvent-dependent term (λ_s) is small, thus requiring only little energy for the adjustment of a generated state (*e.g.*, excited or reduced states) to the new solvent environment. This is a direct consequence of the extended p-electron delocalization within the giant, spherical fullerene framework (diameter > 8 Å).

The sum of these effects bear fundamental consequences upon the classical Marcus treatment of ET, which serves as a valuable guide to control and optimize the efficiency of ET.[5] In particular, the Marcus ET theory treats the dependence of ET rates on the free energy changes of the reaction ($-\Delta G$) as a parabolic curve and predicts an increase in rate with increasing thermodynamic driving force, referred to as the "normal region" of the Marcus curve ($-\Delta G < \lambda$). With increasing exothermicity of the ET reaction optimally thermodynamic conditions are reached, when the driving force equals the overall reorganization energy ($-\Delta G \sim \lambda$). Beyond this maximum the highly exergonic region ($-\Delta G > \lambda$) is entered, where the rate constants actually start to decrease with increasing free energy changes ("inverted region"). Consequently, variation of λ is not only the key to control the maximum of the parabola, but, most importantly, to influence the shape of the parabolic dependence. In principle, smaller λ-values assist in reaching the maximum of the Marcus curve at smaller $-\Delta G$ values and, in turn, in shifting the energy-wasting BET deep into the Marcus "inverted region".

Determination of λ in pulse radiolytic experiments

Our pulse-radiolytic studies, focusing on *inter*molecular ET dynamics between fullerenes and a series of radiolytically generated arene p-radical cations with varying oxidation potentials were designed to shed light onto the small reorganization energy of fullerenes (eq. 1).[4b] Interestingly, parabolic dependence of the rate constants on the thermodynamic driving force was found for *inter*molecular reactions with C_{76} and C_{78}, representing one of the rare confirmations of the existence of the "Marcus-Inverted" region in a truly *inter*molecular forward ET. From these experiments an experimental value of *ca.* 0.6 eV (in dichloromethane) was deduced for the total reorganization energy of C_{76} and C_{78} in oxidative ET processes.

$$(\text{arene})^{\bullet+} + C_{76}/C_{78} \longrightarrow \text{arene} + C_{76}/C_{78}^{\bullet+} \tag{1}$$

We found an even smaller λ-value of 0.48 eV (solvent mixture containing toluene, 2-propanol and acetone; 8:1:1 v/v) in an independent set of experiments, focusing on an *inter*molecular ET between a series of one-electron reduced metalloporphyrins and C_{60}, yielding the fullerene p-radical anion.[7] Despite the fact that the ET rate constants lack the dependence on the driving force for the reaction and are nearly diffusion-controlled, the low reduction potentials of tin-(IV) porphyrins ($Sn^{IV}P$) led to the observation of equilibrium conditions (eq. 2). The equilibrium constant was determined from the kinetic and absorbance plots, from which an average equilibrium constant (K) of 14 ± 3 was derived and used to calculate λ.[6]

$$(Sn^{IV}P)^{\bullet-} + C_{60} \;\; D \;\; Sn^{IV}P + C_{60}^{\bullet-} \qquad\qquad (2)$$

Observation of the Marcus "inverted region" in flash photolysis experiments

The covalent linkage of fullerenes to a number of interesting electro- or photoactive species offers new opportunities in the preparation of materials with the objective to synthesize arrays that exhibit expedited ET kinetics and minimized losses of excited state energy.[7] Consequently, the PET event is converted to a truly *intra*molecular reaction controlled only by the activation energy of the reaction. A fixed and short distance between the two electroactive components prevents the undesired loss of excitation energy *via* alternative radiation and radiationless decay channels. The elimination of the rate-limiting diffusion between a free donor and a free acceptor moiety accelerates also the dynamics of the energy-wasting BET. In some instances, this may even lead to BET processes that are significantly faster than the forward ET (*e.g.*, BET is around the thermodynamic maximum). In this context, the finding by Gust *et al.* is of particular importance, namely, that BET dynamics on a first example of a C_{60}-based systems (**1**) are significantly slower than those of the forward ET.[8] This pioneering work evoked the synthesis of a sheer unlimited number of C_{60}-based donor-acceptor systems incorporating a variety of electron-donor molecules as well as photoluminescence oligomers.[8]

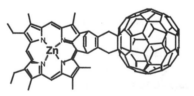

1

Based on the experimentally determined λ-values for fullerenes, which are exceptionally small, the maximum of the Marcus curve should be reached at smaller - ? $G°$, relative to two-dimensional electron acceptors, which in general have less rigid structures and higher reorganization energies than three-dimensional fullerenes. This shifts the more exothermic BET deep into the Marcus "inverted-region", far from the thermodynamic maximum (-? $G° = \lambda$) and inhibits the undesired BET event. At the same time, the "normal-region" is steeper, which leads to a notable acceleration of ET. To illustrate the benefits of incorporating a fullerene rather than a quinone acceptor, which has a similar reduction potential but higher reorganization energy, a fullerene-based porphyrin dyad with a rigid spacer guaranteeing a fixed separation between the two redoxactive moieties, was compared with a quinone-based porphyrin dyad. Remarkably, the fullerene-based dyad gives rise to an accelerated ET (~ 6 times) and decelerated BET process (~ 25 times) relative to the kinetics of the corresponding quinone dyad.[3a]

Thus, at the beginning of our photophysical investigation, we anticipated deceleration of the energy-wasting back electron transfer (BET) in donor-acceptor systems. Recently we accomplished a verification of this assumption by demonstrating, in a series of fullerene-based donor-acceptor systems, that BET occurs clearly in the "Marcus-inverted" region.[9] The selection of a zinc tetraphenylporphyrin (ZnTPP) as a chromophore showed how important the correct choice of the *intra*molecular separation is for the efficiency of light-driven electron transfer, especially when $-?G° \gg \lambda$. The van-der Waals contacts (~ 3.0 Å) in a p-stacked fullerene-porphyrin dyad (**2**) provided an exquisite settings for this study. In particular, the short separation guaranteed that an efficient *intra*molecular ET prevailed over the competing singlet-singlet energy transfer from ^{1}*(p-*p)ZnTPP to C_{60} in all solvents investigated. The change of solvent polarity from non-polar toluene to polar benzonitrile and, thereby, the influence on the free

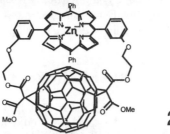

2

energy changes of the associated ET events gives rise to a marked acceleration of the BET with increasing dielectric constant (see Table 1).

Table 1: Photophysical properties of dyad **2** in various solvents.

Solvent	ε dielectric constant	τ radical pair [ps]	?$G_{CR}°$ [eV]
Toluene	2.38	619	1.536
Tetrahydrofuran	7.6	385	1.452
Dichloromethane	9.08	121	1.445
Dichloroethane	10.19	66	1.441
Benzonitrile	24.8	38	1.424

Concluding remarks and outlook

In conclusion, the ability of three-dimensional fullerenes to inhibit BET and still combine it with a fast forward ET renders them unique probes for *inter-* and *intra*molecular ET studies. In addition, the low reduction potential of C_{60} ($E_{1/2}$ = -0.44 V *versus* SCE) appears profitable for their utilisation as novel electron acceptors or relay in multicomponent donor acceptor systems. Remarkably, similar systems based on two-dimensional acceptors (*e.g.*, quinone, etc.) failed to exhibit sufficient lifetimes of the charge-separated states formed, because of the fast occurring BET. The unique delocalization, provided by the three-dimensional structure of the fullerene core, in combination with the small reorganisation energy, on the other hand, prevents a fast BET process in the fullerene-containing systems.

Noteworthy in this context is, last but not least, the recent introduction of elegant and versatile protocols concerning the chemical functionalization of the fullerene core.

ACKNOWLEDGMENT

This work was supported by the Office of Basic Energy Sciences of the Department of Energy and by the European Community (RTN program FUNCARS contract HPRN-CT-1999-00011). This is document NDRL-4221 from the Notre Dame Radiation Laboratory.

REFERENCES

1. a) J. Deisenhofer, O. Epp, K. Miki, R. Huber and H. Michel, *J. Mol. Biol.,* 1984, **180**, 385. b) *The Photosynthetic Reaction Center*; J. Deisenhofer and J.R. Norris, Eds.; Academic Press: 1993. c) G. McDermott, S.M. Prince, A.A. Freer, A.M. Hawthornthwaite-Lawless, M.Z. Papiz, R.J. Cogdell and N.W. Isaacs, *Nature,* 1995, **374**, 517. d) W. Kuhlbrandt and D.N. Wang, *Nature,* 1991, **350**, 130.

2. Examples with leading references: a) D. Gust, T.A. Moore and A.L. Moore, *Acc. Chem. Res.,* 1993, **26**, 198. b) M.R. Wasielewski, *Chem. Rev.,* 1992, **92**, 435. c) M.N. Paddon-Row, *Acc. Chem. Res.,* 1994, **27**, 18. d) N. Sutin, *Acc. Chem. Res.,* 1983, **15**, 275. e) A.J. Bard and M.A. Fox, *Acc. Chem. Res.,* 1995, **28**, 141. f) T.J. Meyer, *Acc. Chem. Res.,* 1989, **22**, 163. g) A.C. Benniston, P.R. Macki and, A. Harriman, *Angew. Chem. Int. Ed.,* 1998, **37**, 354. h) W.B. Davis, W.A. Svec, M.A. Ratner and M.R. Wasielewski, *Nature,* 1998, **396**, 60.

3. a) H. Imahori, K. Hagiwara, T. Akiyama, M. Akoi, S. Taniguchi, T. Okada, M. Shirakawa and Y. Sakata, *Chem. Lett.,* 1996, **263**, 545. b) D.M. Guldi and K.-D. Asmus, *J. Am. Chem. Soc.,* 1997, **119**, 5744.

4. a) M.L. McGlashen, M.E. Blackwood and T.G. Spiro, *J. Am. Chem. Soc.,* 1993, **115**, 2074. b) D.M. Guldi and K.-D. Asmus, *J. Phys. Chem. A,* 1997, **101**, 1472.

5. a) R.A. Marcus, *J. Chem. Phys.,* 1956, **24**, 966. b) R. Marcus and N. Sutin, *Biophys. Acta,* 1985, **811**, 265.

6. D.M. Guldi, P. Neta and K.-D. Asmus, *J. Phys. Chem.,* 1994, **98**, 4617.

7. For recent reviews see a) H. Imahori and Y. Sakata, *Adv. Mater.,* 1997, **9**, 537. b) M. Prato, *J. Mater. Chem.,* 1997, **7**, 1097. c) N. Martin, L. Sanchez, B. Illescas and I. Perez, *Chem. Rev.,* 1998, **98**, 2527. d) A.L. Balch and M.M. Olmstead, *Chem. Rev.,* 1998, **98**, 2123. e) F. Diederich and M. GomezLopez, *Chimica* 1998, **52**, 551. f) D.M. Guldi *Chem. Commun.,* 2000, 321.

8. P.A. Liddell, J.P. Sumida, A.N. McPherson, L. Noss, G.R. Seely, K.N. Clark, A.L. Moore, T A. Moore, D. Gust *Photochem. Photobiol.* 1994, **60**, 537.

9. D.M. Guldi, C. Luo, M. Prato, E. Dietel, A. Hirsch, *Chem. Commun.,* 2000, 373.

Singlet Oxygen Generation by C_{60} and C_{70} - an ESR Study

A. Sienkiewicz*†, S. Garaj†, E. Białkowska-Jaworska*, and L. Forró†

*Institute of Physics, Polish Academy of Sciences, Al. Lotników 32/46, 02-668 Warsaw, Poland,
†Institut de Génie Atomique, Département de Physique, École Polytechnique Fédérale de Lausanne,
CH-1015 Lausanne, Switzerland

Abstract. Under illumination in organic solvents, pristine fullerenes, C_{60} and C_{70}, reveal a high triplet state yield that is close to unity and are known to effectively generate electronically excited singlet oxygen ($^1\Delta_g$). We applied the ESR technique to monitor $^1\Delta_g$ generation efficiency for both fullerenes in toluene solutions that were exposed to the white light. Total signal intensity of TEMPO, a stable nitroxide resulting from $^1\Delta_g$ attack on ESR-silent precursor, TMP (2,2,4,4-Tetramethylpiperidine), was measured to follow the singlet oxygen generation. We also performed a similar experiment using a solid sample of paraffin. The relative efficiency of $^1\Delta_g$ generation in the presence of both C_{60} and C_{70} in solutions and in a solid sample was estimated from the increase rates of TEMPO formation.

INTRODUCTION

The electronically excited state of dioxygene, $^1\Delta_g$ singlet oxygen, is receiving increasing attention in many fields of contemporary science ranging from biomedicine to polymer science [1]. It is also known that pristine fullerenes, C_{60} [60-I_h] and C_{70} [70-D_{5h}], effectively produce $^1\Delta_g$ by absorbing light energy. Since for C_{60} and C_{70} the main deactivation channel of electronically excited singlet states is the Intersystem Crossing (ISC), both pristine fullerenes exhibit close to unity yields of light excited triplet states that can effectively be quenched by a resonant interaction with the ground-state molecular oxygen, O_2 ($^3\Sigma_g$) [2]. These light-excited triplet states are long-lived (~ 1 ms) and energetically sufficiently high (*ca.* 1.5 eV). Therefore, the process of $^1\Delta_g$ generation in solutions is diffusion controlled and occurs with high yield. The mechanism of the process of $^1\Delta_g$ production by C_{60} is schematically shown in Figure 1a. The production of $^1\Delta_g$ can be evaluated from the characteristic near-infrared phosphorescent transition (at 1268 nm) between the lowest excited state, $O_2(a^1\Delta_g)$, and the ground state, $O_2(X^3\Sigma_g)$ [3]. Since such a direct observation requires employment of a sensitive infrared detector, like a liquid nitrogen-cooled germanium photodiode, we explored a simpler, ESR-based method, for monitoring $^1\Delta_g$ generation in toluene solutions containing C_{60} and C_{70}. We also applied the same technique to monitor $^1\Delta_g$ generation in a solid medium. Although it occurred at much lower rate, ESR could easily monitor the production of $^1\Delta_g$ by pristine fullerenes in a sample of solid paraffin. The mechanism of the C_{70}/C_{60}-mediated photosensitization in solutions was also compared to that

CP544, *Electronic Properties of Novel Materials—Molecular Nanostructures*, edited by H. Kuzmany, et al.
© 2000 American Institute of Physics 1-56396-973-4/00/$17.00

occurring in the presence of the well known singlet oxygen generator, a protoporphyrin (PP)-based dye, used for $^1\Delta_g$ generation and tissue inactivation in Photodynamic Therapy of Cancer (PDT) [4-5].

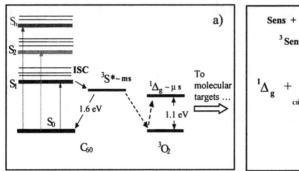
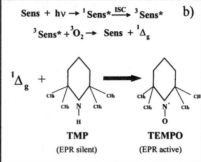

Figure 1. Schematic representation of a) $^1\Delta_g$ generation by C_{60} and b) reaction of $^1\Delta_g$ with TMP.

EXPERIMENTAL

Sterically hindered amines, due to their high chemical affinity to $^1\Delta_g$, can be used as targets to identify singlet oxygen [6-7]. For detecting singlet oxygen production by C_{60} and C_{70} in solutions we used a method introduced first by Y. Lion *et al.* [8]. This technique involves ESR monitoring of a paramagnetic product resulting from $^1\Delta_g$ attack on a diamagnetic substrate. We employed ESR-silent 2,2,6,6-Tetramethylpiperidine (TMP) from Merck KGaA, Darmstadt, Germany, as a molecular target for $^1\Delta_g$. The oxidation product of the diamagnetic TMP is a stable free radical, nitroxide TEMPO. Both photosensitizing agents, C_{60} and C_{70}, have been resublimed before dissolving them in toluene and were of >99.95% and >99.9% purity, respectively. Prof. Alfreda Graczyk (Chemistry Department, Technical Military Academy, Warsaw, Poland) kindly supplied the reference sample, a protoporphyrin-based dye, PP(Ala)$_2$(Arg)$_2$, especially developed for PDT applications. Spectroscopically pure paraffin from Fluka Chemika, Germany was used for preparation of solid-state samples. Bruker ESP300E ESR spectrometer was employed for monitoring the ESR signal growth of TEMPO as a function of illumination time. For performing ESR measurements, aliquots of the toluene solutions containing 2×10^{-4} M/L of C_{60} or C_{70} and 7.8×10^{-2} M/L of TMP were transferred into thin-wall quartz capillaries (0.6 mm ID and 0.84 mm OD) and exposed to the white light from a halogen source (150 W halogen lamp) using four evenly disposed glass fiber-optic light guides. The illumination process was performed outside the ESR cavity at the stabilized temperature of 18 ± 1^0C. The total ESR signal intensity (I_{ESR}) of the stable nitroxide (TEMPO), a paramagnetic product of $^1\Delta_g$ attack on TMP, was measured as a function of light exposure. When the experiment was carried out in nitrogen-bubbled solutions, no nitroxide (TEMPO) formation occurred. The scheme of the major photosensitization process involved in the detection of $^1\Delta_g$ in our experiment is shown in Figure 1b.

RESULTS AND DISCUSSION

The plots of the ESR signal intensity (I_{ESR}) of TEMPO as a function of illumination time for both photosensitizers, C_{60} and C_{70}, as well as for the reference compound, $PP(Ala)_2(Arg)_2$, are shown in Figure 2a & b, respectively. The much slower signal growth of TEMPO against duration of light exposure for the solid sample (paraffin) is shown in Figure 2c. Typical evolution of the ESR signal of TEMPO for illuminated toluene solutions of C_{60} is shown in Figure 2d.

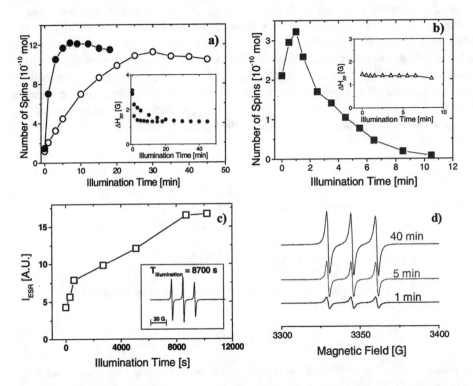

Figure 2. The ESR signal intensity (I_{ESR}) of TEMPO as a function of illumination time for a) C_{60} (open circles) and C_{70} (closed circles) in toluene (inset: progressive narrowing of the TEMPO linewidth as a function of illumination time), b) $PP(Ala)_2(Arg)_2$ in water (inset: ΔH_{pp} of TEMPO against duration of light exposure for water solution of $PP(Ala)_2(Arg)_2$), and c) C_{60} dissolved in solid paraffin (inset: characteristic lineshape of the ESR signal of TEMPO in solid paraffin); d) typical evolution of the ESR signal of TEMPO during illumination of oxygenated toluene solution of C_{60}.

Our comparative study of $^1\Delta_g$ generation in toluene point to higher efficiency of C_{70} as compared to C_{60}. In toluene solutions that were saturated and intentionally non-saturated with oxygen, the initial rates of $^1\Delta_g$ production were approximately 2.5 times greater in experiments in which C_{70} was used. These results are in a qualitatively good agreement with recent experimental data from flash analysis that point to longer lifetimes of excited triplet states in C_{70} molecule. Since $^1\Delta_g$ generation process was performed in sealed capillaries, an important linewidth reduction of the ESR signal of TEMPO was also observed. This phenomenon, the most pronounced in toluene solutions, reflects oxygen consumption within the medium. See inset to Figure 2a.

The ESR method also revealed to be quite useful to monitor the C_{60}-mediated $^1\Delta_g$ production in solid samples. The process of singlet oxygen generation in the solid sample proceeded much slower than in organic solvents due to the considerably lower concentration and diffusion rate of oxygen in solid samples of paraffin. See Figure 2c. However, we believe that there is a rationale for investigating the C_{60}/C_{70}-mediated photosensitization process in solids in view of designing novel biodegradable materials.

The photochemistry of fullerenes also seems to be strikingly different from that of protoporphyrin-based dyes. Our results show that upon longer exposure to light and exhaustion of molecular oxygen from the solvent, the systems containing pristine fullerenes do not turnover from Type II photoreaction mechanism (triplet-triplet energy transfer followed by $^1\Delta_g$ generation) to Type I photoreaction mechanism (free radical-mediated process often involving electron or hydrogen atom transfer). Such a turnover transition is commonly observed for the oxygen-depleted systems containing protoporphyrin-based dyes dissolved either in organic solvents or water. The rapid decay of the TEMPO-related signal in water solution of PP(Ala)$_2$(Arg)$_2$ for longer illumination times can be ascribed to Type I photoreaction mechanism. See Figure 2b.

Comparison of $^1\Delta_g$ generation by pristine fullerenes with the similar process sensitized by the well-established singlet oxygen generator, PP(Ala)$_2$(Arg)$_2$, points to a considerably high efficiency of fullerenes. Therefore, functionalized (water soluble) derivatives of C_{60} and C_{70} seem to be good candidates for biomedical applications, including PDT.

ACKNOWLEDGMENTS

This work has been supported by the Swiss National Science Foundation and partly by the Polish KBN Grant #2-PO3B-018-13 (A.S.).

REFERENCES

1. Frimer, A. A., in *Singlet Oxygen O₂*, edited by R. C. Straight, J. D. Spikes, CRC Press, Boca Raton, FL, 1985, Vol. IV, pp. 92-128.
2. Foote, C. S., "Electron Transfer I", in *Topics in Current Chemistry* **169,** edited by J. Mattay, Springer Verlag, New York, 1994; pp. 347-363.
3. Bernstein, R., and Foote, C. S., *J. Phys. Chem A* **103,** 7244-7247 (1999).
4. Sienkiewicz, A., Bialkowska-Jaworska, E., Kwasny, M., Mierczyk, Z., Graczyk, A., in *Biomeasurements-Optoelectronics in Medical Diagnosis*, edited by H. Wierzba and R. Maniewski, World Health Organization, Warsaw, 1997, pp. 181-190.
5. Zoladek, T., Nguyen Bich Nhi, Jagiello, I., Graczyk, A., and Rytka, J., *Photochemistry and Photobiology*, **66** (2), 253-259 (1997).
6. Rozantsev, E. G., in *Free Nitroxyl Radicals*, edited by H. Ulrich, Plenum Press, London and New York, 1970.
7. Smith, W. F., Jr., *J. Am. Chem. Soc.* **94,** 186-190 (1972).
8. Lion, Y., Delmelle, M., and Van de Vorst, A., *Nature* **263,** 442-443 (1976).

Electrochemical Quartz Crystal Microbalance Study of Redox Active C_{60}/Pd Polymer Films

K. Winkler,[a] K. Noworyta,[b] W. Kutner[b], and A. L. Balch[c]

[a] Institute of Chemistry, University of Bialystok, Pilsudskiego 11/4, 15-443 Bialystok, Poland
[b] Institute of Physical Chemistry, Polish Academy of Sciences, Kasprzaka 44, 01-224 Warsaw, Poland
[c] Department of Chemistry, University of California, Davis, California 95616, USA

Abstract. Properties of conductive C_{60}/Pd polymer films were investigated by simultaneous cyclic voltammetry and piezoelectric microgravimetry at an electrochemical quartz crystal microbalance (EQCM). The films were deposited onto Au electrodes of the EQCM quartz vibrators by concomitant electroreduction of C_{60} and $[Pd^{II}(CH_3COO)_2]_3$ from a 0.1 M tetra(n-alkyl)ammonium perchlorate [alkyl = ethyl (TEA^+), butyl (TBA^+) or hexyl ($THxA^+$)], acetonitrile/toluene (1:4, v/v) solution. The composition of this solution significantly influenced the pattern of the film growth. The size of the counter cation is a major factor determining both the electrochemical properties of the C_{60}/Pd films and their stability with respect to dissolution. The fraction of the film reversibly reduced depends mainly on size of the supporting electrolyte cation and increases in the order: $THxA^+ < TBA^+ < TEA^+$.

INTRODUCTION

Recently, new electro-synthetic procedures for the preparation of two-component polymer films, prepared from fullerenes and transition metal complexes have been developed (1, 2). These two-component polymers are built from fullerene cages covalently bound with transition metal atoms or complexes.

M = Pd, Pt, Ir,
Rh, Au

The polymers are readily prepared by electrochemical reduction of acetonitrile/toluene (1:4, v/v) solutions containing C_{60} and selected transition metal complexes. Properties of these polymers depend on the composition of the polymerizing solution, the nature of solvent and the supporting electrolyte. The rate of charge transport through the polymer film is determined by the rate of diffusion of counter ions within the film.

CP544, *Electronic Properties of Novel Materials—Molecular Nanostructures*, edited by H. Kuzmany, et al.
© 2000 American Institute of Physics 1-56396-973-4/00/$17.00

In the present work, we investigated, under conditions of simultaneous cyclic voltammetry (CV) and piezoelectric microgravimetry, (*i*) the effect of solution composition on formation and electrochemical properties of the C_{60}/Pd films and (*ii*) transport of counter ions of the supporting electrolyte across the film-solution interface during electroreduction and electro-oxidation of the films.

EXPERIMENTAL

Chemicals. The film precursors, $[Pd^{II}(CH_3COO)_2]_3$, and C_{60} from Alfa and MER Corp. (Tucson, AZ), respectively, were used without purification. Tetra(ethyl)-ammonium perchlorate (TEAP), tetra(*n*-butyl)ammonium perchlorate (TBAP) and tetra(*n*-hexyl) ammonium perchlorate (THxAP), from Sigma were dried under reduced pressure at 70°C for 24 h prior to use. Acetonitrile (99.8 %) and toluene (99.8 %) were used as received from Aldrich.

Apparatus. Simultaneous CV and piezoelectric microgravimetry was carried out by using a Autolab electrochemistry system with GPES 4.5 software of Eco Chemie (Utrecht, The Netherlands) and a electrochemical quartz crystal microbalance, EQCM 5510, of the Institute of Physical Chemistry (Warsaw, Poland). The microbalance used was an improved model of that described earlier (3). The 14 mm diameter, AT-cut, plano-convex quartz crystals of the 5 MHz resonant frequencies were from Omig (Warsaw, Poland). The projected area of the Au electrode, vacuum deposited onto the quartz crystal, was $0.24\ cm^2$. Non-polished quartz crystals were used for better adherence of the films. The sensitivity of the mass measurement, was $17.7\ ng\ Hz^{-1}\ cm^{-2}$. Each experiment was performed with a new Au/quartz electrode. In order to minimize the ohmic potential drop, simultaneous CV and piezoelectric microgravimetry was carried out in a two-electrode system of a current follower with a platinum tab serving as the counter and reference electrode.

RESULTS AND DISCUSSION

During electrochemical polymerization of C_{60} in presence of $[Pd^{II}(CH_3COO)_2]_3$, the CV peak currents increase and frequency decreases in simultaneous CV (Fig. 1a) and piezoelectric microgravimetry (Fig. 1b) at EQCM as the C_{60}/Pd polymer film is deposited onto surface of the Au-quartz electrode in consecutive cycles. Both the Pd clusters and the C_{60}/Pd film is co-deposited since deposition proceeds at potentials where both the Pd(II) complex and C_{60} is reduced. In excess of $[Pd^{II}(CH_3COO)_2]_3$, the electrode mass increases much more than it does at a relatively low concentration ratio of the Pd(II) complex to C_{60}.

Presumably, electroreduction of the Pd(II) complex and deposition of the Pd(0)-C_{60} intermediate initiates the growth of the C_{60}/Pd film.

For further investigations, the Au/quartz electrodes coated with the C_{60}/Pd films were removed from the electropolymerization solution, rinsed with acetonitrile, and immersed in a blank acetonitrile solution of the supporting electrolyte.

Electroreduction and electro-oxidation of the film in the blank acetonitrile solution is accompanied by the tetra(n-alkyl)ammonium counter cation ingress and egress, respectively, for charge compensation. At a relatively high potential scan rate, only the outermost layers of the film that are in direct contact with the bathing solution are electrochemically active. At a low scan rate, however, the bulk film material is also electrochemically active. The counter cation exchange between solution and the film is reversible only at a high scan rate while irreversible behavior is observed at a low scan rate.

At relatively large negative potentials, the C_{60}/Pd film is removed from the Au/quartz electrode surface under CV conditions.

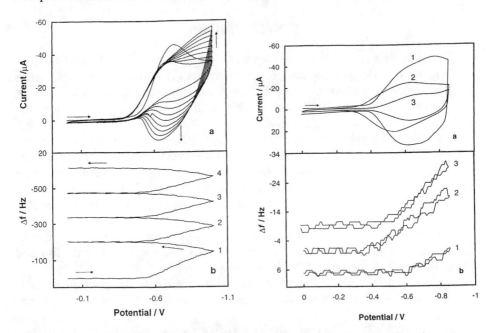

FIGURE 1. Simultaneously recorded at the same Au/quartz electrode (a) multicyclic voltammogram and (b) curves of frequency change vs. potential for 0.31 mM C_{60} and 1.5 mM [$Pd^{II}(CH_3COO)_2$]$_3$ in 0.1 M TBAP, acetonitrile/toluene (1:4, v/v); potential scan rate was 0.1 V s^{-1}.

FIGURE 2. Simultaneously recorded at the same Au/quartz electrode coated with the C_{60}/Pd film (a) multicyclic voltammogram and (b) curves of frequency change vs. potential for 0.1 M supporting electrolytes: TEAP – curve *1*, TBAP – curve *2* and THxAP - curve *3*. The C_{60}/Pd film was prepared under cyclic voltammetry conditions (20 cycles) in 0.31 mM C_{60}, 0.80 mM [$Pd^{II}(CH_3COO)_2$]$_3$ and 0.1 M TBAP, acetonitrile/toluene (1:4, v/v). Potential scan rate was 0.1 V s^{-1}.

The effect of the size of the supporting electrolyte cation on the electrochemical properties of the C_{60}/Pd film is shown in Figure 2. All films were prepared under the same conditions. The voltammograms show (Fig. 2a) that both the reduction and re-oxidation charge is smaller the larger is the size of the cation. The EQCM results allow determination of the mass and charge changes related to the ingress and egress of the cations. These results indicate that the fraction of the film being reduced is markedly lower the larger is the cation. Apparently, smaller TEA$^+$ cations can penetrate the film much more easily than larger THxA$^+$ cations. Assuming that the fullerene density in the C_{60}/Pd film is close to that in the polycrystalline C_{60} film deposited electrochemically, one may conclude that only ca. 2 monolayers of the C_{60}/Pd film are reduced in solution containing THxAP. In contrast, ca. 16 monolayers are reduced in solution containing TEAP. Presumably, these values are underestimated, since it can be expected that density of the fullerene packing in the C_{60}/Pd film is lower than that of the pristine C_{60} crystal.

More details on properties of the C_{60}/Pd films are presented elsewhere (4).

CONCLUSIONS

The C_{60}/Pd film is formed in the potential range of both the C_{60} and Pd(II) complex electroreduction. Composition and properties of the C_{60}/Pd films depend on the electropolymerization conditions. That is, Pd clusters are deposited together with the polymer and the amount of Pd deposited is higher the higher is the concentration ratio of the Pd(II) complex to C_{60} in solution.

Most likely, the polymer is built of the -C_{60}-Pd-C_{60}-Pd- chains which are cross-linked with other chains containing the -C_{60}-Pd-C_{60}- units to form a three-dimensional network. This network is sufficiently permeable since it allows for ingress and egress of the charge balancing counter ions during polymer electroreduction and electrooxidation. Incorporation of counter ions into the film is a slow, rate-determining step of the overall electroreduction.

Mass changes of the Au/quartz electrode are due to (*i*) formation of the C_{60}/Pd film, (*ii*) formation of the Pd clusters, (*iii*) the counter cation incorporation into the film and (*iv*) possible solvent swelling of the polymer.

The fraction of the film reversibly reduced at the same potential scan rate depends mainly on size of the supporting electrolyte cation and increases in order: THxA$^+$ < TBA$^+$ < TEA$^+$.

ACKNOWLEDGMENTS

Financial support of the US-Polish Maria Sklodowska-Curie Joint Fund II (Grant No. PAN/NSF-96-275) and of the Polish State Committee for Scientific Research (Contract No. 7 T07D 077 99 C/4323) to WK as well as the US. National Science Foundation (Grant No. CHE 9610507) to ALB is gratefully acknowledged.

REFERENCES

1. Balch, A. L., Costa, D. A., and Winkler, K., *J. Am. Chem. Soc.*, **120**, 9614 (1998).
2. de Bettencourt-Dias, A., Winkler, K., Hayashi, A., and Balch, A. L., in *Fullerenes. Recent Advances in the Chemistry and Physics of Fullerenes and Related Materials*, Vol. 7, Kamat, P. V., Guldi, D. M., and Kadish, K. M., Eds., The Electrochemical Society, Inc., Pennington (1999), pp. 47-58.
3. Koh, W., Kutner, W., Jones, M. T., and Kadish, K. M., *Electroanalysis*, **5**, 209 (1993).
4. Winkler, K., Noworyta, K., Kutner, W., Balch, A. L., *J. Electrochem. Soc.*, **147**, (2000), in press.

III. FULLERENE POLYMERS

Physical Properties of Highly-Oriented Rhombohedral C_{60} Polymer

M. Tokumoto[a, b], B. Narymbetov[b], H. Kobayashi[b], T. L. Makarova[c], V. A. Davydov[d], A.V. Rakhmania[d], L. S. Kashevarova[d]

[a]Electrotechnical Laboratory, Tsukuba, 305-8568 Japan, [b]Institute for Molecular Science, Okazaki, 444-8585 Japan, [c]Ioffe Physico-Technical Institute, St. Petersburg, 194021 Russia, [d]Veretschagin Institute of High Pressure Physics, Troitsk, 142092 Russia

Abstract. Rhombohedral phase of pressure-temperature treated C_{60} polymer was invesigated by resistivity, X-ray diffraction and ESR measurements. The pure rhombohedral phase of C_{60} polymer displays highly anisotropic electrical properties, and the conductivity in the polymerized 2D plane exhibits a metallic feature with weak localization. X-ray diffraction measurements have revealed that the diffraction patterns are essentially anisotropic. We have carried out the scanning of a reciprocal space in order to reveal the distribution of diffraction intensities. The result testifies to the high degree of mutual orientations of crystallites in the sample not only along the [001] direction but also in the (001) plane. Apparently the observed mosaicity of the crystal is connected with the highly anisotropic electrical properties of the rhombohedral phase of pressure-treated C_{60} polymer. A Dysonian ESR absorption lineshape consistent with the metallic nature was observed.

INTRODUCTION

High-presssure and high-temperature treatment of C_{60} results in various phases of polymerized C_{60}.[1, 2] Recently, the pure rhombohedral phase of C_{60} polymer was found to display highly anisotropic electrical properties.[3, 4] Moreover, the conductivity in the polymerized 2D plane exhibits a metallic feature with weak localization, in contradiction to the theoretical prediction.[5] In this paper, we report the X-ray and ESR studies of the highly anisotropic rhombohedral phase of C_{60} polymer.

EXPERIMENTAL

Sample preparation conditions are described in previous reports.[3, 4] For the dc conductivity measurements six golden contacts were attached with the silver paint in the Montgomery geomerty. Using this configuration in measuring layered systems decreases the chances to misinterpret the resistivty tensor. Current-voltage characteristics were linear in the range of 0.001 – 40 V. In order to decrease the influence of the domain structure of the material, the high pressure – high temperature treated fullerene cylinders were cleaved into small samples with the dimensions of the order of 100 μm. For X-ray measurements, sample was oriented by Laue method and further X-Ray measurements were performed on Rigaku AFC5R diffractometer (MoK$_\alpha$, graphite monochromator) in the powder mode with $\omega/2\theta$ scanning. The sample was oriented with the c-axis lying in the equatorial plane.

CP544, *Electronic Properties of Novel Materials—Molecular Nanostructures*, edited by H. Kuzmany, et al.

RESULTS AND DISCUSSION

Electrical conductivity

Figure 1a shows the temperature dependence of the in-plane resistivity. The samples show a metallic behaviour at high temperatures, but below a certain temperature resistivity increases with decreasing temperature. The low-temperature part of the plot is linear in the ln (T) coordinates. The whole curve in the 4.2 – 300 K temperature range can be described by a simple equation:

$$\rho^{ab} = \rho_0^{ab} - \rho_1^{ab} \ln (T) + \rho_2^{ab} T \tag{1}$$

The first temperature-dependent term can be put down to the quantum corrections of conductivity. The linear temperature dependence of the second mechanism gives strong support to the electron-phonon scattering. This dependence is similar to those found for the alkali-metal intercalated fullerenes.

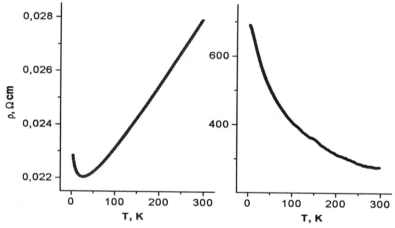

FIGURE 1. Simultaneously measured resistivities of the 2D Rhombohedral phase of C_{60} polymer. (a) in-plane resistivity : ρ^{ab} (left), (b) out-of-plane resistivity : ρ^c (right).

Figure 1b plots the temperature dependent dc resistivity results in the c-axis direction. Resistivity decreases with increasing temperature, as it is typical for semiconductors, but the ratio ρ_{2K}/ρ_{room} is small being less than a factor of 4. The resistivity curves ρ (T) proved to be nearly linear in the ln (T) coordinates in the range 12 – 300 K.

$$\rho^c = \rho_0^c - \rho_1^c \ln (T) \tag{2}$$

These dependencies are usually used to fit the conductivity data to a weak localization or electron-electron interaction models, but due to the low conductivity values these models are unlikely to be applied here. The logarithmic correction in conductivity can appear not through localization but by modifying the 2D density of states. If ρ_c is determined by tunnelling between 2D layers, this modification of DOS can give rise to the same ln (T) behaviour as in the in-plane conductivity.

The distinctive property for the highly-oriented phombohedral C_{60} polymers is their extremely anisotropic resistivity. Figure 2 shows the temperature dependence of the anisotropy ratio. The similarity in the temperature behaviour of the in-plane and out-of-plane conductivity together with enormous anisotropy makes it possible to invoke the model of Kumar and Jayannavar[6] in which the semiconductor-like c-axis resistivity in the layered systems, including graphite, was attributed to anisotropic weak localization. The model involving blocking of coherent interplanar tunnelling by the incoherent in-plane scattering gives the same temperature dependence for the c-axis resistivity as that for the in-plane resistivity, in accordance to our experimental results.

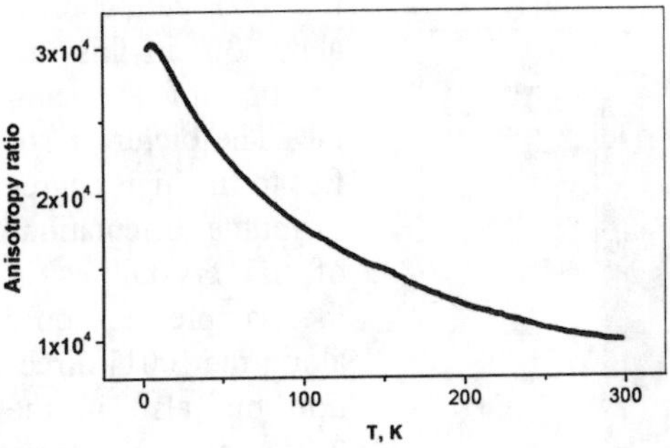

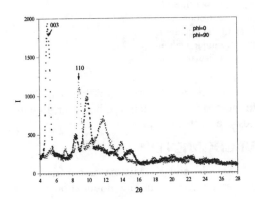

FIGURE 2. Temperature dependence of the anisotropy ratio of resistivity, ρ^c/ρ^{ab}.

X-ray diffraction and ESR

The x-ray diffraction measurements show that the peak positions on the diffraction patterns are in good agreement with the calculated ones for rhombohedral phase with the lattice parameters: $a=9.22$ and $c=24.60$ Å (hexagonal cell). The measurements have revealed that diffraction patterns are essentially changed with the rotation of the sample along the axis perpendicular to the equatorial plane (φ-axis). Figure 3 shows the patterns obtained at the $\varphi=0$ and 90°. It can be seen that at the $\varphi=90°$ orientation of the sample the (00l) reflections practically disappear and the (110) reflection has the highest intensity, that indicates the high texture of the sample.

FIGURE 3. X-ray diffraction patterns obtained at the $\varphi=0$ and 90° orientations of the sample.

To reveal an anisotropy of diffraction intensity distribution we have carried out the scanning of a reciprocal space in this geometry in the wide range of the φ angle 0-180° and set of diffraction patterns was obtained at different values of φ-axis with the interval 10°. The resulted picture of the intensity distribution in the equatorial plane is shown in Figure 4, where the data are drawn as in orthogonal coordinates φ and 2θ to be convenient to compare with the powder diffraction patterns. For non-textured powder sample the distribution of the reflections intensities would be uniform on φ at same values of 2θ but the obtained picture shows the high anisotropy of the intensity distribution. The highest intensity corresponds to the (003) reflection at the φ=0 and 180° that can testify to the preferred-orientation along the (00l) direction. Also, it can be seen in Figure 4 that diffraction spots are separated from each other and we can estimate the mosaicity of the crystal. Observed width at the half maximum (FWHM) for the (003) reflection is about 30° in this orientation of the sample. The picture testifies to the high degree of mutual orientations of the crystallites in the sample not only along the [001] direction but also in the (001) plane. Apparently the observed highly anisotropic electrical properties are connected with the observed mosaicity of the crystal.

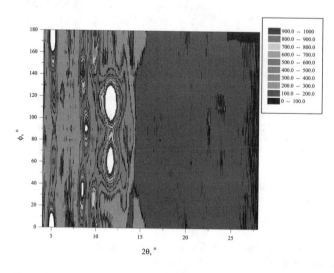

FIGURE 4. X-ray diffraction intensity distribution obtained from the rhombohedral C_{60} polymer sample in the geometry of detection when the [001] and [110] crystallographic directions are in the equatorial plane. The value φ=0° corresponds to the orientation of sample with incident beam parallel to the [110] direction and perpendicular to the [001].

We have also measured ESR spectra of this rhombohedral C_{60} polymer and observed Dysonian lineshape consistent with the metallic conductivity, however, in contrast to the previous ESR study on high-pressure treated C_{60} polymer.[7]

ACKNOWLEDGMENTS

Fruitful discussions with Y. Iwasa, H. Satsuki and K. Miyazawa are acknowledged. T. Makarova is grateful to the Alexander von Humboldt Foundation for the support.

REFERENCES

1) Y. Iwasa et al., Science, **264**, 1570 (1994)
2) M. Nunez-Regueiro et al., Phys. Rev. Lett., **74**, 278 (1995)
3) T.L. Makarova, T. Wågberg, B. Sundqvist, V. Agafonov, V.A. Davydov, A.V. Rakhmanina, L.S. Kashevarova, in *Fullerenes* Vol.7, edited by K.M. Kadish and P.V. Kamat (ECS, 1999) p.628.
4) T.L. Makarova, T. Wågberg, B. Sundqvist, Xiao-Mei Zhu, E.B. Nyeanchi, M.E. Gaevski, E. Olsson, V. Agafonov, V.A. Vadydov, A.V. Rakhmanina, L.S. Kashevarova, Mol. Materials, **13**, 157 (2000)
5) S. Okada and S. Saito, Phys. Rev. B **55**, 4039 (1997)
6) N. Kumar and A. M. Jannavar, Phys. Rev. B **45**, 5001 (1992)
7) M.E. Kozlov, A.A. Zakhidov, K. Yakushi and M. Tokumoto, phys. stat. sol. (b) **197**, 187 (1996)

Dimerization of C_{70} under High Pressure: Thermal Dissociation and Photophysical Properties of the Dimer C_{140}

S. Lebedkin,[1*] W. E. Hull,[2] A. Soldatov,[3] B. Renker,[3]
and M. M. Kappes[1,4]

[1,3] *Forschungszentrum Karlsruhe– Technik und Umwelt, [1]Institut für Nanotechnologie, [3]Institut für Festkörperphysik, P.O.Box 3640, D-76021 Karlsruhe, Germany.*
[2] *Central Spectroscopy Dept., German Cancer Research Center, D-69120 Heidelberg, Germany.*
[4] *Institut für Physikalische Chemie II, Universität Karlsruhe, D-76128 Karlsruhe, Germany*

Abstract. We have recently shown that solid C_{70} efficiently converts into a cap-to-cap dimer C_{2h} C_{140} at pressure of ~1 GPa and temperature of ~200°C [1]. Here we report on measurements of the thermal dissociation of C_{140} (purified by chromatography) at ambient pressure and its photophysical properties, in particular photogeneration of singlet oxygen ($^1O_2$). The kinetics of dissociation of C_{140} in the solid state between 130 and 200°C is well described by a simple first-order process, with an activation energy of 1.6 ± 0.03 eV (compare to 1.75 ± 0.05 eV for the C_{60} dimer C_{120} [2]). In contrast, the thermal dissociation of C_{140} dissolved in o-dichlorobenzene shows different, non-Arrhenius behavior, suggesting a strong influence of the molecular surroundings. The quantum efficiency of $^1O_2$ generation in solutions of C_{140} and C_{120} is close to unity, i.e. similar to that of C_{70} and C_{60}.

INTRODUCTION

Polymerization by [2+2] cycloadditions across reactive double bonds has been well established for C_{60} exposed to high pressure and temperature (p,T) or to UV-visible light [3] and, in principle, is also possible for C_{70} and other fullerenes. Up to now, only a few studies concerning polymerization of C_{70} by irradiation with light [4] and by application of rather high p,T (up to 7.5 GPa and 700°C) [5] have appeared, in part with controversial results. Putative polymeric structures of C_{70} generated in those works remained unclear. Recently, we have shown that C_{70} readily polymerizes like C_{60} at significantly lower p,T of 1-2.5 GPa and 200-300°C [6]. Two dominant structures have been suggested depending on the applied pressure: C_{70} dimers at pressure around 1 GPa and C_{70} "zigzag" chains at 2-2.5 GPa. The latter hypothesis has been very recently supported by X-ray analysis of a polymerized single crystal of C_{70} [7].

In contrast to the "zigzag" chains, the dimer C_{140} could be expected to be soluble like the [2+2] C_{60} dimer C_{120} [8] and 'accessible' to wet chemistry methods. Indeed, we have dissolved and chromatographically separated C_{140} (in ~30-50% yield) from solid C_{70} treated at 1 GPa and 200°C, and thus provided unambigous proof for the formation of the dimer at high pressure [1]. Most remarkably, practically a single C_{2h}

CP544, *Electronic Properties of Novel Materials—Molecular Nanostructures*, edited by H. Kuzmany, et al.
© 2000 American Institute of Physics 1-56396-973-4/00/$17.00

C_{140} isomer (Fig.1) is produced in the solid state reaction out of the five possible [2+2] cycloaddition dimers which have equally low formation energies according to PM3 semiempical calculations [1]. The dimer obtained is the one favored when C_{70} molecules adopt an ordered packing with parallel D_5 axes.

Like other fullerene polymers, C_{140} is metastable at ambient pressure and reverts to C_{70} at elevated temperatures due to the intercage bond breaking. Here, we report on the kinetics of thermal dissociation of C_{140} between 110 and 200°C. Results on efficient photogeneration of singlet oxygen ($^1O_2$) in solutions of C_{140} and C_{120} and a scheme of photophysical proceses in the dimers are presented as well.

FIGURE 1. Cap-to-cap C_{70} dimer C_{2h} C_{140}.

EXPERIMENT

Preparation of C_{140} and C_{120} has been described in detail elsewhere [1,8]. Kinetics of dissociation of C_{140} into C_{70} (powder samples embedded in KBr pellets) was determined from the IR spectra using a Bruker Equinox spectrometer and a temperature-controlled sample oven. The dissociation of C_{140} in o-dichlorobenzene (ODCB) solution (purged with N_2 to minimize oxidation) was followed at 382 nm on a Perkin-Elmer λ900 UV-visible spectrometer. Photogeneration of $^1O_2$ in air-saturated ODCB solutions of C_{70}, C_{60} and their dimers was studied by measuring $^1O_2$ luminescence in the near-infrared (NIR) range with the Equinox spectrometer using a FT-Raman module for detection (900-1700 nm) and a Xe lamp and a Spex monochromator for photoexcitation (300-600 nm). High sensitivity of the FT-NIR-luminescence technique allowed short measurement times; therefore photodissociation of C_{140} and C_{120} (quantum yield $\varphi_{diss} \sim 5 \times 10^{-5}$ in air-saturated solutions [1]) was negligible.

RESULTS

Thermal dissociation. Fig.2a shows characteristic changes in the IR spectrum of solid C_{140} (KBr pellet) decomposing at elevated temperature back into C_{70}. At applied temperatures between 130 and 200°C a decrease/ increase in intensities of the C_{140}/ C_{70} peaks followed a first-order kinetics (Fig.2b). The dissociation rate constant k exhibits a simple activated (Arrhenius) behavior, i.e., $k(T) = \nu_0 \cdot \exp(E_a/k_BT)$. The activation energy determined from a fit of the data in Fig.2c is $E_a = 1.6 \pm 0.03$ eV and the prefactor $\nu_0 = 8.7 \times 10^{14}$ s^{-1}. The value of E_a is similar to that for C_{120} (1.75 ± 0.05 eV [2]). Note that E_a of the depolymerization of "zigzag" chains of C_{70} is also less than E_a of the depolymerization of linear (1D) chains of C_{60} (1.8 and 1.9 eV, respectively [2]). Dissociation of C_{140} in ODCB solution (110-160°C) also proceeds as a first-order process over $\sim 2\tau$, but, in contrast, exhibits different kinetic parameters and non-Arrhenius behavior (Fig.2c,). Evidently, a mechanism of the dissociation of C_{140} (and likely of C_{120} and fullerene polymers) is quite complicated. One can speculate that at first one intercage C-C bond breaks in C_{140}. Molecular dynamics and interactions with the surrounding (likely solvent-specific and strongly temperature-dependent in a solution) can result in the recovery of this bond or in the total dissociation of the dimer. The latter step probably requires a large activation volume in solid C_{140}.

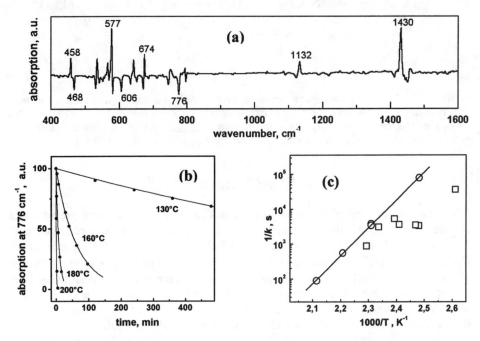

FIGURE 2. (a) Differential IR spectrum of C_{140} heated at 160°C for 35 min. The upward peaks reflectthe appearance of C_{70}; (b) Decrease of 776 cm^{-1} peak of C_{140} vs. time at different temperatures. The curves are exponential fits; (c) Arrhenius plot for the thermal dissociation of C_{140} in the solid state (O) and in ODCB solution (□).

Photophysical properties of C_{140} and C_{120} such as electronic absorption and fluorescence are similar to those of the monomers, indicating that there is only moderate perturbation of the ground (S_0) and excited (*S, *T) electronic states in C_{70} and C_{60} upon dimerization [1,9]. According to our measurements of NIR photoluminescence of $^1O_2$ (Fig.3), the efficiencies of $^1O_2$ generation via energy transfer from excited triplet states (*T) are also similar. The quantum yield of $^1O_2$ is 0.9 ± 0.1 for C_{120} and 0.95 ± 0.05 for C_{140}, being referenced to the yield of ≈ 1 for C_{60} and C_{70} [10]. Consequently, the dimers, like the monomers [10], have quantum yields for the photogenerated triplet state (φ_{ISC}) close to unity. As mentioned above, both dimers photodissociate in a solution. Since the quantum yield φ_{diss} increases

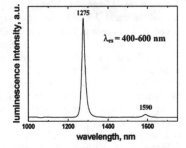

FIGURE 3. Spectrum of $^1O_2$ photoluminescence in air-saturated solutions of C_{60}, C_{70} and their dimers in ODCB.

significantly when dissolved oxygen is removed, the photodissociation likely occurs from the triplet states (efficiently quenched by O_2). This and other common photophysical processes in C_{140} and C_{120} are shown below:

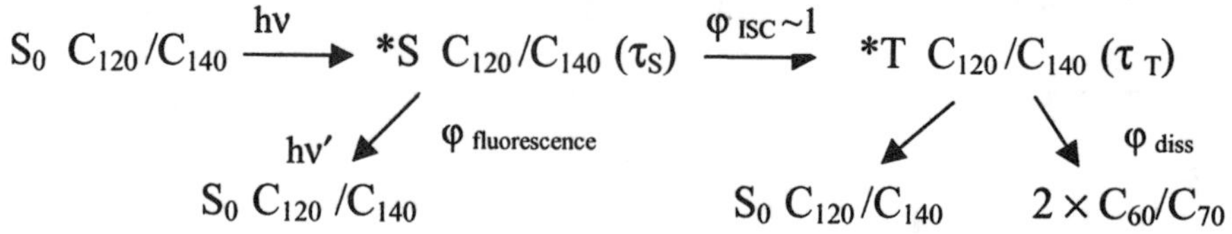

$$S_0 \ C_{120}/C_{140} \xrightarrow{\ h\nu\ } {}^*S \ C_{120}/C_{140} \ (\tau_S) \xrightarrow{\ \varphi_{ISC} \sim 1\ } {}^*T \ C_{120}/C_{140} \ (\tau_T)$$

$$h\nu' \nearrow \quad \varphi \ \text{fluorescence} \qquad\qquad \swarrow \qquad \searrow \quad \varphi \ \text{diss}$$

$$S_0 \ C_{120}/C_{140} \qquad\qquad\qquad S_0 \ C_{120}/C_{140} \qquad 2 \times C_{60}/C_{70}$$

SCHEME 1. Common photophysical processes for the fullerene dimers C_{120} and C_{140} in toluene/ODCB solutions. The lifetimes of excited electronic states: $\tau_S = 1.4$ ns for C_{120} [9]; $\tau_T \sim 10$ and $\sim 50 \mu s$ for C_{120} and C_{140}, respectively [1,9]. The quantum yields: $\varphi_{\text{fluorescence}} = 8 \times 10^{-4}$ and 6×10^{-4} [1,9]; φ_{diss} $= 7 \times 10^{-4}$ and $(2\text{-}3) \times 10^{-3}$ (in deoxygenated solutions) for C_{120} and C_{140}, respectively [1].

CONCLUSIONS

The kinetics of thermal dissociation of solid C_{140} into C_{70} at ambient pressure and elevated temperature follows a simple first-order law and is thermally activated with $E_a = 1.60 \pm 0.03$, similarly to C_{120} (1.75 ± 0.05 eV [2]). In contrast, the dissociation of C_{140} dissolved in ODCB shows non-Arrhenius behavior, suggesting a strong influence of the molecular surroundings. In a solution, the fullerene dimers C_{140} and C_{120} (like C_{70} and C_{60}) exhibit a photogenerated triplet state yield close to unity and high formation efficiency of ${}^1O_2$ via triplet energy transfer. The quantum yield of ${}^1O_2$ is 0.95-1.0 for C_{140} and 0.85-0.95 for C_{120} in air-saturated ODCB solutions.

REFERENCES

1. Lebedkin, S., Hull W. E., Soldatov, A., Renker, B., and Kappes M. M., *J. Phys. Chem. B,* in press.
2. Nagel, P., Pasler, V., Lebedkin, S., Soldatov, A., Meingast, C., Sundqvist, B., Persson, P.-A., Tanaka, T. Komatsu, K., Buga, S., and Inaba, A. *Phys. Rev.* **1999**, *B60*, 16920.
3. Sundqvist, B. *Adv. In Physics* **1999**, *48*, 1, and references cited therein.
4. Rao, A. M., Menon, M., Wang, K.-A., Eklund, P. C., Subbaswamy, K. R., Cornett, D. S., Duncan, M. A., and Amster, I. J. *Chem. Phys. Lett.* **1994**, *224*, 106.
5. Iwasa, Y., Furudate, T., Fukawa, T., Ozaki, T., Mitani, T., Yagi, T., and Arima, T. *Appl. Phys.* **1997**, *A64*, 251; Premila, M., Sundar, C. S., Sahu, P. Ch., Bharathi, A., Hariharan, Y., Muthu, D. V. S., and Sood, A.K. *Solid State Commun.* **1997**, *104*, 237.
6. Soldatov, A., Nagel, P., Pasler, V., Lebedkin, S., Meingast, C., Roth, G., and Sundqvist, B. In *Fullerenes and Fullerene Nanostructures. Proceedings of the International Winterschool on Electronic Properties of Novel Materials 1999*, Kuzmany, H., Fink, J., Mehring, M., Roth, S., Eds., AIP Conference Proceedings 486, Melville, New York, 1999, p.12.
7. Soldatov, A., Roth, G., Lebedkin, S., Meingast, C., Dzyabchenko, A., Johnels, D., Haluska, M., and Kuzmany, H., in preparation.
8. Wang, G.-W., Komatsu, K., Murata, Y., and Shiro, M. *Nature* **1997**, *387*, 583
9. Ma, B., Riggs, J. E., and Sun, Y.-P. *J. Phys. Chem.* **1998**, *B102*, 5999.
10. Arbogast, J. W., Darmanjan, A. P., Foote, C. S., Rubin, Y., Diederich, F. N., Alvarez, M. M., and Whetten, R. B. *J. Phys. Chem.* **1991**, *95*, 11; Arbogast, J. W., and Foote, C. S. *J. Am. Chem. Soc.* **1991**, *113*, 8886.

Chain Orientation and Layer Stacking in the High-Pressure Polymers of C$_{60}$: Single Crystal Studies

R. Moret[*], P. Launois[*], T. Wågberg[¶] and B. Sundqvist[¶]

[*] *Laboratoire de Physique des Solides, UMR8502 CNRS, Bât. 510, Université Paris-Sud, 91405 Orsay, France*
[¶] *Department of Experimental Physics, Umeå University, 90187 Umeå, Sweden*

Abstract. High-pressure polymerisation of C$_{60}$ leads to a variety of new crystalline or amorphous phases which display interesting physical properties. We have prepared one-dimensional (1D, C$_{60}$ chains) and two-dimensional (2D, C$_{60}$ layers) polymers from C$_{60}$ single crystals. The resulting multi-domain crystals have been studied using x-ray diffraction and Raman spectroscopy. The relative orientations of the chains in the "low-pressure" 1D orthorhombic polymer had been characterized previously [1]. We have now determined the specific stacking of the C$_{60}$ layers in the 2D tetragonal and rhombohedral polymers. Using these results we analyze the relations between the different polymers and the intermolecular environments which may play a role in stabilizing the observed polymer structures.

INTRODUCTION

High-pressure high-temperature C$_{60}$ polymers are formed by covalent linking of neighboring C$_{60}$ molecules (via the formation of four-membered rings) along the $\langle 110 \rangle$ directions of the parent monomer cubic structure. As a result, the inter-fullerene distance shortens to 9.1-9.2 Å, as compared to 10 Å in pristine C$_{60}$. The number of covalent bonds per C$_{60}$ molecule increases with pressure and temperature, thus leading to dimers, linear chains (1D polymers) and layers (2D polymers), successively [2]. The structure of these different phases has been investigated by diffraction methods on powder samples and the basic characteristics are already known [2]. However, important information such as the molecular deformation of the C$_{60}$ ball near the inter-molecular bonds, the relative orientation of the chains in the 1D polymers or the stacking of the layers in the 2D polymers is still unknown or disputed. We have been able to characterize the latter two structural properties through the synthesis and study of 1D and 2D polymer samples prepared from C$_{60}$ single crystals. This allows us to discuss the relations and possible transformations between these polymers.

RESULTS

The C$_{60}$ crystals used for the high-pressure high-temperature treatments were grown from purified C$_{60}$ powder by sublimation in a two-zone oven (450-500 °C).

CP544, *Electronic Properties of Novel Materials—Molecular Nanostructures*, edited by H. Kuzmany, et al.

Crystals with negligible twinning contamination were selected. They were compressed in a piston-cylinder device (using silicone oil) and heated in a small furnace made of a Pyrex glass tube wound with Kanthal wire.

The Raman spectra and x-ray diffraction data were collected and analyzed after quenching the crystal to room temperature and ambient pressure. For more details on the experimental procedure and on the analysis of the data we refer the reader to references 1 and 3. The main structural results of these studies are summarized below.

For the 1D polymers it is generally agreed [2,4] that two different orthorhombic phases can be formed, corresponding to "high" and "low" pressure regions which are still ill-defined. We have obtained (at 1-1.2 GPa and 550-585 K) and studied single crystals of the "low" pressure phase [1]. The space group symmetry is found to be Pmnn indicating that the orientation - **around their axes** - of neighboring chains alternates. This is conveniently visualized if one considers the orientation of the four-membered rings joining the C_{60} molecules. As shown in Fig. 1, the orientation angle μ of the four-membered rings relative to the long axis (that originates from a cubic **a** axis of the parent C_{60} structure) of the orthorhombic unit cell is $+\mu$ or $-\mu$, alternatively. This finding is in agreement with the lattice energy minimization performed by Dzyabchenko and coworkers, who also predict an optimal value of ~29° for μ [5]. This value provides a rather satisfactory fit to the diffraction peak intensities but more precise diffraction experiments are needed and planned to ascertain this result.

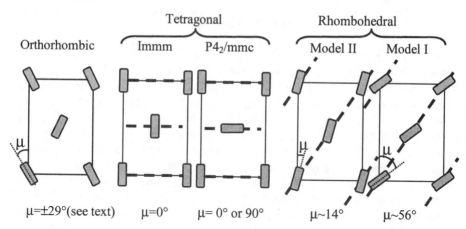

FIGURE 1. Schematic views of the orientation of the polymer chains, running along $\langle 110 \rangle_{cubic}$ directions of the parent monomer lattice, for the different polymer structures. The chains are normal to the figure. The orientation of the four-membered rings is represented by shaded bars which form an angle μ with the $\langle 001 \rangle_{cubic}$ directions. The dotted lines represent the 2D tetragonal and rhombohedral polymer layers which can be obtained, at least schematically, by connecting the C_{60} molecules of the chains by supplementary four-membered rings.

More recently we have succeeded in obtaining two-dimensional polymerisation in a C_{60} single crystal treated at 2 GPa and 700 K (the temperature was established before raising the pressure) [3]. X-ray diffraction and Raman spectroscopy agree on the presence of both and tetragonal and rhombohedral phases in the polymerized single crystal. Furthermore, there is evidence from the Raman data, but not from the x-ray data, for the presence of dimers either in a disordered state or close to the surface. Despite difficulties due to the tetragonal-rhombohedral coexistence, the presence of variants in both cases and a rather poor mosaicity, we have been able to extract information from the diffraction data. We have found that the stacking of the polymer layers in the tetragonal phase is such that successive layers are rotated by 90°, corresponding to the space group $P4_2/mmc$. The Immm model, where successive layers are identical, is thus ruled out, in agreement with the energy minimization results showing that it is slightly less stable [5].

Turning to the rhombohedral phase we have tried to distinguish between the different $R\overline{3}m$ structural models proposed so far. Although our analysis is not fully conclusive we find that model II (in the nomenclature proposed by Davydov et al. [5]) provides a slightly better fit to the diffraction data. We recall that both models I and II result from ABCABC-type stacking of the trigonal C_{60} polymer layers but they differ by a 60° rotation of these layers around the stacking axis.

DISCUSSION

The relation between the parent monomer C_{60} structure and the different polymer phases can be described using the schematic representation of Fig. 2. It shows how the C_{60} chains can be linked by supplementary bonds (four-membered rings) to form the C_{60} layers of the tetragonal and rhombohedral polymers. These layers originate from the {100} and {111} parent cubic planes. In Fig. 1, to visualize the orientation of the C_{60} molecules in these 1D and 2D polymers we show how the chains are oriented around their axes (orientation angle μ, as defined above) using a single type of representation, the chains of Fig. 2 being viewed along their axes.

In the likely hypothesis that 2D polymerisation involves, at some intermediate stage, the formation of chains oriented as in the orthorhombic Pmnn polymer, then Fig. 1 shows that further specific rotations of the chains lead to the 2D structures. For the tetragonal $P4_2/mmc$ structure, the 90° rotation of the successive layers mentioned above corresponds also to an alternation of chains with $\mu=0°$ and 90° (while all chains have $\mu=0°$ (or 90°, equivalently) for Immm). For the rhombohedral structure, it is seen that, for both models I and II, all chains possess the same orientation but with different μ angles of ~ 56.2° and 14.4°, respectively. These remarks should have interesting implications for the understanding of the relation and transformation between the 1D and 2D C_{60} polymers.

It is also worth pointing out that the layer stacking in the rhombohedral [3] (and to a lesser extent the tetragonal) structure is such that near-neighbor C_{60} molecules are close to a double-bond-facing-hexagon type of environment, all the more so as this type of environment is known to be stabilized by pressure in the monomer C_{60} phase.

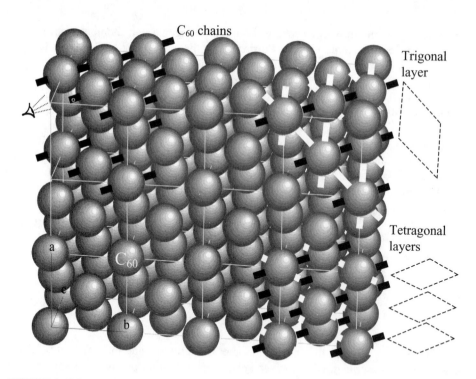

FIGURE 2. Schematic representation of the relation between the 1D and 2D C_{60} polymer structures. The fullerene molecules are represented by spheres (radius 3.5Å). The fcc monomer (**a, b, c**) unit cell is shown on the bottom left. The C_{60} chains (solid black segments) of the orthorhombic polymer are shown on the top left. In Fig. 1 they are viewed along the direction indicated by the eye. The tetragonal and rhombohedral polymer structures are formed by connecting the chains as shown by the white bars, on the right. In this sketch the ~9% contraction of the C_{60}-C_{60} distance due to the formation of the covalent bonds is neglected.

REFERENCES

1. R. Moret, P. Launois, P.-A. Persson and B. Sundqvist, Europhysics, Lett. 40, 55 (1997); P. Launois, R. Moret, P.-A. Persson and B. Sundqvist, in Molecular Nanostructures, ed. H. Kuzmany, J. Fink, M. Mehring and S. Roth. World Scientific, Singapore, 1998, pp. 348-352.

2. For recent reviews, see for instance (a) V.D. Blank, S.G. Buga, G.A. Dubitsky, N.R. Serebryanaya, M. Yu. Popov and B. Sundqvist, Carbon, 36, 319 (1998); (b) B. Sundqvist, Advances in Physics, 48, 1 (1999).

3. R. Moret, P. Launois, T. Wågberg and B. Sundqvist, European Phys. Journal B, 2000, in print. B. Sundqvist, T. Wagberg, P.-A-Persson, P. Jacobsson, S. Stafstrom, R. Moret and P. Launois Proc. IWFAC 99, to be published.

4. V. Agafonov, V.A. Davydov, L.S. Kashevarova, A.V. Rakhmanina, A. Kahn-Harari, P. Dubois, R. Céolin and H. Szwarc, Chem. Phys. Lett. 267, 193 (1997).

5. V.A. Davydov, V. Agafonov, A.V. Dzyabchenko, R. Ceolin and H. Szwarc, J. Solid State Chem. 141, 164 (1998). V.A. Davydov, L.S. Kashevarova, A.V. Rakhmanina, V. Agafonov, H. Allouchi, R. Ceolin, A.V. Dzyabchenko, V.M. Senyavin, H. Szwarc, T. Tanaka and K. Komatsu, J. of Phys. Chem. B 103, 1800 (1999).

Structures of C_{60} Superhard Phases as a Mirror of 3D Polymerization

Nadezhda R. Serebryanaya[a, c], Leonid A. Chernozatonskii[b] and Boris N. Mavrin[c]

[a]Technological Institute for Superhard and Novel Carbon Materials, 7-a Centralnaya St., Troitsk, Moscow Region, 142190 Russia. E-mail: nadya@ntcstm.msk.ru
[b]Institute of Biochemical Physics, 117977 Moscow, Russia.
[c]Institute of Spectroscopy, Troitsk, Moscow Region, 142190 Russia.

Abstract. The crystal structures of superhard C_{60} phases, earlier obtained after high-pressure and high-temperature treatment, were determined by molecular mechanics methods and refined by the profile analysis of X-ray powder diffraction patterns. Three steps of the three-dimensional polymerization (3D) have been found by three types of crystal structures altered in the (9.5 ÷ 13 GPa) pressure range and in the (670 ÷ 820 K) temperature range. The new type of molecular bonding - (3+3) cycloaddition - has been found between neighboring layers of molecules along the space diagonal for all polymerization steps which are characterized by the gradual complication of molecular bonding in the layer itself: (2+2) bonding and/or common four-sided rings. Lattice dynamics simulation of 3D polymerized hardest C_{60} phase has shown that sound velocities, bulk modulus and broad bands in Raman spectra agree with measurements.

INTRODUCTION

Fullerenes are known to undergo polymerization under pressure at high temperatures. A (T, P) C_{60} phase diagram has been reported [1, 2] where is shown the region of parameter synthesis of one-dimensional (1D) orthorhombic chained polymers, 2D tetragonal and rhombohedral polymers and 3D orthorhombic polymers and also several different disordered states. It was concluded that there is a (2+2) cycloaddition in 1D and 2D crystalline C_{60} polymers [3]. Some 3D C_{60} fullerites, obtained after high-pressure (9.5÷13 GPa) and high-temperature treatment (>670 K), are superhard phases with hardness and bulk modulus [4] exceeding those of diamond.. However, the character of C_{60} molecular bonding in superhard phases and their crystal structures were not known up to now. In the present article we have found three main crystal structures of C_{60} superhard phases where the gradual complication of the character of molecular bonding corresponds to the increase of the polymerization degree. Also, we have compared Raman spectra observed with the computed projected density of

CP544, *Electronic Properties of Novel Materials—Molecular Nanostructures*, edited by H. Kuzmany, et al.
© 2000 American Institute of Physics 1-56396-973-4/00/$17.00

vibrational states and have calculated bulk modulus of the hardest phase by lattice dynamics.

CRYSTAL STRUCTURES

The powder diffraction patterns of the C_{60} superhard phases, obtained and published earlier [1] by one of authors, N.R.S., have been used for this investigations. Three types of diffractograms of superhard phases were observed in the quenched samples after treatment at 12÷13 GPa pressure range and 670÷820 K temperature range. These diffractograms have been indexed on the base of the orthorhombic body-centered (*bco*) unit-cell with parameters which are approximately equal to half of a face-diagonal of the face-centered cubic (*fcc*) structure (the C_{60} pristine crystallizes in *fcc*-type of structure). The high value of density (3.4 - 3.2 g / cm ^{3}) calculated from X-ray data for *bco*-phases corresponded to experimental ones. This value has been measured from the whole sample, contained both amorphous and crystalline phases. The inhomogeneity of samples is formed under large gradients (~50 K and ~0.5 GPa) in the pressure chamber [1]. We have determined again the densities of each phase of samples. The density of the crystalline phases is in the 2.2-2.5 g/cm^3 range. Taking into account these values of densities, we will consider structures of superhard phases.

The pseudo-tetragonal *bco* unit cells with parameters c=1.26-1.29 nm, $a \cong b \cong c/\sqrt{2}/2$ was chosen instead of the early *bco*-cell. Then the unit-cell parameters have been refined using DICVOLV and Rietveld programs. Table 1 shows unit-cell parameters and hardness of these three superhard phases of C_{60}. The space group of our *bco* structures is *Immm*. There are 9 independent carbon atoms in asymmetric cell as in the case of the 2D tetragonal phase [3].

Pressure has an enormous effect on C_{60}: bringing the molecules closer together, at high temperatures anisotropic oscillations of atoms should induce an appearance of bonds between molecular layers along the closest [111]-direction of *bco*-structure.

TABLE 1. Unit-cell parameters of 3D superhard body centered orthorombic structures of solid C_{60}. The space group of bco-structures is *Immm*.

Phase	P, GPa	T (K)	a (nm)	b (nm)	c (nm)	$\rho_{X\text{-ray}}$, g/cm^3	ρ_{exp}, g/cm^3	Hardness, GPa
Pristine fcc			1.417			1.68		soft
A bco	13	820	0.867	0.881	1.260	2.48	2.5	150
B bco	12	770	0.869	0.874	1.271	2.43	2.4	120
C bco	13	670	0.873	0.916	1.294	2.30	2.25	60

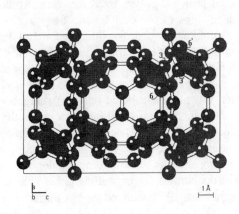

FIGURE 1. The (010) projection of *bco* structure of C superhard phase. (3+3) cycles are formed between shaded hexagons.

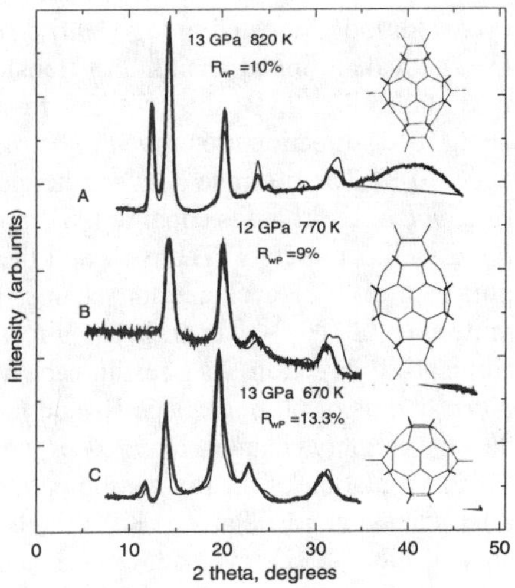

FIGURE 2. Diffraction patterns (thick lines) and simulations (thin lines) of superhard phases. R_{wP}-weighted pattern factor of reliability. (001) projections of C_{60} cages are shown. The amorphous part of pattern is excluded from simulation.

In our work [5], carried out together with the ESRF collaborators, the unusual ellipsoidal Debye-Sherrer diffraction patterns are observed which shows that the giant anisotropic deformation is retained in the quenched samples and the unambiguous proof (as like as the strongly reduced intermolecular distances in all three dimensions) for the existence of 3D polymerized C_{60} phases were yielded.

We have proposed that the shortest intermolecular distance is formed by atom numbers 3 and 6 (Fig. 1) which are located in hexagons oriented along the space diagonal of *bco*-structure. The calculation of atom positions of C_{60} molecule has been carried out in the framework of HYPERCHEM program by the molecular mechanics method: energy optimization of the cluster-geometry of 9 polymerized C_{60} molecules. The calculated coordinates were corrected by Rietveld refinement, all atom positions are listed in our article [6]. The simulation has been carried out with use of the slight preferred orientation (5%) in the [111] direction. The correctness of the chosen orientation has been confirmed by the resemblance of simulated diffraction pattern to that of the first superhard C phase (table 1, Fig. 2) obtained at the first polymerization step (13 GPa, 670 K). Thus, bringing 3rd and 6th atoms closer together is a key to determination of the crystal structure of C_{60} superhard phases. This type of bonding is (3+3) type of cycloaddition which binds molecules of neighboring layers: the central molecule of *bco*-structure is connected with 4 upper and 4 lower neighbor molecules by such (3+3) bonding (Fig.1). The character of the bonding in the (x, y) layer itself is different for these phases.

At the first stage of polymerization the lower hard C structure (Table 1) (2+2) cycloaddition is located along [100] direction. It can be described as the structure of (3+3) covalent linked chains, fast transformed from 1D polymerized structure. At the next stage (Fig. 2, B) the common four-sided rings of the cyclobutane fragment [7] along [010] direction are proved to be more suitable to B-phase, synthesized at 12 GPa and 770 K. The distance between neighbouring molecules of the common four-sided ring type of the chain is equal to 0.87 nm, this value corresponds approximately to a or b unit-cell parameters (Table 1), and these distances are shorter than those (0.91-0.92 nm) in the (2+2) cycloadditon chains. Each C_{60} molecule binds only two neighbor molecules of (001)-layer in [010] direction and there is no bond between molecules along [100] direction. As a result, each molecule binds with 10 neighbors and contains 20 sp^3-atoms of 60 ones. This B- and C-phases can be considered as zeolites because they have empty channels along x- or y-direction, and it is suitable for metal doping.

The simulation of the diffraction pattern of the hardest C_{60} crystalline phase (Fig. 2, A), obtained at 13 GPa, 820 K, is the best fit when all considered types of bonds are applied: the (2+2) cycloaddition reaction takes place along [100] direction, the common four-sided ring bonds are formed along [010] direction, and (3+3)-type bonds connect neighbouring layers along the space diagonal of bco unit-cell. The weighted pattern R_{wP} factor for A-phase (Fig. 2, A) is equal to 10% obtained by the Rietveld refinements. The A phase is formed at the highest stage of polymerization, each molecule has 12 neighbors and contains 24 sp^3 atoms.

RAMAN SPECTRA AND BULK MODULUS

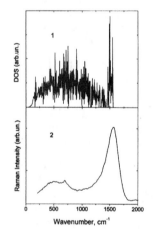

FIGURE 3.Calculated projected DOS for sp^2-bonded atoms (1), (2)- Raman spectrum [1].

The formation of polymerized phases have been confirmed by Raman spectra [1], where the single broad band is observed near 1550 cm^{-1} for the samples synthesized at 9.5÷13 GPa and T > 670 K. It may be interpreted as a structure transition from a molecular structure to a continuous network due to the formation of covalent intermolecular bonds. Fig 3. compares the experimental Raman spectrum of A-phase [1] with the computed projected density of vibrational states. We believe that the very large bandwith (>100 cm^{-1}) of Raman bands gives evidence of the strong disorder of the crystal, in consequence of which the Raman spectrum is proportional to the density of vibrational states. It was shown earlier that the Raman scattering from sp^2-bonded carbon material is ~ 100 times stronger than that of sp^3-bonded carbon atoms, and this is why , in the main, the contribution from sp^2-bonded atoms will be exhibited in the Raman spectra. Fig. 3 shows a good correlation between the calculated projected densities of vibrational states for sp^2-bonded carbons and experimental Raman spectrum, in which is mainly observed the broad intense band near 1550 cm^{-1}. The estimation of the velocity of acoustical waves and bulk modulus of the A-phase has been carried out from lattice dynamics simulation. The potential parameters for 24 sp^3- and 36 sp^2-bonded carbon atoms were found by fitting the calculated dispersion curves of diamond and graphene sheet, respectively, to experimental ones. The experimental values of acoustical velocities (V^{TA} - transverse and V^{LA} - longitudinal), obtained for polycrystalline samples, V^{TA}_{exp} = 7.2–9.6 km/s, V^{LA}_{exp} =17-26 km/s [4], are in the qualitative agreement with the calculated mean velocities: V^{TA} = 10.3 km/s, V^{LA} =21.6 km/s. The value of bulk modulus has been estimated using the calculated mean velocities and it is equal to $K \sim 800$ GPa, that was in accord with the experimental ones (690–1700 GPa) depending on the synthesis [4] and exceeded that of diamond 442 GPa.

The crystal structure of the hardest A-phase is very stable, it remains at 13 GPa up to 1200 K with both some distortions and the gradual increase of amorphous halo. Thus, taking into account the similar Raman spectra of all superhard C_{60} phases we have concluded that the 3D polymerization arises owing to the appearance of (3+3) intermolecular bonding which transforms the molecular structure of pristine into the 3D network, it is getting more complete and rigid in A-phase.

ACKNOWLEDGMENTS

This work was supported by RFFI grants (99-02-17578, 99-02-18193), grant No. 95076 of Russian National Foundation for Intellectual Collaboration and Grant-in-Aid of ICDD (1998-1999 years). We thank V. D. Blank, I. V. Stankevich and M. Menon for fruitful discussions.

REFERENCES

1. Blank, V. D., Buga, S. G., Serebryanaya, N. R., Dubitsky, G. A., Sulyanov, S. N., Popov, M. Yu., Denisov, V. N., Ivlev, A. N. and Mavrin, B. N., *Phys. Letters* A **220**, 149-157 (1996).
2. Blank, V. D., Buga, S. G., Duitsky, G. A., Serebryanaya, N. R., Popov, M. Yu. and Sundqvist, B., *Carbon* **36**, 319-343 (1998).
3. Núñez-Regueiro, M., Marques, L., Hodeau, J.-L., Béthoux, O. and Perroux, M., *Phys. Rev. Lett.* **74**, 278-281 (1995).
4. Blank, V. D., Levin V. M., Prokhorov, V. M., Buga, S. G., Dubitsky G. A. and Serebryanaya, N. R., *JETP* **87**, 741-746 (1998).
5. Marques, L., Mezouar, M., Hodeau, J.-L., Núñez-Regueiro, M., Serebryanaya, N. R., Ivdenko, V. A., Blank, V.D., and Dubitsky, G. A., *Science*, **283**, 1720-1723 (1999).
6. Chernozatonskii, L. A., Serebryanaya, N. R. and Mavrin, B. N., *Chem. Phys. Lett.* **316**, 199-204 (2000).
7. Osawa, S., Osawa, E. and Hirose, Y., *Fullerene Sci. Technol.* **3**, 565-585 (1995).

Density Functional Study of the Phase Diagram of 3D C_{60}-Polymers

Péter Rajczy*, Jenő Kürti*, Sándor Pekker[†] and Gábor Oszlányi[†]

*Dept. of Biological Physics, Eötvös University,
Pázmány P. sétány 1/A, H-1117 Budapest, Hungary
[†]Research Institute for Solid State Physics and Optics,
H-1525 Budapest, P.O.B. 49, Hungary

Abstract. Three-dimensional polymeric structures of C_{60} were investigated with density functional theory using the solid state code Vienna *ab initio* Simulation Package (VASP). Starting from the 2D tetragonal polymer and subsequently reducing the unit cell volume we obtained 8 different possible structures. The calculated bulk moduli were in the range of 150-320 GPa.

INTRODUCTION

As it is well known, covalent bonding between C_{60} molecules can lead to the formation of 1D and 2D polymers with various structures [1–3] Furthermore, it was suggested that applying very high pressure to C_{60} at high temperature transforms it into a 3D polymer. Superhard and ultrahard materials having bulk moduli larger than that of diamond, were produced with this method [4–7]. Their exact structure, however, is still an open question. We carried out first-principles calculations using density functional method, trying to reveal the 3D polymeric structures of C_{60}.

CALCULATIONS

First-principles density functional theory, using Vienna *ab initio* Simulation Package [8,9] (VASP) was applied. VASP uses a finite temperature LDA, and is based upon carefully optimized ultrasoft Vanderbilt-type pseudopotentials. It performs a 3D periodic calculation, using a plane wave basis set.

We started from the 2D tetragonal polymer, and reduced the unit cell volume, hydrostatically. At each step, the volume (but not the shape) of the unit cell V was kept constant during geometry optimization. After reaching the relaxed geometry at a given unit cell volume, the converged hydrostatic pressure p was obtained as well. The bulk modulus B for each total energy minima was calculated with the

CP544, *Electronic Properties of Novel Materials—Molecular Nanostructures*, edited by H. Kuzmany, et al.

formula $B = -V_0 \partial p/\partial V$. For testing purposes, $B = 470$ GPa was obtained for the cubic diamond in good agreement with the experimental value of $B = 445$ GPa.

RESULTS

For the initial geometry two possible arrangements of the 2D tetragonal polymer sheets were taken. In the first case the space group was *Immm* (true *bc* orthorhombic unit cell). In the second case the space group was *P42mmc* (pseudo *bc* tetragonal unit cell).

Compressing the *Immm* unit cell from 1200 Å^3 lead to the formation of a 3D polymer. The resulted structure depended on the way the barrier at ≈ 800 Å^3 was crossed. When this volume was decreased by a small step of $\approx 1\%$ an abrupt change of the structure occurred yielding an energy minimum (1 in Figure 1). A larger step of $\approx 3\%$ from the same point, however, resulted in a different structure (2 in Figure 1). Expanding the latter, lead us again to a new structure (3 in Figure 1).

A similar procedure was carried out in the *P42mmc* case. Two different scenarios were found depending on the way of the compression. Figure 2. shows the case when the exact tetragonal symmetry was maintained during the whole procedure. Two stable structrures were found (1 & 2). Next, the compression starting from the same initial geometry has been carried out in smaller steps. A dramatic change occured due to running into a saddle point at around 830 Å^3: the tetragonal symmetry has been reduced to an orthorhombic one. Choosing different starting points for the subsequent expansions lead us to a manifold of stable 3D polymeric structures. (3, 4 & 5 in Figure 3).

It should be mentioned that the 3D polymeric state obtained by Okada *et al.* [10] is lying higher both in energy and in unit cell volume than our minimum No.2 in Figure 1. This means, together with the usual enthalpy considerations, that the latter is the most favorable among all the structures treated here including that proposed by Okada *et al.*.

The most important parameters for the starting 2D tetragonal structure and for the 8 local energy minima are collected in Table 1. As it can be seen the structure No.2 in Figure 1. has the largest bulk modulus.

SUMMARY

A number of different stable 3D polymeric structures of C_{60} was found using first-principles DFT calculations which were started from the 2D tetragonal polymer. Unit cell volume, under isotropic pressure, was reduced followed by subsequent expansions. Most of the stable structures had a bulk modulus larger than $250 GPa$. The experimental values of the bulk moduli for the superhard and ultrahard materials with unknown structures are even larger than those. This situation can have several possible interpretations. First, we considered only the derivatives of the 2D tetragonal polymer. Second, even for this specific starting point, several additional

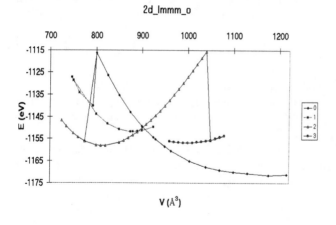

FIGURE 1. Total energy vs unit cell volume for 3D polymers, started with 2D tetragonal polymer having *Immm* space group (*bc* orthorhombic unit cell)

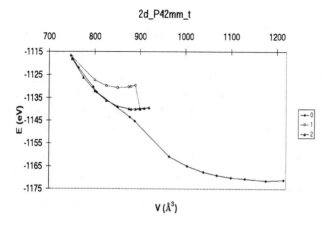

FIGURE 2. Total energy vs unit cell volume for 3D polymers, started with 2D tetragonal polymer having *P42mmc* space group (pseudo *bc* tetragonal unit cell) and preserving the tetragonal symmetry

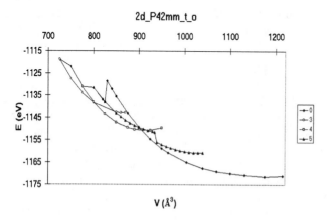

FIGURE 3. Total energy vs unit cell volume for 3D polymers, started with 2D tetragonal polymer having *P42mmc* space group (pseudo *bc* tetragonal unit cell) but loosing its tetragonal symmetry

TABLE 1. Unit cell parameters (a, b, c), unit cell volume (V_0), total energy (E_0) and bulk modulus (B) for the stable structures.

	a(Å)	b(Å)	c(Å)	V_0(Å^3)	E_0(eV)	B(GPa)
2D-tetragonal	8,889	8,889	14,884	1176	-1171,5	30
Immm_o						
1	9,687	9,692	9,373	880	-1151,8	293
2	8,500	7,297	13,140	815	-1158,2	315
3	9.135	8.522	12.844	1000	-1156,8	154
P42mmc_t						
1	9,517	9,517	9,385	850	-1130,7	262
2	9,997	9,997	8,904	890	-1140,2	287
P42mmc_t_o						
3	7,790	8,520	12,958	860	-1142,8	260
4	8,432	8,432	13,011	925	-1150,7	235
5	9,086	9,086	12,477	1030	-1160,9	154

structures must exist which may have larger bulk moduli. Third, a polycrystalline sample as a rule can have much larger bulk modulus than a monocrystalline one. A more detailed analysis of all these results will be published elsewhere.

ACKNOWLEDGMENTS

Work was supported by the grants OTKA T022980, OTKA T032613 and FKFP-0144/2000 in Hungary.

REFERENCES

1. Rao A.M. *et al.*, *Science* **259**, 955 (1993).
2. Pekker S. *et al.*, *Solid State Commun.* **90**, 349 (1994).
3. Iwasa Y. *et al.*, *Science* **264**, 1570 (1994).
4. Blank V.D. *et al. Phys. Lett. A* **220**, 149 (1996).
5. Blank V.D. *et al. Carbon* **36**, 319 (1998).
6. Blank V.D. *et al.* CP442, *Electronic Properties of Novel Materials:* XII International Winterschool, 499 (1998).
7. Marques L. *et al.*, *Science* **283**, 1720 (1999).
8. G. Kresse and J. Hafner, *Phys. Rev. B* **47**, R558 (1993).
9. G. Kresse and J. Furthmüller, *Comput. Mater. Sci.* **6**, 15 (1996); *Phys. Rev. B* **54**, 11169 (1996).
10. S. Okada *et al.*, *Phys. Rev. Lett.* **83**, 1986 (1999).

The phases of Rb$_x$C$_{59}$N from Raman spectroscopy

T. Pichler[1], W. Plank[2], H. Kuzmany[2], N. Tagmatarchis [3], K. Prassides[3]

[1] Institut für Materialphysik der Universität Wien, Strudlhofg. 4, A-1090 Wien and Institut für Festkörper- und Werkstofforschung Dresden, Postfach 270016, D-01171 Dresden.
[2] Institut für Materialphysik der Universität Wien, Strudlhofg. 4, A-1090 Wien.
[3] School of Chemistry, Physics and Environmental Science, University of Sussex, BN1 9QJ Brighton, UK.

Abstract. In this contribution we present resent results on the phases of in situ rubidium intercalated C$_{59}$N. Using Raman spectroscopy six distinct phases were identified by using the charge versus frequency relation of the A_g pinch mode. From additional in situ resistance measurements we find that all phases are semiconducting. Similar to Rb$_x$C$_{60}$ the phase Rb$_3$C$_{59}$N has the minimum resistance.

INTRODUCTION

The electronic properties of fullerenes and their alteration by doping has been in the focus of fullerene research work because of the observed metallic and superconducting fullerides [1,2]. From the beginning vibrational spectroscopy has been shown to be an extremly useful tool to monitor the doping process [1] and to identify the amount of charge transfer of the fullerides [3,4]. Furthermore Raman spectroscopy has been used to predict new phases of the fullerides [5]. In addition to intercalation doping two more doping methods have been used for fullerenes: A) endohedral doping by encageing metal ions (metallofullerenes) and B) on ball doping by substitution of carbon atoms with heteroatoms (heterofullerenes). The most prominent of the heterofullerenes is $(C_{59}N)_2$ [6]. In the solid state this heterofullerene is isostructural and isoelectronic to dimerized Rb$_1$C$_{60}$ [7]. The electronic structure of $(C_{59}N)_2$ has been analysed in detail by a combination of quantum chemical calculations, photoemission and electron energy-loss spectroscopy [8]. It was observed, that the charge distribution of the new highest occupied molecular orbital is predominantly localized at the intermolecular bond and at the nitrogen atoms. The heterofullerene is semiconducting with an energy gap of 1.4 eV about 0.4 eV lower than in C$_{60}$. From Raman spectroscopy of $(C_{59}N)_2$ [9] a strong resemblance to the Raman spectra of C$_{60}$ but a strong splitting of the lines from the degenerate modes was observed. For red laser excitation a strong resonance enhancement of the radial modes and the intercage modes was found. In comparison to the isostructural and isoelectronic compound of the quenched dimerized

CP544, *Electronic Properties of Novel Materials—Molecular Nanostructures*, edited by H. Kuzmany, et al.

Rb_1C_{60} the characteristic Raman modes of these single bond fullerene dimers were identified and can be used as fingerprint modes for these compounds [10].

From this close similarities to C_{60} compounds it is tempting to analyse the intercalation properties of $C_{59}N$ especially regarding the phases, phase separation and electronic transport properties. However, compared to the fullereides much less is known on the phases of the heterofullerene salts. At present only the phase $K_6C_{59}N$ has been fully structurally characterized and was found to be isostructural with bcc K_6C_{60} [11]. Regarding the electronic structure photoemission experiments of heterofullerene salts clearly showed that at least two new semiconducting intermediate phases are formed [12].

In this contribution, we present an *in situ* Raman study of rubidium intercalated thin films of $C_{59}N$ in comparison to the C_{60} fullerides.

EXPERIMENTAL

The preparation of the heterofullerene $(C_{59}N)_2$ has been described previously [6,7]. After the purification process by multi-cycle chromatography the metallofullerene sample was filled into an alumina crucible and degassed in the effusion cell in ultra high vacuum (UHV) at 425 K for 48 hours. For the Raman experiments, thin films of a thickness of about 500 nm were prepared by sublimationat 800 K onto high ohmic Si-wafers with a thin oxide layer as described previously [8]. The samples were then transferred into a purpose built high vacuum Raman cell (p = $2 \cdot 10^{-7}$ mb, T= 80 -500 K) for *in situ* alkali metal intercalation and resistance measurements. The Raman experiments were carried out in a Dilor xy spectrometer at a spectral resolution of 3 cm^{-1} for green (514.5 nm) laser excitations.

Rubidium intercalation was performed in a similar load and equilibrate process as described in Ref. [4]. After intercalation at 425 K and 30 min. equilibration at 500 K the Raman spectra of the majority phases of $Rb_xC_{59}N$ have been collected.

RESULTS AND DISCUSSION

In Fig. 1 the raman spectra of pristine and rubidium intercalated $(C_{59}N)_2$ are depicted. As already mentioned above most of the heterofullerene modes can be related to the characteristic modes from C_{60} [9,10]. For the pristine $(C_{59}N)_2$ a strong splitting of especially the $H_g(1)$ and $H_g(2)$ derived modes are observed. In the left panel of the figure the modes with the more radial character are shown. Upon intercalation these modes show, similar to the C_{60} fullerides, only a very small dependence on the charge transfer to the fullerene cage [13]. The radial breathing mode, for instance, softens by about 1 cm^{-1} for intermediate intercalation and hardens by about 4 cm^{-1} for the fully intercalated $Rb_6C_{59}N$. For the $H_g(1)$ derived mode the stronger crystal fields in the fullerides lead to an additional strong splitting in E_g and T_g derived modes.

Much more sensitive to the charge transfer and the bonding environment are the tangential intramolecular modes. The elongation of the C-C bonds upon in-

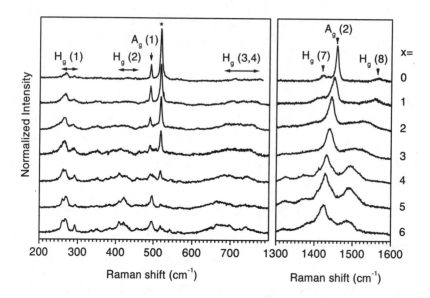

FIGURE 1. Raman spectra of pristine and rubidium intercalated $(C_{59}N)_2$ for green laser excitation (514.5 nm) measured at 425 K. The x numbers are corresponding to the majority phases of $Rb_xC_{59}N$. The asterics denotes the Raman mode from the Si substrate. The Mullikan symbols assign the Raman active symmetry species for C_{60} to the C_{60} derived modes of $C_{59}N$.

tercalation leads to a downshift of these modes, in general [13]. For C_{60} fullerides especially the pentagonal pinch mode has been used as convenient method to characterize the stochiometry of these compounds, because of the approximately linear relationship between softening of the mode and charge transfer. A charge transfer of about 7 cm^{-1} per transferred electron is well established. The right panel of Fig. 1 shows the doping dependence of the pentagonal pinch mode $(A_g(2))$ together with the $H_g(7,8)$ derived modes for $Rb_xC_{59}N$. The strong softening of these modes with intercalation is evident. For the A_g pinch mode we observe, similar to the C_{60} fullerides, a liner softening with increasing intercalation.

Furthermore, and again similar to A_xC_{60}, there is a clear phase separation for $Rb_xC_{59}N$. In Raman measurements conducted during the intercalation (see Fig. 2), we found a change of the Raman intensity of the $A_g(2)$ mode of different phases but no sign of a continuous shift of these mode with intercalation. These two points enable us to use Raman spectroscopy to identify the different phases of these heterofullerene salts. In the left panel of Fig. 3 we show the observed position of the pentagonal pinch mode as a function of charge transfer to C_{60}. Regarding the position of the $A_g(2)$ mode pristine $C_{50}N$ corresponds to C_{60}^-. Interestingly for $Rb_xC_{59}N$ the linear downshift with charge transfer is only slightly smaller (6 cm^{-1}) than for Rb_xC_{60} and thus the frequency versus charge relation in these fullerides

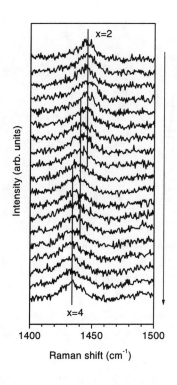

FIGURE 2. Raman spectra conducted during 60 min. of Rb intercalation, as indicated by the arrow. Each individual spectrum was measured for 3 min.. The solid lines are are guidelines for the phases of $Rb_xC_{59}N$ for $x = 2, 3, 4$.

follows a universal behavior. However there are also significant differences between the heterofullerides and the fullerides. In contrast to Rb_xC_{60} not only phases with $x = 1, 3, 4$, and 6 are stable, but additional phases for $Rb_2C_{59}N$ and $Rb_5C_{59}N$ are stable. Regarding the electronic transport properties in a simple charge transfer model one would expect that $Rb_2C_{59}N$, which is isoelectronic to the metallic and superconducting Rb_3C_{60}, has the maximal conductivity. However, the thin film resistance depicted in the right panel of Fig. 3, clearly shows that $Rb_3C_{59}N$ has the maximum conductivity. This is a further indication of the strong localization of the additional electron introduced by the nitrogen substitution. The temperature dependence of the resistance (not shown) is also strongly activated, clearly showing that this phase is semiconducting.

To summarize, we have analysed the intercalation properties of $Rb_xC_{59}N$ by Raman spectroscopy. From the charge vesus frequency relation of the A_g pinch mode we have have identified six phases with $x = 1, 2, 3, 4, 5$ and 6 with a phase separation between them, similar to the C_{60} fullerides. From a comparison with *in situ* resistance measurements we find that all observed phases are semiconducting with the minimum resistance for $Rb_3C_{59}N$. This shows that the extra electron in $C_{59}N$ is strongly localized and has only a small influence on the transport properties.

Aknowledgement: We thank the European Union for funding within the TMR

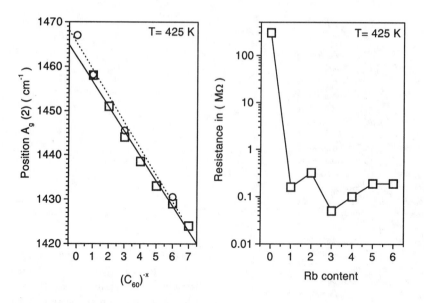

FIGURE 3. Left panel: Position of the Ag(2) mode as a function of charge transfer compared to C_{60} for $Rb_xC_{59}N$ ($\square$) and for Rb_xC_{60} ($\circ$). Right panel: Resistance of the $Rb_xC_{59}N$ film as a function of intercalation level.

Research Network 'FULPROP' (ERBFMRXCRT-970155). N.T. also thanks the European Community for a Marie Curie Fellowship at the University of Sussex. T.P. thanks the ÖAW for funding an APART grant.

REFERENCES

1. R.C. Haddon et al., Nature **350**, 320 (1991).
2. A.F. Hebard et al., Nature **350**, 600 (1991).
3. e.g. T. Pichler et al., Phys. Rev. **B 45**, 1955 (1992)
4. T. Pichler, R. Winkler and H. Kuzmany, Phys. Rev. **B 49** 15879 (1994).
5. J. Winter and H. Kuzmany, Solid State Communic. **84**, 935 (1992).
6. J.C. Hummelen et al., Science 269, 1554 (1995).
7. C.M. Brown et al., J. Am. Chem. Soc. 118, 8715 (1996); C.M. Brown et al., Chem. Mater. **8**, 2548 (1996).
8. T. Pichler et al., Phys. Rev. Lett. 78, 4249 (1997).
9. H. Kuzmany et al., Phys. Rev. B **60**, 1005 (1998).
10. W. Plank et al., (to be published).
11. K. Prassides et al., *Science* **271**, (1996) 1833.
12. M.S. Golden et al., Proc. of the IWEPNM 97, ed. H. Kuzmany, J. Fink, M. Mehring, and S. Roth, Molecular Nanostructures: pp. 86-91, World Scientific (1998).
13. e.g. M.S. Dresselhaus, G. Dresselhaus, and P.C. Eklund, Science of Fullerenes and Carbon Nanotubes, (Academic Press Inc. San Diego 1996).

Single bonded charged fullerene dimers: $(C_{59}N)_2$ versus $(C_{60}^-)_2$

W. Plank, T. Pichler[1], M. Krause, H. Kuzmany

Institut für Materialphysik der Universität Wien, Strudlhofg. 4, A-01090 Wien,
[1] *Institut für Materialphysik der Universität Wien, Strudlhofg. 4, A-01090 Wien and*
Institut für Festkörper- und Werkstofforschung Dresden, Postfach 270016, D-01171 Dresden.

Abstract.
We report on the vibronic structure of the single bonded charged fullerene dimers $(C_{59}N)_2$ and $(C_{60}^-)_2$ in comparison to pristine C_{60}. Raman spectra of $(C_{60}^-)_2$ were measured and compared to spectra from $(C_{59}N)_2$ and C_{60}. Observation and analysis concentrates on the low energy modes. Both dimers exhibit a dramatic resonance below 800 cm^{-1} for red laser excitation. Intercage modes were only observed for the dimers. Their response was lost at the transformation of $(C_{60}^-)_2$ to the polymer.

INTRODUCTION

Polymeric structures of fullerenes are one of the main topics in fullerene research. The fact that the degree of polymerisation is rather uncontrolled has attracted particular attention to the lowest possible oligomeric structures, the dimers. Our interest focused on single bonded charged fullerene dimers, in particular $(C_{59}N)_2$ and $(C_{60}^-)_2$, which we studied using Raman-, FT-Raman-, and IR-spectroscopy.

This work focusses on recent Raman results. We had strong motivation to compare our results on $(C_{59}N)_2$ and $(C_{60}^-)_2$ since they are isostructural and isoelectronic and thus allow to identify dimer typical structures in the spectra. The spectra show a broad resemblance in Raman shift especially below 800 cm^{-1}. Differences and similarities are discussed.

EXPERIMENTAL

The $(C_{59}N)_2$ samples used in this study were prepared by a chemical route as described previously [1]. For the Raman experiments the powdered material was either loosely pressed into a pellet or drop coated from CS$_2$ on a gold coated silicon wafer. The dimeric phase of C_{60} was prepared by doping the surface of a single crystal as described in [2]. By using successive doping and equilibration procedures a thin film of pure RbC$_{60}$ could be obtained. A signature of undoped material appeared during the measurements but could be suppressed by running additional doping-equilibration circles. RbC$_{60}$ is an orthorhombic polymer at room temperature and for temperatures up to about 400 K and exhibits a monomeric fcc lattice

CP544, *Electronic Properties of Novel Materials—Molecular Nanostructures*, edited by H. Kuzmany, et al.
© 2000 American Institute of Physics 1-56396-973-4/00/$17.00

for higher temperatures. The high temperature monomer can be quenched to a low temperature monomer. We achieved this by quenching the monomer from 500 K with rates of about 10 K/s to a monomer at 80 K. The monomer was slowly warmed up to about 150 K where the doped crystal underwent a transition to the dimeric phase. The dimer is stable up to 260 K as it is demonstrated below. Beyond 260 K it transforms into the orthorhombic polymer through an other monomeric state which we assume to be only transient (not shown explicitly). Spectroscopy was always performed in a vacuum better than 10^{-4} Pa. For the experimental setup see also [2].

RESULTS AND DISCUSSION

A comparison between the two dimer spectra and C_{60} is given in Fig.1. Vibrational analysis shows that all modes of the dimers are Raman (89 A_g and 85 B_g) and IR (85 A_u and 89 B_u) allowed. Accordingly compared to C_{60} the number of observable modes is much higher in the case of the dimers. The dimerisation leads to a splitting of the Raman active fundamental modes of the C_{60} sphere. Also new lines appear which are characteristic for the dimer.

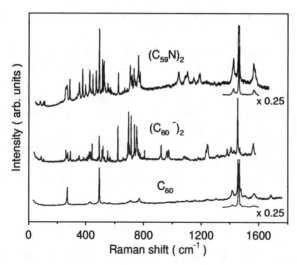

FIGURE 1. Raman spectra of C_{60} (laser excitation with 514.5 nm), $(C_{60}^-)_2$ (647.1 nm) and $(C_{59}N)_2$ (647.1 nm).

The modes at around 100 cm^{-1} which have been assigned to the intercage vibrations [3] are the most significant indicators for dimerisation. We see a strong resonance enhancement for both dimers below 800 cm^{-1}.

The center of gravity for resonance appears to be in the region of the $H_g(3)$ and $H_g(4)$ derived modes for $(C_{60}^-)_2$ ($\approx$700-800 cm^{-1}) and around the $H_g(2)$ derived modes for $(C_{59}N)_2$ (450 cm^{-1}). Peak splitting caused by symmetry reduction compared to C_{60} can clearly be seen on the H_g derived modes of the dimers. This is shown in Fig. 2 for the range from $H_g(1)$ to $A_g(1)$. Resonance enhancement and splitting are particularly dramatic for the $H_g(2)$ derived modes. Correlations between the splitted mode in the dimers can easily be depicted from Fig. 2. The $H_g(1)$ mode is also split with a characteristic component upshifted to 293 cm^{-1} for $(C_{60}^-)_2$ and 288 cm^{-1} for $(C_{59}N)_2$. In the doped or undoped orthorhombic linear polymers this peak is observed at 345 cm^{-1} [4]. It is one of the most significant modes in the Raman spectra of this type of linear polymers.

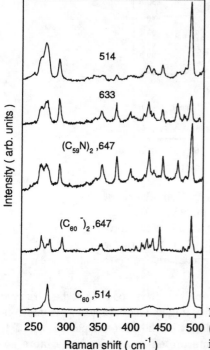

FIGURE 2. Raman spectra of C_{60}, $(C_{60}^-)_2$ and $(C_{59}N)_2$ for excitation with laser wavelength as indicated (in nm).

The very low frequency end of the spectra is given in Fig. 3. The intercage modes for $(C_{60}^-)_2$ have the same pattern as those for $(C_{59}N)_2$ but the line splitting is smaller. Peak positions are 88 cm^{-1}, 98 cm^{-1} and 106 cm^{-1} for $(C_{60}^-)_2$ and 82 cm^{-1}, 103 cm^{-1} and 111cm^{-1} for $(C_{59}N)_2$ respectively.

Contrary to the obvious similarity in the line pattern the temperature dependence of the intercage modes is quite different for $(C_{60}^-)_2$ and $(C_{59}N)_2$. Results for $(C_{60}^-)_2$ are given in Fig. 4. For $(C_{60}^-)_2$, the lowest and the highest component of the three lines become rather strong with increasing temperature and disappear with the phase transition to monomer or polymer. The intercage bonds of the polymer show no detectable response in the pictured energy range. For $(C_{59}N)_2$ the intercage modes appear highly independent of temperature. The bonds do not

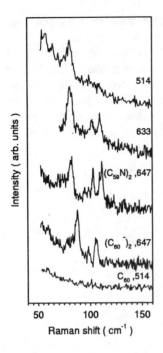 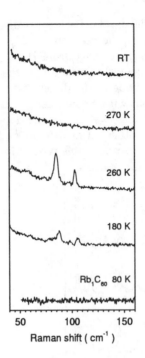

FIGURE 3. Intercage modes of $(C_{60}^-)_2$ (180 K) and $(C_{59}N)_2$ (RT), compared to C_{60} (RT). Excitation with laser wavelength as indicated (in nm).

FIGURE 4. The phases of RbC_{60}: Monomer (80 K), dimer(180, 260 K), polymer(270 K, RT). Spectra for 647.1 nm excitation.

break until the cage itself degrades and a monomer phase could not be observed [5]. In conclusion we can say that the similarity in geometrical and electronic configurations is reflected in the impressing similarity of the Raman spectra of $(C_{60}^-)_2$ and $(C_{59}N)_2$. The small differences can be interpreted as an effect of a more homogeneous distribution of the atomic potential for $(C_{60}^-)_2$.

We thank the European Union for funding within the TMR Research Network 'FULPROP' (ERBFMRXCRT-970155). T. P. thanks the Austrian academy of sciences for an APART grant.

REFERENCES

1. C. M. Brown et al., Chem. Mater. **8**, 2548 (1996)
2. W. Plank et al., (submitted for publication).
3. H. Kuzmany et al., Phys. Rev. B **60**, 1005 (1998).
4. J. Winter , H. Kuzmany, Phys. Rev. B **52**, 7115 (1995).
5. M. Krause et al., (unpublished).

HREELS investigations of adsorbed $(C_{59}N)_2$

J. M. Auerhammer[1], T. Kim[1], M. Knupfer[1], M. S. Golden[1], J. Fink[1],
N. Tagmatarchis[2], K. Prassides[2]

[1]*Institute of Solid State Science and Materials Research Dresden, P.O. Box 270016, D-01171 Dresden*
[2]*School of Chemistry, Physics and Environmental Sciences, University of Sussex, Brighton, BN1 9QJ, United Kingdom*

Abstract. The replacement of one C-atom by an N-atom in C_{60} leads to changes of its geometrical and electronic properties. The resulting heterofullerene radical forms $(C_{59}N)_2$ dimers in the bulk. We have used high resolution electron energy loss spectroscopy (HREELS) to investigate the vibrational and electronic excitations of this novel modified fullerene which we deposited as a multilayer film on single crystal graphite C(0001). The frequencies of the vibrational modes are generally close to those of C_{60}, except for two new modes which can be attributed to the dimer: an intra-dimer vibration at around 18 meV and a dipole-active mode at 104 meV. The electronic excitation spectrum is also similar to that of C_{60} regarding the interband transitions, despite a broadening due to the lower symmetry in $(C_{59}N)_2$. However, in contrast to C_{60}, the first excitonic excitations seen at 1.55 and 1.75 eV are now dipole allowed.

INTRODUCTION

Substitution of a C-atom in the C_{60}-cage by an N-atom - also called on-ball doping - leads to an interesting modification of the fullerenes geometrical and electronic properties. The radical $C_{59}N^{\bullet}$ is not stable, but forms the dimer $(C_{59}N)_2$ [1]. The two molecules are linked at the nearest neighbour of N at the hexagon fusion by one single C-C bond, where, to a large extent, the two extra electrons remain localized [2]. Experiments using EELS in transmission showed further that $(C_{59}N)_2$ is nonmetallic and that the optical gap is reduced by 0.4 eV compared to C_{60} [3].

We investigated the specific effects of the dimerisation with HREELS, since this method can probe excitations from the IR to the soft x-ray regime, whereby dipole active as well as inactive modes and the band gap can be measured.

EXPERIMENTAL

Powder of $(C_{59}N)_2$ was prepared and purified by a chemical route described previously [1]. The material was sublimated from a crucible held at around 830 K under ultra-high vacuum (UHV) conditions onto naturally grown single crystal graphite C(0001) cleaved in UHV and held at 450 K. Vibrational HREELS spectra were taken with primary electron energies of $E_i = 4$ eV in two specular directions ($\theta = 75°$ and $53°$). The electronic excitation spectrum was measured with $E_i = 10$ eV and $\theta = 65°$. All spectra were acquired at room temperature.

CP544, *Electronic Properties of Novel Materials—Molecular Nanostructures*, edited by H. Kuzmany, et al.
© 2000 American Institute of Physics 1-56396-973-4/00/$17.00

RESULTS AND DISCUSSION

Vibrational modes of $(C_{59}N)_2$

The multilayer film of $(C_{59}N)_2$ on single crystal graphite C(0001) showed no low-electron energy diffraction (LEED) pattern, while for the C(0001) itself the usual hexagonal pattern was observed. This implies the growth of disordered layers which is further confirmed by the strong similarity of the specular and off-specular vibrational spectra. A difference, however, is visible when comparing spectra measured at two different angles, $\theta_i = \theta_s = \theta = 53°$ and $75°$, in the specular direction. For larger angles, the cross section for dipole scattering should be higher, i.e. we expect an increase of the dipole-active modes regardless of the disordered surface. In Fig. 1, five peaks at 66, 71, 104, 147 and 178 meV (underlined) appear stronger in the upper ($\theta = 75°$) than in the lower ($\theta = 53°$) HREELS spectrum. The modes at 66, 71, 147 and 178 meV are also found for C_{60} and are attributed to dipole-active T_{1u} modes [4]. The peak at 104 meV cannot be attributed to any mode in C_{60}. The modes at 33, 53, 83, 94, 135, 155, 177 and 193 meV have their counterparts in C_{60} as Raman active $H_g(1)$ to $H_g(8)$ modes. The remaining lines correspond to optically inactive modes.

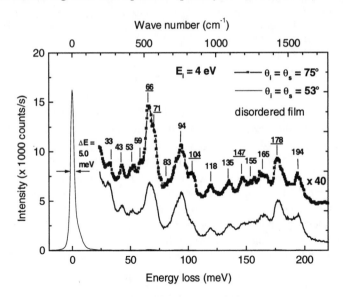

FIGURE 1. HREELS spectrum of the vibrational modes of a multilayer film of $(C_{59}N)_2$ on C(0001) taken with $E_i = 4$ eV taken under specular conditions. The dipole active-modes are underlined.

With improved resolution ($\Delta E = 3.5$ meV) a shoulder on the energy loss and gain side of the elastic peak appears at 18 meV (145 cm^{-1}), (Fig. 2), which we attribute to the gerade intermolecular mode, though it is slightly higher than the lines seen in Raman spectroscopy at 82, 103 and 111 cm^{-1} [5]. They are dipole-forbidden and therefore are expected to be more intense in an off-specular HREELS spectrum or at a disordered surface.

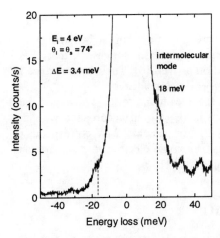

FIGURE 2. HREELS spectrum of the intermolecular mode (marked by a dashed line) of a multilayer film of $(C_{59}N)_2$ on C(0001) taken with $E_i = 4$ eV under specular conditions.

The peak at 104 meV is probably also due to dimerisation since it cannot be found in C_{60}. An even stronger argument is that it does not appear when only a monolayer of $C_{59}N$ is adsorbed on a substrate like copper where it is found to bond covalently as a monomer to the substrate (see [6]). When $C_{59}N$ is adsorbed on graphite the interaction is supposed to be of van der Waals type, and the question still remains whether the first layer consists of monomers and if a good ordering is possible.

Electronic excitations in $(C_{59}N)_2$

Figure 3 shows the electronic excitation spectra of multilayer $(C_{59}N)_2$ adsorbed on graphite taken in the specular direction with $E_i = 10$ eV. The first peaks appear at 1.55 and at 1.75 eV. For C_{60} for the same incident energies, dipole-forbidden lines were found with HREELS [5] which were interpreted to be Frenkel-type molecular excitons: the $^3T_{2g}$ triplet at 1.55 eV and the $^1T_{1g}$ singlet at 1.72 eV [7]. However, in $(C_{59}N)_2$ these lines are dipole-allowed which is corroborated by the data from optical absorption spectroscopy [8].

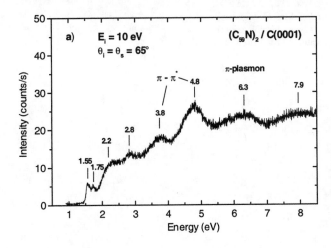

FIGURE 3. HREELS spectrum of the electronic excitations of a multilayer film of $(C_{59}N)_2$ taken with $E_i = 10$ eV on C(0001)

Calculations show that the doping with N leads to a formation of a new HOMO and LUMO in the former band gap of C_{60} [2], where transitions might become dipole allowed, also considering symmetry arguments. It is also suggested that these transitions have to be attributed to the dimerisation as well [6]. Their resemblance in energy with the C_{60} modes would then accordingly be a coincidence.

All further peaks are identical in energy with the ones found for C_{60}; however, they are smeared out due to the symmetry lowering in $(C_{59}N)_2$ in accordance with the results of EELS in transmission of solid $(C_{59}N)_2$ [3]. The peak at 2.2 eV corresponds to the conductivity gap, the ones at 3.7 and 4.8 eV to single electron $\pi-\pi^*$ excitations and the one at 6.3 eV can be assigned to the π–plasmon [3].

CONCLUSION

In summary, HREELS has been applied to measure the vibrational and electronic excitations of multilayer films of the $(C_{59}N)_2$-dimer. Characteristic modes arising from the dimerisation have been found to be the inter-dimer vibration at 18 meV, a further vibrational mode at 104 meV and two dipole-allowed excitonic transitions at 1.55 and 1.75 eV. All remaining vibrational modes and electronic transitions resemble the ones of C_{60} with a broadening due to symmetry reduction in $(C_{59}N)_2$.

ACKNOWLEDGMENTS

We acknowledge financial support from the TMR Research Network 'FULPROP' (ERBFMRXCRT-970155) and thank C. Laubschat for supplying us with the graphite single crystal.

REFERENCES

1. Hummelen, J.C., Knight, B., Pavlovich, J., Gonzalez, R., and Wudl, F., *Science* **269**, 1554 (1995)
2. Andreoni, W., et al., *Chem. Phys. Lett.* **190**, 159 (1992)
3. Pichler, T., et al., *Phys. Rev. Lett.* **78**, 4249 (1997)
4. Gensterblum, G., et al., *Physicalia Mag.* **14**, 239 (1992)
5. Kuzmany, H., Plank, W., and Winter, J., *Phys. Rev. B* **60**, 1005 (1999)
6. Silien, Ch., et al., these proceedings
7. Shirley, E.L., Benedict, L.X., Louie, S.G., *Phys. Rev.* **B54**, 10970 (1996)
8. Plank, W., Pichler, T., Kuzmany, H., submitted to *Eur. J. Phys.* **B**

Electron Spin Density Distribution in the Polymer Phase of CsC_{60}: Assignment of the NMR Spectrum

Thomas M de Swiet

School of Chemistry, University of Nottingham, Nottingham NG7 2RD, United Kingdom

Abstract. I have recently been involved in a project which used high resolution [133]Cs -^{13}C double resonance NMR data and ^{13}C -^{13}C NMR correlation spectra to make a partial assignment of the lines in the ^{13}C NMR spectrum of CsC_{60} to the carbon positions on the C_{60} molecule. (T.M. de Swiet, J.L. Yarger, T. Wagberg, J. Hone, B.J. Gross, M. Tomaselli, J.J. Titman, A. Zettl and M. Mehring *Phys. Rev. Lett.* **84**, 717 (2000)) A completion of the assignment was made on the basis of an *ab initio* calculation. Here I discuss this work with the emphasis on some of the previously unpublished aspects of the data analysis.

INTRODUCTION

The electronic and magnetic properties of the alkali intercalated fullerides, A_nC_{60}, are still only partly understood. The case A=Rb,Cs, $n = 1$ has attracted particular interest [1–22]. NMR has proven a useful probe of structure and electronic properties in AC_{60}. The first NMR evidence for sp^3 carbons and a broad spin density distribution in AC_{60} came from ^{13}C magic angle spinning (MAS) NMR [23] which was followed by a number of detailed investigations [24–31]. In the orientationally ordered structures of AC_{60} that have been proposed, there are 16 inequivalent ^{13}C nuclei, as shown in Fig. 1. A central task is to assign the peaks in the ^{13}C NMR spectrum to the different ^{13}C positions. This step is significant since it yields a map of the hyperfine coupling constant around the fullerene, against which models of the band structure can be tested. Furthermore, such an assignment is a prerequisite to any more detailed NMR study of internuclear distances and improved structural characterisation.

ONE DIMENSIONAL NMR SPECTRA

In order to obtain sufficient sensitivity, samples of CsC_{60} were synthesized using ^{13}C enriched fullerenes. Both 50% [13]C enriched CsC_{60} (sample 1) and 8% [13]C CsC_{60}

CP544, *Electronic Properties of Novel Materials—Molecular Nanostructures*, edited by H. Kuzmany, et al.
© 2000 American Institute of Physics 1-56396-973-4/00/$17.00

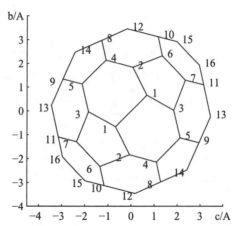

FIGURE 1. Inequivalent carbon positions projected into the crystallographic *bc* plane. The polymer chain runs along the orthogonal *a* axis[3]. Sites 1-14 contain 4 ^{13}C atoms per C_{60} ball, and sites 15-16 contain 2 ^{13}C atoms per C_{60} ball.

(sample 2) were prepared. Fig. 2 shows some one dimensional (1D) ^{13}C magic angle spinning (MAS) spectra of sample 1. The top spectrum at 24 kHz MAS can be fitted well to 11 Lorentzian peaks. Nine of these peaks are characteristic of the polymer phase of CsC_{60} (see Table 1) while the peaks at 179 ppm and 143 ppm are remnants of the cubic phase of CsC_{60} and $\alpha-C_{60}$ respectively [26,27,29]. Previous researchers [29] have been able to identify the shoulder, marked d*, on the side of peak d with a separate Lorentzian. Our data do not permit a meaningful fit of this shoulder. Nevertheless in 2D spectra resolved cross peaks are visible between peak g and region d*, centered on 118 ppm, which is in close agreement to the value of 117 ppm given in [29]. Comparison of the 24 kHz spectrum and the 17 kHz spectrum shows that peak positions move as the spinning speed is varied. This effect probably results from spinning speed affecting sample temperature, and thus Knight shifts and explains why peak h is resolved at 24 kHz but not 17 kHz. In the literature peak h was seen in Fig. 1b of Ref. [27] but not in Ref. [29]. Since there are fewer resolved NMR peaks than ^{13}C positions, some peaks must correspond to more than one position. The poor resolution must be derived from sample disorder, such as incomplete polymerization, and leads to systematic errors in the peak intensities in Table 1, because the individual line shapes are not known *a priori*. This is a particular problem with peaks a and b, and for the assignment analysis they were largely treated as one peak.

The second 17kHz spectrum shows a wider spectral width. The asterisks mark the position of weak spinning side bands visible along the baseline. These spinning side bands show up as ridges parallel to the diagonal in two dimensional spectra. The last spectrum at 19kHz shows the effect of passing the spin polarisation through the POST-C7 based double quantum filter [33] used in the two dimensional spectra. The lack of distortion in the spectrum demonstrates the resonance offset independence of POST-C7. Indeed the main effect of the filter on the 1D spectrum is that the two impurity peaks are removed. This demonstrates that the C60 balls are rapidly rotating in the impurity materials, quenching the spin dipolar couplings and so suppressing the generation of double quantum coherence.

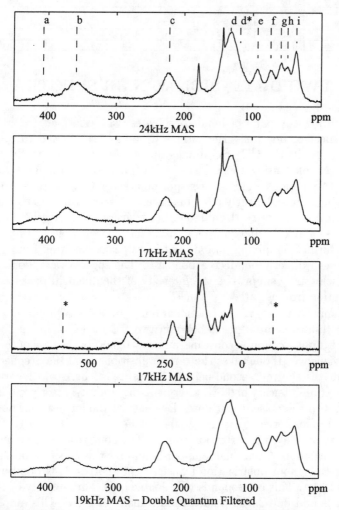

FIGURE 2. MAS NMR spectra of 50% ^{13}C enriched CsC_{60} at different spinning speeds, taken with a Varian/Chemagnetics 3.2mm MAS probe and CMX spectrometer at a field of 7 T, and 2s pulse delay.

TABLE 1. Shift(δ) and intensity, normalised to 60, ($\#/C_{60}$) of the ^{13}C lines in the MAS spectrum of CsC_{60}. R_{CsC} represents the experimental ^{133}Cs -^{13}C REDOR fraction times $(10ms/t)^2$, at short echo times, t. The last line lists the ^{13}C cross peaks which occur between pairs of lines $a, ..., i$(top row) in ^{13}C -^{13}C correlation spectra due to spin polarization transfer between nearest neighbor carbons.

line	a	b	c	d	e	f	g	h	i
δ(ppm) Sample 1	407	359	222	132	93	74	59	49	36
$\#/C_{60}$	1.0	9.6	7.9	24.2	2.9	3.4	3.4	2.6	5.0
δ(ppm) Sample 2	398	359	222	133	97	80	60	46	35
$\#/C_{60}$	4.8	3.8	11.6	18.8	3.0	2.4	7.4	5.2	2.95
δ(ppm) Ref. [29]	397	357	220	126	94	77	55		35
$\#/C_{60}$	7.1	3.5	8	18.2	2.8	1.5	12.6		4
exp. R_{CsC}	2.5	0.8	2.0	2.5	0.5	5.8	2.1	0.9	0.7
cross peaks	d g h	d g h	e f g h	a/b e f h	c d i	c d i	a/b c d*	a/b c d	e f

TWO DIMENSIONAL NMR SPECTRA

In order to identify pairs of bonded ^{13}C sites, we performed ^{13}C MAS dipolar correlation spectroscopy [32]. Here, the ^{13}C spins were allowed to precess freely for a time t_1. Next, the ^{13}C$-^{13}$C dipole-dipole interaction, normally removed by MAS, was recoupled using the POST-C7 [33] pulse sequence, for a mixing time equal to 4 MAS periods. This allows spin polarization to be transferred between ^{13}C nuclei which are sufficiently close in space. After the mixing time, the signal is recorded as a function of t_2. Double Fourier transformation of the data reveals a plot with peaks along the diagonal, $\omega_1 = \omega_2$, resulting from polarization which is not transferred during the mixing period, and cross peaks resulting from polarization transfer between directly bonded carbon sites. The spectra were 'double quantum filtered', which partly suppresses the intensity of the diagonal peaks and removes any contribution from isolated ^{13}C nuclei.

For short mixing times, τ, the fractional build up of double quantum coherence, (which then results in cross peaks between two ^{13}C nuclei) is of the form $\alpha D^2\tau^2$. Here D is the dipolar coupling constant, and $\alpha \sim 0.0287$ in the case of the POST-C7 recoupling sequence. If one takes the carbon positions from the Stephens *et al.* X-ray structure, the shortest second neighbor distance is 2.34Å, which corresponds to a double quantum intensity of 0.5%, after a mixing time of 4 rotor periods at 18kHz, which is beneath the noise in our data. By contrast the longest first neighbor C-C bond is 1.59Å This corresponds to a double quantum intensity of 5%. There is also the possibility of relayed polarization transfer from one site to its second neighbor via the first neighbor. A reasonable upper bound for this effect is the square of the strongest first neighbor polarization transfer. The shortest C-C bond in the X-ray structure is 1.31Å. This corresponds to a double quantum intensity of 16% leading to an upper bound on relayed polarization transfer of 2.4%. This amplitude is on the edge of being detectable in the data. However it can be shown for the majority of spin configurations, that this kind of relayed transfer causes cross peaks of the opposite sign to the ordinary direct transfer cross peaks. Therefore the probability of spurious cross peaks between non-nearest neighbors appears to be small.

These two dimensional techniques, combined with ^{133}Cs-^{13}C REDOR measurements [34,35] enabled the partial assignment of the NMR data which has been presented in [36].

REFERENCES

1. J. Winter and H. Kuzmany *Solid State Commun.* **84**, 935 (1992).
2. Q. Zhu *et al. Phys. Rev. B* **47**, 13948 (1993).
3. P.W. Stephens *et al. Nature* **370**, 636 (1994).
4. O. Chauvet *et al. Phys. Rev. Lett.* **72**, 2721 (1994).
5. S. Pekker *et al. Solid State Commun.* **90**, 349 (1994).
6. G. Oszlanyi *et al. Phys. Rev. B* **51**, 12228 (1995).
7. Q. Zhu, D.E. Cox and J.E. Fischer *Phys. Rev. B* **51**, 3966 (1995).
8. V.A. Atsarkin, V.V. Demidov and G.A.Vasneva *Phys. Rev. B* **56**, 9448 (1997).
9. F. Bommeli *et al. Phys. Rev. B* **51**, 14794 (1995).
10. S.C. Erwin, G.V. Krishna and E.J. Mele *Phys. Rev. B* **51**, 7345 (1995).
11. M. Fally and H. Kuzmany *Phys. Rev. B* **56**, 13861 (1997).
12. A. Janossy *et al. Phys. Rev. Lett.* **79**, 2718 (1997).
13. M. Bennati *et al. Phys. Rev. B* **58**, 15603 (1998).
14. J.R. Fox *et al. Chem. Phys. Lett.* **249**, 195 (1996).
15. D. Bormann *et al. Phys. Rev. B* **54**, 14139 (1996).
16. J.L. Sauvajol *et al. Phys. Rev. B* **56**, 13642 (1997).
17. M.C. Martin *et al. Phys. Rev. B.* **51**, 3210 (1995).
18. H. Schober *et al. Phys. Rev. B* **56**, 5937 (1997).
19. S. Pekker, G. Oszlanyi and G. Faigel *Chem. Phys. Lett.* **282**, 435 (1998).
20. C.H. Choi and M. Kertesz *Chem. Phys. Lett.* **282**, 318 (1998).
21. P. Launois *et al. Phys. Rev. Lett.* **81**, 4420 (1998).
22. K. Tanaka *et al. Chem. Phys. Lett* **272** 189 (1997).
23. T. Kälber, G. Zimmer and M. Mehring *Z. f Physik B* **27** 2 (1995).
24. V. Brouet *et al. Phys. Rev. Lett.* **76**, 3638 (1996).
25. T. Kälber, G. Zimmer and M. Mehring *Phys. Rev. B* **51** 16471 (1995).
26. H. Alloul *et al. Phys. Rev. Lett.* **76**, 2922 (1996).
27. K.-F. Thier *et al. Phys. Rev. B***53**, R496 (1996).
28. R. Tycko *et al. Phys. Rev. B* **48**, 9097 (1993)
29. V. Brouet *et al. Appl. Phys. A* **64**, 289 (1997).
30. V. Brouet *et al. Physica C* **235**, 2481 (1994).
31. P. Auban-Senzier *et al. J. Phys. I (France)* **6**, 2181 (1996).
32. See e.g. H. Geen *et al. J. Magn. Reson.* **125**, 224 (1997).
33. M. Hohwy *et al. J. Chem. Phys.* **108**, 2686 (1998).
34. T. Gullion and J. Schaefer *J. Magn. Reson.* **81**, 196 (1989).
35. T. Gullion and J. Schaefer *J. Magn. Reson.* **92**, 439 (1991).
36. T.M. de Swiet *et al. Phys. Rev. Lett.* **84**, 717 (2000).

Preparation of RbC$_{60}$ by coevaporation

M. Haluška[1], M. Krause[1], P. Knoll[2], H. Kuzmany[1]

[1] Institut für Materialphysik, Universtät Wien, A-1090 Vienna, Austria
[2] Institut für Experimentalphysik, Univeristät Graz, A-8010 Graz, Austria

Abstract. RbC$_{60}$ crystals were prepared directly from the vapour phase by coevaporation in closed evacuated glass tubes with separated source materials using the horizontal five temperature zone furnace. Prepared crystals were characterized by Raman, micro Raman spectroscopy, and optical microscopy. Under certain conditions of growth a considerable concentration of Rb$_1$C$_{60}$ single crystals were detected. The typical crystal size of Rb$_1$C$_{60}$ was 10 μm.

INTRODUCTION

A$_1$C$_{60}$ fullerides can be prepared directly from the corresponding molar ratio of alkali metal and C$_{60}$ compounds either from solution [1], or from the vapour phase [2], or indirectly by mixing and heating well-defined amounts of M$_6$C$_{60}$ and C$_{60}$ powder [3], or by vapour phase doping of pure C$_{60}$ solids [4]. This article will focus on the growth of Rb$_1$C$_{60}$ crystals from the vapour phase using a simultaneous evaporation of crystal components from two separated sources (coevaporation).

For simplification it will be assumed that evaporation of Rb and C$_{60}$ sources and transport of their vapour to the crystal growth region are mutually independent (they behave as congruent vapours) as well as that vapour of Rb immediately reacts with C$_{60}$ in the crystal growth zone.

The important quantity for the vapour transport in the tube is the *Knudsen number* K$_n$.

$$K_n = \frac{l_T}{2r}, \qquad l_T = \frac{k_B T}{\sqrt{2}\pi d^2 p}, \tag{1}$$

where l_T is the mean free path length of molecules, r is the radius of the ampule, d a molecular diameter and p the vapour pressure.

If K$_n$>1 a flow is called *molecular*. Molecule - wall collisions dominate over molecule - molecule collisions. If K$_n$<1 collisions between vapour molecules dominate over collisions with the wall and the flow is called *viscous*. For experimental conditions used during this work K$_n$>1 holds for both Rb and C$_{60}$. The mass transport in the molecular flow regime is described for a congruent vapour by the relation:

$$J_{MF} = \frac{Mr}{4R_g T_a L \epsilon}(p_s^2 - p_c^2), \tag{2}$$

where J_{MF} is the molecular flow, M is the molecular weight, r the radius of the ampule, R_g the gas constant, $\epsilon = p \times \sqrt{m/(2\pi k_B T)}$, L the transport distance between source and growing crystal, p_s and p_c are the equilibrium pressures of the crystallized substance above the source and growing crystal at T_s and T_c, respectively. T_a and p are the average values of T_s and T_c and corresponding p_s and p_c, respectively.

EXPERIMENTAL

Commercially available Rb (Sigma Aldrich 99.6%) and C_{60} crystals prepared from *super gold grade fullerene* from Hoechst purified by vacuum-resublimation technique (purity better than 99.9%) were used as the source materials.

80 to 100 mg of C_{60} (usually 1 crystal) was placed into the reaction tube with ID=8 mm together with a micro pipette with weighted amount of Rb corresponding to $\sim$ 100–110% of stoichiometric $Rb_1 C_{60}$. In some experiments the micro pipette with Rb was deposited into 30 cm long and 2 mm ID glass tube to reduce the transport rate of Rb. The glass tube with the micro pipette and Rb was inserted into the reaction tube. All preparation was carried out under inert condition in an argon dry box. The loaded reaction tube was evacuated and sealed off so that its length was about 50 cm.

The reaction tube was accommodated into a five temperature zone horizontal furnace with windows. Rb and C_{60} were deposited into opposite edges of the tube. The central zone with a window was the crystal growth zone. Another window allowed to observe the diminishing of Rb from its source zone. Two transient zones between the crystal growth zone and both source zones were established to prevent condensation outside the crystal growth zone.

At first a C_{60} nucleation was stimulated. During this stage the part of the reaction tube with Rb was kept at room temperature. When the first small C_{60} crystal was perceived in the crystal growth zone the temperature of the C_{60} source was reduced and the temperature of the Rb source was increased so that the flow of crystallizing substances from both sources to the crystal growth zone had a molecular stoichiometry 1:1. Typical temperature of the crystal growth zone, C_{60} and Rb source zones were 445 °C, 455 °C, and 120 °C, respectively. The temperature profile along the tube is shown in Fig. 1a. After all Rb was evaporated from the pipette the temperature of the tube was slowly decreased to room temperature.

RESULTS

By the procedure described above fulleride crystals and polycrystalline films were grown. From the observed disappearance of Rb from the micro pipette an evaporation rate of 0.8 mg/h was determined for the temperature profile as shown in Fig. 1a. This evaporation rate corresponds to the transport rate predicted from the equation (2). It means that for Rb the assumptions given in the introduction are correct.

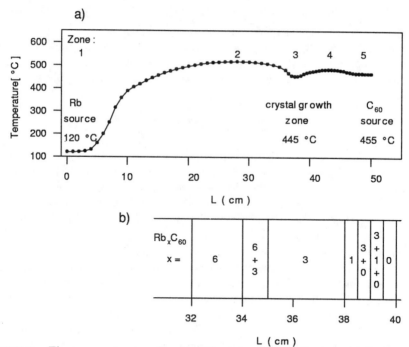

FIGURE 1. The temperature profile along the tube a). Distribution of fulleride phases along crystal growth zone b).

In the case that the micro pipette with Rb was inserted into the 30 cm long thin tube most of the grown material was located in 1.5–3 cm long area in the zone 3. If the micro pipette with Rb was placed directly into the reaction tube the grown material was in a 7–10 cm long area in zone 2 and 3. Larger amount of Rb_1C_{60} was found in the second case.

Solid material in the growth zone was characterized by Raman spectroscopy using the well known relation between the $A_g(2)$ mode frequency and the charge state of C_{60}^{-n}, n=0, 1,...,6. The polymeric Rb_1C_{60} phase was further identified by the Raman linies at ~340 and 630 cm^{-1}, which are absent for all A_xC_{60} stoichiometries with a fcc lattice (x=0, 1, 3, 6). A variation in the stoichiometry was observed. Regions with either single-phase or multiple-phases of Rb_xC_{60} x = 6,3,1 and 0 were detected. The stoichiometry of the grown material depends on its distance from source materials. The typical distribution is shown in Fig. 1b for the temperature profile from Fig. 1a. The best result obtained corresponds to ~50% of Rb_1C_{60} phase from the condensed solids in the crystal growth region.

Figure 2a depicts a spectrum from the zone 3. The strong line at ~340 cm^{-1} is characteristic for orthorombic Rb_1C_{60}. The typical crystal size of Rb_1C_{60} was found to be 10 μm for growth conditions given above.

To check stoichiometry of individual crystallities micro Raman spectroscopy was

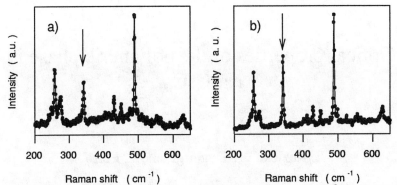

FIGURE 2. Raman spectrum of many crystallities (spot area ~5 mm^2) in the evacuated tube a) and micro Raman spectrum of one crystal (spot area ~5 μm^2) on air b). The arrows indicate a characteristic line for the polymeric Rb_1C_{60}.

used. The measurements were made on air as it is known that Rb_1C_{60} is an air stable phase. Figure 2b depicted a spectrum from an individual crystal. Again the strong line at ~340 cm^{-1} proves that the crystal is Rb_1C_{60} with high purity.

CONCLUSION

Raman and micro Raman spectroscopy showed that single-phase Rb_1C_{60} crystals could by prepared directly from the vapour phase by coevaporation in closed evacuated glass tubes using the horizontal five temperature zone furnace. The typical crystal size of Rb_1C_{60} was 10 μm.

ACKNOWLEDGEMENT

The authors are grateful to the Österreichische Nationalbank Jubiläumsfonds project 7906, to the European Community, TMR project ERBFMRX-CT97-0155 and to the Bundesministerium für Wissenschaft und Verkehr, project GZ 650.346/1-III/2a/99 for financial support and to Hoechst AG for support with C_{60} powder.

REFERENCES

1. T. Saito, Y. Akita, M. Toumoto, P. W. Stephens, K. Tanaka, Solid State Communications **111**, 131 (1999)
2. S. Pekker, A. Janossy, L. Mihaly, O. Chauvet, M. Carrad, L. Forro, Science **265**, 1077 (1994)
3. P. W. Stephens, G. Bortel, G. Faigel, M. Tegze, A. Janossy, S. Pekker, G. Oszlanyi, L. Forro, Nature **370**, 636 (1994)
4. J. Winter, H. Kuzmany, Solid State Communications **84**, 935 (1992)

Optical properties of the polymeric phase in Na$_2$CsC$_{60}$

B. Ruzicka[1], L. Degiorgi[1] and L. Forro'[2]

[1]Laboratorium für Festkörperphysik, ETH-Zürich, CH-8093 Zürich, Switzerland
[2]Department de Physique, IGA, Ecole Polytechnique Federal de Lausanne, CH-1015 Lausanne, Switzerland

Abstract. We present optical data on the (single-bond) polymerized phase of Na$_2$CsC$_{60}$. We find a progressive metal-semimetal crossover, which might indicate an incipient one-dimensional instability. A comparison with the double-bond polymerized phase of AC$_{60}$ is discussed.

INTRODUCTION

Among the families of superconducting alkali fullerides, the Na$_2$AC$_{60}$ (A=Rb, K, Cs) family displays puzzling behaviours. Particularly the much steeper increase of the superconducting transition temperature T_c versus the interfullerene spacing a with respect to the situation in A$_3$C$_{60}$ has not been found a satisfactory and comprehensive explanation yet [1]. Moreover, it is quite well established that the T_c vs. a dependence in Na$_2$AC$_{60}$ cannot be reconciled solely with a much stronger dependence of the density of states (DOS) at Fermi level N(E$_F$) on a than in A$_3$C$_{60}$ [1]. The assumption that alkali atoms act only to expand the lattice is not supported by the experimental results. This also suggests that there is a specific alkali ions effect on N(E$_F$). We recently addressed this issue, as well [2]. We compared the magnetic susceptibility and the optical response between Na$_2$CsC$_{60}$ and K$_3$C$_{60}$. We found that at high temperatures, where both A$_3$C$_{60}$ and Na$_2$AC$_{60}$ are in the fcc phase, Na$_2$AC$_{60}$ is in an insulating phase [2]. This is quite in contrast to the expectation. In fact, Na$_2$AC$_{60}$ has larger bandwidth than A$_3$C$_{60}$ and therefore it should be less prone to correlation effects. The mobility of the small Na$^+$ possibly plays an essential role. We have suggested that a Mott-Jahn-Teller insulating state combined with a (static) charge dissociation in C$_{60}^{2-}$ and C$_{60}^{4-}$ might lead to an insulating state for a nominal C$_{60}^{3-}$ fullerides [2].

Another characteristics of the Na$_2$AC$_{60}$ family is the formation of a single-bond polymerized phase, depending whether one applies pressure and/or from the cooling rate. In this paper we focus our attention on the polymerized phase only, which develops in Na$_2$CsC$_{60}$ by solely applying pressure. The polymerization is supposed to induce a quasi-one-dimensional (1D) electronic structure [3], which will be quite sensitive to one-dimensional instabilities because of the Fermi surface nesting. We

CP544, *Electronic Properties of Novel Materials—Molecular Nanostructures*, edited by H. Kuzmany, et al.
© 2000 American Institute of Physics 1-56396-973-4/00/$17.00

were motivated to optically investigate such phase and compare it with the double-bond polymerized phase encountered in the AC_{60} alkali intercalated fullerides [4,5].

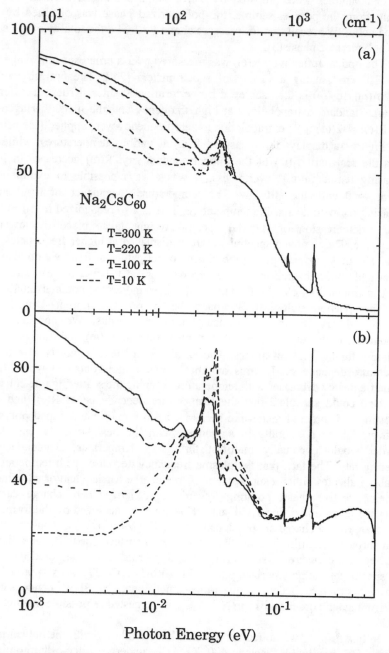

FIGURE 1. Optical reflectivity (a) and optical conductivity (b) as a function of photon energy (wave number) for Na_2CsC_{60} in the polymerized phase.

RESULTS AND DISCUSSION

Our samples were prepared by conventional solid-state reaction method [2]. Starting from the powder sample, the polymerized phase was obtained by forming pressed pellet. We recall that for pressure above 1.12 GPa the totality of the sample is in the polymerized phase [3].

The optical reflectivity $R(\omega)$ was measured over a broad spectral range from 20 up to 10^5 cm^{-1}, using a variety of spectrometers [2,4]. The optical conductivity $\sigma(\omega)=\sigma_1(\omega)+i\sigma_2(\omega)$ is then achieved by performing Kramers-Kronig transformations of $R(\omega)$. Standard extrapolations at high frequency and the metallic Hagen-Rubens (HR) $R(\omega)=1-2(\omega/\sigma_{dc})^{1/2}$ extrapolation to zero frequency were applied (see below) [4].

Figure 1a displays the measured $R(\omega)$ at various temperatures, while Fig. 1b shows the real part $\sigma_1(\omega)$ of the optical conductivity. $R(\omega)$ decreases in FIR with decreasing temperature. Nevertheless, it was always possible to extrapolate $R(\omega)$ towards $\omega=0$ with the HR law. The temperature dependence of $R(\omega)$ implies a decreasing σ_{dc} with decreasing temperature. This is also recognized in $\sigma_1(\omega)$ (Fig. 1b) where a suppression of (Drude) spectral weight takes place by lowering the temperature. The removed spectral weight is displaced at higher frequencies between 20 and 100 meV. Overlapped to the plasma edge feature of $R(\omega)$ we do also observe several additional absorptions: two sharp ones are at 0.1 and 0.2 eV and a broader feature is around 30 meV. We tend to ascribe these modes to phonon excitations [5].

The phenomenological approach based on the Lorentz-Drude classical dispersion theory [6] can reproduce the optical results. We consider a Drude component for the effective metallic contribution to $\sigma_1(\omega)$ and Lorentz harmonic oscillators for the phonon modes. The Drude component is mostly affected by the temperature dependence. It turns out that the best fit is obtained when the Drude plasma frequency decreases with temperature, keeping the scattering rate constant.

One could speculate that the temperature dependence of $R(\omega)$ and $\sigma_1(\omega)$ is suggesting a disordered metal-like scenario, as in non-oriented doped polymers [2]. However, this is not totally in agreement with the best Drude-Lorentz fit result. Disorder would eventually manifest through a temperature dependence of the scattering rate. The fact, that the plasma frequency decreases with temperature, most probably indicates either a suppression of free charge carriers and/or an enhancement of their effective mass, pointing towards a scenario where charge carriers are condensed in a static or localized state. This could be achieved by the formation of a condensate (like a charge or spin density wave, CDW or SDW) as consequence of the Fermi surface instability (i.e., nesting of the Fermi surface due to the one-dimensional character of the electronic structure). Recently, Arcon et al. measured the high-field ESR spectra for the polymerized phase of $Na_2Rb_{1-x}Cs_xC_{60}$ [7]. As x approaches 1, the electronic structure becomes quasi-one-dimensional. A SDW condensate or an antiferromagnetic ordering like in $NH_3K_3C_{60}$ is suggested as possible magnetic ground state [7].

In this respect, we notice the quite striking similarity with the optical properties of the AC_{60} family [4]. Below 400 K, AC_{60} undergo a fcc-orthorhombic phase transition due to the formation of double bond polymers [5], leading to a quasi-one-

118

dimensional electronic structure. It is believed that the Rb and Cs compounds are the most 1D of the whole series [5]. Our optical and microwave results [4] confirm that KC_{60} is a metal at any temperatures while RbC_{60} and CsC_{60} undergo a metal-insulator phase transition. The insulating state is fully developed below 50 K, where also the magnetic susceptibility displays a sharp drop, indicating a suppression of DOS at the Fermi level [4,5]. This was ascribed to the onset of a magnetic ground state, possibly due to a SDW condensate [4].

As far as our optical results on Na_2CsC_{60} are concerned, it remains to be seen whether the scenario of the one-dimensional instability can comprehensively describe our data. In fact, $\sigma_1(\omega)$ (Fig. 1b) at best shows a progressive metal-semimetal crossover rather than a metal-insulator phase transition. Obviously, our data could be considered as a manifestation for an incipient 1D instability. We might speculate that the polymerization is not completely realized and a kind of phase separation with residual metallic component occurs.

ACKNOWLEDGMENTS

The work in Lausanne and Zurich was supported by the Swiss National Science Foundation. BR and LD wish to acknowledge the technical help of J. Müller.

REFERENCES

1. C.M Brown et al., Phys. Rev. **B59**, 4439 (1999) and Ref. therein
2. N. Cegar et al., submitted to Phys. Rev. Lett.
3. S. Margadonna et al., J. Solid State Chem. **145**, 471 (1999)
4. F. Bommeli et al., Phys. Rev. **B51**, 14794 (1995)
5. L. Degiorgi, Adv. in Physics **49**, 207 (1998) and Ref. therein
6. F. Wooten, in *Optical Properties of Solids*, (Academic Press, New York, 1972)
7. D. Arcon et al., preprint (1999)

^{13}C and ^{23}Na NMR Study of Na$_2$KC$_{60}$ in Polymer Phase

T. Saito[1,*] , V. Brouet[1], H. Alloul[1], L. Forró[2]

1. Laboratoire de Physique des Solides, UMR8502, Univesité Paris-Sud, Bât 510, 91405 Orsay, France
2. Departement de Physique, Ecole Polytechnique. Fédérale Lausanne, 1015 Lausanne, Switzerland

Abstract. ^{13}C and ^{23}Na NMR investigations on fulleride that contain sodium, Na$_2$KC$_{60}$, with polymer structure are reported. Temperature dependence of ^{13}C and ^{23}Na NMR spectra show that the polymerization starts around 310K and the amount of polymer phase are estimated to be 80% below 200K. In polymer phase, a gradual change of electric field gradient at Na site is observed by the analysis of ^{23}Na spectrum.

INTRODUCTION

The ternary fullerene compounds Na$_2$AC$_{60}$ (A = K, Rb, Cs) have a face centered cubic (*fcc*) structure in high temperature and the phase transition to simple cubic (*sc*) structure occurs around room temperature. When *sc* structure is maintained by quenching, superconductivity is observed. The loss of superconductivity and cubic symmetry were reported by modest pressure[1,2] or slow cooling[3] in Na$_2$RbC$_{60}$. In this phase, C$_{60}$ molecules are linked by singly bond and form one-dimensional *polymer* structure.[4,5] This type of bonding was also reported in the two dimensional Na$_4$C$_{60}$. Although the existence of ground state instability and a formation of band gap at E_F induced by CDW is predicted by theoretical calculations,[6] it is not observed experimentally in Na$_2$RbC$_{60}$.[7] While the existence of *polymer* phase in Na$_2$KC$_{60}$ is suggested by ESR, structural and electronic properties have not be investigated. In this work, we present ^{13}C and ^{23}Na NMR study of Na$_2$KC$_{60}$. It is shown that slow cooling in Na$_2$KC$_{60}$ also leads to similar phase transition to one-dimensional *polymer* phase. The amount of *polymer* phase in Na$_2$KC$_{60}$ is larger than in Na$_2$RbC$_{60}$. By the analysis of ^{23}Na spectrum, possibility of different environment at Na site is discussed.

EXPERIMENTAL

Sample was prepared by conventional solid-state reaction method. *Polymer* phase was obtained by slow cooling. Phase purity was checked by ESR experiment. It is possible to deconvolute ESR signal into two lines; the broader line corresponds to the

* Present address: Faculty of Integrated Arts and Sciences, Tokushima University, 770-8502, Japan (e-mail address: saito@ias.tokushima-u.ac.jp)

CP544, *Electronic Properties of Novel Materials—Molecular Nanostructures*, edited by H. Kuzmany, et al.

sc phase while the narrow component to the *polymer*. The ratio for *polymer* phase is larger than that for Na_2RbC_{60}.

^{13}C and ^{23}Na NMR experiment was done in magnetic field at 7.5 T (above 77 K) and at 7 T (below 77 K). To produce *polymer* phase, sample was cooled at a rate of -1 K/min. [13]C and [23]Na spectra are obtained by FFT and temperature variation of spectra are shown in Fig. 1. Difference of shift in [23]Na spectrum gives an outline of the phase transition from *fcc* phase at high temperature to *sc* and *polymer* phase. Above 325 K, only *fcc* line is observed at 159 ppm and it disappears completely at 300K. Below 320 K, *sc* line appears at 80 ppm and it remains until 12 K. Below 310 K, *polymer* line appears at ~260 ppm and its intensity become dominant with decreasing temperature.

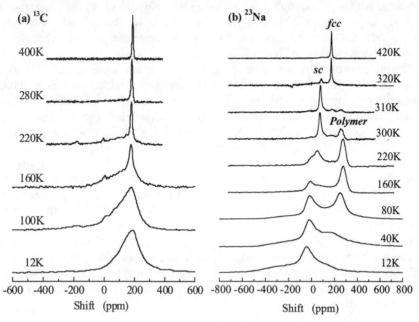

FIGURE 1. Spectra of ^{13}C and ^{23}Na NMR at various temperatures. (a) ^{13}C spectra are observed at 80.2 MHz (above 77K) and 74.9 MHz (below 77K). (b) [23]Na spectra are observed at 84.4 MHz (above 77K) and 78.8 MHz (below 77K).

RESULTS AND DISCUSSION

Polymerization in Na_2KC_{60}

When Na_2KC_{60} is cooled slowly, broad ^{13}C line (line width of 145 ppm) appears just below 300K in addition to ^{13}C sharp line (line width of 10 ppm). Broad line was not observed in quenched *sc* phase until 120K because C_{60} molecules in *sc* phase rotate faster than the frequency of spectrum and the motional narrowing occurs above 120 K[8]. Therefore the broad line in slow cooled sample indicates that the C_{60} molecular rotation is restricted suddenly and it is naturally assigned to the *polymer* phase and sharp line to the *sc* phase. Some fraction of sharp line below 300K indicates that the *sc* phase coexists in addition to *polymer* phase. Area of ^{13}C NMR spectrum directly gives the number of atoms because ^{13}C have magnetic moment $I = 1/2$, while ^{23}Na with $I = 3/2$ might lose FFT spectrum intensity by quadrupole effect. From the deconvolution of ^{13}C spectrum into broad line and narrow line, it is possible to know the ratio of *polymer* phase in Na_2KC_{60}. Just below 300K, *polymer* phase increases rapidly and it gradually increases below 240K. The ratio becomes constant (about 80 %) below 200K indicating that the polymerization saturates at this temperature. Below 120K, sharp *sc* line becomes broad because of freezing of C_{60} molecular rotation as also observed in quenched *sc* line, and it is not possible to deconvolute spectrum into two lines. Although cooling rate -1 K/min in this experiment is rather fast, the ratio of 80 % is large compared with the ratio of 50% in Na_2RbC_{60} with extremely slow cooling rate -0.03 K/min. These results clearly indicate that the polymerization easily occurs in Na_2KC_{60} than in Na_2RbC_{60}.

The progress of polymerization of C_{60} molecule is also observed in ^{23}Na spectrum. In addition to *sc* line at 80 ppm, sharp line is observed at ~260 ppm below 300K. This temperature is coincident with that for ^{13}C broad line. Therefore the line at ~260 ppm is assigned to *polymer* phase. With decreasing temperature, the increase of *polymer* line and the decrease of *sc* line is observed until 220 K which is a similar result in ^{13}C spectrum. Strangely enough, *polymer* line at 260 ppm gradually decreases below 100K and the broad line with a long tail in lower frequency side (lower shift) appears gradually. At 12 K, sharp *polymer* line disappears and the broad line is dominant with minor (~20 %) line coming from *sc* phase. Considering that the ^{13}C *polymer* line does not show any change in spectrum below 100 K, this broad line is also assigned to *polymer* phase. These features are analyzed in next section. Assuming that the line at 80 ppm (which shifts to -50 ppm at 12K) originate from *sc* phase and the rest of the spectrum (sharp 260 ppm line + broad line) from *polymer* phase, it is also possible to estimate the ratio of *polymer* phase. The ratio of *polymer* line obtained by ^{23}Na spectrum gives a similar temperature variation to ^{13}C results. Above assumption would be assured by this consistency.

^{23}Na Spectrum Analysis

One of the features of the broad ^{23}Na line is that it has a long tail in lower frequency side. Another feature is that the line width in frequency is inversely proportional to NMR frequency f_0 which was checked at 28.7 MHz. These results are characteristic of 2nd order quadrupole broadening of central transition (1/2, -1/2); broadening width in frequency is proportional to v_Q^2/f_0, where $v_Q \equiv 3e^2qQ/2I(2I-1)$ is a quadrupole frequency ($eq=V^{33}$ denotes the maximum component of the electric field gradient). Because 12 K spectrum does not contain sharp line, it is possible to estimate the quadrupole parameter for broad line; $v_Q = 2.2$ MHz and asymmetric field gradient parameter $\eta \equiv (V^{11}-V^{22})/V^{33} = 0.85$ which is shown in Fig. 2 (a). This large v_Q in low temperature is consistent with the report of displacement of Na from the center of T site (0, 1/4, 1/2) to (-0.018, 0.266, 0.482) at 2.5K in Na_2RbC_{60} *polymer* by neutron diffraction measurement.[5]

On the other hand, the sharp line at ~260 ppm observed at higher temperature has small quadrupole effects; $\nu_Q < 0.4$ MHz is estimated by the frequency dependence. Therefore it is necessary to introduce that the Na ions have some sites with different electric field gradient. Temperature variation of ^{23}Na spectrum is qualitatively explained by the change of ratio in small (or zero) and large ν_Q. For example, a simulation for the ^{23}Na spectrum at $T = 80$ K is shown in Fig. 2 (b). The spectrum is almost explained by the addition of sharp *polymer* line, broad *polymer* line and *sc* line. However, the simulation intensity between *sc* line and sharp *polymer* line is smaller than observed spectrum. To improve the fitting, it might be necessary to introduce more than two sites; for example the distribution of ν_Q. The change of ν_Q is the direct evidence for the change in Na local environment. This might originate from the displacement of Na position as reported in Na_2RbC_{60} of *polymer* phase.[5]

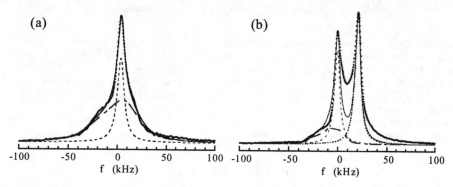

FIGURE 2. Simulation results of ^{23}Na spectrum. Closed circle indicates experimental result. Short dash line indicates *sc* phase, dot line indicates *polymer* phase with $\nu_Q = 0$ MHz, dash line indicates *polymer* phase with $\nu_Q = 2.2$ MHz and solid line indicates addition of each simulation curve. (a) $T = 12$K. $\nu_Q = 2.2$ MHz is used. (b) $T = 80$K. $\nu_Q = 0$ and 2.2 MHz are used.

CONCLUSIONS

^{13}C and ^{23}Na NMR measurements have shown that 1) polymerization of ~80% is observed by slow cooling in Na_2KC_{60}. 2) A gradual change of electric field gradient at Na site is observed in *polymer* phase.

REFERENCES

1. Q. Zhu *et al.*, Phys. Rev. B **52** R723 (1995)
2. J. E. Schirber *et al.*, Physica C **260** 173 (1996)
3. K. Prassides *et al.*, J. Am. Chem. Soc. **119** 834 (1997)
4. G. M. Bendele *et al.*, Phys. Rev. Lett. **80** 736 (1998)
5. A. Lappas et al. J. Phys. Cond. Matt. *in press*
6. P. R. Surján *et al.*, Phys. Rev. B **58** 3490 (1998)
7. D. Arcon *et al.*, Phys Rev. B **60** 3856 (1999)
8. T. Saito *et al.*, J. Phys. Soc. Jpn. **64** 4513 (1995)

X-ray Diffraction of Na_4C_{60} at Low Temperature under High Pressure

Yasuhiro Takabayashi,[a] Yoshihiro Kubozono,[a] Satoshi Fujiki,[a]
Yoshihiro Iwasa[b] and Setsuo Kashino[a]

a) Department of Chemistry, Okayama University, Okayama 700-8530, Japan
b) Japan Advanced Institute of Science and Technology, Ishikawa 923-1292, Japan

Abstract. X-ray diffractions of two-dimensional Na_4C_{60} are studied at various temperatures above 11 K in a pressure region up to 67 kbar. All X-ray diffraction patterns measured in this study can be analyzed by using a body-centered monoclinic (bcm) structure of space group $I2/m$. New Bragg reflections which cannot be assigned to $I2/m$ are not observed in all X-ray diffraction patterns. The β at 11 K and 67 kbar is the smallest ($93.66(8)°$) among those of Na_4C_{60} lattice of bcm. In this crystal, the center-to-center distance, $9.57(2)$ Å, between the C_{60} molecules associated with van der Waals contacts approaches to that, $9.20(2)$ Å, between the C_{60} molecules in a polymer chain, suggesting the realization of three-dimensional polymeric structure in Na_4C_{60} at a higher pressure than 67 kbar.

INTRODUCTION

Dimeric and polymeric phases were recently discovered in alkaline-metal intercalated C_{60} [1-7]. In AC_{60} (A: K, Rb and Cs), one-dimensional (1D) polymeric phase is formed by cooling slowly the face-centered cubic (fcc) phase realized at high temperature to temperature below 280 K [1]. In this phase, the C_{60} molecules connect through [2+2] cycloaddition to form four-membered ring. On the other hand, when cooling rapidly the fcc AC_{60} phase to temperature below 280 K, the metastable dimeric phase is realized through the formation of a single C-C bond between the C_{60} molecules [2,3]. The cubic superconducting Na_2RbC_{60} (superconducting critical temperature $T_c = 3.5$ K) loses the cubic symmetry to form the 1D polymeric phase by applying the modest pressure below 3 kbar and/or by cooling slowly the cubic phase to temperature below 230 K [4-6]. In this polymeric phase, the C_{60} molecules connect through single C-C bonds.

Recently, Oszlanyi et al. first found two-dimensional (2D) polymeric phase in Na_4C_{60} at 300 K under ambient pressure [7]. In this phase, the C_{60} molecules connect through single C-C bonds. This phase shows a metallic behavior, and transforms to the monomeric body-centered tetragonal (bct) phase above 500 K which also shows a metallic behavior [8]. Very recently, we studied the structure of Na_4C_{60} in a pressure region up to 53 kbar at 300 K [9]. The pressure dependence of the C-C bond between the polymerized C_{60} molecules along <111> direction and the van der Waals C...C contact between the monomeric C_{60} molecules along <111> suggests that a three-

CP544, Electronic Properties of Novel Materials—Molecular Nanostructures, edited by H. Kuzmany, et al.

dimensional (3D) polymeric phase would be formed above 150 kbar. On the other hand, the temperature dependence of spin susceptibility, line width and g-factor estimated from ESR shows the magnetic transition below 100 K [9]. Further, the temperature dependence of lattice constants suggests the structural phase transition below 100 K [9], although new Bragg reflection are not observed. Consequently, the magnetic transition may be ascribed to either Peierls (CDW) or spin-Peierls transition accompanied by the structural transition (dimerization), which is typical behavior of low-dimensional instability.

Realization of various crystal structures such as 3D polymerization under high pressure and dimerization at low temperature as well as monomerization at high temperature is characteristic of 2D polymeric Na_4C_{60}. Therefore, it is of importance to search new phases of Na_4C_{60} under various environments. In the present paper, we report the crystal structure of Na_4C_{60} at low temperature under high pressure determined from X-ray diffraction.

EXPERIMENTAL

The Na_4C_{60} samples was prepared by annealing stoichiometric amounts of C_{60} and Na metal for 100 h at 723 K under a vacuum of 10^{-5} Torr; a trace of benzene contained in C_{60} powder was removed before the annealing. The composition of the sample was determined to be $Na_{3.82(7)}C_{60}$ on the basis of the Rietveld refinement of X-ray diffraction pattern at 300 K and 1 bar. The sample was introduced into a diamond anvil cell with small amounts of NaCl for X-ray diffraction measurements at low temperature under high pressure. X-ray diffraction pattern was measured with synchrotron radiation at the BL-1B in the Photon Factory of High Energy Accelerator Research Organization (KEK-PF). The wavelength λ of 0.6904(1) Å was used in the X-ray diffraction measurement. The temperature was regulated within ±1 K with an Iwatani He closed cycle cryostat (M310). The pressure was estimated from the lattice constant of NaCl [10].

RESULTS AND DISCUSSION

The polymer chains between the C_{60} molecules in Na_4C_{60} are formed along the <111> direction in the (101) plane, as is shown in Fig. 1 [7,9]. The center-to-center distance between the C_{60} molecules along the <111> direction, which is associated with the polymer chain, is 9.28(1) Å at 300 K and 1 bar [9] This value is slightly larger than that of 1D polymeric RbC_{60}, 9.14 Å [1]. On the other hand, the center-to-center distance between the C_{60} molecules along <111> direction in the

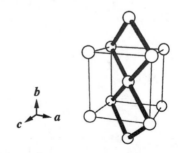

Fig. 1. Unit cell of Na_4C_{60} lattice; only C_{60} molecules are pictured by white balls. Polymer chain are drawn by thick lines.

(101) plane, which is associated with van der Waals C...C contacts, is 9.91(1) Å. Figs. 2(a) and (b) show the X-ray diffraction patterns of Na_4C_{60} at 200 K and 45 kbar, and at 11 K and 67 kbar, respectively. The Rietveld refinements for the X-ray diffraction patterns were performed in a space group of $I2/m$ according to the previous results [7,9]. No Bragg reflections which cannot be assigned to $I2/m$ are observed in these patterns.

The lattice constants, a, b, c and β, are 11.17(7), 11.55(7), 9.81(6) Å and 94.2(1)° at 200 K and 45 kbar, respectively, while 11.13(2), 11.50(2), 9.83(2) Å and 93.66(8)° at 11 K and 67 kbar. It was found that the values of β approached to 90° with an increase in pressure at 300 K [9]. Further, the β decreased monotonously with a decrease in temperature at 1 bar (β = 96.08(4)° at 300 K, β = 95.88(1)° at 12 K). The β at 11 K and 67 kbar is the smallest (93.66(8)°) observed so far. The center-to-center distances between the C_{60} molecules along <111> and <111> directions are 9.20(2) and 9.57(2) Å at 11 K and 67 kbar, respectively; the difference in distances decreases largely in comparison with that at 300 K and 1 bar (9.28(1) Å along <111>, 9.91(1) Å along <111>), suggesting that the geometry approaches to the 3D one.

Fig. 3 shows the unit-cell volume V of Na_4C_{60} at various temperatures and pressures. The V decreases largely with an increase in pressure, but the temperature dependence under pressure is not clear. The temperature dependence of V at 1 bar was also very small (V = 1349.7(6) Å^3 at 280 K, V = 1345.8(4) Å^3 at 12 K) [9]. In this study, the structural phase transition from the monoclinic phase of $I2/m$ at low temperature under high pressure were not clearly confirmed because new Bragg reflections were not observed. As the next step, the temperature dependent X-ray diffraction should fully be studied at the same pressure to search the structural transition at low temperature under high pressure. It is of interest to study the magnetic and structural transition under high pressure because the low-dimensionality and 3D porimerization reflect directly to the transition.

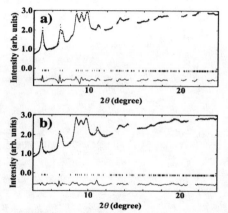

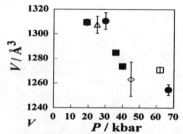

Fig. 3. Pressure dependence of of Na_4C_{60}. ■; 300 K, △; 250 K, ◇; 200 K, □; 101 K, ●; 11 K.

Fig. 2. X-ray diffraction patterns of Na_4C_{60} (a) at 200 K and 45 kbar, and (b) at 11 K and 67 kbar. The observed and calculated data are the symbols + and the solid lines, respectively. The allowed peak positions and the difference between the observed and calculated patterns are drawn by the ticks and the solid lines, respectively. Weighted R-factor R_{wp} = 0.025 and Intensity R-factor R_I = 0.004 in (a) and R_{wp} = 0.019 and R_I = 0.006 in (b).

REFERENCES

[1]   Stephens, P. W. *et al.*, *Nature* **370**, 636-639 (1994).
[2]   Zhu, Q., Cox, D. E., and Fischer, J. E., *Phys. Rev.* **B51**, 3966-3969 (1995).
[3]   Oszlanyi, G. *et al.*, *Phys. Rev.* **B54**, 11849-11852 (1996).
[4]   Zhu, Q., *Phys. Rev.* **B52**, R723-726 (1995).
[5]   Prassides, K. *et al.*, *J. Am. Chem. Soc.* **119**, 834-835 (1997).
[6]   Bendele, G. *et al.*, *Phys. Rev. Lett.* **80**, 736-739 (1998).
[7]   Oszlanyi, G. *et al.*, *Phys. Rev. Lett.* **78**, 4438-4441 (1997).
[8]   Oszlanyi, G. *et al.*, *Phys. Rev.* **B58**, 5-7 (1998).
[9]   Kubozono, Y. *et al.*, *Phys. Rev.* **B** Submitted.
[10]   Menoni, C. S., and Spain, I. L., *High-Temp. & High Pressure*, **16**, 119-125 (1984).

IV. ENDOHEDRALS

Structure and dynamics of endohedral fullerenes

K. Vietze[*,†], G. Seifert[‡], P.W. Fowler[*]

[*] School of Chemistry, University of Exeter, Stocker Road, Exeter EX4 4QD, UK
[†] Institut für Festkörper- und Werkstofforschung, D-01069 Dresden, Germany
[‡] Theoretische Physik, Universität GH Paderborn, D-330998 Paderborn, Germany

Abstract. Semi-relativistic Density-Functional based Tight Binding calculations and molecular dynamics simulations for endohedral metallo-fullerenes (e.g. rare earth endohedrals) indicate a high mobility of the encapsulated atom(s). This results in large amplitude oscillations and travel between minima inside the cage, both of which may lead to a vibrational behaviour which is not always well described by harmonic oscillator approximation approaches for the interpretation of spectroscopic data.

INTRODUCTION

Though there has been much progress in experimental structural analysis of endohedral metallo-fullerenes over the past few years there are still difficulties in the interpretation of spectroscopic data, especially of vibrational spectra. Some of these problems may arise from the insufficient description of the metal–cage interaction by models that involve harmonic-oscillator approximations and are based on the assumptions of harmonic potentials and small vibrational amplitudes. These models do not sufficiently take into account other dynamic effects, such as anharmonic potential contributions, large amplitude oscillations and propagation between several minimum energy configurations with an effective change of structure. Molecular dynamics simulations show that the complex dynamic behaviour of the endohedral atom(s) may lead to an increase or a decrease in effective symmetry, depending on the thermal regime and the time scale of the experimental technique.

A Density-Functional based Tight Binding formalism (DF–TB [1,2]) with a scalar-relativistic extension has been used to investigate structural and especially dynamic properties of a variety of endohedral fullerenes. Results for $Eu@C_{74}$ are presented here as an example to illustrate the important role of the intra-cage dynamics beyond the harmonic approximation. IR spectra were calculated from MD simulations by means of the Fourier transform of the velocity auto-correlation function and the molecular dipole moment.

CP544, *Electronic Properties of Novel Materials—Molecular Nanostructures*, edited by H. Kuzmany, et al.
© 2000 American Institute of Physics 1-56396-973-4/00/$17.00

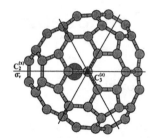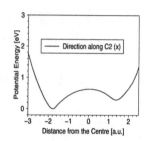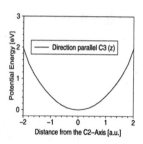

FIGURE 1. Molecular structure of Eu@C$_{74}$ for the IPR isomer of C$_{74}$ (left; the symmetry labels refer to the undisturbed, empty cage) and the potential encountered by the Eu atom in the direction along the main C$_2$ axis (middle) and parallel to the C$_3$ axis (right). Distances are given in atomic units, and the origin is the centre of the cage.

Eu@C$_{74}$ – STRUCTURE AND ENERGETICS

The deformations of the cage upon endohedral doping with the Eu atom are small (of the same estimated order as amplitudes of the cage carbons at a simulation 'temperature' of 140K, i.e. ∼0.08Å), but leading to definite reduction in symmetry to C_{2v} even for the cage itself. Elongation of the cage in the direction of the C_2 axis (Fig.1, left), on which the Eu is located , is about 0.077Å (1%), but since the bonds of the carbon framework are rather rigid the cage is squeezed in the direction of the C_3 axis by 0.08Å (1.1%). Deformations perpendicular to these axes are smaller by a factor of 2. The diagram in the middle of Fig.1 shows the molecular potential encountered by the Eu atom upon diplacement along the main C_2 axis of a frozen cage. The potential along this axis (x) is quite narrow around its well defined minimum, but will become anharmonic for higher vibrational excitations. In contrast, the potential for displacement parallel to the C_3 axis (z; Fig.1, right) is almost harmonic over a wide spatial region, but is much broader, rendering any assumption of small vibrations problematic. Models based on this assumption may fail, because the force constants for the other vibrational directions

TABLE 1. Relative stabilities of the first three non-IPR endohedrals of Eu@C$_{74}$. Isomers are labelled by the position of the empty cage in the spiral sequence [3].

Isomer	Symmetry (empty)	Energy [eV]	Energy [kJ/mol]
14246 (IPR)	$C_{2v}(D_{3h})$	0	0
13393	$C_1(C_1)$	+1.07	+103.2
14049	$C_1(C_1)$	+0.84	+80.8
14227	$C_2(C_2)$	+1.60	+154.6

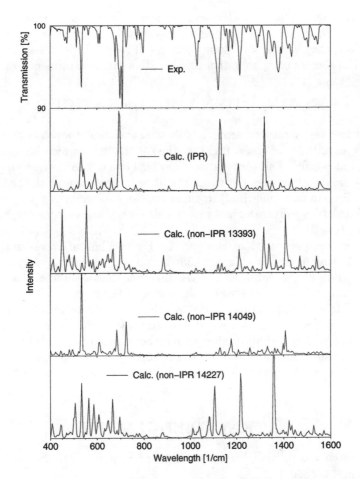

FIGURE 2. IR spectra of the IPR and the first three non-IPR endohedrals of Eu@C_{74}, as calculated from the MD simulations, in comparison to experimental data [4] (top).

are then no longer constant. In molecular dynamics simulations (see next section) the amplitudes of these vibrations can be as large as about 1.7a.u. (0.9Å) along a direction perpendicular to the main axis, depending on the 'temperature' of the simulation. [1]

As experimental results [4,5] point towards an even lower symmetry for the structure of Eu@C_{74}, endohedrals of three non-IPR isomers of C_{74} (those with only one

---

[1] **N.B.** In a microcanonical MD simulation the accessible thermodynamic quantity is the average kinetic energy per particle. To map this internal kinetic energy to temperature, the usual relation $\bar{\varepsilon}_{kin} = 3/2\,kT$ has been used throughout this work. As the exact, system-dependent relation is unknown, comparisons with experimental temperatures must be taken as indicative only.

pentagon-pentagon adjacency each and hence with low energy penalties as empty cages) were also included in the present investigations. However, as can be seen from Table 1, they are far less stable than the IPR-endohedral. In addition, their calculated IR spectra compare rather poorly with with the exprimental data (Fig.2).

Eu@C_{74} – DYNAMIC BEHAVIOUR

Even at low 'temperatures' around 140K the europium atom shows large oscillations with amplitudes of about 0.17Å in the C_3 direction, which increase to more than 0.22Å at $\sim$200K. These oscillations may lead to a dynamic symmetry breaking from C_{2v} down to C_s or even C_1, which would be in agreement with the interpretation of IR, Raman and (polarized) resonant Raman measurements [4,5], and indeed the calculated IR spectrum agrees rather well with the spectrum observed in the experiment (Fig.2).

At higher 'temperatures' of $\sim$600K the Eu oscillations have amplitudes of 1.7a.u. (0.9Å) along z and $\sim$1a.u. (0.53Å) along x, which is clearly outside the harmonic region of the potential in that direction (Fig.1, middle), and shifts the average position of the atom towards the centre of the cage.

At even higher 'temperatures' ($\sim$900K) the static distortions of the cage are completely 'washed out', and the Eu atom starts to cycle randomly between the three (once more degenerate) minima on a rather short time scale of a few picoseconds. This, in turn, increases the effective symmetry to D_3 or higher, which would be relevant for spectroscopic methods with longer characteristic time scales (e.g. NMR).

ACKNOWLEDGEMENT

This research was partially supported by the EU under TMR Network Contract 'Biofullerenes' CT980192.

REFERENCES

1. G. Seifert, H. Eschrig and W. Bieger
   Z. phys. Chemie **267**, 529 (1986)
2. D. Porezag, Th. Frauenheim, Th. Köhler, G. Seifert and R. Kaschner
   Phys. Rev. B **51**, 12947 (1995)
3. P.W. Fowler, D.E. Monolopoulos
   An Atlas of Fullerenes, Clarendon Press, Oxford, 1995
4. P. Kuran, M. Krause, A. Bartl and L. Dunsch
   Chem. Phys. Lett. **292**, 580–586 (1998)
5. M. Krause, P. Kuran, L. Dunsch
   'A vibrational spectroscopic structure analysis of Eu@C_{74}', *Proc. of the XIII Int. Winterschool on Electronic Properties of Novel Materials*, Kirchberg, Austria, 1999

The Encapsulation of Trimetallic Nitride Clusters in Fullerene Cages

H. C. Dorn,[1] S. Stevenson,[1] J. Craft,[1] F. Cromer,[1] J. Duchamp,[1] G. Rice,[1] T. Glass,[1] K. Harich,[1] P. W. Fowler,[2] T. Heine,[3] E. Hajdu,[4] R. Bible,[4] M. M. Olmstead,[5] K. Maitra,[5] A. J. Fisher[5] and A. L. Balch[5]

1) Department of Chemistry, Virginia Tech, Blacksburg, VA 24061
2) School of Chemistry, University of Exeter, Stocker Road, Exeter EX4 4 QD UK
3) Dipartimento di Chemica 'G. Ciamician', Universita di Bologna, via Selmi 2, Bologna I-40126, Italy
4) Searle, 4901 Searle Parkway, Skokie, IL 60077
5) Department of Chemistry, University of California, Davis, California 95616, USA

Abstract: The Kratschmer-Huffman electric-arc generator typically produces endohedral metallofullerenes in low yields with a wide array of different products, but the introduction of nitrogen leads to a new family of encapsulates. A family of endohedral metallofullerenes $A_nB_{3-n}N@C_{2n}$ (n=0-3, x=34, 39, and 40) where A and B are Group III and rare-earth metals is formed by a trimetallic nitride template (TNT) process in relatively high yields. The archetypal representative of this new class is the stable endohedral metallofullerene, $Sc_3N@C_{80}$ containing a triscandium nitride cluster encapsulated in an icosahedron (I_h), C_{80} cage. The $Sc_3N@C_{80}$ is formed in yields even exceeding empty-cage C_{84}. Other prominent scandium TNT members are $Sc_3N@C_{68}$ and $Sc_3N@C_{78}$. The former $Sc_3N@C_{68}$ molecule represents an exception to the well known isolated pentagon rule (IPR). These new molecules were purified by chromatography with corresponding characterization by various spectroscopic approaches. In this paper we focus on the characterization and properties of this fascinating new class of materials.

1. Introduction

For the empty C_{80}-I_h cage, computational results[1-3] suggest significant stabilization upon donation of 6 electrons, $(C_{80})^{6-}$ and experimental evidence supports an icosahedral cage for the $La_2@C_{80}$ endohedral.[3] Recently, we reported the first examples of a new family of stable metal (A,B) endohedral metallofullerenes, $A_{3-n}B_nN@C_{80}$ (n=0-3) that are stabilized by donation of 6 electrons to the C_{80}-I_h cage.[4] The endohedral nature of $Sc_3N@C_{80}$ was confirmed via a single crystal X-ray diffraction study of $(Sc_3N@C_{80})\cdot Co^{II}(OEP)\cdot 1.5$ chloroform$\cdot 0.5$ benzene.[4] The $Sc_3N@C_{80}$ molecule is in close proximity but not covalently bound to the $Co^{II}(OEP)$, which makes face-to-face contact with another $Co^{II}(OEP)$ moiety. The N-Sc distance and the closest Sc-C bond distance are 0.198 ± 0.002 and, 0.220 ± 0.002 nm, respectively. In the solid, the scandium ions face three pentagons within the C_{80} cage.

CP544, *Electronic Properties of Novel Materials—Molecular Nanostructures*, edited by H. Kuzmany, et al.
© 2000 American Institute of Physics 1-56396-973-4/00/$17.00

2. $A_{3-n}Sc_nN@C_{80}$ (n=0-3) Group III Endohedral Metallofullerenes

The $A_{3-n}Sc_n@C_{80}$ family members were prepared utilizing the TNT approach (presence of nitrogen) with a typical example illustrated, $Y_3N@C_{80}$ (Fig. 1a). The facile formation of scandium and yttrium TNT members, allows competitive formation conditions for comparisons with other Group III and rare-earth elements. Cored graphite rods were packed with A_2O_3 and Sc_2O_3 (constant A/Sc, 3/2% atomic ratio), powdered graphite mixture, and cobalt oxide. As previously reported for nanotube production, low levels of cobalt (and nickel) enhance nanotube formation.[5,6] The rods were subsequently vaporized in a Krätschmer-Huffman generator (He/N_2 mixture).[7] We have observed similar enhancements of both the TNT and non-TNT endohedral metallofullerenes by factors of 3-6 relative to empty-cage fullerenes with the inclusion of low levels (100-180 mg) of cobalt oxide. The soot obtained from the generator was extracted with carbon disulfide and the soluble fraction (fullerenes and endohedral metallofullerenes) was analyzed by negative-ion mass spectrometry.

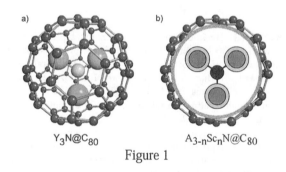

a)　　　　　　　　　　　b)

$Y_3N@C_{80}$　　　　　　$A_{3-n}Sc_nN@C_{80}$

Figure 1

The NI-DCI mass spectra for "mixed" Group III and scandium, $A_{3-n}Sc_nN@C_{80}$ TNT members are shown in Fig. 2. This data clearly confirms the higher yield advantage for A_3N cluster formation of Group III (relative to the non-TNT members and empty-cages) by the TNT process. As illustrated in Fig. 2, the yield enhancement for $Sc_3N@C_{80}$ is at least an order of magnitude greater than the usually prominent non-TNT $Sc_2@C_{84}$ formed under non-TNT conditions (absence of N_2). The high yields suggest favorable spatial and/or bonding interactions for the YSc_2N and Y_2ScN clusters inside the cage. Although all members of the $A_nSc_{3-n}N@C_{80}$, (n=0-3) Group III family are observable in the soluble fraction, $La_3N@C_{80}$ is formed at a very low level (~10% of $Sc_2@C_{84}$). The mixed lanthanum TNT members $LaSc_2N@C_{80}$ and $La_2ScN@C_{80}$ are also formed in relatively lower quantities. The lower yields for the $La_nSc_{3-n}N@C_{80}$ family members are consistent with the significant increase in the ionic radii for La (0.1045 nm) relative to Y (0.0900 nm) and Sc (0.0745 nm). For all Group III examples, we have assumed a trivalent state for each encapsulated metal atom (La, Y, and Sc). This suggests a contribution of one electron (per metal atom) for bonding to the central nitrogen atom and two electrons for cage stabilization, $(C_{80})^{6-}$ *vide supra*.

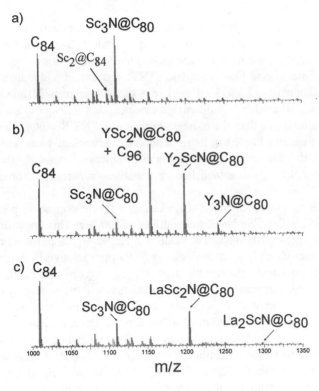

Figure 2
NI-DCI mass spectra for soluble extract: a) graphite rods
packed with Sc_2O_3; b) graphite rods packed with 3/2%
Y_2O_3/Sc_2O_3; c) graphite rods packed with 3/2% La_2O_3/Sc_2O_3.

In summary, Group III and rare-earth trimetallic nitride template (TNT) formation
for the A_3N cluster in the icosahedral C_{80} cage is consistent with two chief factors: 1)
an optimum A_3N cluster size, and 2) a requisite trivalent character of the encapsulated
metal ions. The high stability of these new TNT members will allow isolation of
numerous purified samples in the near future.

3. $Sc_3N@C_{68}$: A Violation of the Isolated Pentagon Rule

One of the sacrosanct rules in the evolving field of fullerenes, nanotubes, and
endohedral metallofullerenes has been the isolated-pentagon rule (IPR).[1,8-9] Although
exceptions have been predicted,[10-12] there are no well characterized violations of this
rule. Most endohedral metallofullerenes isolated and characterized to date[13-15] have
carbon cages of seventy or more carbon atoms (e.g., $La_2@C_{72}$, $Sc_2@C_{74}$, $Er_2@C_{82}$,
$Sc_2@C_{84}$, $Sc_3@C_{82}$, $Sc_3N@C_{80}$) with IPR allowed structures. Of all possible carbon
cages with less than seventy atoms, only the well recognized C_{60}-I_h can satisfies the
isolated pentagon rule. Scandium encapsulates, $Sc_3N@C_{68}$, $Sc_3N@C_{78}$, and $Sc_3N@C_{80}$

are formed in a Krätschmer-Huffman generator by the trimetallic nitride template (TNT) process (in the presence of nitrogen) in soluble extract yields of ~ 1%, 1%, and 10%, respectively. We observe enhanced production of three members of the series by the trimetallic nitride template (TNT) process in a dynamic mixture of nitrogen and helium gas (20/300 ml/min) into the Krätschmer-Huffman apparatus.[7] Besides $Sc_3N@C_{80}$, the other homologues, $Sc_3N@C_{68}$ and $Sc_3N@C_{78}$ are produced in slightly higher abundance than the most prominent non-TNT $Sc_2@C_{84}$, the endohedral metallofullerene usually formed in larger amounts in scandium soots under non-TNT conditions. After the usual chromatographic isolation protocol, the purity of $Sc_3N@C_{68}$ and $Sc_3N@C_{78}$ was established by negative-ion mass spectrometry.

The ^{45}Sc NMR spectrum for $Sc_3N@C_{68}$ exhibits a single symmetric peak in carbon disulfide at 296 K with a ^{45}Sc NMR linewidth of ~5600 Hz. This linewidth is slightly greater than the previously reported value for $Sc_3N@C_{80}$ and the chemical shift corresponds to considerably greater shielding (92.5 ppm relative to external $ScCl_3$).[4] On the ^{45}Sc NMR timescale, the results for $Sc_3N@C_{68}$ suggests that the three Sc atoms are equivalent at this temperature, which is consistent with three-fold symmetry. The $Sc_3N@C_{68}$ species represents a clear departure from previous endohedral metallofullerenes, and encapsulation of a four-atom molecular cluster in a relatively small carbon cage of only sixty-eight carbon atoms gives rise to a clear violation of the isolated-pentagon rule (IPR). For fullerenes smaller than C_{70} only C_{60} has a cage with a classic IPR allowed icosahedral cage. For a C_{68} cage, the spiral algorithm[1] finds 6332 distinct fullerenes. In neutral fullerene cages it is well established that each fused pentagon pair carries an energy penalty of 70-90 kJ/mole.[16] The qualitative preference for low-N_p isomers (N_p=number of fused pentagons) is confirmed by model DFTB calculations[17] that treat the cage as an empty fullerene capable of accepting electrons from a central reservoir. A highly symmetric structure (D_3, N_p=3) is proposed for $Sc_3N@C_{68}$ that is consistent with this computational approach.

4. Chromatographic Retention Behavior of $Sc_3N@C_{68}$, $Sc_3N@C_{78}$, and $Sc_3N@C_{80}$

The characteristic colors bluish-purple, green, and reddish-orange for carbon disulfide solutions of $Sc_3N@C_{68}$, $Sc_3N@C_{78}$, and $Sc_3N@C_{80}$, respectively, illustrate changes in the electronic structure as a function of carbon cage differences (C_{68}, C_{78}, and C_{80}). To date, we have found no chromatographic evidence for additional isomers of $Sc_3N@C_{68}$, $Sc_3N@C_{78}$, and $Sc_3N@C_{80}$, but the chromatographic behavior of this TNT family still provides information regarding the charge distribution and polarity of these three species. It has been previously established that less polar chromatographic stationary phases, such as PBB (pentabromobenzyl) generally exhibit weaker intermolecular interactions and give chromatographic retention times proportional to the polarizability/ π-electron count of the fullerene cage.[4,18] We find that $Sc_3N@C_{68}$, $Sc_3N@C_{78}$, and $Sc_3N@C_{80}$ when injected onto a PBB chromatographic column (carbon disulfide solvent) have elution times corresponding to empty cages C_{74}-C_{75},

C_{82}-C_{83}, and C_{84}-C_{85}, respectively, (Fig. 3). This suggests significant transfer of π electron population to the cage surface, although possibly attenuated by the presence of the central electronegative nitrogen atom in the encapsulated cluster.

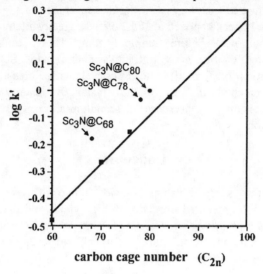

Figure 3

Chromatographic retention data for TNT members and empty-cage fullerenes PBB/CS$_2$, solvent (capacity factor, k'=t$_r$-t$_o$/t$_o$) versus carbon cage number.

5. Air Oxidation Study of Sc$_3$N@C$_{80}$

A black film of Sc$_3$N@C$_{80}$ sample was evaporated from a carbon disulfide solution directly onto a Au pad. The corresponding photoelectron spectroscopy (XPS) spectrum is shown in Fig. 4b. The observed spectrum suggests an absorption centered at ~400.9 eV for the 2p$_{3/2}$ core level and a second peak due to spin-orbital coupling 4.8 eV higher energy than the 2p$_{3/2}$ peak. Although a much smaller nitrogen peak is observed at 396.4 eV, the scandium and nitrogen XPS signal areas (corrected for relative sensitivities) provide good agreement for a 3 to 1 ratio of atoms for the Sc$_3$N cluster. In addition, the binding energy for the nitrogen peak (396.4 eV) compares favorably with the value reported for scandium nitride,[19] 396.2 eV, as well as the 2p$_{3/2}$ core level Sc peak, 400.7 eV. The binding energy for the nitrogen in Sc$_3$N@C$_{80}$ (396.4 eV) also disfavors a structure of an encapsulated Sc$_3$N cluster with a rapidly inverting electron lone pair located on a sp^3 hybridized nitrogen atom. Takahashi and coworkers[20] have reported XPS results for Sc$_2$@C$_{84}$ with a 2p$_{3/2}$ peak at ~401 eV that we have also repeated for reference purposes (Fig. 4a) illustrating a typical scandium endohedral metallofullerene (without encapsulated nitrogen).

The black $Sc_3N@C_{80}$ film prepared for the XPS experiments described in Fig. 4 was slowly heated in air. At a temperature of 663-673 K, the black film was converted to a white crystalline film and the subsequent XPS spectrum (Fig. 4c) for the scandium $2p_{3/2}$ core level shifted to 402.9 eV with the disappearance of the nitrogen peak. In addition, the carbon 1s peak centered at 285.2 eV was greatly attenuated relative to the corresponding carbon peak before sample heating. These results suggest nearly complete conversion of the $Sc_3N@C_{80}$ film into scandium oxide, Sc_2O_3. A SEM image of the scandium oxide film (Fig. 4d) indicates crystals with dimensions in the range of 0.1 - 0.6 μm. Elemental analysis of the crystals by energy dispersive X-ray spectroscopy indicates only the presence of scandium and oxygen with significantly lower levels of carbon.

6. Conclusions

The results of this study illustrate a wide range of new TNT endohedral metallofullerenes can be prepared in relatively high yields and purity . The limited physical and chemical properties suggest a wide range of applications in the emerging nano-science field. The unique structural, chemical, and reactivity features for these new encapsulates will clearly provide new directions in host-guest chemistry.

7. Acknowledgments

AIB thanks the National Science Foundation and HCD thanks LUNA Innovations for supporting phases of this study. HCD also acknowledge technical support from P. Phillips. PWF and TH acknowledge support from contract 'FMRX CT96 0126 USEFULL' under the European Union TMR Network scheme.

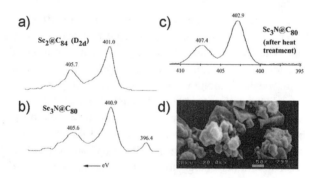

Figure 4
XPS of: a) $Sc_2@C_{84}$ (D_2d) film; b) $Sc_3N@C_{80}$ film;
c) $Sc_3N@C_{80}$ film after heat treatment 673K;
and d) SEM of $Sc_3N@C_{80}$ after heat treatment.

References

1. Fowler, P.W., and Manoloupoulos D.E., An atlas of fullerenes. Oxford Univ. Press, *Oxford* (1995).
2. Kobayashi, K., and Nagase, S. *Chem. Phys. Lett.* **262**, 227-232 (1996).
3. Kobayashi, K., Nagase, S.,and Akasaka, T. *Chem. Phys. Lett.* **261**, 502-506, (1996).
4. Akasaka, T., et al. *Angew. Chem. Int. Ed. Engl.* **36**, 1643-1645, (1997).
4. Stevenson, S., Rice, G., Glass, T., Harich, K., Cromer, F., Jordan, M. R., Craft, J., Hajdu, E., Bible, R., Olmstead, M. M., Maitra, K., Fischer, A. J., Balch, A. L. and Dorn, H. C. *Nature, 401*, **55**(1999).
5. Thess, A., Lee, R., Nikolaev, P., Dai, H., Petit, P., Robert, J., Xu, C., Lee, Y. H., Kim, S. G., Rinzler, A. G., Colbert, D. T., Scuseria, G. E., Tomanek, D., Fischer, J. E., Smalley, R. E., *Science*, 1996, 273-483.
6. Guo, T., Nikolaev, P., Thess, A., Colbert, D. T., Smalley, T. E., *Phys. Lett.* 1995, 243-49.
7. Krätschmer, W., Fostiropoulos, K., and Huffman, D. R. *Chem. Phys. Lett.* **170**, 167-170 (1990).
8. Kroto, H. W. *Nature*, **329**, 529-531 (1987).
9. Schmalz, T. G., Seitz, W. H., Klein, D. J., and Hite, G. E. *J. Amer. Chem. Soc.* **110**, 1113-11127 (1988).
10. Kobayashi, K., Nagase, S., Yoshida, M. and Osawa, E. *J. Amer. Chem. Soc.* **119**, 12693-12694 (1997).
11. Dorn, H. C. et al. in Fullerenes: Recent Advances in the Chemistry and Physics of Fullerenes and Related Materials (eds. Kadish, K. M. and Ruoff, R. S.) 990-1002, The Electrochemical Society, Pennington, 1998.
12. Wan, T. S. M. et al. *J. Amer. Chem. Soc.* **120**, 6806-6807 (1998).
13. Bethune, D. S., Johnson, R. D., Salem, J. R., de Vries, M. S., and Yannoni, C. S. *Nature*, **366**, 123 (1993),
14. Nagase, S., Kobayashi, K., and Akasaka, T. The electronic properties and reactivities of metallofullerenes, fullerenes recent advances in the chemistry and physics of fullerenes and related materials," ed. by K. M. Kadish and R. S. Ruoff, The Electrochemical Society, Inc., Pennington, 747-76 (1995).
15. Nagase, S., Kobayashi, K., and Akasaka, T. *Bull. Chem. Soc. Jpn.* **69**, 2131-2142 (1996).
16. Albertazzi, E., Domene, C., Fowler, P. W., Heine, T., Seifert, G., Van Alsenoy, C., and Zerbotto, F. *Phys. Chem. Chem. Phys.* **1**, 2913-9 (1999).
17. Seifert, G., Porezag, D., and Frauenheim, T. *Int. J. Quantum Chem.*, **58**, 185-192 (1996).
18. Fuchs, D., et al. *J. Phys. Chem.* **100**, 725 (1996).
19. Wagner, C.D., Riggs, W.M., Davis, L.E., Moulder, J. F., Muilenberg (ed.) "Handbook of X-Ray Photoelectron Spectroscopy", Perkin-Elmer Corporation, 6509 Flying Cloud Drive, Eden Prairie, Minn. 55344 (1978).
20. Takahashi, A., Ito, M., Inakuma, M., Shinohara, S. *Phys. Rev.* B52 (19) 13812-13814 (1995).

Photoemission spectroscopy of Gd@C_{82}

T. Schwieger[1], H. Peisert[1], M. Knupfer[1], M. S. Golden[1], J. Fink[1], T. Pichler[2], H. Kato[3], H. Shinohara[3]

[1]*Institute of Solid State and Materials Research Dresden, P.O. Box 270016, D-01171 Dresden*
[2]*Institute of Material Science, University of Vienna, Strudlhofgasse 4, 1090 Vienna, Austria*
[3]*Department of Chemistry, Nagoya University, Nagoya 464-8602, Japan*

Abstract. In this contribution we present a study of the electronic properties of Gd@C_{82} using high energy spectroscopy. From UPS we find a close similarity to La@C_{82}. From the analysis of the valence band and core level spectra excited using x-ray radiation, we conclude that the Gd-ion has transferred three electrons to the C_{82} cage, although there are signs of hybridisation between the C_{82} cage and the rare earth ion.

INTRODUCTION

Several metallofullerenes have been produced in macroscopic quantities in the last few years [1]. Recently, Gd based metallofullerenes have received considerable attention as possible tracers in medical MRl applications [2]. Analysis of the endohedral-C_{82} vibrational mode in Raman [3], or the magnetic properties of Gd@C_{82} [4] have led to the conclusion that the Gd ion present in Gd@C_{82} is trivalent, although a more direct proof of this has been missing up to now. In this regard, photoemission spectroscopy offers the possibility of directly probing the electronic structure of metallofullerenes such as Gd@C_{82}, and is particularly powerful in determining the 4f count - and hence the formal valency - in rare earth metallofullerenes [5]. In Gd@C_{82}, ultraviolet photoemission spectroscopy (UPS) has been used to study the charge transfer from the encapsulated Gd ions to the C_{82} cage [6], and it has been shown that this electron transfer also modifies the C 2p-derived electronic states.

EXPERIMENTAL

The preparation and separation of Gd@C_{82} has been described previously [7]. After purification by multi-cycle chromatography, recrystallization from CS_2 and degassing in UHV at 160°C for 100 hours, thin films of the Gd@C_{82} were prepared for photoemission by sublimation onto freshly sputtered gold foil substrates. The samples were then transfered under UHV conditions into the spectrometer, where they were studied using monochromatic Al Kα radiation (1486.6 eV) with an energy resolution of 350 meV. The UPS was carried out using the He Iα-line (21.2 eV) from a helium

CP544, *Electronic Properties of Novel Materials—Molecular Nanostructures*, edited by H. Kuzmany, et al.

discharge lamp with an overall energy resolution of 150 meV. The spectra of the Gd@C_{82} film contained no contribution from the substrate, remaining solvent or any contamination. All photoemission lines present could be attributed to either Gd or C.

RESULTS AND DISCUSSION

As mentioned above, valence band photoemission spectra can give information about the charge transfer from the Gd ion to the carbon π^* molecular orbitals. As a first step we measured the valence band of Gd@C_{82} by UPS and compared these spectra to those of the empty fullerene C_{82} (from Ref. [8]), and the monometallofullerenes La@C_{82} (from Ref. 8) and Tm@C_{82} (see Fig. 1). Our valence band spectrum of Gd@C_{82} is similar to that shown in Ref. [6]. The onset of the highest occupied molecular orbital (HOMO) in C_{82} is at about 1.2 eV (BE), which is slightly higher than in Tm@C_{82} (0.9 eV BE). Tm in Tm@C_{82} has been proven to be divalent [5], and has therefore, like C_{82}, a closed-shell configuration: C_{82} and Tm@C_{82} are consequently semiconductors. In a simple one-electron picture, a charge transfer of three electrons to the carbon cage should lead to metallic behaviour. However, as can be seen from Fig. 1, even with the essentially trivalent La ion, the monometallofullerene La@C_{82} is also semiconducting, with a spectral onset of 0.35 eV (BE).

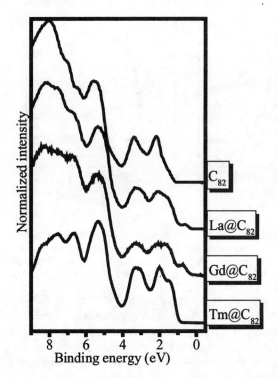

FIGURE 1. Valence band photoemission spectra obtained with He Iα-radiation for C_{82} (from Ref. [8]) and three different rare earth metallofullerenes. The spectra have been normalised to the feature at about 5 eV, where the first σ-derived states occur.

The additional structure in the valence band spectrum of La@C_{82} centred at about 0.5 eV (BE) is thus attributed to a singly occupied molecular orbital (SOMO) due to the charge transfer of the third electron to the C_{82}-cage [8].

Upon inspection of Fig. 1, it is immediately clear that the Gd@C_{82} UPS spectrum is very similar to that of La@C_{82}. The onset energy is 0.4 eV and the small peak attributed to the SOMO as well as the other σ-

and π-states of the C_{82} are well reproduced. Therefore, the presence of a SOMO-feature in the UPS spectra gives us first indirect evidence for the trivalent nature of the Gd-ion inside the C_{82} cage. Turning now to the core level spectra, the left panel of Fig. 2 depicts the photoemission profile of the Gd 4d core level line of both metallic gadolinium and of Gd@C_{82}. Because of the strong interaction between the core-hole and the 4f electrons in the photoemission final state the lineshape is complex, being dominated by multiplet effects [9]. The multiplet fine structure in the metallic Gd spectrum is clearly visible. In contrast, the fine structure of the first peak of 4d line of Gd@C_{82} is nearly completely washed out. This broadening might be caused by hybridisation between the electronic levels of the Gd with those of the fullerene, as has been shown to be the case in La@C_{82} [10]. Additionally, the screening environments for the 4d core-hole could be radically different in the metallic Gd and semiconducting endofullerene, which could also account for the shift of the 4d line in Gd@C_{82} to higher binding energies in comparison to the metallic Gd. As metallic gadolinium is well known to be trivalent ($4f^7$ ground state), the close similarity in the Gd 4d lines for both systems gives concrete evidence for the trivalency of the encapsulated Gd inside the C_{82} carbon cage.

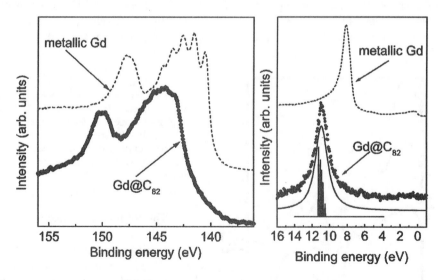

FIGURE 2. Photoemission spectra of Gd 4d multiplet (left panel) and the Gd 4f multiplet (right panel). In both panels the upper dashed curves show the corresponding spectra from metallic Gd, whereas the lower lines belong to Gd@C_{82}. For the Gd 4f data from the metallofullerene, the multiplet distribution calculated from an atomic model for a $4f^6$ final state is depicted under the data as thin solid lines.

The right panel of Fig. 2 shows the Gd 4f photoemission spectrum of Gd@C_{82}. Because of the much higher photon energy in XPS, the relative intensity of the Gd 4f emission relative to the carbon derived electronic states is dramatically increased, thus enabling a straight forward subtraction of the background from the carbon cage photoemission [5]. No fine structure of the multiplet splitting in the Gd 4f line could be resolved. Compared with the 4f line of the metallic Gd, we found a shift of 2.7 eV to higher binding energy, which is the same value as the shift for the 4d line.

The formal trivalency of the Gd ion inside the C_{82} cage can be finally proven by the comparison of the Gd 4f line in the XPS valence band spectra of the Gd@C_{82} with atomic multiplet calculations, which reproduce the photoemission 4f lines of the rare earth metals very accurately [11]. As can be seen from Fig. 2, the excellent agreement between the experimental Gd 4f spectrum of the monometallofullerene and the multiplet distribution for a $4f^6$ photoemission final state delivers the ultimate confirmation of the trivalent nature of the Gd in this system. No assymetry of the line shape of the multiplet was obtained by the fitting procedure, as can be expected from the nonmetallic nature of the endohedral. Nevertheless, the substantially increased width of the individual 4f multiplet lines in the endofullerene (0.7 eV) compared to Gd metal (0.30 eV) also points to the effects of hybridisation or altered screening in the former.

CONCLUSIONS

In summary, photoemission spectroscopy has been used to determine the formal valency of the Gd ion inside the C_{82} cage of Gd@C_{82}. From both the UPS and XPS measurments we can conclude that the Gd is trivalent, having transferred three electrons to the C_{82} cage. Nevertheless, there are signs that, in contrast to the case of Tm@C_{82}, but in keeping with that of La@C_{82}, hybridisation occurs between the levels of the metal ion and its host fullerene cage.

ACKNOWLEDGMENTS

We acknowledge financial support from the TMR Research Network 'FULPROP' (ERBFMRXCRT-970155). T.S. thanks the BMBF (05 5F8BD11) and T.P. the ÖAW for funding.

REFERENCES

1. Bethune D. S., Johnson R. D., Salem J. R., de Vries M. S., Yannoni C. S., Nature **366**, 123 (1993)
2. Shinohara, H., these proceedings
3. Krause M, Kuran P., Kirbach U., Dunsch L., Carbon **37**, 113 (1999)
4. Dunsch L., Eckert D., Fröhner J., Bartl A., Kuran P., Wolf M., Müller K.-H., Electrochemical Society Proceedings, **98-8**, 955 (1998)
5. Pichler T., Golden M.S., Knupfer M., Fink J., Kirbach U., Kuran P., Dunsch L., Phys. Rev. Lett., **79**, 3026 (1997)
6. Hino S., Umishita K., Iwasaki K., Miyazaki T., Miyamae T., Kikuchi K., Achiba Y., Chem. Phys. Lett., 281, **115** (1997)
7. Shinohara H., Kishida M., Nakane T., Kato T., Bandow S., Saito Y., Wang X. D., Hashizume T., Sakurai T., Recent Advances in the Chemistry and Physics of Fullerenes and Related Materials, The Electrochemical Society, Inc., Vol.1, pp.1361-1381 (1994).
8. Poirier D. M., Knupfer M., Weaver J. H., Andreoni W., Laasonen K., Parrinello M., Bethune D. S., Kikuchi K., Achiba Y., Phys. Rev. B, **49**, 17403 (1994)
9. Ogasawara H., Kotani A., Thole B. T., Phys. Rev. B, **50**, 12332 (1994)
10. Kessler B., Bringer A., Cramm S., Schlebusch C., Eberhardt W., Suzuki S., Achiba Y., Esch F., Barnaba M., Cocco D., Phys. Rev. Lett., **79**, 2289 (1997)
11. Gerken F., J.Phys. F : Met. Phys. 13, 703 (1983)

Sc₂@C₈₄/Au(110) : A Synchrotron Radiation Study

I. Marenne[1], P. Rudolf[1], M.R.C. Hunt[1]*, J. Schiessling[2], L. Kjeldgaard[2],
P. Brühwiler[2], M.S. Golden[3], T. Pichler[3]○, M. Inakuma[4], H. Shinohara[4]

1. Laboratoire Interdisciplinaire de Spectroscopie Electronique, Facultés Universitaires Notre-Dame de
    la Paix, Rue de Bruxelles 61, B-5000 Namur, Belgium
2. Department of Physics, Uppsala University, Box 530, S-75121 Uppsala, Sweden
3. Institut für Festkörperforschung, IFW Dresden e.V., P.O.Box 270016, D-01171 Dresden, Germany
4. Department of Chemistry, Nagoya University, Nagoya 464-01, Japan

Abstract. The electronic properties of Sc₂@C₈₄ deposited on Au(110) were investigated by X-ray absorption spectroscopy , valence band and core-level photoemission spectroscopy. We compared the bulk spectra of Sc₂@C₈₄ to those of a monolayer produced by annealing the thick film to about 750 K. Strong evidence is found for chemisorption of the first layer of Sc₂@C₈₄: the C1s photoemission spectrum shows a shift in binding energy and an asymmetric broadening towards higher binding energies pointing to metallic screening and/or an increase in the number of non-equivalent carbon sites. The valence band spectra present a broadening and non-rigid shifts indicating hybridization of the molecular electronic states with those of the substrate. The X-ray absorption spectra recorded at the Sc 2p edge does not change with film thickness. We can therefore conclude that the scandium electronic structure is not influenced by the chemisorption of the endofullerene.

INTRODUCTION

Since the first endohedral metallofullerenes were successfully synthesised, this novel type of material has attracted much interest because the insertion of an atom inside the cage is expected to confer extremely interesting electronic properties on the fullerene. Questions of particular importance are the interaction and charge transfer between the metal ion and the fullerene cage, and the existence of metallic, superconducting or ferromagnetic compounds. So far, different species like La, Y, Tm and Sc and most of the lanthanides have been encapsulated in various hollow fullerenes, and many characterisation techniques have been applied to La@C₈₂, La₂@C₈₀, Y₂@C₈₄, Sc₂@C₈₄, Sc₃@C₈₄, Tm@C₈₂ [1-5]. These metallofullerenes have been found to be very stable even in air, especially for compounds with Sc.

In this study we investigate the reactivity of Sc₂@C₈₄ when deposited on a gold surface, a substrate on which C₆₀ was found to chemisorb with hybridisation between the adsorbate and substrate electronic states and charge transfer from the substrate to the fullerene [6-8]. We deposited a multilayer and a monolayer of Sc₂@C₈₄ (isomer III) on the Au(110) surface and probed the electronic properties and the influence of the substrate on the endofullerene by X-ray photoemission spectroscopy (XPS), ultraviolet photoemission spectroscopy (UPS) and X-ray absorption spectroscopy.

CP544, *Electronic Properties of Novel Materials—Molecular Nanostructures*, edited by H. Kuzmany, et al.
© 2000 American Institute of Physics 1-56396-973-4/00/$17.00

EXPERIMENTAL

Sc$_2$@C$_{84}$ was synthesised and purified as described elsewhere [3,9,10] and loaded into an alumina crucible. A thick Sc$_2$@C$_{84}$ film (visible by eye) was sublimated at 710°C in UHV onto a clean (1x2) reconstructed Au(110) single crystal surface held at room temperature. After characterisation, the sample was annealed at about 480°C to desorb the multilayer and retain only the chemisorbed molecules on the surface. Thus we obtained approximately a monolayer (ML) of Sc$_2$@C$_{84}$ as verified from the ratio of the Au 4f$_{7/2}$ to C1s XPS intensities in comparison to C60 qdsorption on the same surface.

The photon emission and x-ray absorption experiments were performed at the VUV beamline at the Elettra synchrotron radiation facility in Trieste, Italy. This beamline is equipped with a spherical grating monochromator of the DRAGON type [11]. The photoemission spectra were collected with an angle integrated (20° acceptance cone) hemispherical analyser, for the X-ray absorption spectra the photocurrent from the sample was recorded. The photon energy was calibrated by measuring Au 4f$_{7/2}$ or the Fermi edge with first and second order light. The experimental resolution for the X-ray absorption, the valence band and the core level photoemission spectra was 80, 50 and 150 meV, respectively.

RESULTS AND DISCUSSION

Figure 1.(a) displays the valence band photoemission spectra recorded for the multilayer and the ML of Sc$_2$@C$_{84}$. The spectrum of the clean gold is also plotted for

(a)   (b)

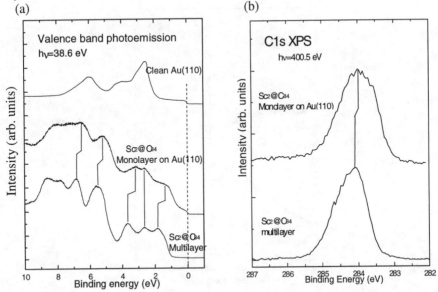

FIGURE 1. (a) Valence band photoemission spectra and (b) C1s core level spectra of the multilayer and monolayer of Sc$_2$@C$_{84}$

comparison. The multilayer spectrum, in total agreement with previously published data [12], shows the onset of the highest occupied molecular orbital (HOMO) at 1 eV, consistent with the theoretical prediction [13] of a closed shell configuration. The monolayer spectrum shows the same features as that of the multilayer, however, non-rigid shifts (ranging from 0 to 400 meV) of the peaks derived from those π and $\pi+\sigma$ molecular orbitals which energetically overlap with the Au bands are observed. This is a clear signature of a hybridisation of the molecular electronic states with those of the substrate and hence of chemisorption. Similar hybridisation effects have been observed when C_{60} is adsorbed onto the same surface [6]. Close inspection of the ML spectrum also reveals an increase in intensity next to the Fermi level compared to the spectrum of clean Au(110). We interpret this as a strong indication for a charge transfer from the substrate to $Sc_2@C_{84}$ and conclude that similarly to the ML's of smaller fullerenes [14], the monolayer of $Sc_2@C_{84}$ is likely to be metallic.

A comparison of the C1s core level for the multilayer and the ML is shown in figure 1 (b). The ML spectrum is shifted towards lower binding energy, E_B, compared to the multilayer, and is broader. Based on previous results for C_{60} adsorption on Au(110) [6], this is in perfect agreement with a chemisorption of the first layer . In fact, the metallic character of the film allows for screening of the core-hole through creation of electron-hole pairs, resulting in an asymmetric broadening of the peak towards higher binding energies. Due to the charge transfer, the peak shifts towards lower E_B and possibly a charge carrier plasmon satellite which would also broadens the peak. Finally, hybridisation with the substrate states creates non-equivalent carbon sites which can contribute to the broadening of the C1s line. The latter factor seems to

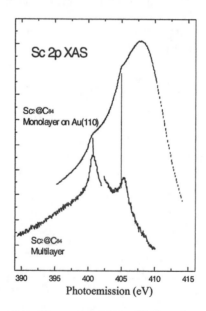

FIGURE 2. X-ray absorption spectra of the multilayer and monolayer of $Sc_2@C_{84}$

be more important in $Sc_2@C_{84}$ than in C_{60} [14] given the more symmetric lineshape observed here.

We also looked for possible changes in the Sc electronic structure induced by chemisorption onto the Au(110) surface. Figure 2 shows the X-ray absorption spectrum recorded at the Sc 2p edge for the multilayer and the ML $Sc_2@C_{84}$. The background shape reflects that of the undulator harmonic. Nevertheless, one can not seen any differences in the peak position in this spectrum. This seems to indicate that the fullerene cage protects the Sc from the surrounding environment. Such a protective function of the cage also agrees with the result reported in the literature [15] that $Sc_2@C_{84}$ films are stable in air and the Sc ion does not oxydise.

In summary, we have studied the adsorption of $Sc_2@C_{84}$ on Au(110) and found that the electronic structure of the first layer is changed by the interaction with the substrate, both due to hybridisation and charge transfer from, the gold to the fullerene. X-ray absorption spectra recorded at the Sc 2p edge reveal no difference in energy position between the ML and the thick film suggesting that the Sc electronic structure is not affected by the chemisorption.

This work was performed within the EU-TMR "FULPROP" network, contract n° ERBFMRX-CT97-015, with additional funding from the EU-TMR Large Scale Facility Programme and from the Belgian National Fund for Scientific Research (FNRS). I.M. and T.P. acknowledges for financial support the FRIA and the Osterreichische Akademie der Wissenschaften, respectively.

REFERENCES

* Present address: School of Physics and Astronomy, The University of Nottingham, University Park, Nottingham NG7 2RD, U. K.
° Also at Institut für Materialphysik, Universität Wien, Strudlhofg. 4, A-1090 Wien, Austria.
[1] Y. Chai et al., J. Phys. Chem. 95, 7564 (1991)
[2] H. Shinohara et al., J. Phs. Chem. 96, 3571 (1992)
[3] H. Shinohara et al., J. Phs. Chem. 97, 500 (1993)
[4] H. Shinohara et al., Nature (London). 357, 52 (1992)
[5] U. Kirbach and L. Dunsch, Angew. Chem. Int. Ed. Engl. 35, 2380 (1996)
[6] A. J. Maxwell et al., Phys. Rev. B49, 10717 (1994)
[7] S. Modesti, S. Cerasari and P. Rudolf, Phys. Rev. Lett. 71, 2469 (1993)
[8] M. Pedio et al., J. Electron. Spectrosc. Relat. Phenom. 76, 405 (1995)
[9] E. Yamamoto et al., J. Am. Chem. Soc. 118 2293 (1996)
[10] H. Shinohara et al., J. Am. Chem. Soc. 97, 4259 (1993)
[11] C. Quaresima et al., Nucl. Instr. Meth. Res. A 364, 374 (1995)
[12] T. Pichler et al, in Electronic Properties of Novel Materials Science and Technology of Molecular Nanostructures, eds. H. Kuzmany, J. Fink, M. Mehring, and S. Roth, AIP Conference Proceedings 486, Melville, N.Y., 1999, p.132.
[13] K. Laasonen, W. Andreoni and M. Parinello, Science 258, 1916 (1992); S. Nagase, K. Kobayashi, Chem. Phys. Lett. 214, 57 (1993); ibid. 31, 319 (1994)
[14] P. Rudolf, M. S. Golden and P. A. Brühwiler, J. Electron. Spectrosc. Relat. Phenom. 100, 409 (1999)
[15] T. Takahashi et al., Phys. Rev. B52, 13812 (1995).

Bonding and Dynamics of Diatomic Metal Encapsulates in Carbon Cages

M. Krause[1], M. Hulman[1], H. Kuzmany[1], P. Georgi[2], L. Dunsch[2], M. Inakuma[3], T.J.S. Dennis[3], H. Shinohara[3]

[1]Institut für Materialphysik, Universität Wien, Strudlhofgasse 4, A-1090 Wien, Austria
[2]Institut für Festkörper- und Werkstofforschung Dresden, D-01171 Dresden, Germany
[3]Department of Chemistry, University of Nagoya 464-8602, Japan

Abstract. Raman scattering of D_{2d}-Sc_2@C_{84} and D_{2d}-Y_2@C_{84} is compared to the response of C_{60}, empty D_{2d}-C_{84}, and Y@C_{82}. By a detailed analysis of the metal-cage vibrations we found that the unit cell of D_{2d}-Y_2@C_{84} contains at least 2 molecules. No decoupling of the yttrium ions from the C_{84} cage was observed up to 550 K. The vibrational line widths are governed by a thermally activated process starting above 140 K that could correspond to an order/disorder phase transition.

INTRODUCTION

In previous vibrational studies on endohedral metallofullerenes the redox state of Gd^{3+}@$C_{82}{}^{3-}$ [1] and Eu^{2+}@$C_{74}{}^{2-}$ [2] as well as details of the charge transfer in the Sc_2@C_{84} isomer III [3] were clarified. Now we extend our studies on D_{2d}-Y_2@C_{84}, the diyttrium counterpart of Sc_2@C_{84} (III). Based on the assignment of Y-C_{84} modes we analyze its solid state structure. Temperature dependent Raman data are used to study the dynamics of the material with respect to a decoupling of the yttrium ions from the cage and an order/disorder phase transition. We focus our analysis on the energy range below 300 cm^{-1}, which is less crowded with lines and supplies all information needed.

EXPERIMENTAL

The fullerene soot was prepared by a modified Krätschmer arc method using metal- or metal oxide-graphite mixtures as raw material. Its soluble components were extracted with CS_2 in a Soxhlet and separated by two stage HPLC (high performance liquid chromatography) using toluene as the mobile phase [4,5]. The sample purity of Sc_2@C_{84} and Y_2@C_{84} was at least 98% as confirmed by HPLC and mass spectrometry.

For Raman studies 50 μg Y_2@C_{84} and 200 μg Sc_2@C_{84} fullerene were dissolved in 1 ml toluene and as such dropcoated on a gold covered silicon substrate. The resulting polycrystalline films were dried under ambient conditions. Only Sc_2@C_{84} was then heated in high vacuum at 350°C to remove residual solvent molecules or other impurities. For the spectroscopic analysis the samples were placed into a cryostat and kept in vacuum better than 5x10^{-7} mbar. A Lakeshore model 330 was used for temperature control. Raman spectra were excited with 647 and 514 nm radiation of a Kr/Ar mixed

CP544, *Electronic Properties of Novel Materials—Molecular Nanostructures*, edited by H. Kuzmany, et al.

gas laser stabilite 2018 (Spectra-Physics, USA) using a line focus of 0.05 x 2.0 mm^2 area. A premonochromator and adapted interference filters were applied to eliminate laser plasma lines. The Raman radiation was collected in a back scattering geometry by a triple spectrometer XY 500 (Dilor, France) equipped with a CCD detector.

RESULTS AND DISCUSSION

In Fig.1 we display the low energy Raman spectra of C_{60}, D_{2d}-C_{84}, D_{2d}-$Sc_2@C_{84}$, and D_{2d}-$Y_2@C_{84}$ at 100 K. C_{60} shows a single line group of the $H_g(1)$ mode at 270 cm^{-1} and exhibits no feature below 200 cm^{-1}. Empty C_{84} has likewise only one structure peaking at 219 cm^{-1}. The metallofullerenes exhibit a comparable structure around 230 cm^{-1}, but additional metal induced lines are observed at lower frequencies and attributed to vibrations between the C_{84} cage and the encaged metal ions. In total 8 Sc-C_{84} and even 12 Y-C_{84} lines were resolved. The wave numbers are listed in Tab.1 together with the data of $Y@C_{82}$. The frequencies of $Y@C_{82}$ and $Y_2@C_{84}$ are in good agreement. Only the $Y_2@C_{84}$ line group around 85 cm^{-1} has no counterpart in the $Y@C_{82}$ spectra.

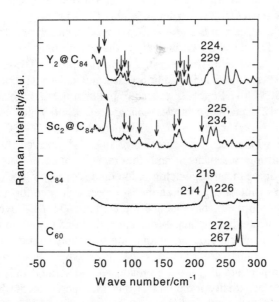

FIGURE 1. Low energy Raman spectra of C_{60}, D_{2d}-C_{84}, D_{2d}-$Sc_2@C_{84}$, and D_{2d}-$Y_2@C_{84}$, 100 K, 647 nm excitation, 2 (C_{60},C_{84}) and 5 mW laser power, 1.5 cm^{-1} spectral bandpass, 0.5 (C_{60}) and 2 h integration time. Metal-cage modes are marked by arrows. Spectra are shifted along the intensity scale.

TABLE 1. Wave numbers [cm^{-1}] of metal-cages modes

Sample	Stretching modes	Deformation modes
Sc$_2$@C$_{84}$	210, 175, 167	138, 110, 95, 87
		60
Y$_2$@C$_{84}$	190, 184, 177, 173	94, 88, 82, 77
		54, 48, 42, 39
Y@C$_{82}$	183 [6], 177 [5]	54 [6]

From x-ray analysis of D$_{2d}$-Sc$_2$@C$_{84}$ the Sc atoms are accommodated on the fourfold inversion axis (S$_4$) of the D$_{2d}$-C$_{84}$ cage [7]. Assuming an identical structure for D$_{2d}$-Y$_2$@C$_{84}$ a group theoretical treatment for the free molecules leads to 4 metal-cage modes with the following symmetry species:

$$\Gamma_{vib,M-C_{84}} = 1A_1(Ra) + 1B_2(Ra,IR) + 2E(Ra,IR).$$

The A$_1$ and the B$_2$ mode are metal-cage vibrations along the molecular S$_4$ axis which have a distinct bond stretching character. The E modes correspond to metal displacements perpendicular to S$_4$ and are metal-cage deformation vibrations. For one of them the C$_{84}$ cage is moving whereas it is silent for the other. Only the former one has a counterpart in the Y@C$_{82}$ spectrum at 54 cm^{-1} (Tab.1).

To understand the large number of metal-cage modes the crystal structure has to be taken into account. D$_{2d}$-Sc$_2$@C$_{84}$ crystallizes in the space group P2$_1$ with two formula units per unit cell [7]. The site symmetry of the molecule is C$_1$. Therefore the E modes can be split by the crystal field and in addition each molecular vibration can be doubled by the interaction of the two Sc$_2$@C$_{84}$ molecules in the unit cell, known as factor group splitting. In total 12 metal-cage modes can appear, 8 of which are observed for Sc$_2$@C$_{84}$. Our data reveal a similar crystal structure for D$_{2d}$-Y$_2$@C$_{84}$. Compared to Sc$_2$@C$_{84}$ even the whole set of 12 metal-cage lines is observed. In Fig.2 we have illustrated the splitting mechanism and the related energies for the metal-cage stretching and one of the metal-cage deformation modes of D$_{2d}$-Y$_2$@C$_{84}$.

The observation of metal-cage vibrations is a clear evidence for the existence of a chemical bond between the metal ions and the carbon cage. Moreover the small line widths at 100 K indicate that the metal ions are rigid bonded and no change between different positions of the cage occurs within the vibrational time scale. In Fig.3 we show the low energy Raman spectrum of Y$_2$@C$_{84}$ for temperatures from 32 to 550 K. With increasing temperature all lines are broadened and some details of the spectrum are no longer resolved. Really important is that the energy of the Y-C$_{84}$ stretching modes remains constant over the whole range and that the splitting in two stretching modes is present up to 550 K. Apparently the yttrium ions are bonded in an identical fashion at all temperatures studied.

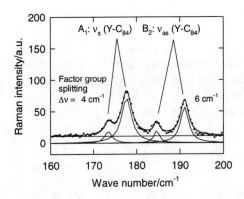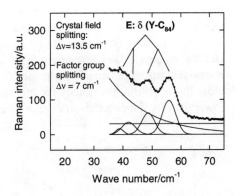

FIGURE 2. Crystal field and factor group splitting of the metal-cage stretching modes and one of the metal-cage deformation modes in the low frequency Raman spectrum of solid D_{2d}-$Y_2@C_{84}$ at 100 K

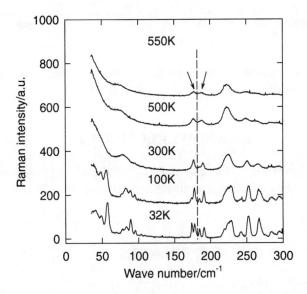

FIGURE 3. Low frequency Raman spectrum of solid D_{2d}-$Y_2@C_{84}$ at temperatures between 32 and 550 K, recording parameters as given for figure 1

Informations on the solid state dynamics can be obtained from the line width evolution with temperature. In Fig.4 we compare the data of the Y-C_{84} stretching mode at 177 cm^{-1} and the cage mode at 250 cm^{-1}. The widths of both lines are constant up to 140 K. According to predictions for a two phonon decay of a given excited phonon they should start to increase from 0 K on and reach 3-5 times higher values at 550 K than observed. Obviously a two phonon decay model is not suited to fit the experimental data. Instead the measured behavior can be understood in terms of a thermally activated coupling of phonons to rotational modes starting above 140 K. This process can be understood as transition from an ordered rotational state to a disordered rotational state of the metallofullerene. The transition temperature of 140-150 K is very close to 160-180 K previously reported for empty C_{84} [8] and $Sc_2@C_{84}$ [9] and much lower than for the order/disorder phase transition of C_{60} at 255K.

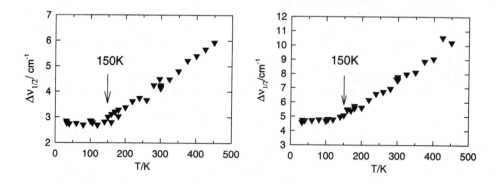

FIGURE 4. Temperature dependence of the line widths $\Delta v_{1/2}$ for the symmetric Y-C_{84} stretching mode of D_{2d}-$Y_2@C_{84}$ (left side) and the cage line at 250 cm^{-1} (right side)

ACKNOWLEDGMENT

Financial support by the EU, TMR-network ERBFMRX-CT97-0155 and the Fonds zur Förderung der Wissenschaftlichen Forschung in Austria, project P11943 is gratefully acknowledged.

REFERENCES

1. M. Krause, P. Kuran, U. Kirbach, L. Dunsch, Carbon **37,** 113 (1999),
2. P. Kuran, M. Krause, A. Bartl, L. Dunsch, Chem. Phys. Lett. **292,** 580 (1998)
3. M. Krause et al., J. Chem. Phys. **111,** 7976 (1999)
4. E. Yamamoto et al., J. Amer. Chem. Soc. **118,** 2293 (1996)
5. M. Krause et al., in preparation
6. S. Lebedkin et al., Appl. Phys. A **66,** 273 (1998)
7. M. Takata et al., Phys. Rev. Lett. **78,** 3330 (1997)
8. S. Margadonna, private communication
9. M. Hulman et al., Physica B Cond. Mat. **244,** 192 (1998)

Temperature and Field Dependent ESR Measurements on $Sc_3@C_{82}$

J. Rahmer*, A. Grupp*, M. Mehring*,
A. Bartl†, U. Kirbach†, and L. Dunsch†

*2. Physikalisches Institut, Universität Stuttgart,
Pfaffenwaldring 57, 70550 Stuttgart, Germany
†Institut für Festkörper- und Werkstofforschung Dresden,
Helmholtzstr. 20, 01171 Dresden, Germany

Abstract. ESR measurements in X band (9 GHz) and W band (94 GHz) have been performed on $Sc_3@C_{82}$ dissolved in trichlorobenzene. In X band, in addition to the hyperfine line-pattern which is due to the coupling of the electron spin to the scandium nuclei, the spectra of the cooled sample exhibit a line which is not observed in other solvents. In W band, upon cooling, the spectrum develops a strong asymmetry of the hyperfine pattern. These unusual effects are discussed in terms of changes in the electronic state of the molecule and internal dynamics of the scandium atoms.

INTRODUCTION

$Sc_3@C_{82}$ is an endohedral fullerene consisting of a scandium trimer forming an equilateral triangle inside a fullerene cage. X-ray diffraction experiments established that the molecule has the C_{3v} symmetry of the cage, but that the trimer is displaced from the center along the symmetry axis [1]. A view perpendicular to the C_{3v} axis and a view from "above" the molecule (along the symmetry axis) are shown in the left and right part of Fig. 1, respectively.

The nucleus of ^{45}Sc (natural abundance 100 %) carries a spin of $I=7/2$. In an ESR experiment the hyperfine coupling of the nuclear spins of the 3 scandium atoms to the electron spin leads to a splitting of the resonance into a pattern of 22 lines. The relative intensities of the individual lines are determined by the degeneracy of the hyperfine levels involved in the transition.

X-BAND MEASUREMENTS

The lower plot in Fig. 2 shows the room-temperature spectrum of $Sc_3@C_{82}$ dissolved in trichlorobenzene (TCB) in X band. From a fit to the data one obtains the parameters displayed in Table 1.

CP544, *Electronic Properties of Novel Materials—Molecular Nanostructures*, edited by H. Kuzmany, et al.

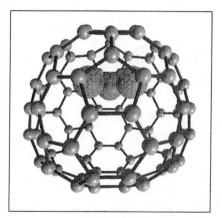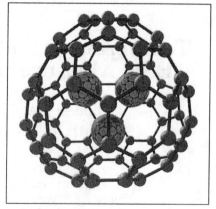

FIGURE 1. "Front" and "top" view of the $Sc_3@C_{82}$ molecule.

TABLE 1. Parameters obtained from a fit to the X band data. The fit assumes that the line intensities are determined by the statistical weight of the hyperfine levels, and that the linewidth is the same for all line components.

$Sc_3@C_{82}$ in TCB at room temperature	
linewidth FWHM [G]	1.90
isotropic g factor	1.9991
hyperfine constant a_{iso} [G]	6.29

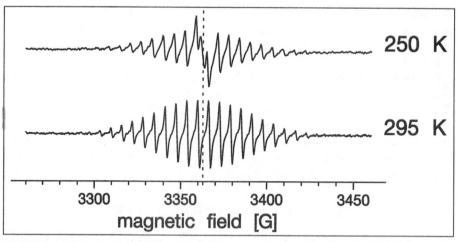

FIGURE 2. X-band spectra of $Sc_3@C_{82}$ in TCB at room-temperature and at 250 K. The solvent freezes at about 289 K.

In general, the linewidth found for $Sc_3@C_{82}$ is much larger than for $Sc@C_{82}$ [2]. This is attributed to the internal dynamics of the scandium trimer [3]. In our case, using TCB as a solvent, the linewidth of $\Delta B_{FWHM} = 1.90$ G is also large in comparison to linewidths reported for $Sc_3@C_{82}$ in other solvents, e.g., 0.8 G in decaline [3] or even 0.4 G in CS_2 [4]. This indicates a strong influence of the solvent on the dynamics of the system.

In fact, upon cooling, the X-band spectra of the frozen solution (Fig. 2, upper plot) exhibit a new peak in the center of the hyperfine pattern which is not observed in other solvents. From a fit to the hyperfine pattern and the central line it can be shown that in the temperature range between room-temperature and 200 K ESR intensity is transferred from the pattern to the central line. This means that under the influence of the solvent the molecule passes into a new electronic state. Due to the comparatively small width of the central line ($\Delta B_{FWHM} = 6.6$ G at 230 K), the hyperfine interaction in the new electronic state must be significantly reduced.

W-BAND MEASUREMENTS

Like in X-band, the room-temperature spectra in W band (94 GHz) exhibit the 22-line hyperfine pattern displayed in the lower plot of Fig. 3. It is only upon cooling that a striking new feature is observed. In the range between 260 K and 200 K the amplitude of the low-field hyperfine components decreases strongly in comparison to the high-field components, thus leading to the asymmetric spectrum

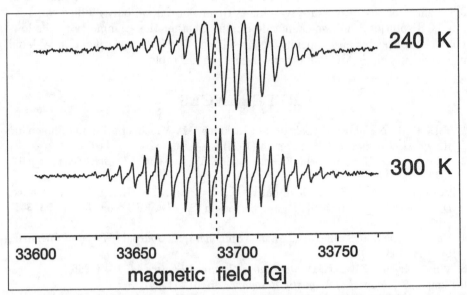

FIGURE 3. W-band spectrum of $Sc_3@C_{82}$ in TCB at 300 and 240 K.

shown in the upper plot of Fig. 3. As the population of the various hyperfine levels does not differ significantly, the change in line amplitudes must stem from a variation of the linewidths. Dynamic effects can account for a line-broadening that depends on the nuclear magnetic quantum number M_I [5,6].

We assume that the line-broadening is due to a dynamic Jahn-Teller-effect that results from the equivalence of the three scandium positions in the cage. The magnitude of the effect depends on the strength of the hyperfine coupling and the product of g anisotropy and applied B_0 field. Thus it can be understood that the effect is only visible in W band, where the B_0 field is about ten times the field in X band.

We furthermore attempted to obtain information about relaxation times, using pulsed ESR experiments. Due to the fast relaxation, pulse echo signals can only be obtained at very low temperatures. At 10 K a homogeneous T_2 relaxation time of 2 μs was found using a two-pulse echo-decay sequence.

CONCLUSION

We find that $Sc_3@C_{82}$ in a TCB solution exhibits unusual ESR behavior. The appearance of an additional line in the center of the hyperfine pattern can only be understood in terms of a different electronic state that the molecule forms at low temperatures in the presence of TCB as a solvent.

In W band, in addition to the central line, the hyperfine pattern develops a significant asymmetry when the sample is cooled. This must be due to a dynamic line-broadening effect. We assert that this behavior is caused by a dynamic Jahn-Teller effect of the degenerate scandium trimer. So far we only studied $Sc_3@C_{82}$ in a TCB solution, thus we cannot determine whether the asymmetry in the low-temperature spectra also depends on the solvent. For further clarification W-band measurements on $Sc_3@C_{82}$ in other solvents are desirable.

REFERENCES

1. Takata, M., Nishibori, E., Sakata, M., Inakuma, M., Yamamoto, E., and Shinohara, H., *Physical Review Letters* **83**, 2214 (1999).
2. Kato, T., Suzuki, S., Kikuchi, K., and Achiba, Y., *Journal of Physical Chemistry* **97**, 13425 (1993).
3. van Loosdrecht, P. H. M., Johnson, R. D., de Vries, M. S., Kiang, C.-H., Bethune, D. S., Dorn, H. C., Burbank, P., and Stevenson, S., *Physical Review Letters* **73**, 3415 (1994).
4. Kato, T., Bandou, S., Inakuma, M., and Shinohara, H., *Journal of Physical Chemistry* **99**, 856 (1995).
5. Wilson, R. and Kivelson, D., *Journal of Chemical Physics* **44**, 154 (1966).
6. Zimpel, Z., *Journal of Magnetic Resonance* **85**, 314 (1989).

The Samarium Fullerene Family

Petra Georgi, Silvio Lieb, Josué Navarro de San Pio,
Pavel Kuran[*] and Lothar Dunsch

*IFW Dresden, Abt. Elektrochemie und leitfähige Polymere
Helmholtzstr. 20, D-01069 Dresden, Germany*
*[*]Institut für Analytische Chemie, TU Dresden,
Zellescher Weg 17, 01217 Dresden*

Among the rare earth metals samarium is expected to form an ion of the two-valent redox state in endohedral fullerene structures. As compared to thulium and europium metallofullerenes a similar distribution of carbon cage structures is expected for samarium.
The samarium fullerene structures produced by the Krätschmer-Huffman method were studied with respect to the influence of the Sm-carbon ratio on the type and yield of the metallofullerenes. The chromatographic separation carried out by a two step HPLC resulted in a larger family of Sm-fullerenes (C_{2n}, $2n = 74, 78, 82, 84, 86, 88, 90, 92$) encapsulating one ion as detected by mass spectrometry. Only the $Sm-C_{76}$ structure was missing. No dimetallofullerenes of Sm were found.
The samarium structures of C_{74} and C_{82} were characterised by UV-Vis and IR spectroscopy. The results were compared with those of $Eu@C_{74}$ and $Tm@C_{82}$. The redox state of the metal ion in the metallofullerene was shown to be Sm^{2+}

INTRODUCTION

The carbon cage size and the isomeric structure distribution of endohedral fullerenes is influenced by the type and redox state of the metal included. Among the rare earth metals the valence state of the metals is preferably +3 (La, Ce, Y, Gd) and only in some cases +2 (Tm and Eu). Samarium is expected to be two-valent in endohedral fullerene structures. As compared to thulium and europium fullerenes a similar distribution of cage structures is expected for samarium. Therefore samarium was selected to study the influence of the type of metal on the carbon cage size and the distribution of the cage isomers.

EXPERIMENTAL

The samarium fullerenes produced by a modified form [1] of the Krätschmer arc vaporisation method were extracted with a Soxhlet-extractor for 20 hours with CS_2. The resulting fullerene mixture was dissolved in xylene for HPLC separation which was carried out in two steps. In the first step the fullerene mixture was separated on a

CP544, *Electronic Properties of Novel Materials—Molecular Nanostructures*, edited by H. Kuzmany, et al.

preparative HPLC (Gilson Abimed) with a Cosmosil Buckyprep column (Nacalai, 250 x 20 mm) and toluene as eluent. Fig. 1 shows the mass spectra of the fractions containing endohedral samarium fullerenes. The analytical HPLC (HP series 1050) as the second step was carried out on a Buckyclutcher column (SES, 250 x 10 mm) with toluene as mobile phase.

The mass spectrometric measurements were done on a MAT 95 (Finnigan, Germany) with methane as reagent gas and negative ion detection after chemical ionisation. The

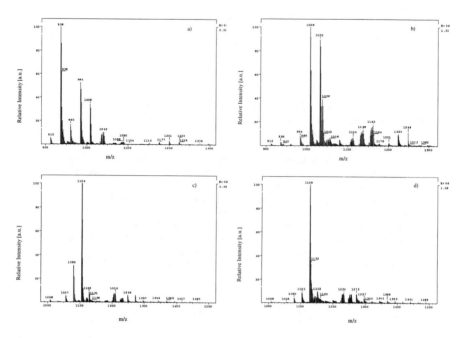

Fig. 1: Mass spectra of HPLC fractions containing endohedral samarium fullerenes after prep. HPLC
a) C_{80}- fraction with $Sm@C_{74}$ (m/z = 1042)
b) C_{84}- fraction with $Sm@C_{80}$ (m/z = 1114), $Sm@C_{82}$ (m/z = 1138) and $Sm@C_{84}$ (m/z = 1162)
c) C_{92}- fraction with $Sm@C_{88}$ (m/z = 1210) and $Sm@C_{90}$ (m/z = 1234)
d) C_{94}- fraction with $Sm@C_{90}$ (m/z = 1234) and $Sm@C_{92}$ (m/z = 1258)

UV-Vis spectra of the fullerenes in a solution of CS_2 were taken by an UV 3101 PC spectrometer (Shimadzu, Japan) in Suprasil 300 quartz cells of 10 mm optical path length.

Room temperature infrared spectra of $Sm@C_{82}$ and $Sm@C_{74}$ were obtained from polycrystalline films of ca. 20µg material drop-coated on KBr disks. The IR spectra were measured in transmission by a FTIR spectrometer IFS 66v (Bruker, Germany) under vacuum (1 mbar pressure) with 2 cm^{-1} resolution and a baseline correction using a scattering correction available within the spectrometer software.

RESULTS AND DISCUSSION

1. Sm@C₇₄

The UV-Vis-NIR spectrum of Sm@C_{74} (fig. 2) shows absorption bands at 501 and 763 nm as the highest ones in this spectrum, followed by the absorption bands at 598, 643 and 697 nm of medium intensity and the three NIR transitions at 1192, 1549 and 1972

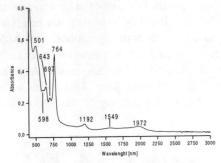

Fig. 2: UV-Vis-NIR spectrum of Sm@C_{74} in carbon disulfide solution

Fig. 3: IR spectrum of Sm@C_{74}

nm of weak intensity. The onset of this spectrum is at 2600 nm. This low energetic onset indicates a very small band gap of the fullerene.

The IR spectra of Sm@C_{74} (fig. 3) gives vibrational modes between 400 and 1800 cm^{-1} which are attributed to cage vibrations. These modes are compared with the published data of endohedral Eu@C_{74} [2] indicating that the Sm@C_{74} and the Eu@C_{74} are endohedral twins. A detailed analysis of the modes by Raman spectroscopy is under way.

2. Sm@C₈₂ – Isomer II and III

During the HPLC separation three isomers of Sm@C_{82} were observed by different retention times. Two isomers of Sm@C_{82} were isolated with a purity of 98%. The third isomer was unstable in the separation. Because of the two-valent redox state of

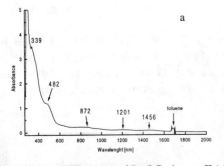
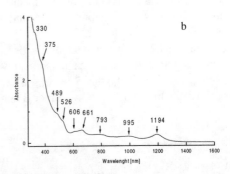

Fig. 4: UV-Vis-NIR spectra of Sm@C_{82} isomer II (a) and Sm@C_{82} isomer III (b)

samarium in the endohedral structure, the UV-Vis-NIR spectra were compared with the spectra of the three Tm@C$_{82}$ isomers A, B and C [3] in which Tm was shown to be two-valent [4].

The UV-Vis-NIR spectra of these isomers (fig. 4) show large differences in the number of absorption bands and the onset of the spectra. The Sm-isomer II (fig. 4a) shows five absorption bands with strong ones at 339 and 482 nm and an onset at 1800 nm. This Sm@C$_{82}$ II has almost the same UV-Vis-NIR spectrum as Tm@C$_{82}$ B isomer. Therefore a similar cage isomer (C$_s$) is expected.

The Sm-isomer III (fig. 4b) with nine UV-Vis-NIR absorption bands shows an onset at a wavelength of 1400 nm indicating a larger band gap for this isomer in comparison to isomer II. The Sm@C$_{82}$ isomer III spectrum has similar UV-Vis absorption bands as that of the Tm@C$_{82}$ isomer A. As the sample coding in both separations was made with respect to the retention times the isomers of Sm and Tm seems to be opposite in its cage structure and retention time. Thus the counterpart to Tm@C$_{82}$ isomer A in the cage structure is Sm@C$_{82}$ isomer III with the largest retention time. This seems to be caused by a different influence of the metal ion on the polarity of the endohedral

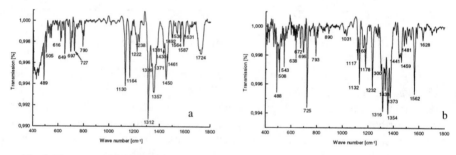

Fig. 5: IR spectra of Sm@C$_{82}$ isomer II (a) and Sm@C$_{82}$ isomer III (b) on KBr disks

carbon cage due to the difference of their ionic radii. The influence is to be studied in more detail with other metal ions of the some redox state.

The IR spectra of the two stable Sm@C$_{82}$ isomers II and III (fig. 5) results in vibration modes between 400 and 1800 cm^{-1} which are identified as internal cage vibrations. A hint for the identification of the different isomers is given by the modes in the range between 800 and 1100 cm^{-1}. In the spectrum of Sm@C$_{82}$ isomer III (fig. 5b) additional cage vibrations are visible at 890 and 1031 cm^{-1}. For Tm@C$_{82}$ isomer A these vibrations were also found.

Comparing all available data of the isomers of Sm@C$_{82}$ and Tm@C$_{82}$ it is to be concluded that the same cage isomers were found for both metals. The Tm@C$_{82}$ isomer A as well as the Sm@C$_{82}$ isomer III consist of a carbon cage with C$_{3v}$-symmetry while the Tm@C$_{82}$ isomer B and the Sm@C$_{82}$ isomer II have C$_s$-symmetry.

CONCLUSIONS

The successful preparation, purification and characterisation of one $Sm@C_{74}$ isomer and two stable $Sm@C_{82}$ isomers is presented for the first time. The samarium endohedral fullerene structures show large similarities with $Eu@C_{74}$ in the case of $Sm@C_{74}$ and with $Tm@C_{82}$ isomers in the case of the $Sm@C_{82}$ isomers. By comparing the vibrational spectroscopic data it is concluded that $Sm@C_{82}$ isomer II has a C_{3v}-symmetry (like $Tm@C_{82}$ A) and $Sm@C_{82}$ isomer III a C_s-symmetry (like $Tm@C_{82}$ B).

ACKNOWLEDGMENTS

The financial support by *BMBF, DFG (Schwerpunktprogramm Grundlagen der elektronischen Nanotechnologie)* and *EU (TMR-contract ERBFMRX-CT97-0155)*. As well as experimental work of Heidi Zöller are duly acknowledged.

REFERENCES

1   L. Dunsch, F. Ziegs, U. Kirbach, K. Klostermann, A. Bartl and U. Feist, in "Electronic properties of fullerenes (H. Kuzmany, J. Fink, M. Mehring and S. Roth, Eds.) Sprnger series in Solid State Sciences 117 (1993) 39
2   P. Kuran, M. Krause, A. Bartl, L. Dunsch, Chem. Phys. Lett. 292 (1998) 580 – 586
3   L Dunsch, P. Kuran, U. Kirbach, D. Scheller, ECS Proc. 97-14 (1997) 523 – 536
4   U. Kirbach and L. Dunsch, Angew. Chem. Int. Ed. Engl. 35 (1996) 2380; T. Pichler et al. Phys. Rev. Lett. 79 (1997) 3026

Thin films of Li@C$_{60}$

R. Ehlich, C. Stanciu and K. Buttke

Max Born Institut, Max Born Straße 2A, D- 12489 Berlin, Germany

Abstract. Thermal stability of endohedral Li@C$_{60}$ molecules was studied with high performance liquid chromatography (HPLC), laser desorption time of flight mass spectrometry and double sublimation thermal desorption spectrometry. HPLC purified material with a 90% nominal content of Li@C$_{60}$, was used. The material is transformed by heat into a non-soluble and non-sublimable state. High heating rates (10 K/s) are appropriate to sublime Li@C$_{60}$. Thin films of Li@C$_{60}$ were investigated by atomic force microscopy, optical absorption and second harmonic generation (SHG). First evidence for the photopolymerization of Li@C$_{60}$ films was obtained from SHG measurements.

INTRODUCTION

The endohedral fullerene Li@C$_{60}$ is the first alkali fullerene produced in macroscopic quantities [1]. Interesting electronic properties due to the charge transfer from alkali atom to C$_{60}$ and the off center position of the atom in the C$_{60}$ cage were predicted theoretically [2]. Electronic and optical property investigations are largely based on the availability of solid state material or thin films. The thermal stability of Li@C$_{60}$, the sublimability and the thin film structure are investigated. First results on optical properties and phototransformation are presented here.

EXPERIMENT

The endohedral fullerene Li@C$_{60}$ is produced by low energy (30 eV) Li$^+$ implantation in C$_{60}$ films following the method introduced by Campbell and coworkers [1,3,4]. Macroscopic quantities can be produced due to a conversion efficiency of up to 40% using a rotatable configuration for continuous deposition. The Li$^+$ - irradiated C$_{60}$ films are dissolved in CS$_2$. A first step of enrichment of Li@C$_{60}$ to about 70% is achieved using anisol, which dissolves mainly C$_{60}$. Solutions in o-dichlorbenzene (oDCB) are used for chromatography through a Cosmosil 5PBB column. The HPLC analysis of the final product shows a fullerene concentration of up to 95% Li@C$_{60}$ [1,3]. Purging the material in hexane reduces solvent impurities. Dried Li@C$_{60}$ was kept under N$_2$ atmosphere at 5°C for 1 year without any significant decay.

CP544, *Electronic Properties of Novel Materials—Molecular Nanostructures*, edited by H. Kuzmany, et al.
© 2000 American Institute of Physics 1-56396-973-4/00/$17.00

Thermal stability of Li@C$_{60}$ was investigated by heating small quantities in a quartz crucible under high vacuum at given temperatures for one hour. The heated material was ultrasonicated with oDCB and the soluble fraction was analyzed by HPLC. Sublimability tests were performed under UHV by double sublimation thermal desorption. The Li@C$_{60}$ films are produced by sublimation from small evaporators at a distance of 5 mm to a Ni(111) crystal with a constant heating rate of 10 °C/s up to 800 °C. The pressure during deposition was in the 10^{-8} mbar range. The neutral desorbed molecules from either the evaporator or the Ni(111) crystal are ionized by electron-impact and detected with a Balzers QMS 421 quadrupole mass spectrometer. The C$_2$ fragmentation of fullerenes is negligible at an electron energy of 45 eV [5].

A Nd:YAG laser at 1064 nm (pulse width 30 ps, fluence 7 mJ/cm^2), incident on the sample at 45° was used for the SHG measurements. The second harmonic signal is recorded via a gated integrator, and normalized to a SHG reference from a quartz crystal. It is important to note that the fundamental wavelength of 1064 nm does not induce photopolymerization. The SH signal from the film was measured after each step of Ar$^+$ laser irradiation.

RESULTS AND DISCUSSION

The integrals over the HPLC peaks identified as Li@C$_{60}$ (Li@C$_{60}$)$_2$ and C$_{60}$, which are given in figure 1 show that Li@C$_{60}$ is fairly stable for temperatures up to 175 °C. A similar behavior was found for anisol enriched compared to HPLC purified material. There is no clear correlation which would indicate a thermal decay of Li@C$_{60}$ into C$_{60}$.

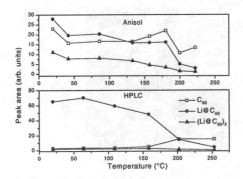

FIGURE 1. HPLC analysis of the soluble fraction of Li@C$_{60}$ materials heated for 1 hour

The soluble fraction of Li@C$_{60}$ decreases significantly between 175 and 200° C. A slight increase of the C$_{60}$ fraction was found for temperatures between 150 and 180 °C which could be due to a release of some Li atoms from C$_{60}$ cages catalyzed by impurities [6]. In the case of the anisol material the C$_{60}$ fraction is reduced for higher temperatures similar to the Li@C$_{60}$. The dimer fraction is also reduced with increasing temperature, but the drop at about 175 °C is less pronounced. The loss of endohedral material is considered to be due to a transformation into an insoluble, polymerized fraction. Measurements of the kinetics at 220°C have shown that about 80 % of the Li@C_{60} HPLC material and about 40 % of the Li@C_{60} from anisol material is lost within five minutes. The thermal transformation appears to be correlated to some extend with the concentration of Li@C_{60} in C_{60}.

The sublimability of Li@C_{60} requires good thermal stability at higher temperatures. Time of flight mass spectroscopy has shown that at least 78 % of the fullerenes in the HPLC material desorbed with 337 nm excimer laser pulses is Li@C_{60}. Thermodesorption measurements gave a percent ratio R = $I_{Li@C_{60}}$/ ($I_{Li@C_{60}}$+$I_{C_{60}}$) between the Li@C_{60} and C_{60} of R = 9.2 % for sublimation directly from the evaporator. A similar result was obtained using Li^+ ion irradiated C_{60} films [5]. The double sublimation of films deposited on a Ni substrate gave R = 9.8 % which proofs that the decay of Li@C_{60} to C_{60} which was found for single sublimation is due to the heating of the molecule by electron impact in the ionizer of the mass spectrometer. Double sublimation and irradiation of the film on the Ni crystal with $4.3 \cdot 10^{21}$ photons/cm^2 from a 514 nm Ar^+ laser at 350 mW/cm^2 gave R = 9.9 % Li@C_{60}. We can thus conclude that the endohedral material is not destroyed by thermal sublimation or by laser irradiation at 514 nm .

Thin films of Li@C_{60} were deposited under high vacuum on Si(100), Ni(111), quartz and Au(111) on mica substrates. The material organizes on the surface in islands with different size and density. Similar structures were found for Si(100) (Fig.2) and quartz substrates - islands of around 1.5 µm diameter and 300 nm height. On Ni(111) these islands had typical diameters of less than 0.6 µm and a height of about 50 nm, indicating a stronger interaction with the metallic substrate.

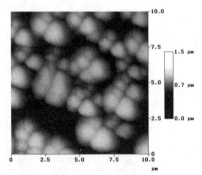

FIGURE 2. AFM Image of Li@C_{60} islands grown by sublimation on an oxidized Si(100) substrate

The gold surface with a rough topography of terrace edges shows a more homogeneous coverage and only very few islands comparable to deposits on other substrates were found. However, we can not exclude that a few islands were deposited as droplets. The films deposited on quartz were also measured by optical absorption. The absorption shows very broadened features in the UV- vis range up to 650 nm, with some similarity to the C_{60} films. A strong and very broad absorption maximum is found at around 1.1 µm. An absorption at this wavelength is characteristic for the C_{60} anion and thus it may be due to charge transfer from the Li to the C_{60} cage [7].

The C_{60} films after irradiation with Li^+ ions showed a significant increase in the second harmonic response due to the formation of $Li@C_{60}$ [8]. The phototransformation of $Li@C_{60}$ films by Ar^+ Laser irradiation was measured for the first time by second harmonic generation. The photopolymerisation of C_{60}, which was measured in comparison, was showing a double exponential decrease of the SH signal as a function of the irradiation time [9]. The fast exponential was attributed to the loss of monomers and the slow exponential was interpreted as the formation of oligomers Saturation was observed after a dose of $4 \cdot 10^{20}$ photons/cm^2, which is more than one order of magnitude faster than results obtained with Raman spectroscopy [10,11]. SHG is particularly sensitive for the early stages of photopolymerisation. A similar dependence of SHG on Ar^+ laser irradiation was found for $Li@C_{60}$ and interpreted as being due to photopolymerization. Photopolymerized $Li@C_{60}$ films can be separated from the substrate by solvent treatment and recovered as free films.

ACKNOWLEDGMENTS

The authors would like to thank I. V. Hertel, MBI Berlin, E. E. B. Campbell, Uni. Gothenburg and N. Krawez, IBM for fruitful discussions regarding production and stability of the endohedral $Li@C_{60}$. The AFM image was measured by C. S. Daróczi from MTA Budapest.

REFERENCES

1. Tellgmann R., Krawez N., Lin S.H., Hertel I. V., Campbell E. E. B., Nature **382**, 407 (1996)
2. Tomanek, D., Li, Y. S., Chem.Phys. Lett. 243, **42** (1995)
3. Gromov A., Krätschmer W., Krawez N., Tellgmann R., Campbell E. E. B., Chem.Commun., 2003 (1997)
4. Krawez N.,. Tellgmann R, Gromov A., Krätschmer W., Hertel I. V., J. Mol. Mat. **10,** 19 (1998)
5. Kusch Ch., Krawez N., Tellgmann R., Gomes Silva A., Winter B., Campbell, E. E. B. Appl. Phys. A **66**, 293 (1998)
6. Krawez, N., Gromov, A., Buttke, K., Campbell, E. E. B., Eur. Phys. J. D9, 345 (1999)
7. Krawez N., Dissertation Freie Universität/ Max Born Institut Berlin (1998)
8. Campbell E. E. B., Fanti M., Hertel I.V., Mitzner R., Zerbetto F., Chem.Phys.Lett. **288**, 131 (1998)
9. Ehlich R., Stanciu C., Buttke K. J. Mol. Mat. submitted (1999)
10. Wang Y., Holden J. M., Bi X. X. Bi, Eklund P. C., Chem. Phys. Lett. **211**, 341 (1993)
11. Park S., Han H., Kaiser R., Werninghaus T., Schneider A., Drews D., Zahn D. R. T., J. Appl. Phys. **84**, 1340 (1998)

Metal Doped Fullerenes: Endohedral and Networked Dopants

W. Eberhardt, R. Klingeler, P. S. Bechthold, M. Neeb

Forschungszentrum Jülich GmbH, Institut für Festkörperforschung, D-52425 Jülich, Germany

Abstract. We report on a systematic study on the production and relative abundancies of fullerenes doped with group IIIB transition metal atoms as well as Cerium and Calcium. The smallest stable endohedral cage sizes observed are $n = 30$ atoms for Sc and $n = 36$ for Y, La, and Ce. Fullerenes doped with two metal atoms are also observed. In this case one of the metal atoms is proposed to be integrated into the carbon network, whereas the other one is endohedral. The smallest cage sizes are $n = 27$ carbon atoms for 2xSc, $n = 31$ for 2xCe, and $n = 33$ for 2xY and 2xLa. Starting with $n = 70$ (72) carbon atoms, both metal atoms are endohedral for Y (La). The metal reactivity and ability to induce metallofullerene formation is discussed.

INTRODUCTION

Immediately following the discovery of endohedrally doped fullerenes about 15 years ago [1] the question about the smallest possible endohedral fullerene came up without saying. Among others, the discovery of $U@C_{28}$ raised excitement in this field of research [2], and so did $La@C_{36}$ as smallest La containing fullerene [3]. This has stimulated extensive research activity on endohedrally doped fullerenes, since they were supposed to exhibit unique properties determined by the dopant atom [4]. However even now, 15 years after the initial discovery, many of the intrinsic properties are not yet completely understood, e. g. the exact properties of the electronic structure and the charge transfer in these systems [5]. This gap in knowledge might partially be due to the fact that only a few endohedrally doped fullerenes are available in quantities amenable to conventional solid state spectroscopic methods.

EXPERIMENTAL SETUP

The experimental setup used in the present work has been developed for mass selected soft-landing of cluster ions onto a substrate under UHV conditions. The clusters are grown in a 100 Hz laser vaporization source driven by the 2nd harmonic of a Nd:YAG solid state laser. Careful attention was paid to the source design in order to seal the reaction chamber and optimize pressure conditions within the source. This

CP544, *Electronic Properties of Novel Materials—Molecular Nanostructures*, edited by H. Kuzmany, et al.

plasma source is coupled to a magnetic field mass analyzer. Ion abundance spectra are obtained by scanning the magnetic flux density and using a secondary electron multiplier behind the exit slit. Due to the high impact energies, constant detector sensitivity for all masses is expected. The ion transmission kinetic energy is 4 keV, and the resolution used in the current spectra is $m/\Delta m \approx 250$.

Due to the condensation process inherent to a laser plasma cluster source, the cluster formation slows down when a high stability cluster is formed. Accordingly the intensity distribution of the mass spectra reflects the most stable isomers. In particular, the smallest possible cage size capable of encapsulating a metal atom can be obtained.

RESULTS

Geometries Of Doped Fullerenes

Singly Doped Fullerenes

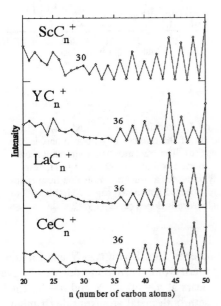

Figure 1. Intensity of singly doped carbon cluster cations as a function of the number of carbon atoms in the cluster (n). The intensity oscillations are characteristic for cage structures. From the onset of the oscillation the smallest possible endohedrally doped cage sizes $M@C_n^+$ can be deduced.

As shown in Fig. 1, the intensities of singly doped carbon cluster cations MC_n^+ reveal the onset of a distinct even-odd alternation as a function of n at a certain cluster size. This onset is interpreted to indicate the formation of stable endohedral fullerenes. The smallest observed cage sizes are $n = 30$ for Sc and $n = 36$ for Y, La, and Ce. The intensities have been evaluated from the corresponding mass spectra which will be published elsewhere [6].

Twofold Doped Fullerenes

Concerning the intensities of twofold metal-doped carbon cluster cations, the most interesting feature is found to be the onset of an odd-even alternation in the same size region as the MC_n^+ even-odd alternation, namely $n = 27$ for Sc, $n = 33$ for Y and La, and $n = 31$ for Ce (Fig. 2). Again, fullerene structures are most reasonable to explain this kind of intensity pattern. Since odd-numbered clusters are more abundant with respect to the even-numbered ones, one of the metal atoms is supposed to occupy the defect site introduced by the missing carbon atom, and the other metal atom is trapped within the cage. This kind of structure was first proposed by Jarrold et *al.*, and it will be referred to as "networked (nw) endohedral fullerene", $M@(MC_n)^{nw}$ [7]. Thus, the smallest possible networked cage size capable of encapsulating a metal atom is obtained.

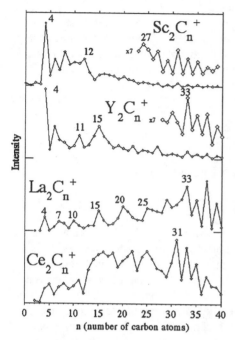

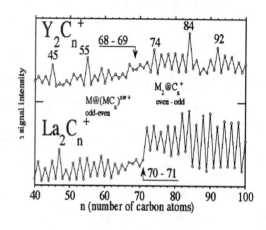

Figure 2. Intensity of twofold doped carbon cluster cations as a function of the number of carbon atoms in the cluster (n). The intensity oscillations are characteristic for cage structures. From the onset of the oscillations the smallest possible endohedeally doped networked cage sizes $M@(MC_n)^{nw+}$ can be deduced.

Figure 3. Intensity of $Y_2C_n^+$ and $La_2C_n^+$ as a function of the number of carbon atoms in the cluster (n). The transition from alternation odd-even to even-odd at $n = 69$ and $n = 71$, respectively, indicates a structural transition from endohedrally doped networked cages $M@(MC_n)^{nw}$ to biendohedrally doped cages $M_2@C_n$.

There is another interesting observation concerning fullerenes doped with two Y and La atoms, respectively, that is a distinct change in the intensity pattern from alternation odd-even to even-odd at a certain cluster size (Fig. 3). This transition appears at $n = 70$ for Y and $n = 72$ for La, respectively, and it lables the onset of cage structures with both metal atoms inside. It might be explained by the simple model that cages exceeding a certain size provide sufficient space to encapsulate both metal atoms inside, $M_2@C_n$ (see Fig. 4).

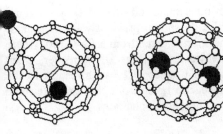

FIGURE 4. Proposed structures for twofold doped fullerenes: endohedrally doped networked cage $M@(MC_n)^{nw}$ (left sketch) and bi-endohedrally doped cage $M_2@C_n$ (right sketch).

Smallest Cage Size Of Doped Fullerenes

Summarizing the results on the smallest possible cage sizes of endohedrally doped pure and networked carbon cages, a major spread of the smallest possible n (n_{min}) is not observed. The hypothesis that the smallest endohedral fullerene that can be formed for a given element is a function of its ionic radius, as formulated by Guo et *al* [8], is thus supported, since the ionic radii of the metal atoms studied in the present work are close to each other. Taking into consideration the reduced n_{min} for Sc and its smaller ionic radius with respect to the other metal ions, the general trend of increasing n_{min} with increasing ionic radius is confirmed. The differences in the smallest possible sizes between pure carbon cages and networked cages are two atoms (three carbon atoms minus the networked metal atom) for the group IIIB transition metals, and four atoms for cerium. The smallest possible endohedrally doped networked cage size is smaller than the smallest possible endohedrally doped pure carbon cage size. Thus, networking seems to stabilize the endohedrally doped cluster with respect to disintegration upon shrinkage, and the 4f metal cerium is more successful in doing so than the group IIIB transition metals. When taking away one carbon atom of the pristine cage (with its four valence electrons), and replacing it by a lanthanum atom (which has got three valence electrons), there is one electron less in the system. The lanthanum atom needs to form three bonds with the neighbouring carbon atoms, and three electrons are required. As a matter of fact, the networked cage lacks one π valence electron compared to the undoped network. This reduced charge density inside and outside the carbon sphere might be the decisive feature for encapsulating a metal atom at the smaller cage sizes. Since the cerium valence region contains strongly localized 4f character, a reduced ability to contribute to the Ce-C bonds may be expected, and thus cerium should be more effective concerning this effect.

Doped Carbon Cluster Growth

Reactivity

Concerning the ability of metal atoms to induce doped fullerene cation growth, a systematic behaviour can be deduced from the mass spectra. The relative abundance of pristine fullerenes compared to doped fullerenes (which depends on the particular source conditions, but seems not to depend proportionally on the metal-to-carbon ratio in the sample rod) decreases when going from Sc to Y to La and the rare earth metal Ce, as already shown previously [9]. The least relative abundance of doped fullerenes is observed in the spectra taken from a Ca-C rod, where comparable intensities for doped and pristine fullerenes are found (Fig. 5). Moreover, when using a mixed metal-graphite rod, the relative abundance of C_{50}^+ and C_{60}^+ with respect to the other pristine carbon clusters is much higher than in the spectra taken from pure graphite rods. These seem to be the only "empty" fullerenes still present under these conditions.

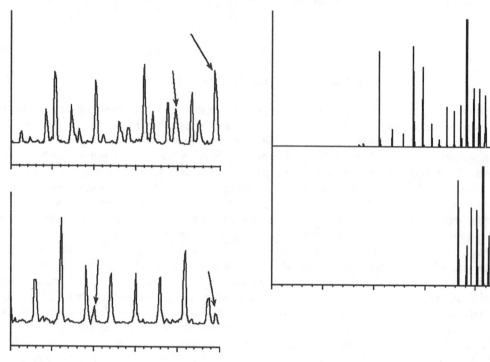

FIGURE 5. Comparison of the relative abundancies of pure and endohedral fullerenes for Ca and Y, respectively. The $M@C_{60}^+$ to C_{60}^+ ratio is $\approx$ 1 and 4 for Ca and Y, respectively.

Figure 6. Comparison of the mass spectra taken from a pure and metal-doped carbon rod, respectively. In the latter one no pure carbon cluster intensity is detected below the metal monomer.

Growth Mechanism

One of the most striking feature in the mass spectra taken from mixed metal-carbon rods is the complete lack of pure carbon cluster cation intensity in the mass range below the metal monomer. The metal monomer lables the onset of any ion signal intensity (Fig. 6). This indicates a metal induced cluster growth where the metal atom serves as core for carbon atom or fragment aggregation.

CONCLUSIONS

In the present work the formation and relative stability of endohedrally and networked endohedrally doped fullerenes is investigated. Examining the group IIIB transition metals and cerium as dopants, we find the smallest possible cage sizes capable of encapsulating a metal atom to be $n = 30$ for Sc, and $n = 36$ for Y, La, and Ce, as inferred from cation mass spectra. The metal-networked species are capable to accommodate a metal atom at a smaller size than the pure cages. Concerning twofold doped species, $Y_2C_n^+$ and $La_2C_n^+$ show structural transitions from endohedrally doped networked cages to bi-endohedrally doped cages at $n = 69$ and $n = 71$, respectively. As proposed earlier, the smallest pure cage sizes capable of encapsulating a metal atom depend on the metal ionic radii. The cluster growth process is supposed to be driven by a metal seed.

ACKNOWLEDGMENTS

The Sonderforschungsbereich 341 of the Deutsche Forschungsgemeinschaft has supported this work. Technical support by H. Pfeifer and J. Lauer is gratefully acknowledged. The group of Y. Achiba placed doped carbon rods to our disopsal.

REFERENCES

1. Heath, J. R., O'Brian, S. C., Zhang, Q., Liu, Y., Curl, R. F., Kroto, H. W., Tittel, F. K., and Smalley, R. E., *J. Am. Chem. Soc.* 107, 7779-7780 (1985)
2. Guo, T., Diener, M., Cai, Y., Alford, M. J., Haufler, R. E., McClure, S. M., Ohno, T., Weaver, J. H., Scuseria, G. E., and Smalley, R. E., *Science* 257, 1661- (1992).
3. Chai, Y., Guo, T., Jin, C., Haufler, R. E., Chibante, L. P. F., Fure, J., Wang, L., Alford, J. M., and Smalley, R. M., *J. Phys. Chem.* 95, 7564-7568 (1991)
4. Bethune, D. S., Johnson, R. D., Salem, J. R., de Vries, M. S., and Yannoni, C. S., *Nature* 366, 123-128 (1993)
5. Kessler, B., Bringer, A., Cramm, S., Schlebusch, C., Eberhardt, W., Suzuki, S., Achiba, Y., Esch, F., *Phys. Rev. Lett.* 79, 2289-2292 (1997)
6. Klingeler, R., Bechthold, P. S., Neeb, M., and Eberhardt, W., *J. Chem. Phys.*, submitted
7. Shelimov, K. B., and Jarrold, M. F., *J. Am. Chem. Soc.* 117, 6404-6405 (1995)
8. Guo, T., Smalley, R. E., and Scuseria, G. E., *J. Chem. Phys.* 91, 352-359 (1993)
9. Kietzmann, H., Dissertation Universität zu Köln (1996)

Synthesis and EPR studies of N@C$_{60}$ and N@C$_{70}$ radical anions

P. Jakes[†], B. Goedde[†], M. Waiblinger[‡], N. Weiden[†], K.-P. Dinse[†], and A. Weidinger[‡]

[†]Technische Universität Darmstadt, Petersenstr. 20, 64287 Darmstadt, Germany
[‡]Hahn-Meitner Institut, Glienickerstr. 100, 14109 Berlin, Germany

Abstract. Dilute solutions of anion radicals produced from enriched endohedral N@C$_{60}$ and N@C$_{70}$ samples were prepared by reduction with lithium in tetrahydrofuran (THF). EPR spectra were measured at room temperature with a Bruker X-Band spectrometer. Although EPR signals from mono- to penta-anions of empty fullerenes gave characteristic signals, no signals could be detected for these reduction states originating from endohedral nitrogen. We believe that increased spin relaxation rates of states of higher spin multiplicity cause a line broadening impeding detection. Continuing reduction finally resulted in the diamagnetic hexaanions, detected by vanishing EPR signals of empty fullerenes and a reappearence of the characteristic three line spectrum of endofullerenes. The chemical shift was quite similar to those measured by the point-like ^{3}He.

I INTRODUCTION

EPR is an essential tool to identify diamagnetic states of fullerene poly-anions with EPR. For non-degenerate levels, diamagnetic and paramagnetic states alternate with reduction. In case of C$_{60}$, the three-fold degenerate LUMO let us only predict the hexaanion and the unreduced molecule to be diamagnetic. Due to the Jahn-Teller distortion the tetraanion as well as the dianion could be diamagnetic. Labeling C$_{60}$ internally with nitrogen gives the opportunity to probe the effective spin of the cage in various reduction states, because only diamagnetic states are expected to lead to well resolved nitrogen EPR spectra. The fact that the nitrogen atom has no chemical interaction with the cage, its excellent chemical stability, and extremely narrow signals make it possible to use it as a probe and measure chemical shifts in the range of ppm. The high reactivity against impurities and the possibility to polymerize especially of the higher fullerene anions make it difficult to obtain reproducible results. Although electrochemical reduction of C$_{60}$ and C$_{70}$ in the group of Echegoyen [1] gave evidence for a reversible six-electron transfer to C$_{60}$ and C$_{70}$, fullerene anions of C$_{60}$ and C$_{70}$ are difficult to characterize.

CP544, *Electronic Properties of Novel Materials—Molecular Nanostructures*, edited by H. Kuzmany, et al.
© 2000 American Institute of Physics 1-56396-973-4/00/$17.00

Chemical reduction with lithium [2] or potassium [3] as reducing agent produced a variety of anions detected by EPR and UV/VIS. Only recently Saunders and coworkers [4] succeeded in the generation of hexaanions of C_{60} and C_{70} by using lithium and corannulene as sensitizer at -78°C. The ^{3}He inside the C_{60} hexaanion is found to be more strongly shielded than in the neutral C_{60}, whereas the ^{3}He inside the C_{70} hexaanion is more deshielded than in the neutral C_{70}. These results stand in complete accord with earlier predictions that the magnetic properties of C_{60} and C_{70} would be altered dramatically, and in opposite directions, by reduction to their hexaanions.

II   EXPERIMENTAL

We used two different ways to create the anions. First, lithium and ultrasound, second, lithium in combination with corannulene as sensitizer were used. $N@C_{60}$ and $N@C_{70}$ were prepared at the Hahn-Meitner-Institut (HMI Berlin) and enriched at the Technical University Darmstadt (TUD). The enrichment factor is in the order of ten.   Tetrahydrofuran (THF) was degassed several times before being transfered into a flask containing Na-K 5:1 alloy and a magnetic stirrer. The solvent was then stirred for about two days until a permanent blue color had developed. A 4 mm glass tube with a constriction to seperate two chambers was used. 100 μg of the enriched endohedral material dissolved in toluene was introduced. In the case of corannulene as additional sensitizer, 100 μg of corannulene in carbondisulfide (CS_2) were added (molar ratio ca. 1:3). After several times coevaporating with CS_2 to remove toluene the solid material in the glass tube was transported in a glove box where a lithium wire was inserted. After connecting the glass tube to a high vacuum line, THF was transfered and the sample finally sealed-off on a high vacuum line. EPR spectra have been recorded on a Bruker ESP 300 E spectrometer at X-Band frequency (9.63 GHz) with 1.56 kHz field modulation. For chemical shift determination, $N@C_{60}$ and $N@C_{70}$ in small tubes outside the probe were used.

A   EPR Spectra of Poly-Anions of $N@C_{60}$ and $N@C_{70}$

A stepwise reduction of $N@C_{60}$ to produce poly-anions where done with lithium and the aid of ultrasound at room temperature [2]. If mono-anions of C_{60} cage appear no signal for the nitrogen could be detected anymore. After appearence of the characteristic signals for the mono- up to the penta-anion the signal for the nitrogen is well resolved. We believed that increased spin relaxation rates of states of higher spin multiplicity cause a line broadening impeding detection. A similar experiment with $N@C_{70}$ failes, but a mixture of $N@C_{60}$ and $N@C_{70}$ in a ratio of ca. 3:1 shows the hexaanions of both species (figure 1). We think that C_{60} reacts with the impurities or prevents polymerisation of C_{70} poly-anions.

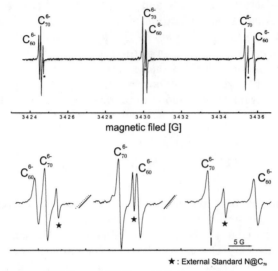

FIGURE 1. Reduction of N@C_{60} and N@C_{70} with lithium and ultrasound at room temperature.

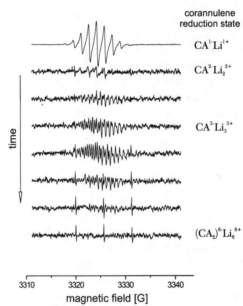

FIGURE 2. Sequence of EPR spectra of various reduction levels of N@C_{60} and corannulene as sensitizer.

Reduction of N@C_{60} with lithium and corannulene as sensitizer was performed at -78°C (12 h). No paramagnetic C_{60} mono-anions signal could be observed. Due to the experimental setup we suppose, that C_{60} is reduced very qickly to C_{60}^{4-}. The broad signal of C_{60}^{5-} (10 G) is not observed because of the small modulation amplitude of 13 mG. The signals of N@C_{60}^{6-} are well resolved.

III   RESULTS AND CONCLUSIONS

It is noteworthy, that sampling the internal field in fullerenes by extended atomic charge distribution leads to relative chemical shift values similar to those measured previously with the ^{3}He nuclear spin. Evaluation of the data shows, however, that the differences in internal fields "seen" by the nitrogen atom increases in the highly reduced cages for both fullerenes: The g value of the N@C'_{60} hexa-anions decreases by 35.1(5) ppm (compared to a chemical shift change of +42.4 ppm), the g value shift of the N@C_{70} hexa-anion amounts to +41.0(5) ppm (compared with -37.1 ppm). (The different sign quoted for changes in g value and chemical shift data originates from an anti-linear definition of the latter parameters). It seems, that the difference in g values of both molecules in their ground state (Δg = 19 ppm) and reduced state (Δg = 57 ppm) is nearly identical to those values measured by ^{3}He-NMR. The isotropic hyperfine coupling constant (hfcc) of N@C_{60} hexa-anion increases to 15.907(5) MHz from 15.877(5) MHz, whereas it decreases from 15.133(5) MHz (N@C_{70}) to 15.077(5) MHz in N@C_{70} hexa-anion, measured under identical conditions, indicating small change in the spin density at the nucleus. Because the observed variations have different sign in C_{60} and C_{70} cages, no simple explanation is possible, however. Finally, the increase of homogeneous EPR linewidth of the hexa-anion of N@C_{60} was tentatively related to a decrease of its spin relaxation time from 50 μs to 5 μs.

IV   ACKNOWLEDGEMENTS

We thank Prof. L. T. Scott (Boston College) for providing samples of corannulene

REFERENCES

1. Qingshan Xie, Eduardo Perez-Cordero, Luis Echegoyen, *J. Am. Chem. Soc.*, **1992**, *114*, 3978-3980.
2. J. W. Bausch, G. K. S. Prakash, G. A. Olah, D. S. Tse, D. C. Lorentz, Y. K. Bae,R. J. Malhotra, *J. Am. Chem. Soc.*, **1991**, *113*, 3205-3206.
3. M. Baumgarten, A. Ggel, L. Gherghel, *Adv. Mat.* **1993**, *5*, 458-461.
4. E. Shabtei, A. Weitz, R. C. Haddon, R. E. Hoffman, M. Rabinovitz, A. Kong,R. J. Cross, M. Saunders, Pei-Chao Cheng, L.T. Scott, *J. Am. Chem. Soc* **1998**, *120*, 6389-6393.

Squeezing and Shielding of Atoms
in Fullerene Traps

K.-P. Dinse*, N. Weiden*, B. Goedde*, P. Jakes*, M. Waiblinger**, and
A. Weidinger**

*Phys. Chem. III, TU Darmstadt, Petersenstr. 20, D-64287 Darmstadt, Germany
**Hahn-Meitner-Institut, Glienickerstr. 1000, D-14109 Berlin, Germany

Abstract. Ideal trapping behaviour was recently discovered for nitrogen incased in C_{60} and C_{70}, although significant increase of Fermi contact interaction gave evidence for compression of the atomic spin distribution. Fullerene cages therefor are candidates for the study of atomic orbital interaction with inert confinements. Prolate deformation of the spin and charge distribution of nitrogen in C_{70} was detected by high resolution ENDOR spectroscopy. Magnetic shielding averaged over the extended electronic orbitals in poly anions of fullerenes was found to be similar but not identical to the values measured for helium atoms at the cage center of the cages.

INTRODUCTION

From early on fullerenes have been envisaged as candidates for chemical traps. Molecular super structures could be of considerable interest for the generation of quantum level systems with controlled de-coherence behavior as required for the realization of quantum computers. Ideal cage behavior was discovered for paramagnetic group V elements like nitrogen and phosphorus incased in C_{60} and C_{70}. EPR and ENDOR data gave evidence that neutral atoms are encapsulated in their quartet spin ground state [1]. A 50% increase of the Fermi contact interaction in case of nitrogen gave evidence for significant compression of the atomic spin distribution. Fullerenes could be expected to be candidates for the study of the deformation of atomic charge and spin distributions by an "inert" confinement, which is rigid enough to give measurable effects even considering the "stiffness" of atomic orbitals. Using C_{60} and C_{70}, cages with nearly identical volume but of different symmetry are available. ^{14}N (I=1) could sense charge and spin distributions in C_{70} via dipolar and quadrupolar hyperfine interaction (hfi).

RESULTS AND DISCUSSION

ENDOR Investigation Of N@C_{70}

A natural extension of the investigation of the local symmetry around the encased nitrogen atom was to use C_{70} instead of C_{60} as confinement, thus imposing D_{5h}

CP544, *Electronic Properties of Novel Materials—Molecular Nanostructures*, edited by H. Kuzmany, et al.
© 2000 American Institute of Physics 1-56396-973-4/00/$17.00

boundary conditions. Intrinsic Zero Field (ZFS) interaction as well as nuclear quadrupole interaction (nqi) is expected if the long axis orientation is not isotropically averaged on the time scale of the experiments. Nqi, being not influenced by crystal imperfections, can only be measured by ENDOR. In Fig. 1, pulsed W-band ENDOR spectra narrow ENDOR lines for the solid state $N@C_{70}$ sample. Only under higher spectral resolution it became apparent (see Fig. 2) that in contrast to $N@C_{60}$ with its extremely narrow nitrogen ENDOR transitions of 4 kHz width (FWHM) [2], typical powder line shapes are detected for $N@C_{70}$.

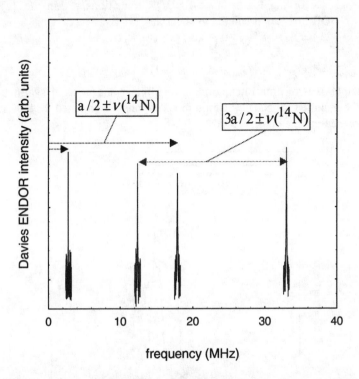

FIGURE 1. W-band (95 GHz) ENDOR transitions of $N@C_{70}$ measured at 80 K.

Because of limited material, all studies were performed at a microwave (MW) frequency of 95 GHz. At this frequency, a 100 μg sample of 10^{-4} relative concentration was sufficient to give ENDOR signals of reasonable intensity. For the quartet electron spin coupled to the nuclear spin I = 1 a total of eight "allowed" ENDOR transitions are expected which are pair-wise degenerate. Selective MW excitation of the m_I = +1 or −1 hfcc enables undistorted observation of a particular subset of four non-degenerate transitions, important for an elucidation of the source of line broadening. Allowed ENDOR transitions are predicted at

$$\nu_{ENDOR} = \left| \left\{ m_s (a_{iso} + A_{zz}(\Theta, \Phi)) - \nu_n \right\}(m_I - m_I') + \tfrac{3}{2} Q_{zz}(\Theta, \Phi)(m_I^2 - m_I'^2) \right| \quad (1)$$

Here, ν_ε and ν_n denote electronic and nuclear Larmor frequencies, m_s and m_I describe electronic and nuclear spin magnetic quantum numbers, a_{iso} denotes the isotropic nitrogen hyperfine coupling constant (hfcc), $A_{zz}(\Theta,\Phi)$ and $Q_{zz}(\Theta,\Phi)$ are elements of the traceless dipolar and quadrupolar hfi after transformation to the laboratory frame. In its eigenframe, Q_{zz} can be related to the nuclear quadrupole coupling constant eQq/h by Q_{zz} = eQq/2h in case of ^{14}N with I = 1. Because dipolar interactions depend on terms linear in m_I and m_I', whereas quadrupolar terms depend on the difference of squared m_I values, it is in general possible to determine both coupling constants separately. In our case (I = 1) it was sufficient to tune the EPR excitation (using "soft" MW pulses) to the outer hfcc corresponding either to m_I = +1 or to m_I = -1. Recording for instance ENDOR transitions within the $|m_s|$ = 1/2 manifold, ENDOR lines are predicted to occur at

$$\nu_{(m_I=+1,\, m_I'=0)} = \left| \left(\pm \tfrac{1}{2}A_{zz} + \tfrac{3}{2}Q_{zz} \right)\tfrac{1}{2}(3\cos^2\Theta - 1) \pm \tfrac{1}{2}a_{iso} - \nu_n \right|$$

$$\nu_{(m_I=-1,\, m_I'=0)} = \left| \left(\pm \tfrac{1}{2}A_{zz} - \tfrac{3}{2}Q_{zz} \right)\tfrac{1}{2}(3\cos^2\Theta - 1) \pm \tfrac{1}{2}a_{iso} - \nu_n \right|$$

(2)

if the principal axes of both interactions are collinear and of axial symmetry, which can safely be assumed for N@C$_{70}$. Similar expressions are obtained for the $|m_s|$ = 3/2 manifold, by scaling the dipolar terms by 3.

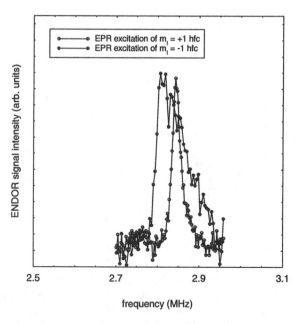

FIGURE 2. Low frequency ENDOR line of N@C$_{70}$ showing partial cancellation of dipolar and quadrupolar interactions.

In Fig. 2, W-band ENDOR lines are shown demonstrating the predicted partial cancellation of powder line broadening originating form dipolar and quadrupolar interactions for one of the ENDOR transitions. Although limited signal-to-noise ratio does not allow an accurate line shape analysis, the observed pattern is consistent with that predicted for an interaction of axial symmetry. From the intensity maximum at the low frequency edge of the low frequency ENDOR line one directly can deduce the negative sign of the principle element of the anisotropic part of the dipolar coupling constant, because the dominant isotropic hfcc a_{iso} = 15.12 MHz can safely be assumed to be positive. We deduce A_{zz} = -94(5) kHz and Q_{zz} = +21(3) kHz, giving eQq/h = +42(6) kHz. The determination of the sign of the quadrupole coupling tensor element relative to A_{zz} (and a_{iso}) is unambiguous when using the expressions given in Eq (2).

A simple interpretation of the observed set of signs is as follows. The ground state of neutral nitrogen ^{14}NI ($^4S_{3/2}$) in cubic site symmetry must exhibit vanishing EFG and A_{zz}. Confinement of nitrogen in C_{70} could lead to a relative increase of the expectation value $\langle r \rangle$ for the electron in the orbital parallel to the principal symmetry axis of C_{70} and a decrease for the "perpendicular" electrons. As a result we expect inverse changes of $\langle 1/r^3 \rangle$. When comparing with data of $^{14}NIII$ ($^2P_{3/2}$) with its singly occupied 2p orbital, sign inversion of A_{zz} and eqQ/h for N in C_{70} is predicted (see Eq (3)), if equal population of the three 2p orbitals persists.

$$A_{zz}, Q_{zz} \propto \left\langle \frac{1}{r^3} \right\rangle_{2p_z} - \sum_{i=x,y} \frac{1}{2} \left\langle \frac{1}{r^3} \right\rangle_{2p_i} \tag{3}$$

For the free ion, *positive* A_{zz} and *negative* eQq were measured [3], consistent with a positive spin density and a negative charge density in the 2p orbital (positive Q and positive magneto-gyric ratio given for ^{14}N). If alternatively a significant change in orbital population would be assumed as a result of lifting the degeneracy of perpendicular and parallel orbitals, an increase of charge and spin density in the preferred parallel orbital would lead to an opposite set of signs. We therefor conclude that the dominant effect of the interaction of the cage with the atomic charge distribution is *deformation* and not *re-population*.

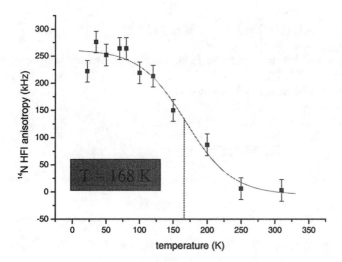

FIGUR 3. Temperature dependence of the ENDOR line axial fit parameter. At 168 K orientational melting of the long axis orientation is observed.

Orientational melting of the long axis orientation leads to a collaps of the powder line shape. Fitting the ENDOR lines with a line shape function corresponding to axial symmetry, a direct measure of remaining anisotropy is obtained. In Fig. 3, the temperature dependence of the axial fit parameter is depicted. The measured melting temperature is close to the value reported earlier from an analysis of μSR data obtained from exohedral myonium adducts of C_{70} [4]. One can conclude that long axis reorientation fast on the EPR time scale occurs well below the reported phase transition temperatures of C_{70}.

Chemical Shielding Detected By EPR

Previous work of Cross, Saunders and coworkers [5] had shown that the capability of the fullerene cage to shield magnetic fields depends on its reduction state. Diamagnetic shielding of C_{60}, for instance, increases by 42.4 ppm in the hexa anion state, whereas the shielding capability of C_{70} is reduced by 37.1 ppm under the same conditions. By using nitrogen instead as probe for the internal field, the following questions could be answered: Does different spatial extend of the probing spin change the value of shielding, and second, is the electronic state of the encapsulated atom "inert" to such an extend that g value shifts are equivalent to nuclear spin chemical shifts.

Experimental results, which are presented in detail in a further contribution [6] gave compelling evidence for inert encapsulation, i. e., g value shifts (-35(1) ppm and +41(1) ppm) were in very good qualitative agreement with the NMR data.

ACKNOWLEDGMENTS

This work was supported by the Deutsche Forschungsgemeinschaft with special grants DFG Di182/19-1, Di182/22-1, Ro2159/1-1, promoting the use of high frequency EPR in Physics, Chemistry and Biology.

REFERENCES

1. T. Almeida Murphy, T. Pawlik, A. Weidinger, M. Höhne, R. Alcala, and J.-M. Spaeth., *Phys. Rev. Lett.*, **77**, 1075 (1996).
2. N. Weiden, H. Käß, and K.-P. Dinse, *J. Phys. Chem.*, B **103**, 9826 (1998).
3. A. Schirrmacher, H. Winter, H. J. Andrä, Y. Ouerdane, J. Désesquelles, G. DoCao, and A. Denis, *J. Physique*, **48**, 905 (1987).
4. T.J. S. Dennis, K. Prassides, E. Roduner, L. Cristofolini, and R. DeRenzi, *J. Phys. Chem.*, **97**, 8663 (1993).
5. E. Shabtai, A. Weitz, R. C. Haddon, R. E. Hoffman, M. Rabinovitz, A. Khong, R. J. Cross, M. Saunders, P.-C. Cheng, and L. T. Scott, *J. Am. Chem. Soc.*, **120**, 6389 (1998).
6. P. Jakes et al., these proceedings

Thermodynamic Properties of N@C$_{60}$

Filip Uhlík*, Zdeněk Slanina¶, and Eiji Ōsawa¶

*Department of Physical and Macromolecular Chemistry, Charles University,
Albertov 6, 128 43 Prague, Czech Republic
¶Laboratories of Computational Chemistry & Fullerene Science,
Department of Knowledge-Based Information Engineering,
Toyohashi University of Technology, Tempaku-cho, Toyohashi, Aichi 441-8580, Japan

Abstract. Contributions to thermodynamic functions U, S, and C_v of the endohedral N@C$_{60}$, resulting from the motion of the N atom inside the cage, are calculated using spherical well approximation with *ab initio* potential. The contribution to C_v has a maximum at temperature around 230 K.

INTRODUCTION

Endohedral complex N@C$_{60}$ can be prepared by bombarding C$_{60}$ with nitrogen ions from a conventional plasma discharge ion source [1, 2]. An atom in a small cavity represents interesting system from thermodynamic point of view, e. g., hydrogen atoms dissolved in palladium [3] or He@C$_{60}$ [4].

At not very high temperatures due to the I_h symmetry of the C$_{60}$ cage the N atom "sees" only an almost spherical potential near the cage center. In this approximation we calculate the energy eigenvalues of the stationary Schrödinger equation, the partition function and the contributions to the thermodynamic functions U, S, and C_v.

THEORETICAL

After separating translation of the center of mass and rotation of the cage and transforming stationary Schrödinger equation into spherical coordinates we obtain [5]

$$-\frac{\hbar^2}{2\mu}\frac{d^2}{dr^2}\Phi(r) + \left(\frac{\hbar^2}{2\mu r^2}L(L+1) + V(r) - \epsilon\right)\Phi(r) = 0, \qquad (1)$$

where μ is the reduced mass of the system (due to the much larger cage mass, μ is essentially the mass of the N atom), $V(r)$ is the potential, $\hbar$ is Planck's constant divided by 2π, $L = 0, 1, 2, \ldots$, $\Phi(r)$ is the radial wave eigenfunction, and ϵ is the $(2L+1)$ times degenerate energy eigenvalue.

Let's define

$$Q_j \equiv \sum_{L=0}^{\infty}\sum_{i=0}^{\infty}(2L+1)\left(\frac{\epsilon_{L,i}}{kT}\right)^j \exp\left(-\frac{\epsilon_{L,i}}{kT}\right), \qquad (2)$$

CP544, *Electronic Properties of Novel Materials—Molecular Nanostructures*, edited by H. Kuzmany, et al.
© 2000 American Institute of Physics 1-56396-973-4/00/$17.00

where k is the Boltzmann constant and T the thermodynamic temperature. The thermodynamic functions can be then conveniently expressed as

$$\frac{U}{kT} = \frac{Q_1}{Q_0}, \quad \frac{S}{k} = \frac{Q_1}{Q_0} + \log Q_0, \quad \text{and} \quad \frac{C_v}{k} = \frac{Q_2}{Q_0} - \frac{Q_1^2}{Q_0^2}. \tag{3}$$

COMPUTATIONAL

The potential was calculated *ab initio* at the UHF/3-21G level using the Gaussian [6] program package. The C_{60} cage was fully optimized and then, with the C atoms held fixed, the energy was calculated for several different positions of the N atom on the lines between the cage center and C atom (V_1) and the cage center and both pentagonal (V_5) and hexagonal (V_6) face centers. The resulting values were interpolated using cubic splines [7] with second derivatives at the end points approximated by the second derivatives of the second order Lagrange interpolation polynomial. Finally, an effective radial potential $V(r)$ was formed as a weighted average

$$V(r) \equiv \frac{60}{92}V_1(r) + \frac{12}{92}V_5(r) + \frac{20}{92}V_6(r). \tag{4}$$

All four potential functions are displayed in Figure 1.

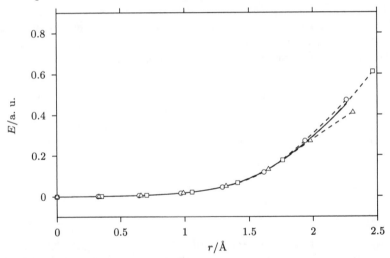

FIGURE 1. Potential energy functions (energy in Hartrees, distance in Ångstroms). Solid line corresponds to effective potential V, dashed ones correspond to potentials V_1 (squares), V_5 (triangles), and V_6 (circles).

Equation 1 was solved on interval $(0, r_{\max})$ using the Numerov $O(h^4)$ method, where h is the space discretization step. This method leads to a generalized unsymmetrical tridiagonal eigenvalue problem. According to [8] a symmetric tridiagonal matrix with the same eigenvalues can be constructed and its eigenvalues can be exactly bracketed. After the bracketing we isolate each eigenvalue by a fail-safe combination of the Newton-Raphson method and bisection [7] and apply "Richardson's deferred approach to the limit" and extrapolate each calculated energy eigenvalue to both $h = 0$ and $r_{max} = \infty$, that is, to the exact solution.

After calculating all eigenvalues such that

$$(2L + 1) \exp \left(-\frac{\epsilon_{L,i}}{kT} \right) < \varepsilon, \tag{5}$$

where ε is a predetermined accuracy, we calculate the sums Q_0, Q_1, and Q_2. Using these sums we finally calculate the contributions to the thermodynamic functions.

RESULTS AND DISCUSSION

Several lowest energy levels for the ^{14}N isotope are shown in Figure 2. The energy levels in one column have the same values of i and $L = 0, 1, 2, \ldots$ according to their increasing energies. The leftmost column represents the complete energy spectrum.

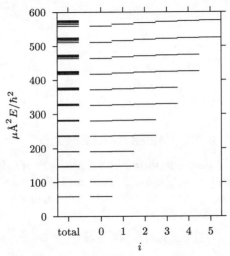

FIGURE 2. Calculated energy levels (energy in units of $\hbar^2/(\mu \text{Å}^2)$).

The temperature dependence of the contributions to the thermodynamic functions for the ^{14}N isotope are shown in Figure 3. The C_v contribution have a temperature maximum around 230 K.

Our treatment produces only approximate contributions to the thermodynamic functions resulting from the motion of the N atom inside the C_{60} cage, not the overall values. A calorimetric experiment would yield the overall values. The reported temperature maximum of the C_v function need not be present in the overall value owing to the compensation by other contributions. It can be supposed, however, that all other contributions to C_v (overall translation, rotation, and vibrations) are essentially identical for an empty and a filled cage. Hence, in order to test our results, a differential calorimetry should be performed, giving the C_v difference for the filled and the empty cage. This approach would, however, require much higher yield of the endohedral form than presently available. Some endohedral metal fullerenes can be prepared with higher yields, but the above treatment cannot be applied to them due to the much stronger interaction of the metal with cage.

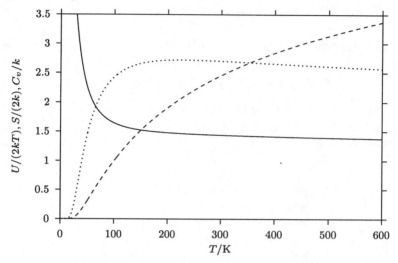

FIGURE 3. The temperature dependence of $U/(2kT)$ (solid line), $S/(2k)$ (dashed line), and C_v/k (dotted line).

ACKNOWLEDGMENTS

The research has in part been supported by the Ministry of Education, Science and Culture in Japan and by the Japan Science and Technology Corporation. Authors thank The Supercomputer Centers of Charles University, Prague, and Masaryk University, Brno, for computer time.

REFERENCES

1 Murphy, T. A., Pawlik, Th., Weidinger, A., Höhne, M., Alcala, R., and Spaeth, J.-M., *Phys. Rev. Lett.* **77**, 1075–1078 (1996).
2 Knapp, C., Dinse, K.-P., Pietzak, B., Waiblinger, M., and Weidinger, A., *Chem. Phys. Lett.* **272**, 433–437 (1997).
3 Slanina, Z., *Thermochim. Acta* **156**, 285–290 (1989).
4 Slanina, Z., Uhlík, F., and Ōsawa, E. in *Recent Advances in The Chemistry and Physics of Fullerenes and Related Materials*, volume 7, edited by Kamat, P. V., Guldi, D. M., and Kadish, K. M., The Electrochemical Society, Inc., Pennington, 1999, pp. 816–821.
5 Flügge, S., *Practical Quantum Mechanics*, second edition, Springer-Verlag, 1974, p. 144.
6 Gaussian 98, Revision A.7, Gaussian, Inc., Pittsburgh PA, 1998.
7 Press, W. H., Teukolsky, S. A., Vetterling, W. T., and Flannery, B. P., *Numerical Recipes in C*, second edition, Cambridge University Press, 1994, pp. 113, 362.
8 Lindberg, B., *J. Chem. Phys.* **88**, 3805–3810 (1988).

The Exchange Coupling between the N Atom and the Fullerene Cage in N@C$_{60}^{n-}$

L. Udvardi

Dept. of Theoretical Physics, Institute of Physics, Technical University of Budapest,
H-1521 Budapest, Budafoki út 8, Hungary

Abstract. The electronic structure of N@C$_{60}$, N@C$_{60}^{-1}$, N@C$_{60}^{-3}$, N@C$_{60}^{-6}$ have been calculated at MCSCF level. In all cases considered the extra electrons occupy the LUMO of the C$_{60}$ molecule according to the Hund's rule and the nitrogen keeps its atomic state. The coupling between the electrons of the cage and the nitrogen is estimated from the difference of the energies of the system with different multiplicity in the case of the paramagnetic ions. The calculations resulted ferromagnetic coupling which is strong enough to make the ESR signal of the nitrogen undetectable.

I INTRODUCTION

Since the discovery of the La@C$_{60}$ a great deal of experimental effort has been invested into the research of the endohedral fullerenes. The N@C$_{60}$ was first reported in 1996 by Almeida Murphy *et al.* [1]. ESR experiments [1–3] have been showed that surprisingly the nitrogen keeps its atomic quartet state inside the cage and it remains intact even if the fullerene takes part in a chemical reaction [4,5]. This property makes the encapsulated nitrogen good ESR signaling agent in the processes involving the C$_{60}$ molecule. It is natural to apply the endohedral nitrogen to follow the changes in the structure of the systems during the charging of the molecule. However, the ESR signal of the nitrogen in electro-chemically charged N@C$_{60}$ anions [6] and Rb doped C$_{60}$ [7] could be detected only in the case of diamagnetic cage: N@C$_{60}$, N@C$_{60}^{-6}$.

In our study the electronic structure of the N@C$_{60}^{-n}$: n=0,1,3,6 were calculated using multi-configurational self-consistent field method (MCSCF). The coupling between the electrons of the nitrogen and the electrons of the fullerene molecule was described by means of a Heisenberg type Hamiltonian. The coupling constant was extracted from the results of the MCSCF calculations on the different spin states of the complex.

CP544, *Electronic Properties of Novel Materials—Molecular Nanostructures*, edited by H. Kuzmany, et al.
© 2000 American Institute of Physics 1-56396-973-4/00/$17.00

II   RESULTS AND DISCUSSIONS

The geometry of the systems were calculated by applying restricted open–shell Hartree–Fock (ROHF) method using Dunning's double zeta basis [8]. In all the systems we considered the extra electrons were accepted by the LUMO of the C_{60} molecule and the lowest energy belonged to the state with the highest multiplicity, according to the Hund's rule. In the case of the $N@C_{60}^{-1}$ this is a triply degenerate $^3T_{1u}$ state and the system undergoes a Jahn–Teller distortion. The distorted system has d_{2h} symmetry. The details of the result of the geometry optimization are summarized in Table 1. The nitrogen remains intact in the anions and its valence orbitals are very similar to the atomic ones. The weak interaction between the systems implies the introduction of a simple Heisenberg type model in order to describe the coupling between the electrons of the nitrogen and the electrons of the fullerene cage:

$$H = H_{C_{60}} + H_N + J\vec{S}_{C_{60}}\vec{S}_N \qquad (1)$$

where H_N and $H_{C_{60}}$ denote the Hamilton operator of the nitrogen and the C_{60} and $S_{C_{60}(N)} = \sum_{j,\alpha,\alpha'} a_{j\alpha}^+ \vec{\sigma}_{\alpha\alpha'} a_{j\alpha'}$ where $a_{j\alpha}^+$, $a_{j\alpha}$ creates and annihilates a state on the C_{60} (N), respectively. This form of the Hamiltonian has the same symmetry as the original Hamiltonian has and its eigen vectors are also eigen states of the square of the spin of the complex.

N@C_{60}	N@C_{60}^{-1}
I_h	D_{2h}
a = 1.4561	a = 1.4540
b = 1.3801	b = 1.3790
	c = 1.3787
N@C_{60}^{-3}	d = 1.4539
I_h	e = 1.4541
a = 1.4495	f = 1.4539
b = 1.4032	g = 1.3787
	h = 1.4539
N@C_{60}^{-6}	i = 1.4540
I_h	j = 1.3787
a = 1.4472	k = 1.3788
b = 1.4292	l = 1.4539
	m = 1.4540

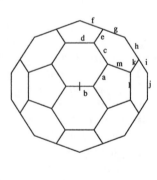

Table 1. Bond lengths in the different N@C_{60} anions resulted by ROHF calculations. The lengths are given in Å.

In the lack of the coupling the energy of the singly ionized anion in the state S = 2 : $|\Psi_{S=2}>=|\uparrow\uparrow\uparrow,\uparrow>$ and in the state S = 1:

$|\Psi_{S=1}>= \frac{1}{\sqrt{2}}(|\uparrow\uparrow\uparrow,\downarrow> +|\downarrow\downarrow\downarrow,\uparrow>)$ would be the same. The arrows represent the occupation of the p–orbitals of the nitrogen and the occupation of the LUMO of the C_{60}, respectively. Due to the interaction the degeneracy of the two states will be resolved and the splitting will be proportional to the J exchange coupling.

$$E_{S=2}^{-1} - E_{S=1}^{-1} \tag{2}$$

$$= J <\uparrow\uparrow\uparrow, \uparrow | \vec{S}_{C_{60}} \vec{S}_N | \uparrow\uparrow\uparrow, \uparrow> - $$
$$J \frac{1}{2} (<\uparrow\uparrow\uparrow, \downarrow | + <\downarrow\downarrow\downarrow, \uparrow>) \vec{S}_{C_{60}} \vec{S}_N (| \uparrow\uparrow\uparrow, \downarrow> + | \downarrow\downarrow\downarrow, \uparrow>) = \frac{3}{2} J \tag{3}$$

In order to determine the energy of the different spin states MCSCF calculations have been done with the complete active space of $N(p_x \ p_y \ p_z) + C_{60}(LUMO \ t_{1u})$ orbitals. Since the p–orbitals of the nitrogen have no density at the nucleus it is clear that the inclusion of the 2s,3s orbitals of the nitrogen in the active space would be necessary if other properties, like the hyper fine coupling tensor, were calculated. However, the extension of the active space implies the appearing of such excitations in the wave-functions for which the expectation values of $H_{C_{60}}$ and H_N are not canceled in Eq. 2.

In the case of the triply ionized anion the ground state is a S = 3 heptet state: $| \uparrow\uparrow\uparrow, \uparrow\uparrow\uparrow>$. In the case of the present active space the ROHF and the MCSCF methods supply the same ground state. The purest state with lower multiplicity is turned out to be the S=1, S_z=0 state with the largest CI coefficients $\alpha = 0.6700612$: $|\Psi_{S=1}> = \alpha (| \uparrow\uparrow\uparrow, \downarrow\downarrow\downarrow> - | \downarrow\downarrow\downarrow, \uparrow\uparrow\uparrow>)$ All the other CI coefficients were neglected and α was substituted by $\frac{1}{\sqrt{2}}$ in order to keep the wave-function normalized. The exchange coupling can be derived from the difference of the energy of the two states as:

$$E_{S=3}^{-3} - E_{S=1}^{-3} \tag{4}$$

$$= J <\uparrow\uparrow\uparrow, \uparrow\uparrow\uparrow | \vec{S}_{C_{60}} \vec{S}_N | \uparrow\uparrow\uparrow, \uparrow\uparrow\uparrow> - $$
$$J \frac{1}{2} (<\uparrow\uparrow\uparrow, \downarrow\downarrow\downarrow | - <\downarrow\downarrow\downarrow, \uparrow\uparrow\uparrow>) \vec{S}_{C_{60}} \vec{S}_N (| \uparrow\uparrow\uparrow, \downarrow\downarrow\downarrow> - | \downarrow\downarrow\downarrow, \uparrow\uparrow\uparrow>) = -9J \tag{5}$$

The energy of the different states and the exchange coupling constants are summarized in Table 2.

	E_0	ΔE	J	J/μ_B
N@C_{60}^{-1}	-2325.525633 a.u	-3.45 meV	-2.30 meV	-39.7 T
N@C_{60}^{-3}	-2325.302800 a.u.	-5.19 meV	-0.58 meV	-10.0 T

Table 2. Energies of the different states of the N@C_{60} anions and the exchange couplings.

In both cases the calculations resulted ferromagnetic coupling between the electrons of the nitrogen and the electrons of the charged fullerene cage. The coupling is rather strong and it is comparable to the exchange coupling in some ferromagnetic systems [9]. In the case of paramagnetic fullerene cages the ESR spectrum is dominated by the broad lines of the valence electrons of the C_{60}. Since the concentration of the encapsulated nitrogen is very shallow its ESR signal can be detected only in the case if the broadening of the lines due to the spin–lattice relaxation is small. Our conclusion is that the large coupling we got between the nitrogen and the fullerene molecule inhibits the detection of the ESR signal of the endohedral complex with partially filled LUMO of the

C_{60}. Our assumption is confirmed by experiments on electro-chemically produced radical anions by Jakes et al. [6]. They found the disappearance of the nitrogen signal as the charge of the complex was increased. The typical three line of the nitrogen appeared again only if the LUMO of the C_{60} was completely filled forming a six fold ionized anion. Similar results has been found by Janossy in the case of alkali doped C_{60} [7]. The endohedral nitrogen could be detected only in the diamagnetic Rb_6C_{60}.

This research was supported in part by the Hungarian National Research Fund (OTKA) under Grants No. T24137.

REFERENCES

1. Almeida Murphy, T. Pawlik, A. Weidinger, M. Höhne, R. Alcala, J.M. Spaeth, Phys. Rev. Lett. **77**, 1075 (1996)
2. C. Knapp, K.-P. Dinse, B. Pietzak, M. Waiblinger, A. Weidinger, Chem. Phys. Lett. **272** (1997) 433-437
3. B. Pietzak, M. Waiblinger, T. Almeida Murphy, A. Weidinger, M. Höhne, E. Dietel, A. Hirsch, Chem. Phys. Lett. **279** (1997) 259-263
4. B. Pietzak, M. Waiblinger, T. Almeida Murphy, A. Weidinger, M. Höhne, E. Dietel, A. Hirsch, Molecular Nanostructures; eds H. Kuzmany, J. Fink, M. Mehring and S. Roth, (World Scientific, Singapore, 1998), 180-183
5. E. Dietel, A. Hirsch, B. Pietzak, M. Waiblinger, K. Lips, A. Weidinger, A. Gru, K.-P. Dinse, J. Am. Chem. Soc. **121** (1999)
6. P. Jakes B. Goedde M Waiblinger K.-P. Dinse, this issue
7. A. Janossy, S. Pekker, L. Korecz, F. Simon, L. Forró, this issue
8. T.H.Dunning, Jr., P.J.Hay Chapter 1 in "Methods of Electronic Structure Theory", H.F.Shaefer III, Ed. Plenum Press, N.Y. 1977, pp 1-27.
9. J.A.Hofmann, A. Paskin, K.J. Tauer, R.S. Weiss, J.Phys. Chem. of Solids 1 45 (1956)

Electron Spin Relaxation Rates T_1^{-1} and T_2^{-1} in Diluted Solid N@C$_{60}$

S. Knorr*, A. Grupp*, M. Mehring*, M. Waiblinger[†], and
A. Weidinger[†]

*2. Physikalisches Institut, Universität Stuttgart, Pfaffenwaldring 57, 70550 Stuttgart, Germany
[†]Hahn-Meitner-Institut Berlin, Glienicker Straße 100, 14109 Berlin, Germany

Abstract. We report on pulsed electron-spin-resonance (ESR) investigations of endo-hedral N@C$_{60}$. The measurements were performed in X band (9.5 GHz) and W band (94 GHz). Applying a two-pulse sequence with a variable flip angle of the second pulse, we could not only separate the contributions of the central and the satellite transitions, but also monitor the phase transition in C$_{60}$ at $T \approx 250$ K. Analysing the temperature dependence of the longitudinal relaxation rate T_1^{-1}, the hyperfine relaxation, which is caused by the damped quantum oscillator of the nitrogen atom inside C$_{60}$, was found to play the most important role. From a comparison of the experimental data with our relaxation model, the oscillator energy was determined to be $E_{osc} \approx 11(2)$ meV.

INTRODUCTION

Nitrogen inside the C$_{60}$ fullerene (N@C$_{60}$) is almost freely suspended at the centre of the fullerene, and the whole compound is stable at room temperature [1]. The nearly perfect shielding of the encaged N atom from the outside keeps the electronic spin with $S = 3/2$. Depending on the nitrogen isotope, hyperfine coupling to the nuclear spin of $I = 1$ (^{14}N) or $I = 1/2$ (^{15}N) must be taken into account. In ^{14}N@C$_{60}$, we find at room temperature a three-line ESR spectrum with linewidths of $\Delta B_{pp} = 0.08$ G (X band) and 0.15 G (W band) respectively. A detailed analysis of the line shape exhibits a second broader contribution with $\Delta B_{pp} = 0.40$ G which arises from a non-vanishing zero-field splitting due to lattice disorder.

EXPERIMENTAL

In order to separate the contributions of the central transition $m_S = +1/2 \leftrightarrow m_S' = -1/2$ (E_M) and of the satellite transitions $m_S = \pm 3/2 \leftrightarrow m_S' = \pm 1/2$ (E_Q) to the total echo amplitude, we applied a $\frac{\pi}{2}$-τ-β pulse sequence, where β is the rotation angle of the second rf pulse [2]. These experiments were performed in X band. In addition, we determined the temperature dependence of the longitudinal

CP544, *Electronic Properties of Novel Materials—Molecular Nanostructures*, edited by H. Kuzmany, et al.
© 2000 American Institute of Physics 1-56396-973-4/00/$17.00

and transverse relaxation rates in W band with standard Hahn-echo and inversion-recovery pulse sequences.

RESULTS AND DISCUSSION

The $\frac{\pi}{2}$-τ-β experiment

In Fig. 1, the theoretically expected E_M and E_Q contributions to the echo amplitude are displayed on the left side. The maximum echo amplitude is at $\beta \approx 65°$. On the right side of Fig. 1, experimental data at two selected temperatures below and above the phase transition are shown. This phase transition was first characterized by Tycko *et al.* with ^{13}C NMR measurements [3]. Tycko concluded that above the transition, the molecules undergo continuous rotational diffusion, whereas below the transition, they can only jump between symmetry-equivalent orientations.

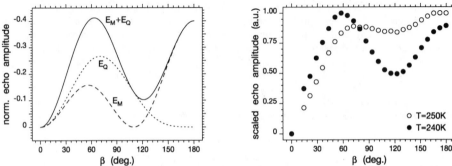

FIGURE 1. Theoretically expected E_M and E_Q contributions to the echo amplitude (left) and experimental results in X band (right). The phase transition in C_{60} at 250 K shows up as a flattening of the β-dependence caused by the molecular rotations above 250 K.

At $T \leq 240$ K, i.e., below the phase transition, we find that the measurements agree very well with the theoretical curve (see Fig. 1). This implies that the icosahedral symmetry of the C_{60} is lowered due to cage deformations leading to separate E_Q and E_M transitions. At $T \geq 250$ K, the peak at $\beta \approx 65°$ tends to vanish due to the rotational averaging of the molecular distortions. In this case, the β-dependence approaches the $\sin^2(\beta/2)$ behaviour expected for an undistorted spin $S = 3/2$ system.

Relaxation rates T_1^{-1} and T_2^{-1} in W band

Investigating the longitudinal relaxation rate T_1^{-1} in W band, we find a pronounced increase with temperature as can be seen in Fig. 2, left. In principle, there are two processes causing longitudinal relaxation: hyperfine and fine-structure relaxation. Hyperfine relaxation is caused by the damped quantum oscillator of the

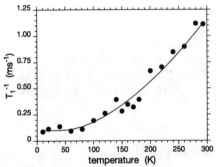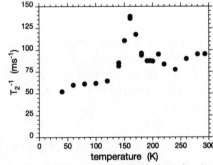

FIGURE 2. Temperature dependence of the longitudinal (left) and transverse (right) relaxation rates. Measurements were performed in W band with the standard inversion-recovery and Hahn-echo pulse sequences respectively.

nitrogen atom inside C_{60}, whereas fine-structure relaxation originates from intermolecular strain which fluctuates due to molecular rotation and phonon scattering.

Analysing the temperature dependence, we conclude that hyperfine relaxation plays the dominant role. The corresponding characteristic relaxation rate R_{1a} can be expressed as

$$R_{1a} = \Delta a^2 \left[\coth\left(\frac{\hbar\omega_0}{k_BT}\right) \right]^\alpha \text{Im}\{\chi(\omega_{0S})\}. \tag{1}$$

Details will be published elsewhere. Here, Δa^2 represents the mean square amplitude of the hyperfine fluctuation, ω_0 is the vibrational frequency of the N atom, $\text{Im}\{\chi(\omega_{0S})\}$ is the imaginary part of the dynamic susceptibility at the Larmor frequency of the electron spin, and α is an adjustable parameter to model the change in hyperfine interaction with the displacement of the N atom. The latter manifests itself as an increase of a_{iso} with temperature [4].

The fit to the experimental data (dashed line in Fig. 2, left) yields a vibrational frequency of $\omega_0/(2\pi) \approx 2.4(4) \times 10^{12}$ Hz corresponding to an oscillator energy of $E_{\text{osc}} \approx 11(2)$ meV. Our interpretation is supported by a comparison with preliminary ^{15}N@C_{60} data. The gyromagnetic ratio of the ^{15}N nuclear spin is larger by a factor of 1.40 which leads to an increase of the isotropic hyperfine coupling constant by the same factor. Taking Eq. (1) into account, we expect the T_1^{-1} rates to increase by a factor of about 2 (according to $^{15}a_{\text{iso}}^2/^{14}a_{\text{iso}}^2$) which is in agreement with our experimental findings.

Recently, Uhlík _et al._ calculated the potential of the N atom inside the cage and also the lowest energy levels of the oscillator [5]. Although the potential turns out to be anharmonic, the energy levels and their degree of degeneracy resemble very much those of the harmonic oscillator. In this calculation, the first excited state is found to be separated from the ground state by $\Delta E \approx 13$ meV in accordance with our results.

The transverse relaxation rates are only weakly temperature dependent and increase slowly with rising temperature (see Fig. 2, right). However, we find a rather sharp maximum at $T = 160$ K. We ascribe this maximum in T_2^{-1} to a resonance between the molecular reorientation jump frequency of the C_{60} cage and the size of the zero-field splitting. From ^{13}C NMR data and μSR measurements, the molecular orientational correlation rate τ_c^{-1} was determined to be $\tau_c^{-1} = 16 \times 10^6$ s^{-1} at $T = 200$ K [3] and $\tau_c^{-1} = 8 \times 10^6$ s^{-1} at $T = 160$ K [6], which is in good agreement with the measured residual linewidth due to zero-field splitting of $\omega_{zfs} \approx 6.9 \times 10^6$ s^{-1}.

ACKNOWLEDGMENTS

We acknowledge partial support of this project by the Deutsche Forschungs-gemeinschaft (Schwerpunkt "Hochfeld-EPR in Biologie, Chemie und Physik") and the Fonds der Chemischen Industrie. We are grateful to B. Gödde (TU Darmstadt) for the HPLC purification of the sample.

REFERENCES

1. Almeida Murphy, T., Pawlik, T., Weidinger, A., Höhne, M., Alcala, R., and Spaeth, J. M., *Phys. Rev. Lett.* **77**, 1075–1078 (1996).
2. Mehring, M., and Kanert, O., *Z. Naturforsch.* **24a**, 768–774 (1969).
3. Tycko, R., Dabbagh, G., Fleming, R. M., Haddon, R. C., Makhija, A. V., and Zahurak, S. M., *Phys. Rev. Lett.* **67**, 1886–1889 (1991).
4. Grupp, A., Pietzak, B., Waiblinger, M., Almeida Murphy, T., Weidinger, A., and Roduner, E., "Vibration of Atomic Nitrogen Inside C_{60}," in *Molecular Nanostructures*; edited by Kuzmany, H., Fink, J., Mehring, M., and Roth, S., World Scientific, Singapore, 1998, pp. 224–226.
5. Uhlík, F., Slanina, Z., and Ōsawa, E., "Thermodynamic Properties of N@C_{60}," in this volume.
6. Kiefl, R. F., Schneider, J. W., MacFarlane, A., Chow, K., Duty, T. L., Estle, T. L., Hitti, B., Lichti, R. L., Ansaldo, E. J., Schwab, C., Percival, P. W., Wei, G., Wlodek, S., Kojima, K., Romanow, W. J., McCauley, Jr., J. P., Coustel, N., Fischer, J. E., and Smith, III, A. B., *Phys. Rev. Lett.* **68**, 1347–1350 (1992).

Magnetic Interaction in Diluted N@C$_{60}$

M. Waiblinger[*], B. Goedde[**], K. Lips[***], W. Harneit[*], P. Jakes[**],
A. Weidinger[*], K.-P. Dinse[**]

* Hahn-Meitner-Institut Berlin, Glienicker Straße 100, 14109 Berlin, Germany
**Phys. Chemie III, TU Darmstadt, Peterstrenstr. 20, 64287 Darmstadt, Germany
*** Hahn-Meitner-Institut Berlin, Kekulestr. 5, 12489 Berlin, Germany

Abstract. Nitrogen in C$_{60}$ is a paramagnetic atom with a magnetic moment corresponding to the S = 3/2 electronic spin. The dipolar interaction of two N@C$_{60}$ at closest distance, i.e. 1 nm apart, reaches a maximum value of 1.85 mT or 52 MHz, more than three times larger than the hyperfine interaction with the nuclear spin. For diluted N@C$_{60}$ in a C$_{60}$ matrix, the distribution of interactions leads to a line broadening in EPR experiments. We report here on a EPR line width measurement for N@C$_{60}$ in C$_{60}$ with concentrations varying by a factor of 1000. For the first time a concentration in the percent range was reached by repeated HPLC enrichment. A surprising and completely unexpected result is that sharp lines appear superimposed on the broad lines. We attribute the sharp line to motional narrowing of N@C$_{60}$ diffusing on the surface of C$_{60}$ grains. It can be assumed that the diffusion seen here is a general phenomenon of C$_{60}$, the doped fullerene being just the indicator.

INTRODUCTION

Nitrogen encapsulated in C$_{60}$ is an almost free atom with a magnetic moment corresponding to the S = 3/2 ground state of atomic nitrogen with g ≈ 2 [1-3]. The as-produced samples are highly diluted and have usually concentrations of N@C$_{60}$ in C$_{60}$ in the order of 10^{-4}. In this situation the magnetic moments are so far apart that the mutual interaction can be neglected. However, if N@C$_{60}$ is enriched the magnetic moments come to a distance where the interaction becomes appreciable. This leads to a dipolar broadening of the electron paramagnetic resonance (EPR) lines with a predictable increase of the line width with concentration.

C$_{60}$ together with N@C$_{60}$ condenses in the fcc structure. The maximum magnetic dipolar interaction at closed distance (1nm) is 1.85 mT or 52 MHz in EPR. In a diluted polycrystalline sample both the distance and orientation are statistically distributed leading to a broadening of the EPR lines. This broadening has been studied in this work as a function of the concentration of N@C$_{60}$ in C$_{60}$.

CP544, *Electronic Properties of Novel Materials—Molecular Nanostructures*, edited by H. Kuzmany, et al.
© 2000 American Institute of Physics 1-56396-973-4/00/$17.00

EXPERIMENTAL DETAILS AND RESULTS

N@C$_{60}$ was produced by ion implantation with the method described in Ref. 1. The as-prepared sample had a concentration of N@C$_{60}$ in C$_{60}$ in the order of 10^{-4}. The concentration was determined by comparing the EPR intensity of the sample in solution with that of a calibration standard (weak pitch). The total amount of N@C$_{60}$ plus C$_{60}$ in the sample was determined by high performance liquid chromatography (HPLC). Due to known problems in absolute spin number calibrations, the absolute concentrations are rather uncertain but the relative concentrations of different enriched samples are already accurate.

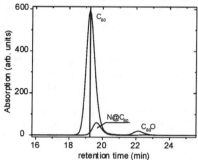

FIGURE 1. HPLC retention diagram for a C$_{60}$ sample. For comparison the calculated N@C$_{60}$ diagram is shown for a concentration of 10 % and the experimentally known retention time differences between C$_{60}$ and N@C$_{60}$ [4].

C$_{60}$ and N@C$_{60}$ are chemically very similar and therefore elute in chromatography at only slightly different retention times as shown in Fig. 1 (HPLC column: Cosmosil Buckyprep / 5PYE) [4]. This effect can be used to separate N@C$_{60}$ from C$_{60}$ by repeated fractionation.

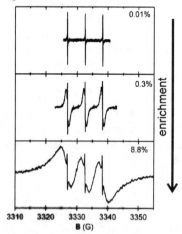

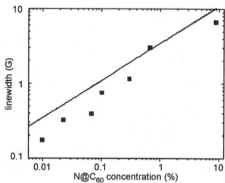

FIGURE 3. Linewidth measured by EPR for samples with different concentrations. The straight line shows the predicted linewidth calculated according to equation 2.

FIGURE 2. Powder EPR spectra of N@C$_{60}$ for different concentrations. Note the increase of the broadening of the lines with increasing concentration. The sharp lines are attributed to motional narrowing by N@C$_{60}$ diffusing on crystallite surfaces.

The separation was achieved by cutting the C_{60} peak in half and collecting the second part for further treatment. Thus, in each step the enrichment was approximately a factor of two. The loss of $N@C_{60}$ per step was less than 5%. In the present case 12 enrichment steps were performed and a concentration enrichment by approximately a factor of 1000 was obtained (the actual enrichment factor per step was somewhat less than 2). The concentrations of $N@C_{60}$ and C_{60} after each step were determined by EPR and HPLC as described above.

Representative powder EPR spectra are shown in Fig. 2. Clearly the broadening of the lines with increasing concentrations is seen. The experimental line width was determined by assuming a Gaussian line-shape. The results are shown in Fig. 3.

The sharp lines superimposed on the broad lines in Fig. 2 are attributed to a small fraction (≈ 0.1 %, estimated from integrated intensities), of rapidly diffusing $N@C_{60}$, thereby averaging out the dipolar field distribution by motional narrowing. We assume, that this diffusion occurs at the surface of the crystallites of the powder sample.

DISCUSSION

Figure 4 shows schematically two adjacent $N@C_{60}$ molecules with a specific spin of the enclosed nitrogen atoms.

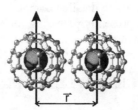

FIGURE 4. Schematic view of two nearest $N@C_{60}$ neighbors.

For a diluted system, many different configurations with different separation and orientation occur. In the second moment approximation, an effective magnetic field distribution at each spin from the surrounding spins is predicted [5] given by second moment σ^2,

$$\sigma^2 = \frac{3}{4} \frac{\gamma_e^2 \hbar^2 \mu_0^2}{16\pi^2} S(S+1) \sum_j \frac{(1-3\cos^2 \vartheta_j)^2}{r_j^6} c, \text{ for like spins.} \quad (1)$$

The summation is over all sites of the fcc lattice and the concentration c gives the probability that these sites are occupied. γ_e is the electron gyromagnetic ratio, in the present context it is sufficient to use that of the free electron, μ_0 is the vacuum permeability and $S = 3/2$ the spin of the system. Averaging out the angular dependant part for polycrystalline samples gives 4/5 and the sum over $1/r_j^6$ is for the fcc lattice is

14.45 in units of the nearest neighbor distance which in the case of C_{60} is 1 nm. One may argue that the contribution from $N@C_{60}$ molecules in the first shell leads to a completely different EPR frequency and therefore can be left out in the summation for not too high concentrations. Then the $1/r^6$ term reduces to 2.45. Due to the hyperfine splitting only 1/3 of the spins are like and therefore a modified pre-factor in (1) with a value between ¾ for like spins and 1/3 for unlike spins must be taken. This finally leads to

$$\sigma = 34.4\,G \times \sqrt{c} \qquad (2)$$

The straight line in Fig. 3 shows the square root of c dependence of the predicted line width. The predicted absolute value is somewhat too large compared to the experiment, but considering the crudeness of the model and the experimental uncertainty of the concentrations the agreement is acceptable.

CONCLUSIONS

Enrichment of $N@C_{60}$ leads to a broadening of the EPR lines as expected from the dipolar interaction of the magnetic moments. The concentration dependence and the magnitude of the line broadening are in fair agreement with a simple model based on the momentum method. For a more detailed comparison of the linewidth and also the lineshape with theory we are presently preparing Monte Carlo calculations, which do not have the limitations of the momentum method.

ACKNOWLEDGMENTS

This project was financially supported by the Deutsche Forschungsgemeinschaft (DFG)

REFERENCES

1. T. Almeida Murphy, T. Pawlik, A. Weidinger, M. Höhne, R. Alcala, J.M. Spaeth, *Phys. Rev. Lett.* **77**, 1075 (1996)
2. B. Pietzak, M. Waiblinger, T. Almeida Murphy, A. Weidinger, M. Höhne, E. Dietel, A. Hirsch, *Chem. Phys. Lett.* **279** (1997) 259
3. C. Knapp, K.-P. Dinse, B. Pietzak, M. Waiblinger, A. Weidinger, *Chem. Phys. Lett.* **272** (1997)
4. A. Weidinger, M. Waiblinger, B. Pietzak, T.A. Murphy, *Appl. Phys. A,* **66: (3)** 287 (1998)
5. A. Abragam, The Principles of Nuclear Magnetism, Oxford University Press (1961)

Electron spin resonance of $N@C_{60}^{6-}$ in the fulleride salt Rb_6C_{60}

A. Jánossy[1], S. Pekker[2], F. Fülöp[1], F. Simon[1], G. Oszlányi[2]

[1]Technical University of Budapest, Institut of Physics H-1521 Budapest POBox 91 Hungary
[2]Research Institut for Solid State Physics and Optics, H-1525 Budapest, Hungary

Abstract. The use of $N@C_{60}$ as a spin probe is demonstrated in the ionic fulleride salt Rb_6C_{60}. The salt contains a few ppm of $N@C_{60}$ and was synthesised using a low temperature reaction. The electron spin resonance of $N@C_{60}^{6-}$ of this compound at 225 GHz (g=2 at 8.1 T) shows a small diamagnetic shift of 88±15 ppm with respect to pure $N@C_{60}$.

INTRODUCTION

N atoms encapsulated into fullerenes has a great potential for magnetic probes in Electron Spin Resonance (ESR) spectroscopy[1]. Since the N atoms are only weakly interacting with the surrounding cage, ESR lines are narrow. Recently, it has been shown that the endohedral N atom can be a sensitive indicator of molecular distortions[2]. In this paper we show that $N@C_{60}$ may be incorporated into fulleride salts. We successfully synthesised Rb_6C_{60} starting from dilute $N@C_{60}:C_{60}$. The observation of the $N@C_{60}^{6-}$ ESR proves that the endohedral compound survives ionization to the hexaion with the N atom remaining at the center of the cage.

EXPERIMENTAL

Endohedral $N@C_{60}$ was produced in an electric discharge tube following the method of Pietzak et al.[3]. The typical concentration of $N@C_{60}$ in C_{60} after purification is between 10^{-6} and 10^{-5} as measured by ESR at X band. During the discharge a nitrogen gas pressure of 1 mbar was maintained and typically a current of 0.2 mA was applied between electrodes placed about 10 cm apart. The tube is heated above 500 °C and the C_{60} powder placed at the bottom is continuously sublimated into the discharge. The endohedral $N@C_{60}$ is produced in the gas and is collected at a water cooled surface. C_{60} is heavily damaged at that part of the cathode where discharge is intense. However, damage to C_{60} is small at cooled surfaces a few centimeters away from the visible discharge and here most of the C_{60} is recovered after purification. We showed that it is not neccessary to collect the product at the cathode; similar N concentrations are obtained at other cooled surfaces. N ions were not implanted in significant amounts into C_{60} which has been condensed onto the surface before the discharge was turned on. This also shows that $N@C_{60}$ is predominantly formed in the gas.

CP544, *Electronic Properties of Novel Materials—Molecular Nanostructures*, edited by H. Kuzmany, et al.

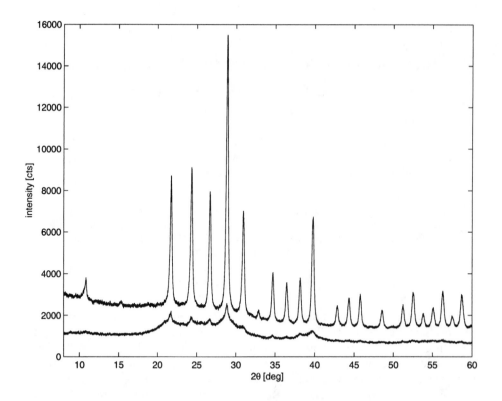

Figure 1. X-ray diffraction spectra of Rb_6C_{60} produced at low temperature. Lower trace: after synthesis and drying at 120 °C. Upper trace: after anneal at 300 °C.

Since N atoms escape the cage at elevated temperatures, a low temperature method was devised to produce the Rb_6C_{60} salt doped with $N@C_{60}$. 0.1 mmol of $N@C_{60}$ doped C_{60} was dissolved in 70 ml of toluene (purified by trap to trap distillation from liquid Na/K alloy). Stoichiometric amount of Rb was added to the solution in a reaction tube and the mixture was sealed in vacuum. The reaction was carried out at 55 °C with sonication for 1 hour. The solution became colorless after sedimentation of the black precipitate formed. The material was filtered in a glove box and dried in dynamic vacuum at 120 °C for 4 hours.

X ray diffraction verified that the stoichometry of the material was indeed Rb_6C_{60} (Figure 1). The spectra were recorded with a Huber G670 Image Plate Guinier camera and Cu Kα1 radiation. The crystallinity of the as received material was poor, weak and broad diffraction lines of Rb_6C_{60} were sitting on an amorphous background. An overnight annealing at 300 °C dramatically improved crystallinity. Rietveld analysis proved that the stoichometry corresponded well to the chemical formula and no trace of a minority phase was present.

The electron spin resonance was observed in the poorly crystalline material at 225.00 GHz between 2 and 25 K. The spectra were taken with the superconducting magnet in the persistent mode and the millimeter wave Gunn diode oscillator locked to a quartz 100 MHz reference oscillator. To sweep the field a current was added to the modulation coil wound around the sample. Although the pure $N@C_{60}$ and $N@C_{60}^{6-}$ spectra were not taken simultaneously, the magnet remained in the persistent mode while the samples were exchanged to ensure precision in determining the g-factor.

RESULTS AND DISCUSSION

Figure 2 shows a spectrum at 10 K of $N@C_{60}^{6-}$ in the fulleride salt Rb_6C_{60} compared to $N@C_{60}:C_{60}$. As expected, the N hyperfine splittings of the neutral and charged systems are equal (0.56 mT) within the experimental accuracy of a few %. The intrinsic linewidths of the pure system are very narrow and the ESR is distorted by saturation at 10 K in spite of a mm wave power less than a milliwatt and that there is no cavity. Under the same conditions the $N@C_{60}^{6-}$ ESR in Rb_6C_{60} is not saturated. The N lines of the doped Rb_6C_{60} system are relatively broad, about 0.2 mT and independent of temperature. The ESR of the N atom in the hexaion is slightly shifted to higher fields by 0.7 mT at 8 T thus the relative shift is $\Delta B/B_0 = 88\pm15$ ppm with

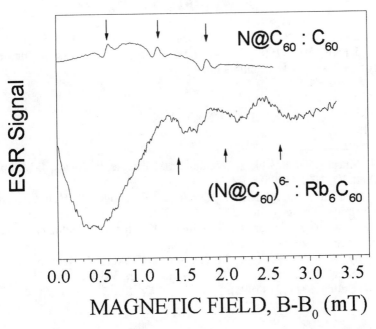

FIGURE 2. Comparison of the ESR spectra of $N@C_{60}$ and $N@C_{60}^{6-}$ at temperature of 10 K and exciting frequency 225 GHz. B_0=8.07 T.

respect to the neutral N@C_{60}. This corresponds to a diamagnetic shielding of the external magnetic field at the N atom by the sixfold charged cage with respect to the shielding due to the neutral cage. The temperature independent linewidth is explained by dipolar hyperfine coupling of the $S=3/2$ nitrogen electrons to ^{85}Rb and ^{87}Rb nuclei. We find for the second moment of the linewidth $M_2 = 0.22$ mT using the Van Vleck formula and the known structure[4] of Rb_6C_{60}. The 24 first Rb neighbours are responsible for most of the linewidth.

CONCLUSIONS

Since N atoms encapsulated in neutral C_{60} cages and in C_{60}^{6-} fulleride hexaions are stable and remain at the center of the fullerene molecule it is almost certain that N@C_{60}^{n-} is stable in solids for all n between 0 and 6. Thus N encapsulated C_{60} may be an attractive ESR probe in both neutral and ionic fullerenes to measure distortions, molecular motion and internal fields. For example it may be interesting to study the A_4C_{60} compounds which are close to a metallic state. In paramagnetic compounds, like A_3C_{60} superconductors in the normal state, interaction of the $S=3/2$ N spin with the electrons on the cage (or with conduction electrons) is expected to be strong and it is unlikely that the ESR of the N atom is observable.

ACKNOWLEDGEMENTS

This work was supported by the Hungarian State Grants OTKA T 029150, T 032613, T 029931 and FKFP 0352/1997. We are indebted to R. Kökényessi for numerical calculation of the Van Vleck second moment.

REFERENCES

[1] Almeida T. Murphy, Th. Pawlik, A. Weidinger, M. Höhne, R. Alcala, J.-M. Spaeth, Phys. Rev. Lett. **77**, 1075 (1996).

[2] B. Pietzak, M. Waiblinger, Almeida T. Murphy, A. Weidinger, M. Höhne, E. Dietel, A. Hirsch, Chemical Physics Letters **279**, 259 (1997).

[3] E. Dietel, A. Hirsch, B. Pietzak, M. Waiblinger, K. Lips, A. Weidinger, A. Gruss, K.-P. Dinse, J. Am. Chem. Soc. **121**, 2432 (1999).

[4] D. W. Murphy, M. J. Rosseinsky, R. M. Fleming, R. Tycko, A.P. Ramirez, R. C. Haddon, T. Siegrist, G. Dabbagh, J. C. Tully, and R. E. Walstedt, J. Phys. Chem Solids, **53**, 1321 (1992).

N@C$_{120}$ - The First Step To Endohedral Polymers

B. Goedde[†], M. Waiblinger[‡], P. Jakes[†], N. Weiden[†], K.-P. Dinse[†]
and A. Weidinger[‡]

†Technische Universität Darmstadt, Petersenstr. 20, 64287 Darmstadt, Germany
‡Hahn-Meitner Institut, Glienickerstr. 100, 14109 Berlin, Germany

Abstract. Introducing an endohedral atom in one of the cages of the dimer opens the possibility to get information about changes in the electronic structure of the molecule due to polymerization. Nitrogen is known to form a stable endohedral fullerene and can be used for further chemical processing. In this contribution we will present first EPR results of the dimer being 'doped' with nitrogen in one of the cages. Zero Field Splitting induced by cage distortion is significantly larger than that found in the monoadduct (C_{2v}-N@C$_{61}$(COOCH$_2$CH$_3$)$_2$).

I INTRODUCTION

Dimerization and polymerization of fullerenes has gained attention in the last years. It is known that polymers can be synthesized by photopolymerization, polymerization under high pressure and mechanochemical synthesis. Due to the limited stability of nitrogen encapsulated in fullerenes under irradition and high temperature the photopolymerization and the high-pressure methods seem to be unsuitable for the synthesis of N@C$_{120}$. Therefore we used the mechanochemical approach, first discovered by Wang *et al.* [1]. In this reaction solid C$_{60}$ is milled in a high-speed vibration mill together with an inorganic additive, to give C$_{120}$ which is separated from the educts by washing and HPLC.

We believe that N@C$_{60}$ behaves chemically similar to C$_{60}$, so we first optimized the reaction parameters, including mill load, milling time and frequency with 'empty' cages. We also tested different inorganic additives to obtain a better insight in the reaction mechanism.

N@C$_{60}$ was synthesized at the Hahn-Meitner-Institut (Berlin) by bombardment of C$_{60}$ with nitrogen ions. The concentration of N@C$_{60}$ in C$_{60}$ is between 10^{-5} and 10^{-4}. The material also contains about 10% C$_{70}$ which was removed together with other impurities originating from bombardment with HPLC. The material was dried overnight on a high-vacuum line to remove remaining toluene.

CP544, *Electronic Properties of Novel Materials—Molecular Nanostructures*, edited by H. Kuzmany, et al.
© 2000 American Institute of Physics 1-56396-973-4/00/$17.00

II EXPERIMENTAL

The fullerene samples were mixed with $CaCl_2$ in molar excess of 5:1 and milled in a stainless steel mortar with grinding balls. Our high-speed vibration mill resembles commercially available mills for IR-sample preparation.

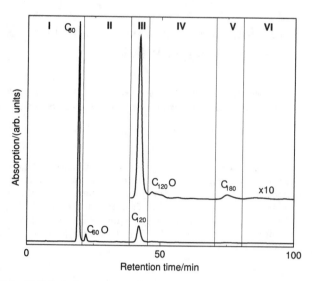

FIGURE 1. HPLC chromatogram of a reaction mixture with $CaCl_2$ as additive.

Milling time was 20 minutes. The mixture was dissolved afterwards in o-dichlorobenzene and micro-filtrated. Purification was done in a 3-step HPLC process using a Cosmosil Buckyprep (250x10mm) column running toluene as eluent (flow rate of 4 ml/min). Figure 1 shows a HPLC chromatogram of the primary reaction mixture. As it can be seen in the inset, not only C_{120} is formed but also small amounts of higher oligomeres are present. Due to slow decomposition of $N@C_{120}$, the sample still contained some $N@C_{60}$. The sample was sealed off on a high vacuum line and stored at 77 K.

III DRIFT SPECTROSCOPY

Identification was done using UV/vis-Spectroscopy which results in a good agreement with Wang et $al.$. In order to confirm that fraction III (in figure 1) consists of C_{120}, the DRIFT (Diffuse Reflectance Infrared Fourier Transform) spectrum was compared with reference spectra of C_{120} and $C_{120}O$ [2]. Plotted is the Kubelka-Munk function $f(R) = (1 - R^2)/2R$, where R is the diffuse reflectance. One can see that our spectrum is in good agreement with the published C_{120} spectrum. The indexed peak (*) arises from remaining o-dichlorobenzene.

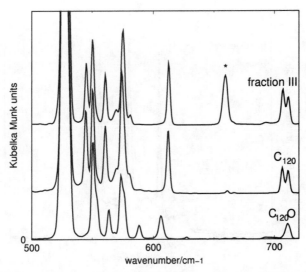

FIGURE 2. DRIFT spectrum of fraction III together with two reference spectra, traces of o-dichlorobenzene are indicated by (*).

IV EPR SPECTROSCOPY

Standard c.w. EPR spectra taken at room temperature showed only weak structures in addition to the three line main pattern. Because of limited signal to noise ratio, no reliable fit with an expected quartet powder pattern was possible. In figure 3 an echo-detected spectrum taken at 80 K is shown displaying the quartet powder pattern clearly. The main advantage of this detection method compared to c.w. EPR is that the spectra are free of distortion by saturation and modulation broadening and broad structures are not suppressed.

The three line structure can be attributed to nitrogen hfi $a_{iso} = 0.55$ mT. The additional features can be successfully simulated assuming Zero-Field-Splitting with the following parameters: D=0.5 mT, E=0.02 mT. Non-vanishing D and E values reflect lowering of symmetry by dimerization at the nitrogen site from T_h to C_s due to a small drop shaped distortion. Not astonishing, the measured D value is larger than that observed for the monoadduct caused by a single sp^3 substitution and much larger than that caused by long range order in the low temperature phase in polycrystalline C_{60}. The reduction of isotropic ^{14}N-hfs from 15.76 MHz for N@C_{60} to 15.55 MHz for N@C_{120} (obtained by simulation) agrees with the similar trend observed for the monoadduct N@$C_{61}(COOC_2H_5)_2$. An impurity seems to be responsible for the non-symmetric shoulder on the central line.

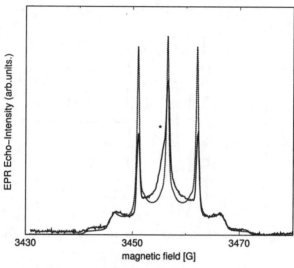

FIGURE 3. Echo detected EPR spectrum at 80 K and simulation (dotted line) Total acqusition time was 240 hours. (*) Indicates an EPR active impurity.

V CONCLUSION

We could show that nitrogen stays confined in the C_{60} cage during dimerization. The atom in one of the C_{60} cages of the dumbbell shaped molecule experiences a ZFS significantly larger than those observed in single and multiadducts of C_{60}. The small observed reduction of $a_{iso}(^{14}N)$ indicates that the neutral atom is not interacting differently with the modified cage.

VI ACKNOWLEDGEMENT

We thank Frank H. Hennrich and Manfred M. Kappes for their support and for recording the DRIFT spectra.

REFERENCES

1. G.-W. Wang, K. Komatsu, Y. Murata, M. Shiro, *Nature*, **1997**, *387(5)*, 583-586.
2. H.J. Eisler, F.H. Hennrich and M.M. Kappes, in: *Electronic properties of novel materials*, H. Kuzmany, J. Fink, M. Mehring and S. Roth, Eds. (World Scientific Singapore 1998), 180.

N@C_{60} for Quantum Computing

W. Harneit, M. Waiblinger, K. Lips[*], S. Makarov, A. Weidinger

Hahn-Meitner-Institut, Glienicker Str. 100, D-14109 Berlin, Germany
[*]Hahn-Meitner-Institut, Kekuléstr. 5, D-12489 Berlin, Germany

Abstract. Nitrogen in C_{60} (N@C_{60}) has properties which could be favourable for spin quantum computing. The spin system has an electron spin S = 3/2 and a nuclear spin I = 1. Due to the good shielding of the nitrogen atom from its environment, the spin relaxation times are relatively long compared to other spin systems. Therefore, the decoherence time (i.e. time available for computing) is fairly long. It has been proposed that endohedral fullerenes can be arranged on the surface of a substrate in such a way that the spins of different N@C_{60} interact with one another and thus the number of bits can be augmented. Specific arrangements and possibilities of addressing individual spins will be discussed.

SPIN QUANTUM COMPUTING

One of the most successful approaches in quantum computing [1] is bulk nuclear magnetic resonance (NMR) quantum computing in liquids [2-4]. The basic is to use a large number of identical molecules (or, quantum computers) with inequivalent nuclear spins (the qubits) which are controlled by rf pulses. The different local environment of the spins induces a chemical shift in the resonance frequency that allows one to address the spins individually. Using many identical molecules enhances the signal intensity to useful levels, and the use of a liquid phase ensures motional averaging of spurious couplings between the molecules. Several quantum algorithms using NMR computers have been demonstrated [3] involving up to 7 qubits [4].

For practically useful quantum computers, the number of qubits has to be raised substantially higher [5]. Increasing the number of nuclear spins and hence the number of qubits will get increasingly harder for two reasons : (i) as the size of the molecule increases, the signal drops rapidly due to insufficient spin polarization (the thermal probability to find n spins in the ground state scales with 2^{-n}), and (ii) the pulse sequences increase in length since one has to distinguish between very similar spin resonance frequencies [1,5]. Bulk NMR quantum computing is therefore eventually limited by the realizable number of qubits.

To overcome these difficulties, several solid state based spin quantum computing schemes have been proposed. The first proposal by Seth Lloyd dates back to 1993 [6] and requires neither addressing the spins individually nor to manipulate their coupling. Instead, spins of only three different types (i.e. with distinguishable resonance frequency) are lined up in a linear or two-dimensional array and one uses more complex addressing pulses which make use of the local (dynamic) environment of the spins to be addressed. We will show below how N@C_{60} could be used for this purpose.

CP544, *Electronic Properties of Novel Materials—Molecular Nanostructures*, edited by H. Kuzmany, et al.
© 2000 American Institute of Physics 1-56396-973-4/00/$17.00

A recent, quite more elaborate, scheme for solid state spin quantum computing has been proposed by Kane [7]. Here, the qubit is a phosphorus atom embedded in silicon beneath a gate electrode (A-gate). It can be tuned into and out of resonance, and thus addressed individually, by varying the hyperfine coupling of the phosphorus atom. Coupling between adjacent spins is achieved by varying the electronic wave function overlap with an electrode (J-gate) between the atoms. The great drawback of this proposal, however, is the necessity of positioning the phosphorus atoms precisely in order for the elaborate gate structure to work. We will discuss how $N@C_{60}$ could be used for this scheme, too.

$N@C_{60}$ PROPERTIES

In this paper, we investigate the possibility to build a solid-state quantum computer based on the *electronic* spin of the endohedral fullerene $N@C_{60}$. An electron spin quantum computer would take advantage of the higher gyromagnetic ratio of the electron, the detection limits and spin polarization being more favorable several orders of magnitude. On the other hand, the resonance frequencies of electron spins are more difficult to manipulate since the g-factor is pinned to a value of about two. The hyperfine interaction can be useful for this.

The $N@C_{60}$ molecule has quite unique properties that make it suitable for quantum computing. The nitrogen atom is in a highly symmetric environment exactly at the center of the fullerene cage, leading to a very good shielding of the spin. At the same time, the cage provides a useful "handle" to the electron spin. Thus, one can use the numerous schemes to nano-position C_{60} molecules that have been developed over the past years in order to build elaborate geometric arrangements of the spins.

$N@C_{60}$ [8] has electronic spin S = 3/2, a very sharp ESR triplet (Fig.1, left panel) and very long relaxation times $T_1 \approx 100\text{-}1000$ ms and $T_2 \approx 20$ µs [9]. The excellent chemical shielding of the usually rather reactive nitrogen atom is evidenced by the

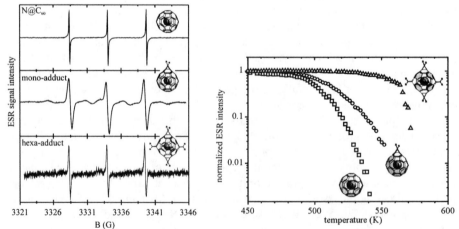

FIGURE 1. ESR signal of solid $N@C_{60}$ and some of its chemical modifications (left panel, ref. [11]) and thermal stability of these molecules in an isochronal experiment (right panel, ref. [12])

possibility to attach molecular addends on the C_{60} cage without losing the atomic character of nitrogen [10], as shown in Fig. 1 for mono- and hexa-adducts of $N@C_{60}$.

The ESR signals of the adducts are very similar to those of pure $N@C_{60}$ [11] confirming that the nitrogen atom is still in its quartet ground state. The additional line for the monoadduct arises from a fine structure due to a drop-like deformation of the cage which is again absent for the hexa-adduct. $N@C_{60}$ is stable at room temperature; Fig. 1 (right panel) shows that the thermal stability is even enhanced for the adducts, probably because the addends block the natural escape path of the nitrogen atom [12]. In summary, the wave function of nitrogen inside the cage can be tuned by chemical modification of the cage itself which at the same time enhances the thermal stability.

$N@C_{60}$ QUANTUM COMPUTING SCHEMES

In as-produced $N@C_{60}$, only about one C_{60} molecule in ten thousand contains a nitrogen atom. At this low spin density, the ESR remains sharp in the solid phase. Perturbing nuclear spins, e.g. due to a ^{13}C atom in the cage, get averaged out statistically and, at temperatures higher than the rotational melting transition, also due to the rotational motion of the cages on site.

Upon increasing the spin density (i.e. the ratio of N-filled to empty C_{60} molecules), the ESR lines show broadening characteristic of dipolar coupling [13]. The coupling constant J can be estimated from the z-component of the magnetic dipole field

$$B_{dip} \approx 9.3\,G \times \left(3\cos^2\theta - 1\right)\left(r/nm\right)^{-3}$$
(1)

induced at a neighbouring site ($r \approx 1$ nm, $\theta = 0$) to be about $B_{max} \approx 18.5$ G or

$$J = 2\pi / \left(\gamma_e B_{max}\right) \approx 52\,MHz$$
(2)

where we have used the gyromagnetic ratio of the electron, $\gamma_e \approx e / m_e = 1.76 \times 10^{11}$ Ts.

Based on this maximal coupling and the relaxation time T_2 which is equivalent to the coherence time one can calculate a *figure of merit* η_{pr} which is roughly the number of elementary clock cylces of the quantum computer that elapse before one has to apply quantum error correction schemes [5]. Table 1 shows that a $^{14}N@C_{60}$ linear array as proposed by Lloyd [6] would be favourable in these terms.

The different resonance frequencies needed for Lloyd's computer can be realized in endohedral group-V fullerenes [8]. The naturally abundant ^{14}N isotope (99.63 %) has a nuclear spin I = 1 yielding an ESR triplet whereas the ^{15}N isotope (abundance 0.37 %, commercially available in pure form) shows a doublet due to its nuclear spin I = 1/2 [14]. Furthermore, $P@C_{60}$ has been produced [15] showing properties similar to that

TABLE 1. The possible performance of spin quantum computation schemes.

Spin system	Decoherence time T_2	Coupling time $T_C = 1 / 2\,J$	Figure of merit $\eta_{pr} = T_2 / T_C$
liquid NMR[a]	500 ms	5 ms	100
solid NMR[a]	20 ms	200 μs	100
solid ESR[a]	10 μs	200 ns	50
Kane's ^{31}P array[a]	10 μs	10 ns	1000
$^{14}N@C_{60}$ array	> 20 μs[b]	10 ns[c]	>2000

[a] numbers taken from ref. [5]. [b] measured at 300 K in dilute $N@C_{60}$ [9]. [c] estimated from Eq. (2) [13].

of $^{15}N@C_{60}$. Finally, it is possible to incorporate nitrogen in C_{70} [10], making yet another set of spin resonance frequencies available for quantum computing.

As for Kane's silicon-based quantum computer [7], a variation might be possible using $N@C_{60}$ as the spin system (Fig. 2). The realization of A gates (i.e. resonance frequency tuning) is now only hypothetical and relies on the possibility to polarize the nitrogen atom or its surrounding cage, a question that is only being investigated. With a complete charge transfer, the spin ground state would change from quartet ($N@C_{60}$) to triplet ($N^-@C_{60}$). The coupling of spins through J-gates should be similar to Kane's original proposal, although the distances between gates might be larger since a charge transfer to the conduction band would be necessary.

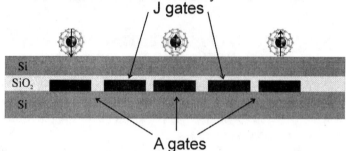

FIGURE 2. A possible solid state quantum computer based on $N@C_{60}$. The fullerene molecules are placed on top of buried electrodes (A gates) which modify the resonance frequency of the electron spin and its coupling to the silicon conduction band. A seperate set of electrodes (J gates) in between the molecules modifies the spin coupling due to quenching of the silicon conduction channel [16].

ACKNOWLEDGMENTS

This work was supported by the Information Societies Technology programme of the European Union under contract no. IST-1999-11617 (QIPD-DF project).

REFERENCES

1. for a recent review see, e.g. C. H. Bennett and D. P. DiVincenzo, *Nature* **404**, 247 (2000).
2. N. Gershenfeld and I. L. Chuang, *Science* **275**, 350 (1997).
3. L. M. K. Vandersypen *et al.*, *Nature* **393**, 143 (1998); M. Mosca *et al.*, *Nature* **393**, 344 (1998);
4. R. Marx *et al.*, *quant-ph¹/9905087* (1999); E. Knill *et al.*, *Nature* **404**, 368 (2000).
5. M. Mehring, Appl. Magn. Res. 17, 141 (1999)
6. S. Lloyd, *Science* **261**, 1569 (1993); *quant-ph¹/9912086* (1999).
7. B. E. Kane, *Nature* **393**, 133 (1998).
8. A. Weidinger *et al.*, *Appl. Phys.* **A 66**, 287 (1998)
9. A. Grupp *et al.*, *this issue*
10. E. Dietel *et al.*, *J. Am. Chem. Soc.*, (1999)
11. B. Pietzak *et al.*, *Chem. Phys. Lett.* **279**, 259 (1997)
12. M. Waiblinger *et al.*, *submitted to Phys. Rev.* **B**
13. M. Waiblinger *et al.*, *this issue*
14. T. Almeida-Murphy *et al.*, *Phys. Rev. Lett.* **77**, 1075 (1996)
15. C. Knapp *et al.*, *Molecular Physics* **95**, 999 (1998)
16. J. Twamley, *private communication*

V. NANOTUBE PREPARATION, NANOTUBE CHARACTERIZATION

Single-Wall Carbon Nanotubes: Study of Production Parameters using cw CO_2-Laser Ablation Technique

W.K. Maser[1*], E. Muñoz[1], A.M. Benito[1], M.T. Martínez[1],
G.F. de la Fuente[2], E. Anglaret[3], A. Righi[3], and J.-L. Sauvajol[3]

[1]Instituto de Carboquímica, CSIC, María de Luna 12, E-50015 Zaragoza (Spain)
[2]ICMA, CSIC-Universidad de Zaragoza, María de Luna 3, E-50015 Zaragoza (Spain)
[3]GDPC, Université de Montpellier II, F-34095 Montpellier Cedex (France)

Abstract. We report the production of single-wall carbon nanotubes (SWNTs) using a continuous wave (cw) 10.6 μm CO_2-laser. A wide range of experimental parameters and their influence on the formation of SWNTs has been investigated. It turns out that the temperature environment near the zone of evaporation plays a key role in the growth process of SWNTs. This result has important consequences for the optimization of the production of SWNTs.

INTRODUCTION

The availability of „tailored" samples of single wall carbon nanotubes (SWNTs) with determined diameters and chiralities in large amounts certainly would boost the exploitation of their promising application potential in various areas of technology ranging from electronics up to new systems for the storage of energy. However, the production of SWNTs in a controlled way and in large amounts currently represents a major challenge. Today, one technique to obtain large lab-scale amounts of SWNT material is the laser ablation method. This technique allows to vary parameters in a systematic way, and can give valuable information not only about possibilities for a controlled growth and a future up-scale production but also about the still not well understood formation mechanism.

In this article, we describe and discuss results obtained by using a cw-CO_2 Laser technique employed under various experimental conditions. Parameters like catalyst composition, type of gas, its pressure and its flow, laser power, and the effects of the laser-operating mode, cw versus pulsed, as well as the temperature have been investigated. Here the temperature is a unique parameter in the whole formation process of SWNTs. Many of the changes of sample characteristics can be related to this parameter. Effects induced by varying other parameters lead, more or less directly, to changes of the temperature environment necessary for the assembling of SWNTs. Therefore, the temperature plays a key role during the whole production process.

*Corresponding author: wmaser@carbon.icb.csic.es

CP544, *Electronic Properties of Novel Materials—Molecular Nanostructures*, edited by H. Kuzmany, et al.
© 2000 American Institute of Physics 1-56396-973-4/00/$17.00

EXPERIMENTAL

A CO_2-laser operating in continuous wave mode at a wavelength of 10.6 μm is used in order to produce SWNTs [1]. The laser beam is focused onto the lateral side of a graphite/bi-metal composite target rod placed inside a stainless steel chamber (Fig. 1). The evaporation chamber is filled from its bottom side with inert gas. Under best conditions, about 200 mg/hour of the target material gets evaporated. The following parameters have been employed:

i)   Metal composition of the graphite target (in at%): Single metals: Ni (2), Co (2), Y (0.5), Fe(2); bi-metallic mixtures: Ni/Y (4/1), (2/0.5), (1/0.25), (0.6/0.6), (0.5/0.13); Ni/Co (4/1), (2/2), (2/0.5), (1/0.25), (0.6/0.6), (0.5/0.13); Ni/Fe (4/1), (2/0.5), (0.6/0.6); Co/Y (2/0.5), Ni/La (2/0.5).

ii) Gas, pressure and flow-rate: argon, nitrogen at 50, 100, 200, 300, 400 and 500 Torr. Flow-rates: 1, 0.6, 0 l/min. Helium at 400 Torr at a flow-rate of 9 l/min.

iii) Laser power density and cw versus pulsed mode: 12 kW/cm^2, 9 kW/cm^2, 6kW/cm^2 (pulsed mode at a frequency of 2kHz).

iv) Temperature: Introduction of an additional graphite pipe around target rod (Fig. 2).

The as-produced carbonaceous materials have been investigated by scanning (SEM) and transmission (TEM) electron microscopy as well as by Raman spectroscopy performed at excitation energy of 2.41 eV. Typical microscopy images and Raman spectra can be found in [1,2,3].

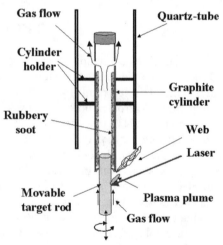

Figure 1. Experimental set-up: The laser beam inside the evaporation chamber is guided by a plane mirror to a parabolic mirror from where it is focused onto the target rod. About 1 cm above the target rod a quartz tube, in which SWNT material gets collected, can be seen.

Figure 2. Experimental detail showing an additional graphite pipe around the target rod. The graphite pipe, which is fixed inside the quartz tube, leaves a 1mm channel between target rod and pipe for the flowing buffer gas. During the evaporation process this channel creates a hot zone and changes the temperature environment.

RESULTS AND DISCUSSION

Effect of target composition: without metals, no SWNTs have been found in the produced soot material. The same result was observed with rods containing Y or Fe. However, adding metals like Ni and Co to the pure graphite target, a powdery soot was formed which contained a few and in most cases only isolated SWNTs. Here Ni led to highest yields. Using graphite/bi-metal targets, high amounts of SWNTs with diameters between 1.2-1.6 nm organized in bundles of typical length of 1 μm were found [1,2]. Highest yields were achieved with Ni/Y (4/1), (2/0.5) and Ni/Co (2/2) and substantially lower amounts belong to the lowest bi-metal concentrations. In general, Ni/Y and Co/Y mixtures lead to major contribution of large diameter tubes while in samples produced with Ni/Co, Ni/Fe and Ni/La contributions of small diameter tubes are dominating.

Gas, pressure and flow: No difference in using argon and nitrogen can be observed. Both gases allow the formation of high yields of SWNTs in an optimum pressure range between 200 and 400 Torr and both gases show a similar cut-off threshold at about 100 Torr where samples get dominated by amorphous carbon. On the other hand, this optimum pressure range for argon and nitrogen is not favorable in the case of helium. At 400 Torr, here only samples of amorphous carbon have been produced. No differences in the sample material could be observed by changing the flow conditions in the range indicated above.

Power and laser operating mode, cw versus pulsed: lowering the power from 12 to 9 kW/cm^2 first leads to changes in the evaporation rate (from 200 to 90 mg/h). However, no differences in the produced SWNT material could be noticed. Lowering the power to 6 kW/cm^2 (pulsed mode), no heating of the rod can be observed, the evaporation rate is very low (4 mg/h), and the samples are dominated by amorphous carbon.

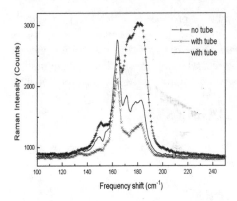

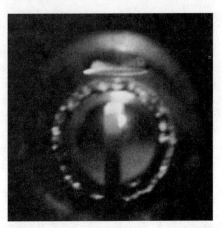

Figure 3. Raman spectra of the radial breathing modes of SWNTs. Samples produced with the graphite pipe show a systematic shift to lower frequencies (larger tube diameters).

Figure 4. Heated target rod during the evaporation process. Different grey tones represent different temperature zones of the rod.

Temperature gradient: In order to keep the temperature around the target rod a graphite pipe around the top of the target rod has been introduced (Figure 2). The heat radiation from the target rod heats the walls of the graphite pipe and an additional hot channel, through which the gas is flowing, is created. The soot material contains large amounts of SWNTs whose diameters are systematically larger than those in samples produced without graphite pipe (Fig. 3).

The key to explain all these results is related to the temperature. The energy of the CO_2-laser beam not only is used for evaporating material, but also to heat the target rod up to 1200°C in a distance of 1 cm around the focal spot (Fig. 4). This heat acts like a furnace and therefore, the temperature environment around the focal spot is at sufficient high temperature in order to form SWNTs. If the laser energy falls below a certain threshold, the target gets not heated anymore, consequently the temperature gets too low and no SWNT can be found in the soot. On the other hand, keeping the temperature gradient small over a longer distance around the evaporation zone helps to increase the yield of SWNTs in the samples. Furthermore, higher temperatures clearly lead to the formation of larger tube diameters. Gases also strongly influence the temperature environment [4]. Light gases like helium lead to increased cooling rates and to rapid lowering of the temperature. Therefore, the optimum pressure range for helium lies above the values found for argon and nitrogen. Flow rates certainly have an influence on the temperature, however in our experiments they are far too small to induce significant changes. Beside that they are absolutely necessary for forming SWNTs, metals also can change the heat distribution of the target rod as well as of the gas phase and therefore lead to changes of the temperature environment, resulting in samples with different diameter distributions.

CONCLUSIONS

Temperature plays a key role in the formation process of SWNTs. It has to be sufficient high in order to provide the suitable thermal energy for nanotube formation. It also determines the diameter of the tubes. Therefore, for any controlled and up-scaled production it becomes absolutely necessary to create and to maintain a sufficient and controlled hot temperature environment around the evaporation zone.

ACKNOWLEDGMENTS

One of us (EM) acknowledges funding from the Departamento de Educación y Cultura de la Comunidad Autónoma de Aragón.

REFERENCES

1. Maser, W.K., Muñoz, E., Benito, A.M., Martínez, M.T., de la Fuente, G.F., Maniette, Y., Anglaret, E., Sauvajol J.L., Chem.Phys. Lett., **292**, 587 (1998)
2. Muñoz, E., Maser, W.K., Benito, A.M., Martínez, M.T., de la Fuente, G., Righi, A., Sauvajol, J.L., Anglaret, E, Maniette, Y, Appl. Phys. A **70**, 145 (2000)
3. http://www.icb.csic.es/nanotubos/first.html
4. Gamaly, E.G., Rode, A.V., Maser, W.K., Muñoz, E., Benito, A.M., Martínez, M.T., de la Fuente, G.F., Appl. Phys. A **70**, 161 (2000)

Mechanism of Metal Catalysts Controlling Formation of Single-Wall Carbon Nanotubes

Masako Yudasaka[1], Yohko Kasuya[1], Morio Takizawa[1], Shunji Bandow[1], Kunimitsu Takahashi[2], Fumio Kokai[2], and Sumio Iijima[1,3,4]

[1] Nanotubulites Project, ICORP-JST, c/o NEC Corporation, 34 Miyukigaoka, Tsukuba, Ibaraki 305-8501, Japan
[2] Institute of Research and Innovation, 1201 Takada, Kashiwa, Chiba 277-0861, Japan
[3] NEC Corporation, Corporation 34 Miyukigaoka, Tsukuba, Ibaraki 305-8501, Japan
[4] Meijo University, Shiogamaguchi, Tempaku-ku, Nagoya, Japan

Abstract. In the formation of single-wall carbon nanotubes (SWNTs) by laser ablation and arc discharge, metal catalysts are necessary. We found that NiCo was the most effective in catalyzing SWNT formation and that other metals followed the order NiCo>Ni~NiFe>>Co~Fe. We confirmed that all these metals were effective catalysts of graphitization, indicating their equal potential-ability for catalyzing SWNT formation. The reason Co and Fe could not well catalyze SWNT formation in laser ablation and arc discharge is discussed based on the segregation manner of graphite on Co and Fe.

INTRODUCTION

Single-wall carbon nanotubes (SWNTs) were discovered by Iijima in 1993 [1]. Since then, their structures, properties, and applications have been studied and summarized in journals and books, e.g., ref. 2. The progress of such studies advanced enormously when sufficient quantities of SWNTs for research became available from pulsed Nd:YAG laser ablation [3]. Later on arc discharge [4] and CO_2 laser ablation [5] were also found to produce SWNTs. Despite this progress, however, the formation mechanism of SWNTs still remains unclear. Since we think that the understanding the formation mechanism helps the further development of the production methods that can supply large quantity of SWNTs with high yield, we have carried out the in-situ observation for the SWNT growth in the laser ablation, and performed structural analyses of the products. Our experimental results and growth models are summarized as follows [6-11]. Carbon and metal evaporate from the target as a result of the laser beam irradiation, and their frequent collisions generate carbon-metal droplets. During these processes, which take several milliseconds, the emitted materials move about several centimeters away from the target, and the temperature of the droplets changes from above 3000 K [12, 13] to the atmosphere temperature in the chamber of above 1200 K that is optimum for SWNT growth [9,14,15]. We think that metal particles with diameters of about 1 to 2 nm segregate in the carbon-metal droplets when these droplets cool to 1400–1500

CP544, *Electronic Properties of Novel Materials—Molecular Nanostructures*, edited by H. Kuzmany, et al.
© 2000 American Institute of Physics 1-56396-973-4/00/$17.00

K, and these small metal particles catalyze SWNT formation.

SWNTs were produced by Nd:YAG laser ablation with high yields when NiCo or Ni was used, but when Co or Fe was used, the yield was extremely small [9]. We inferred the reason for this based on experimental results as follows. When Co particles with sizes of 1 to 2 nm segregated, the temperature of the Co-C droplets was too high for carbon to form SWNTs. The Fe particles segregated in the Fe-C droplets at a temperature that was too low for carbon to react. The validity of these explanations is reconsidered in this report: we clarified the activity of the metal catalysts in SWNT formation in different formation methods, and studied the interaction of metal and carbon by our own method.

EXPERIMENTAL

The lowest temperatures (T(L)) above which carbonaceous materials were transformed into graphite as a result of dissolution of carbon in metal and recrystallization of carbon as graphite were found as follows. The carbonaceous materials we chose to study were diamond-polycrystalline films with a thickness of about 10 μm formed on a SiC crystal, diamond-like amorphous carbon (DLC) formed on a quartz glass plate by vacuum deposition, highly oriented pyrolytic graphite, and graphite-like amorphous carbon (GLC) deposited on a quartz glass plate by chemical vapor deposition (CVD) at a temperature of 1270 K and pressure of 1×10^{-6} Torr using anthraquinone as a carbon source. The films of Fe, Co, Ni, NiFe, and NiCo were deposited on these carbonaceous materials and heat treated at various temperatures for 1 hour at 1×10^{-6} Torr. Hereafter, we use abbreviations, such as NiCo/diamond to mean diamond covered by NiCo film. After the heat treatment, the structural changes of the carbonaceous materials and metal films were checked by scanning electron microscopy (SEM) and Raman spectrum measurement.

The quantity of SWNTs formed by Nd:YAG laser (0.5 to 6 W irradiating within an area of 0.1 cm^2, pulse width: 6-7 ns, 10 Hz, 532 nm) ablation (Ar 600 Torr, 1470 K, 600 pulses) using a carbon target containing metal catalysts (Fe, Co, Ni, NiFe, and NiCo with concentrations of 0.3 to 9 atom%) were investigated in a similar manner as we reported previously [7].

The quantity of SWNTs formed by DC arc discharge was estimated roughly from the weights and Raman spectra of soot existing on the chamber wall, around the cathode, and around the anode. DC arc discharge was performed at DC 100 A for 10 s in 500 Torr of He atmosphere. On top of the anode was placed the carbon target containing metal particles; it was the same type of the target used in the Nd:YAG laser ablation. The cathode was a carbon rod. The diameter of the electrodes was 1 cm.

Results and Discussion

Most of the SWNTs formed in our Nd:YAG laser ablation were contained in the web-like deposit, and its weight proved to be a good measure of the quantity of

SWNTs [9]. The quantities of web-like deposits formed using various metal catalysts are shown in Fig. 1a. The largest quantity was when NiCo was used, and the quantities for other metals followed the order NiCo > Ni ~ NiFe >> Co ~ Fe. When Co or Fe was used, almost no web-like deposits but only amorphous carbon (a-C) were formed on the chamber wall.

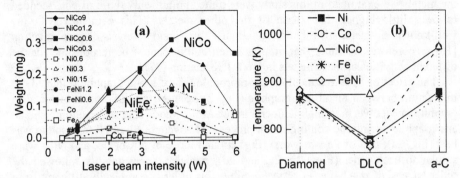

FIGURE 1. Quantities of web-like deposits formed by Nd:YAG laser ablation (a), and values of T(L) above which various carbonaceous materials were graphitized by metals (b).

TABLE 1. Quantities of soot produced by arc discharge using graphite anode containing metal catalysts (shown as concentrations in atom%). Numbers in front of (a-C) indicate the ratio of Raman peak intensity of a-C at 1350 cm^{-1} to that of SWNT at 1591 cm^{-1}.

Catalyst		Chamber soot		Cathode soot		Anode soot	
Ni 0.3	Weight	11.37		2.61		0.61	
	Raman	SWNT	0.07(a-C)	SWNT	0.02(a-C)	SWNT	0.11(a-C)
Ni 0.6	Weight	12.53		2.12		5.3	
	Raman	SWNT	0.02(a-C)	SWNT	0.07(a-C)	SWNT	0.03(a-C)
Ni 0.15	Weight	2.8		1.01		0.17	
Co 0.15	Raman	SWNT	0(a-C)	SWNT	0.005(a-C)	(No SWNT)	a-C
Ni 0.3	Weight	15.93		6.18		1.84	
Co 0.3	Raman	SWNT	0.03(a-C)	SWNT	0.03(a-C)	SWNT	0.09(a-C)
Ni 0.3	Weight	6.49		0.75		1.66	
Fe 0.3	Raman	SWNT	0.03(a-C)	SWNT	0.15(a-C)	(No SWNT)	a-C
Fe 0.6	Weight	2.64		0.91		0.24	
	Raman	SWNT	0.5(a-C)	(No SWNT)	a-C	SWNT	0.07(a-C)
Co 0.6	Weight	9.42		2.03		1.21	
	Raman	SWNT	0.44(a-C)	(No SWNT)	a-C	(No SWNT)	a-C

The quantity of SWNTs formed by arc discharge was also influenced by the metal catalyst. To compare the SWNT quantities formed by using various metal catalysts, we measured the weight of the soot and observed its Raman spectrum (Table 1). We judged that the SWNT quantity formed by arc discharge was largest when NiCo was used and the quantities formed by other metals followed the order NiCo > Ni > NiFe >> Co ~ Fe. This order was almost identical to the order obtained for Nd:YAG laser ablation. A similar order was also found for SWNT formation by CO$_2$ laser ablation.

To understand the way in which the metals control the SWNT formation in laser ablation and arc discharge, we studied the interaction of metal and carbon. Raman spectra showed that the diamond of NiCo/diamond was transformed into graphite by heat treatment at and above 870 K, so T(L) was 870 K. Values of T(L) for the other combinations of carbonaceous materials and metals are shown in Fig. 1b. Those for the metals we examined in this study were quite similar; only those of NiCo tended to be a little higher than those of the other metals. (This coincides with the well-known characteristics that Ni, Co, and Fe are good catalysts of graphitization [16].) Therefore, we think that graphitization ability cannot be responsible for the different activities of these metals in SWNT formation [17].

The effect of heat treatment on the morphologies of metal/graphite reflected the interaction between metal and graphite (Fig. 2) because the dissolution of metal in graphite or vice versa at high temperature was followed by the segregation of metal and graphite during the cooling-down process. The wettability of Ni to graphite was best (Fig. 2a), that of Fe was worst (Fig. 2e) and those of NiCo, Co, and NiFe were similar to each other (Figs. 2b, 2c, and 2d); namely, the wettability followed the order Ni >> NiCo ~ NiFe ~ Co >> Fe. Since this order is completely different from the activity of metal as an SWNT formation catalyst, we think that wettability cannot control SWNT formation.

When we observed the metal surface carefully, we found that the segregated graphite formed film-like layers on islands of Ni, NiCo, and NiFe after heat treatment (Figs. 2a, 2b and 2c), but not on the islands of Co and Fe (Figs. 2d and 2e). In the case of Fe, the segregated graphite formed islands that existed laterally to the side of the Fe islands (Fig. 2e). The Co islands laterally touched the side of steps on the graphite substrate (Fig. 2d). The X-ray diffraction data showed that (111) of Ni, NiCo, and NiFe grew preferentially on the graphite surface, while Co grew with its (111) and (100) faces parallel to the graphite surface. We tentatively think that the preferential crystallographic orientations of metal and graphite in contacting with each other controls the catalytic activity of the metal in SWNT formation as follows.

A metal particle with a diameter of 1 to 2 nm segregates in the metal-carbon solution and a graphite shell can form around it, but the diameter of the graphite shell is too small to keep a spherical structure because graphite prefers flat extension. This would cause lifting off of the graphite cap from the metal particle followed by continuous growth of the graphite layer around the metal particle, leading to SWNT formation. If Co or Fe segregated, they would not form a graphite shell around themselves, but would aggregate by themselves and the metal particles with sizes of 1 to 2 nm would quickly become larger. [1] This coincides with the former explanation of why Co cannot catalyze SWNT formation, which was introduced above. In the case of Fe, the high solubility of Fe in C would also prevent graphite formation, which was also introduced above.

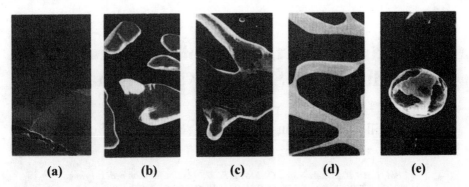

FIGURE 2. SEM images (a) Ni/graphite, (b) NiCo/graphite, (c) NiFe/graphite, (d) Co/graphite, and (e) Fe/graphite after heat treatment at 1370 K.   (━━━━ : 5 µm)

REFERENCES

1. Iijima, S., Ichihashi, T., *Nature*, **363**, 603(1993).
2. Saito, R., Dresselhaus, G., Dresselhaus, M. S., *"Physical Properties of Carbon Nanotubes"*, Imperial College Press, London, 1998.
3. Guo, T., Nikolaev, P., Thess, A., Colbert, D. T., Smalley, R. E., *Chem. Phys. Lett.*, **243**, 49(1995).
4. Journet, C., Maser, W. K., Bernier, P., Loiseau, A., Chapelle, M. L., Lefrant, S., Deniard, P., Lee, R., Fisher, J. E., *Nature*, **388**, 756(1997).
5. Maser, W. K., Munz, E., Beito, A. M., Martinez, M. T., Fuente, G. F., Maniette, Y., Anglaret, E., Sauvajor, J. –L., *Chem. Phys. Lett.*, **587**, 292(1998)
6. Yudasaka, M., Komatsu, T., Ichihashi, T., Achiba, Y., Iijima, S., *J. Phys. Chem. B*, **102**, 4892(1998).
7. Yudasaka, M., Ichihashi, T., Iijima, S., *J. Phys. Chem . B*, **102**, 10201(1998).
8. Yudasaka, M., Ichihashi, T., Komatsu, T., Iijima, S. *Chem. Phys. Lett.*, **299**, 91(1999).
9. Yudasaka, M., Yamada, R., Sensui, N., Wilkins, T., Ichihashi, T., Iijima, S. *J. Phys. Chem . B*, **103**, 6224(1999).
10. Kokai, F., Takahashi, K., Yudasaka, M., Iijima, S., *Appl. Phys. A.*, **69**, S229(1999).
11. Kokai, F., Takahashi, K., Yudasaka, M., Iijima, S., *J. Phys. Chem.*, in press (2000).
12. Steinback, J., Braunstein, G., Dresselhaus, M. S., Ventatesan, T., Jacobson, D. C., J. Appl. Phys., **58**, 4347(1988).
13. Ishigaki, T., Suzuki, S., Kataura, H., Kraetschmer, W., Achiba, Y., *Appl. Phys. A*, **70**, 121(2000).
14. Bandow, S., Asaka, S., Saito, Y., Rao, A. M., Grigorian, L., Richter, E., Eklund, P. C., Phys. Rev. Lett., **80**, 3779(1998).
15. Kataura, H., Kumazawa, Y., Maniwa, Y., Ohtsuka, Y., Sen, R., Suzuki, S., Achiba, Y., *Carbon*, in press.
16. Marsh H., Warburton, A. P., *J. Appl. Chem.*, **20**, 133(1970).
17. Since a poor graphitization catalysts, Pd, showed high T(L), above 1270 K, and produced little SWNTs, we think that the graphitization ability needs to be high for a metal to catalyze the SWNT formation.

† A recent report on SWNT formation by CVD shows that Fe or FeCo is more effective than Ni in catalyzing the SWNT formation [Colmer et al., Chem. Phys. Lett., **317** 83(2000)]. Does this suggest that the lateral growth of graphite to the Fe or Co islands is preferable to grow SWNTs by CVD? It is difficult to comment further on the effect of metal catalysts in SWNT formation by CVD or gas phase growth because studies on CVD formation have only just started and little information is available on metal catalyst activities.

Investigation of the Role of Yttrium in the Production of SWNT by arc Discharge

J. Gavillet[1], A. Loiseau[1], O. Stephan[2], S. Tahir[3], P. Bernier[3]

[1] LEM, UMR 104 Onera-Cnrs, ONERA B.P. 72, 92322 Châtillon cedex, France
[2] LPS, Université Paris-Sud, Bât.510, 91405 Orsay, France
[3] GDPC, Université Montpellier 2, Place E. Bataillon, CC 26, 34095 Montpellier Cedex 05, France

Abstract. We present a TEM study of single wall nanotubes (SWNTs) produced with the electric arc discharge technique using a Ni/Y catalyst (of various relative proportions). The study shows that SWNT ropes grow from metallic nanoparticles which are found to have definite Ni-Y compositions. This link has been   confirmed for the different Ni/Y anode compositions investigated.

INTRODUCTION

Carbon single-wall nanotubes (SWNTs) can be produced with different synthesis techniques (arc discharge [1], pulsed laser ablation [2], continuous laser ablation [3] or catalytic decomposition [4], etc...). In all cases, SWNTs can only be produced when metallic catalysts are used. The role of the catalyst is still not clearly understood; this strongly limits our understanding of the SWNT growth mechanism. We have studied the composition of the nanoparticles produced during arc discharge synthesis of SWNTs using a Ni/Y catalyst using Transmission Electron Microscopy (TEM). We show that the nanoparticle composition is strongly related to the nature of the carbon structures emerging from them and that long bundles of SWNTs grow from nanoparticles whose Ni/Y compositions fall within a specific range.

EXPERIMENTAL

Sample production was performed with the arc-discharge generator using the optimized conditions described in [1]. Nickel and yttrium were used as co-catalysts, the total catalyst concentration in the anode was ~1-5 at.% and different Ni/Y proportions have been tested: 100/0, 80/20, 50/50, 20/80, 0/100.
We focused our TEM analysis to the cathode collarets which contained the highest density of SWNT. High Resolution TEM (HRTEM, Jeol 4000 FX) was used to identify the different nanotube morphologies (diameter size, bundle diameter, length, nanoparticle size, etc...). The metallic composition of the nanoparticles was analyzed using Energy Dispersive X-Ray (EDX, Philips CM20). The minimum probe that could be used was ~20 nm in diameter. Electronic Energy Loss Spectroscopy (EELS) in Scanning TEM mode (VG Hb 501 STEM) with a probe size of 1 nm was used to detect the presence of carbon inside the catalyst nanoparticles.

CP544, *Electronic Properties of Novel Materials—Molecular Nanostructures*, edited by H. Kuzmany, et al.
© 2000 American Institute of Physics 1-56396-973-4/00/$17.00

RESULTS

HRTEM characterization shows the presence of different structures in the samples (single-wall nanotubes, nanoparticles, graphitic shells, amorphous carbon etc...). Most SWNTs are self-assembled in crystalline bundles. A systematic study of their extremities has revealed that all the bundles have one free end and the other attached to a metallic nanoparticle with a diameter smaller than 40 nm. These observations lead to the conclusion that SWNT bundles are emerging from small catalyst nanoparticles.

Two kinds of SWNTS rope morphologies were observed in various proportions depending on the nominal yttrium concentration in the anode (Fig. 1, Tab.1): long SWNTs bundles and sea urchin like structures.

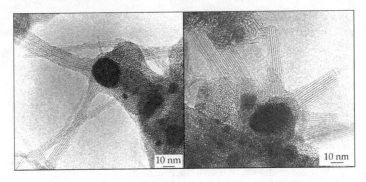

FIGURE 1. The two kinds of SWNT morphologies: long SWNT bundles (left), sea-urchin (right).

In long nanotubes bundles, the length of theSWNTs exceeds one micron and the average nanotube diameter is 1.35 nm . They were observed in all samples except for that synthesized with the pure yttrium catalyst. In general,only 1 or 2 such bundles are emerging from the same catalyst nanoparticle (Fig. 1a). On the contrary, in sea-urchin features, many bundles are irradiating from a given nanoparticle (Fig. 1b). In these bundles, SWNTs are much shorter (length < 200 nm) and their average diameter is larger (1.8 nm) than in long bundles. This second kind of bundles was only present when the Y/Ni ratio in the anode was higher than 20%. Tab.1 gives a summary of the relative abundance of these structures in the five samples.

TABLE1. Summary of the relative abundance of the structures in samples.

Anode composit. (Ni/Y at. %)	Observed structures / Abundance
100:0	Long SWNTs bundles : low yield / No sea-urchin
80:20	Long SWNTs bundles : High yield / No sea-urchin
50:50	Long SWNTs bundles : High yield / Sea-urchins : low yield
20:80	Long SWNTs bundles : High yield / Sea-urchins : low yield
0:100	Sea-urchins : low yield / No long SWNTs bundles

The compositions of the nanoparticles giving rise to these different structures were determined by EDX. We observed that nanotube morphology was linked to the composition of the nanoparticles. Nanoparticles related to long bundles have a low yttrium content within a specific range (0-15%) (Fig. 2) whereas sea-urchins grow from nanoparticles relatively richer in yttrium (>11%) (Fig. 2). In contrast, nanoparticles surrounded by amorphous carbon or graphitic shells have very dispersed Ni/Y compositions without any particular tendency (Fig. 2).

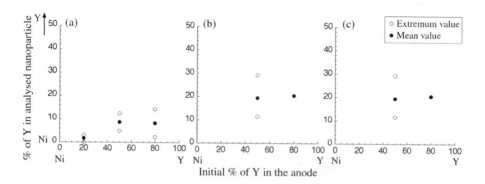

Initial % of Y in the anode

FIGURE 2. Results of EDX analyses for different anode Ni/Y proportions and for the different carbon structures linked to nanoparticles a) long SWNT ropes b) sea-urchin c) amorphous carbon or graphitic shell.

These experiments also show that only a small amount of yttrium condenses in the collaret (Fig. 2). Whatever the carbon structure surrounding the nanoparticles, the Y/Ni ratio is always smaller than that of the anode. Yudasaka et al. [5] have shown that most of the yttrium is contained in the hard cathode deposit.

EELS line scan spectra were done across nanoparticles related to long SWNT bundles (Fig. 3a). Carbon and nickel profiles are found to be anti-correlated, which means that the metallic nanoparticles do not contain any carbon (Fig. 3b). Moreover, theperfect similarity between the carbon edge fine structures of the border and of the middle of the nanoparticles (Fig. 3c) shows that the carbon contribution observed in the center of the line scan is only due to the disordered carbon surrounding the nanoparticles.

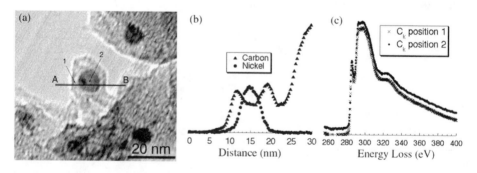

FIGURE 3. EELS analysis results: a) picture of the analyzed nanoparticle b) Carbon and nickel profiles recorded through the nanoparticle between the poin ts A and B. c) Comparison of the carbon edge fine structures at positions 1 and 2 of the nanoparticle.

DISCUSSION

The present results show that SWNT are actually linked to catalytic metallic nanoparticles and that there is a direct relationship between the composition of a nanoparticle and the type of nanotubes emerging from it. The addition of yttrium in the nickel particles up to a maximum concentration of 15% is found to improve the growth of long bundles. One the other hand, a higher proportion of yttrium leads to the formation of sea-urchin-like structures composed of nanotubes with larger diameters.

These results strongly support the hypothesis of a SWNT growth initiated at a particle surface according to a root-growth mechanism based on a vapor-liquid-solid model. This model was first proposed by Saito et al [6] to explain the formation of sea-urchin-like structures. We believe that it is also applicable to the formation of long SWNT ropes. In this model, SWNT nucleation occurs during the solidification of metal-carbon (M-C) nanoparticles via the segregation of carbon to the particle surface. The occurence of such a segregation process is attested by the absence of carbon inside the nanoparticles in our analyses. Since the equilibrium feature corresponds to a nanoparticle surrounded by a graphitic shell, SWNTs result from a particular kinetic process, in which carbon diffusion, solidification temperature of the M-C liquid and local temperature gradient are critical parameters. Previous works have already pointed out the importance of the solidification point [7, 8]. The present observations suggest that both the morphology of the bundles and the diameter of the nanotubes are closely determined by the composition and the solidification conditions of the nanoparticle acting as a SWNT seed.

Long ropes seem to be produced in a stationary regime that allows for a relatively long growth process whereas sea-urchin formation seems to be more rapid, with a stronger driving force for carbon segregation. Moreover, the nanotubes involved in sea-urchins are relatively short, which implies an interrupted growth and their diameter are larger suggesting a higher temperature of formation. These differences have to be linked to the composition difference between the nanoparticles producing the two kinds of nanotubes.This composition difference implies different solidification temperatures and different segregation conditions, that we will analyse in detail with the help of the ternary Ni-Y-C phase diagram in a forthcoming paper.

REFERENCES

1. C. Journet, W. K. Maser, P. Bernier, A. Loiseau, M. L. De La Chapelle, S. Lefrant, P. Deniart, R. Lee and J. E. Fisher, *Nature* **388**, 756 (1997)
2. A. Thess, R. Lee, P. Nikolaev, H. J. Dai, P. Petit, J. Robert, C. H. Xu, Y. H. Lee, S. G. Kim, A. G. Rinzler, D. T. Colbert, G. E. Scuseria, D. Tomanek, J. E. Fisher and R. E. Smalley, *Science* **273**, 483 (1996)
3. W.K. Maser, E. Munoz, A.M. Benito, M.T. Mart•nez, G.F. de la Fuente,Y. Maniette, E. Anglaret, J.-L. Sauvajol, *Chem. Phys. Letters 292, 587 (1998)*
4. J-F. Colomer, G. Bister, I. Willems, Z. Konya, A. Fonseca, G. Van Tendeloo and J. B. Nagy, *Chem. Phys. Letters*, 1343 (1999)
5. M. Yudasaka, N. Sensui, M. Takizawa, S. Bandow, T. Ichihashi, S. Iijima, *Chem. Phys. Letters,* **312**, 155 (1999)
6. Y. Saito, M. Okuda, N. Fujimoto, T. Yoshikawa, M. Tomita and T. Hayashi, *Jpn. Appl. Phys.* **33**, L526 (1994)
7. H. Kataura, Y. Kumazawa, Y. Maniwa, Y. Ohtsuka, R. Sen, S. Suzuki and Y. Achiba in press
8. Alvarez et al., to be published

Simple Energetics of Single-Walled, Open-Ended, Finite Carbon Nanotubes

M.W. Radny and P.V. Smith

School of Mathematical and Physical Sciences, The University of Newcastle, Callaghan, Australia 2308

Abstract. In this paper we examine how the total energies of finite, open-ended, single-walled carbon nanotubes are affected by the presence of chemically active dangling bonds at the ends of the tubes. Results have been obtained by carrying out empirical Brenner potential molecular dynamics total energy calculations on tubes with different symmetries. The total energy curves for the finite tubes are shown to have a well-defined minimum. These minima result from a balance between the chemical energy arising from the dangling bonds at the ends of the tubes and the mechanical energy due to the stress induced by the curvature of the tubes. A simple relationship has been derived for the radius of the minimum energy nanotube containing a given number of atoms. This relationship shows that as the number of atoms in the tube increases, the length of the minimum energy tube with that number of atoms increases much faster than the radius of the tube. The values of the total energies and radii defining these energy minima are found to be nearly identical for all symmetries.

Extended carbon tubes, because of their curvature, have an excess of energy (strain energy) compared with the energy of the flat infinite graphite sheet from which the extended tubes can be formed. This energy scales as R^{-2}, where R is the radius of the tube [1]. In this paper we examine how this well known behaviour is affected by the presence of chemically active dangling bonds at the ends of finite, open-ended, single-walled carbon tubes.

To this end, we have performed empirical Brenner potential molecular dynamics total energy calculations [2] for a number of single-walled, open-ended tubes with armchair [(m,0)k], zigzag [(m,m)k], and chiral [(m,0.5m)k] symmetries [3]. All of the calculations have been carried out for tubes comprised of 840 and 480 carbon atoms. Calculations were also performed for infinite graphite strips and extended tubes having the same number of atoms in the periodic cell.

The variations of the strain energies of extended tubes with armchair, chiral and zigzag symmetries as a function of their radii are shown in Figures 1a, 1b and 1c, respectively. Each of these curves can be described by the expression

$$E(extended\ tube) = (N \times A / x^2) + N \times C,\qquad(1)$$

where x is the number of atoms at the edges of the tubes, and A and C are constants with the values given in Table 1. As expected, the first term represents the excess

· CP544, *Electronic Properties of Novel Materials—Molecular Nanostructures*, edited by H. Kuzmany, et al.

a)

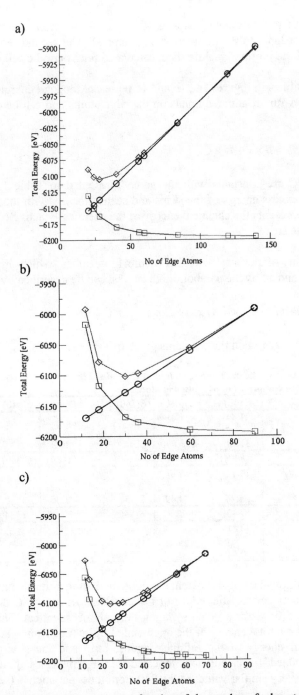

b)

c)

FIGURE 1. Variation of the total energy as a function of the number of edge atoms for extended single-walled carbon tubes (squares), infinite graphite strips (circles), and finite, open-ended carbon tubes (diamonds) for a) armchair, b) chiral and c) zigzag tubes comprised of 840 carbon atoms.

227

strain energy of the tubes and scales as R^{-2} (this follows from the fact that the tube radii are proportional to the number of edge atoms). The second term represents the total energy of the infinite graphite sheet calculated for a periodic cell with N atoms.

The straight lines in Figures 1a, 1b and 1c represent the total energies of the infinite graphite strips with unsaturated bonds on the edge atoms and can be described by the expression

$$E(strip) = B \times x + N \times C,$$
(2)

where B and C are constants with the values presented in Table 1. The first term represents the excess energy of the strips and scales linearly with the number of edge atoms (x). It represents the chemical energy of the chemically active edge atoms of the infinite graphite strips.

The variations of the total energies of the finite tubes, which are represented in Figures 1a, 1b and 1c by the parabolic-like curves, can be described by the expression

$$E(finite\ tube) = (N \times A / x^2) + B \times x + N \times C,$$
(3)

where A, B and C are again the constants given in Table 1.

TABLE 1. Values of the parameters A, B, and C obtained from fitting to equations (1)-(3) for the tubes comprised of 840 and 480 atoms.

ARMCHAIR	A(840)	A(480)	B(840)	B(480)	C(840) & C(480)
Equation 1.	31.44	31.45	-	-	7.38
Equation 2.	-	-	2.13	2.13	7.38
Equation 3.	31.44	31.69	2.11	2.12	7.38
ZIGZAG	A(840)	A(480)	B(840)	B(480)	C(840) & C(480)
Equation 1.	24.14	24.17	-	-	7.38
Equation 2.	-	-	2.58	2.62	7.38
Equation 3.	24.24	24.49	2.55	2.56	7.38
CHIRAL	A(840)		B(840)		C(840)
Equation 1.	30.81		-		7.38
Equation 2.	-		2.27		7.38
Equation 3.	30.64		2.26		7.38

Comparing the values of the parameters obtained from fitting equations (1)-(3) to the calculated data reveals the values of the constants A, B, and C derived from the different curve fittings for a given symmetry are almost identical. This suggests that one can predict the total energy of the open-ended finite tube from a knowledge of the excess chemical energy of the infinite graphite strip and the strain energy of the infinite tube. Indeed, the calculations repeated for the armchair and zigzag tubes with 480 atoms give very similar values to the 840 atom tube parameters (see Table 1).

The total energy curves for the finite tubes are seen to be characterised by deep minima. These minima arise from a balance between the chemical energy arising from the dangling bonds at the ends of the tubes and the mechanical energy due to the stress induced by the curvature of the tubes. This relationship holds for all three of the symmetries that we have studied. From the formulas presented above, this balance is achieved when the radius of the finite tube is given by

$$R = const(N \times A/B)^{1/3}. \tag{4}$$

Calculating the length of the minimum energy tube one finds that the length of the tube increases much faster than its radius as the number of atoms in the tube increases. For instance, the most stable finite armchair tube for 840 carbon atoms is predicted to be the (7,7)30 tube with $R = 0.48$nm and a length of 7.38nm, while for 480 atoms the most stable tube is the (6,6)20 tube with $R = 0.41$nm and a length of 4.92nm.

There are also only relatively small differences between the minimum energy values of tubes with different symmetry. The most stable 840 atom tube is the armchair finite tube (-6105.0 eV) followed by the almost degenerate in energy chiral (-6101.87 eV) and zigzag (-6101.82 eV) tubes. The radii of these minimum energy tubes are also similar: 0.48nm (armchair), 0.47nm (zigzag), and 0.51nm (chiral). These results are consistent with recent experimental studies on single-walled carbon nanotube bundles [4] which led to the conclusion that within a bundle no tube chirality (symmetry) is favoured, and the diameters of the tubes are almost uniform. On the other hand, the experimentally observed distribution of the radii of single tubes is clearly sensitive to the type of catalytic metal used in their production [5]. This suggests that while kinetic processes dominate the process of tube formation, maintaining a balance between the chemical and mechanical energies of the tubes plays an important role.

ACKNOWLEDGMENTS

We would like to thank the University of Newcastle for financial support during the course of this work. It is also a pleasure to thank A.Loiseau, W.Maser and D.Walters for very useful discussions about the present results.

REFERENCES

1. Saito, R., Dresselhaus, G., and Dresselhaus, M.S., „Phonon Structure and Raman Effect of Single-Walled Carbon Nanotubes," in *The Science and Technology of Carbon Nanotubes,* edited by Tanaka, K., Yamabe, T., and Fukui, K., Oxford: Elsevier Science, 1999, pp. 51-63
2. Brenner, D.W., *Phys.Rev. B* **42**, 9458-9466 (1990); and erratum *Phys. Rev. B* **46**, 12592-12594 (1992).
3. Yu, J., Kalia, R. K., and Vashishta, P. , *Chem. Phys.* **103**, 6697-6703 (1995).
4. Henrard, L.,Loiseau, A., Journet, C., and Bernier, P., *Eur.Phys.J. B* **13**, 661-669 (2000).
5. Ugarte, D., Stockl, T., Bonard, J.M., Chateland, A., and De Heer, W.A., „Capillarity in Carbon Nanotubes," in *The Science and Technology of Carbon Nanotubes,* edited by Tanaka, K., Yamabe, T., and Fukui, K., Oxford: Elsevier Science, 1999, pp. 128-142

Template-Based Plasma Synthesis of Carbonaceous Nanostructures

A. Huczko[*], H. Lange[*], J, Tyczkowski[†], P. Kazimierski[†], and P. Tomassi[¶]

[*]Department of Chemistry, Warsaw University, Pasteura 1, 02-093 Warsaw, Poland
[†]Centre for Molecular and Macromolecular Studies, PAS, Sienkiewicza 112, 90-363, Łódź Poland
[¶]Institute of Precision Mechanics, Duchnicka 3, 00-967 Warsaw, Poland

Abstract. The preliminary results of the experimental study on non-equilibrium plasma (glow and barrier discharges in Ar, He, H_2, and NH_3) decomposition of hydrocarbons (methane, pentane, acetylene and benzene) are presented. The resulting carbonaceous solid products were deposited onto the anodic aluminum oxide membranes with nanometer scale channels which served as templates. Various techniques (SEM, AFM, elemental analysis) were used to characterize obtained nanostructures.

INTRODUCTION

A plasma is an excited and ionized gas characterized by glow discharge-light emitted by the excited species that make up the plasma as they return to lower energy states. Non-equilibrium plasma (also called cold or low-pressure plasma) is of a primary interest for deposition processes. Energetic species and the UV radiation present in the plasma initiate them.

Carbon nanotubes (CNTs) are known to be very good electron emitters [1]. This is why since their discovery there has been a lot of speculation about the use of CNTs in the construction of field emission displays. However, a prerequisite for a major break-through in this area would be their perfect alignment on a flat surface. Many techniques resulted in growing large arrays of the well-aligned CNTs [2-4]. However, these methods mostly involve the use of catalysts at relatively high temperatures. Only recently Ren et al. [5] succeeded in using plasma–enhanced CVD for the aligned nanotubes synthesis. This technique allowed for the deposition to proceed at temperatures below 666°C, which corresponds to the strain point of the best display glass. This result can be considered as a good step toward the construction of the flat panel displays with CNTs.

In this paper we present the results of the exploratory study on the plasma decomposition of organic compounds over templated supports. The low-temperature synthesis of aligned carbon nanostructures seems to be by far superior by comparison with laser and arc plasmas or catalytic techniques. The presented method may help to overcome the main drawback of the catalytic thermal deposition of nanocarbons (temperatures at least within a few hundreds of °C) since the deposition from non-equilibrium plasma occurs at much lower temperature.

CP544, Electronic Properties of Novel Materials—Molecular Nanostructures, edited by H. Kuzmany, et al.
© 2000 American Institute of Physics 1-56396-973-4/00/$17.00

EXPERIMENTAL

The deposition of carbonaceous products onto the substrate was performed in two reactors with (i) glow discharge (system A) and (ii) barrier discharge plasma (system B).

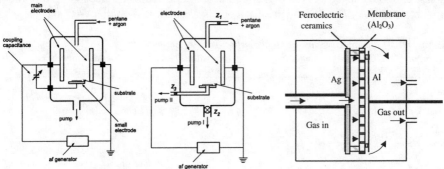

FIGURE 1. Schematic view of an audio-frequency glow discharge reactor (mode of operation: A1 - left, A2 – middle) and a dielectric barrier discharge (DBD) – right.

Low-pressure glow discharge plasma was used in a 3-electrode reactor (A1) or a 2-electrode reactor (A2) [6] - Fig. 1. The reactors worked at 20 kHz frequency and 100 W power. A mixture of pentane vapor and argon (C_5H_{12} content from 20 to 80 vol. %) at a pressure of several Pa was flown through the system. The deposition process was carried out for 5-15 min.

A dielectric barrier discharge (DBD) reactor (system B) used in this study is also sketched in Fig. 1. The key feature of DBDs is, unlike other plasma sources, to produce non-thermal plasma at atmospheric pressure. The flow of reactants (various plasma gases and carbon-bearing reactants) was within 20 sccm/min and a plasma was generated at ferroelectric sample surface applying a.c. 200 Hz frequency and 800 V voltage.

Anodic aluminum oxides (AAO) 60-100 μm thick with channels of several nm diameter were used as substrates. The application of AAO membranes in production of aligned CNTs by high-temperature catalytic decomposition of hydrocarbons was reviewed recently [7].

The morphology of deposited carbonaceous nanostructures were determined using an ultra-high resolution field emission SEM (LEO 1530) with a unique GEMINI column. Very high (3 nm) resolution was permitted at extremely low operating voltage (1 kV). Also the contact mode AFM (Digital Instruments NanoScope IIIa) observations were performed. The chemical composition of electrode deposits was evaluated by elemental analysis.

RESULTS AND DISCUSSION

In the system A1 the electric field was supposed to promote a flow of charged decomposition products through a membrane while in the system A2 a differential pressure was a driving force promoting the gas flow. For all tests performed in this reactor a homogeneous deposits were formed on the AAO template surface. The operational conditions of the experiments carried out in the system B are shown in Table 1. The presence of AAO within the plasma zone did not disturb the discharge itself, i.e. plasma species could penetrate the template channels. Depending on process parameters more or less homogeneous deposits were formed on the AAO surface.

Table 1. DBD Plasma Decomposition of Carbon-Bearing Reactants

Test No.	Gas Plasma	Reactant	Deposition duration [h]	Deposit membrane characterization
F-1	NH_3	C_2H_2	1	
F-2	H_2	C_2H_2	1	brownish homogeneous with
F-3	NH_3	CH_4	1	carbonaceous islands
F-4	H_2	CH_4	1	
F-5	NH_3	C_6H_6	0.5	
F-6	He	C_6H_6	1	fractal structure of carbonaceous
F-7	Ar	C_6H_6	1	islands
F-8	H_2	C_6H_6	1	
F-9	NH_3	C_6H_6	1	Only brownish homogeneous
F-10÷12	He	$C_6H_6+CCl_4$ $+ C_4H_4S$	1-3	deposit

A plasma gas used influences the characteristics of the discharge, e.g. the most uniform plasma was observed for He while for Ar plasma separated sparks and streamers were clearly visible. It is reflected by the morphology of the deposits, which were more homogeneous for helium plasma. Black fractal "flowers" were observed in deposits when the plasma was non-homogeneous. Thus, the local spark discharge resulted in total decomposition of hydrocarbons. It was confirmed by the elemental analysis of solid deposits scraped off the metallic electrode, e.g. for the test No. F-6 the C/H atomic ratio was higher (1.25) comparing to the starting reactant (C_6H_6). Also the kind of hydrocarbon influenced the morphology of the deposit. Products of methane decomposition were solid while for C_2H_2 and C_6H_6 a sticky and viscous deposit was formed, which later solidified. The deposition process is not a time-independent. The deposition rate is higher at the beginning and stabilizes with evolving time. It may suggest that the formed deposit influenced somehow the characteristics of the deposition. Such conclusion was confirmed by SEM observations of the deposits obtained – Figs 2 and 3. The double structure was revealed with thin homogeneous film covering and clogging the template surface and „cauliflower-like" regular islands of nanometer-size on it. Such morphology of the layer was also confirmed by AFM observation. Due to the excessive flow of reactants and a high deposition rate in those preliminary tests the penetration of channels by decomposition products was poor. Also, the carbonization was not high enough. We have shown, however, that plasma deposition by means of non-equilibrium decomposition of hydrocarbons offers a good way to deposit carbonaceous nanomaterials onto the AAO membranes. The studies are under way to deposit nanocarbons within the pores of AAO (and also track membranes) by plasma technique. This would be of a particular importance in search for the low temperature technique of aligned CNTs production.

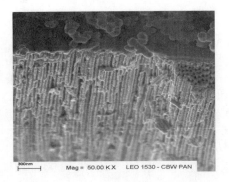

FIGURE 2. SEM of AAO template structure and of the deposits (system A2): left - cross-section, right – surface.

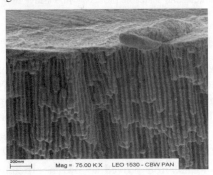
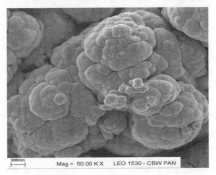

FIGURE 3. SEM of AAO template structure and of the deposits (system DBD, test F-6): left - cross-section, right – surface.

ACKNOWLEDGMENTS

The work was supported by the Committee for Scientific Research (KBN) through the Department of Chemistry, Warsaw University (grant No. 3 T09A 05816).

REFERENCES

1. Terrones, M., Hsu, W. K., Kroto, H. W., and Walton, D. R. M., *Topics in Current Chem.*, **199**, 190-234 (1999).
2. Huang, S., Dai, L., and Man, A. W. H., *J. Phys. Chem. B*, **103**, 4223-4227 (1999).
3. Kyotami, T., Tsai, L., and Tomita, A., *Chem. Commun.*, 701-705 (1997).
4. Che, G., Lakshmi, B. B., Martin, C. R, Fischer, E. R., and Ruoff, R. S., *Chem. Mater*, **10** (1), 260-265 (1998).
5. Huang, Z. P., Xu, J.W., Ren, Z. F., Wang, J. H., Siegal, M. P., and Prorencio, P. N., *Appl. Phys. Lett.*, 73, 1998, 3845-3850.
6. Tyczkowski, J., *J. Vac. Sci. Technol.* **A 17(2)**, 470-478, (1999).
7. Huczko, A., *Applied Physics A*, in print.

Iron Catalyzed Growth of Carbon Nanotubes

Frank Rohmund, Lena K.L. Falk, Eleanor E.B. Campbell

School of Physics and Engineering Physics, Gothenburg University and Chalmers University of Technology, SE-412 96 Gothenburg, Sweden

Abstract: The catalytic activity of iron particles is employed to grow carbon nanotubes in various modifications. Single-walled nanotubes (SWNT) are obtained from catalytic disproportionation of CO in the gas phase in a continuous and scaleable process, while arrays of aligned multi-walled nanotubes (MWNT) are grown efficiently on deposited iron particles from acetylene. In both cases $Fe(CO)_5$ serves as iron-containing precursor. These nanotube synthesis methods are efficient and simple and therefore interesting for technical applications.

Transition metal-catalyzed chemical vapor deposition (CVD) provides a way for the production of carbon nanotubes at moderate temperatures as demonstrated in many studies. The technique employs the catalytic disproportionation of a carbon-containing gas (hydrocarbon molecules, CO) on small metal particles, which are supported on inert substrates. Both MWNT[1] and SWNT[2,3] can be obtained via this method. In the former case it is possible to synthesize films of aligned MWNT by various CVD-related processes[4-7]. Among the transition metals, iron proved to be an efficient catalyst for CVD production of carbon nanotubes. We study the production of carbon nanotubes on iron clusters in the gas phase as well as on supported iron catalyst particles from CO and acetylene. In the case of the CO carbon feedstock, SWNT grow in the gas phase with no nanotube growth on supported iron particles[8] while MWNT are obtained on supported iron particles from acetylene. The latter process provides a simple means for production of large arrays of aligned MWNT [9].

Both gas phase production of SWNT and synthesis of MWNT on supported iron particles is performed in a horizontal tube furnace apparatus as discussed in detail elsewhere[8]. SWNT growth from CO in the gas phase was demonstrated initially by Nikolaev et al.[10]. In our experiment, a mixture of CO and $Fe(CO)_5$ and, optionally, H_2 is fed into the hot (1100 °C) zone of the furnace through a water-cooled injector. The thermal dissociation of the carbonyl leads to formation of iron clusters, which serve as catalysts for the nanotube growth via the Boudouard reaction. This process produces only SWNT and iron particles that are encapsulated in graphitic carbon shells. No MWNT and no or little formation of amorphous carbon are observed[8,10]. Addition of hydrogen to the $CO/Fe(CO)_5$ mixture increases the SWNT yield[8]. Since the commercial CO used in our experiments contains a contamination of iron pentacarbonyl, SWNT growth can be achieved without any additional $Fe(CO)_5$. Fig. 1

CP544, *Electronic Properties of Novel Materials—Molecular Nanostructures*, edited by H. Kuzmany, et al.

shows a TEM micrograph of SWNT material, which was produced from a CO/H_2 mixture with 10 % hydrogen and no additional carbonyl at 1100 °C and 1 atm.

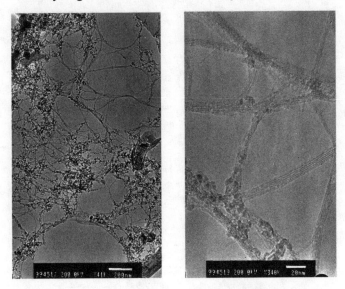

Figure 1. SWNT obtained from iron-catalyzed disproportionation of CO in the gas phase

The lower magnification image in the left part of the figure shows entangled filamentous material. The higher magnification image on the right side reveals that the filaments consist of ropes of SWNT. The ropes are not covered with amorphous carbon debris in contrast to raw SWNT material obtained by laser vaporization[11] or arc discharge[12]. The SWNT are not clean, however, but are decorated with iron particles. These iron particles are encapsulated with several layers of graphitic carbon as observed in high resolution TEM micrographs[8]. The amount of decoration with encapsulated particles relative to the amount of SWNT can be reduced by efficient mixing of the reactive gases and application of high pressure CO in the growth region[10]. From the observations that a) only SWNT but no MWNT are formed in this process, b) the decorating particles have diameters larger than approx. 2.5 nm, c) the SWNT have a broad diameter distribution and d) the particles are encapsulated in *multiple* layers of carbon, the nanotube growth scenario depicted in fig. 2 can be proposed[2,8]. Small iron particles grow from iron atoms supplied by the carbonyl and catalyze the Boudouard reaction, accompanied by the dissolution of a large fraction of carbon in the metal particle. If the particle diameter reaches that of a C_{60} molecule (approx. 7 Å) it can be capped with a carbon hemisphere. In the diameter range between 0.7 and approx. 2.5 nm the nucleation of a SWNT is energetically favored compared to complete encapsulation of the iron (carbide) particle. The situation is reversed for particles larger than 2.5 nm which prefer to encapsulate. Multiple encapsulation is the consequence of iron-carbon phase separation of the particle.

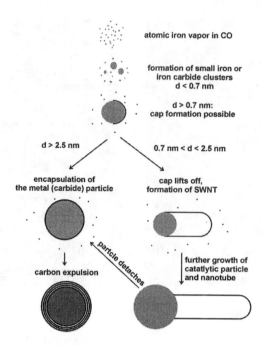

atomic iron vapor in CO

formation of small iron or
iron carbide clusters
d < 0.7 nm

d > 0.7 nm:
cap formation possible

d > 2.5 nm   0.7 nm < d < 2.5 nm

encapsulation of   cap lifts off,
the metal (carbide) particle   formation of SWNT

carbon expulsion   particle detaches   further growth of
catatlytic particle
and nanotube

Figure 2. Schematic scenario of the iron catalyzed growth of SWNT in the gas phase from CO

Addition of acetylene has no impact on the SWNT growth process in the gas phase besides producing amorphous carbon coatings but leads to growth of MWNT on iron particles which are deposited in the hot zone of the tube furnace[8]. No SWNT are synthesized on the supported iron particles from acetylene, probably because the metal particles have too large diameters. Instead, large arrays of aligned MWNT can be grown in the hot zone of the furnace on substrates like silicon[9]. In fig. 3 two SEM images of an aligned MWNT film, grown at 750 °C on a piece of silicon wafer, are depicted. The iron coating on the substrate was obtained by dissociation of $Fe(CO)_5$ at 200 °C with subsequent nanotube growth from acetylene in the absence of carbonyl at 750 °C in the presence of an excess of Ar buffer gas. The material consists entirely of MWNT with a high degree of alignment. Similar films can be obtained under simultaneous supply of carbonyl and acetylene at 750 °C. This method for aligned MWNT synthesis is extraordinarily efficient and simple compared to other methods considering that no separate surface coating process is necessary and only very simple reactants are used.

The presented iron catalyzed CVD processes for carbon nanotube production provide methods for the synthesis of two different classes of nanotubes, SWNT and aligned MWNT. The gas phase method for SWNT production is easily up-scaleable and therefore represents a possible way for mass production of SWNT. The process for making films of aligned MWNT makes simple production of nanotube coatings for technical applications feasible.

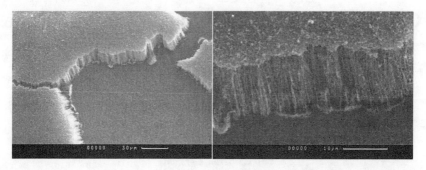

Figure 3. SEM images of aligned MWNT grown on silicon by catalytic disproportionation of acetylene on deposited iron particles

REFERENCES

[1]A. Fonseca, K. Hernadi, P. Piedigrosso, J.-F. Colomer, K. Mukhopadhyay, R. Doome, S. Lazarescu, L.P. Biro, Ph. Lambin, P.A. Thiry, D. Bernaerts, and J.B. Nagy, Appl. Phys. A **67**, 11 (1998)

[2]J.H. Hafner, M.J. Bronikowski, B.R. Azamian, P. Nikolaev, A.G. Rinzler, D.T. Colbert, K.A. Smith, and R.E. Smalley, Chem. Phys. Lett. **296**, 195 (1998)

[3]J.-F. Colomer, G. Bister, I. Willems, Z. Konya, A. Fonseca, G. Van Tendeloo, and J.B. Nagy, Chem. Comm. 1343 (1999)

[4]W.Z. Li, S. Xie, L.X. Qian, B.H. Chang, B.S. Zou, W.Y. Zhou, R.A. Zhao, and G. Wang, Science **274**, 1701 (1996)

[5]Z.F. Ren, Z. P. Huang, J.W. Xu, J.H. Wang, P. Bush, M.P. Siegal, and P.N. Provencio, Science **282**, 1105 (1998)

[6]S. Fan, M.G. Chapline, N.R. Franklin, T.W. Tombler, A.M. Cassel, and H. Dai, Science **283**, 512 (1999)

[7]H. Kind, J.-M. Bonard, Ch. Emmenegger, L.O. Nilsson, K. Hernadi, E. Mailard-Schaller, L. Schlapach, L. Forro, and K. Kern, Adv. Mat. **11** (15), 1285 (1999)

[8]K. Bladh, L.K.L. Falk, and F. Rohmund, Appl. Phys. A **70** (3) (2000)

[9]F. Rohmund, L.K.L. Falk, and E.E.B. Campbell, submitted for publication (2000)

[10]P. Nikolaev, M.J. Bronikowski, R.K. Bradley, F. Rohmund, D.T. Colbert, K.A. Smith, and R.E. Smalley, Chem. Phys. Lett. **313**, 91 (1999)

[11]A. Thess, R. Lee, P. Nikolaev, H. Dai, P. Petit, J. Robert, C. Xu, Y. H. Lee, S.G. Kim, A.G. Rinzler, D.T. Colbert, G.E. Scuseria, D. Tomanek, J.E. Fischer, and R.E. Smalley, Science **273**, 483 (1996)

[12]C. Journet, W.K. Maser, P. Bernier, A. Loiseau, M. Lamy de la Chapelle, S. Lefrant, P. Deniard, R. Lee, and J.E. Fischer, Nature **388**, 756 (1997)

Calculation of Single-wall Carbon Nanotube Diameters from Experimental Parameters

H. Kanzow, C. Lenski, and A. Ding

Technische Universität Berlin, Optisches Institut, Sekr. P 1-1, Str. d. 17. Juni 135,
10623 Berlin, Germany
kanzow@physik.tu-berlin.de

Abstract. The growth of bundles of single-wall carbon nanotubes is explained by assuming a transition state, in which precipitated graphene sheets detach from the surface of the liquid catalyst particle forming caps thereby avoiding dangling bonds at the graphene edges. By considering the energetic situation of the transition state we are able to calculate the diameter distribution of the tubes depending on their formation temperature and the composition of the catalytic metal and carbon containing particles.

INTRODUCTION

SWCNTs have been produced for some years by co-vaporisation of pure carbon and a metal catalyst either in an electric arc discharge [1-4] or by high power laser irradiation [5-9]. However, the mechanism of the tube growth is still unclear. The occurrence of sea urchin morphologies in the experiments [1,2] strongly favour a root growth model, where the formation takes place at a large catalytic particles: The carbon and metal atoms condense at high temperatures forming liquid droplets. Fullerene like caps attached to them are prolonged by segregated carbon atoms, which are incorporated leading to a single-wall growth [1,2,10,11]. The temperature range, where SWCNT growth takes place in the arc experiment, was determined by Saito et al. to be between 650 to 1100°C [3]. Heating of the area around the plasma led to an increase of the SWCNTs diameters [4,6,7,9]. The diameter distribution was also strongly dependent on the metal used as catalyst in the plasma methods [2,3]: Mean diameters of 1.0 - 1.1 nm were achieved with iron, 1.1 - 1.2 nm with nickel and 1.3 - 1.4 nm using cobalt as catalyst in the arc.

Under experimental conditions melting temperatures of the catalytic particles are drastically reduced because of the oversaturation of the particles with carbon and the surface energy associated with the small dimensions [12]. Carbide formation is not possible for Fe, Ni and Co at these reaction temperatures [13].

CP544, *Electronic Properties of Novel Materials—Molecular Nanostructures*, edited by H. Kuzmany, et al.
© 2000 American Institute of Physics 1-56396-973-4/00/$17.00

Growth Model

The formation of SWCNTs is thought to occur at liquid metal and carbon containing particles at temperatures between 650 and 1100°C [3]. The reaction scheme was presented in more details in [11]. In short, a flat small graphene sheet is attached to the surface of a metal and carbon containing particle. The segregation of carbon atoms near the graphene sheet and therefore the sheet growth is thought to take place slowly compared with average time needed that a special oscillation state occurs. This state is characterised by a collective vibrational movement of all carbon atoms in the sheet (except at the sheet border) away from the sheet particle interface. This will happen, if there is enough kinetic energy to overcome the work of adhesion and the work of bending of the sheet. Other oscillating states are not considered, because they do not lead to the cap formation. From a certain angle of the unsaturated sp^2 orbitals on the edges of the graphitic plane to the particle surface (which we arbitrarily set to 90° in our calculations), the cap is stabilised by the interaction with the metal orbitals. Now we consider the energetic situation of the transition state (Fig. 1, Tab. 1):

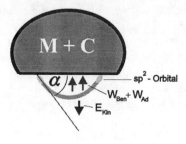

Figure 1. Model of the transition state: A fullerene cap is formed from a graphene sheet on a catalytic particle.

$$W_{Ben} \approx \frac{\Delta H_f}{A_C} - 2 \cdot E_{S,G} = \frac{e}{\pi \cdot d^2}\left[A + B \cdot \ln\left(\frac{\pi \cdot d^2}{A_C}\right)\right] - 2 \cdot E_{S,G} \tag{1}$$

$$W_{Ben} = E_{kin} - W_{Ad} \tag{2}$$

$$E_{Kin} = \frac{k \cdot T}{A_C} + \frac{k \cdot T}{A_{MC}} \tag{3}$$

$$A_{MC} \approx \left[X_C \cdot a_{C-C} + (1 - X_C) \cdot a_{M-M}\right]^2 \tag{4}$$

$$W_{Ad} \approx X_C \cdot W_{Ad,G} + (1 - X_C) \cdot W_{Ad,M} \tag{5}$$

$$W_{Ad,G} \approx 2 \cdot E_{S,G} \tag{6}$$

Table 1. Equations to calculate tube diameter distributions: A_C, Area of a carbon atom in graphite; A_{MC}, Area of an atom in the surface of the catalytic particle; a_{M-M}, Atom distance in the pure metal M, a_{C-C}, Bond length in graphite; d, Cap (= tube) diameter; A,B, Constants from the fit in fig. 3 of [14]; $E_{S,G}$, Free surface energy of graphite per area; ΔH_f, Heat of formation for an atom in a fullerene cage; E_{Kin}, Kinetic energy per area at the interface between the graphene sheet and the particle; X_C, Molfraction of carbon in the surface of the catalytic particle; Temperature: T, Work of adhesion per area: W_{Ad}, Work of adhesion per area of graphite to liquid carbon: $W_{Ad,G}$, Work of adhesion per area of graphite to the metal M: $W_{Ad,M}$, Work of bending per area: W_{Ben}

The necessary equations (3) to (6) to calculate (2) have been discussed in our previous paper in details [11]. The free energy available for bending the sheet into a fullerene like cap is given by (2). Note that we had to correct (3) [11]. The parameter related to the type of metal used is the temperature dependent function of the work of adhesion per area of graphite to the liquid metal $W_{Ad,M}$ (5). If mixed metals are considered, a linear combination of the $W_{Ad,M}$ - functions of the single metals is used. The eq. (1) is new and combines the available free energy for bending with the diameters of the fullerene caps. It derives from the fit function of the elastic theory of Thersoff [14] to heat of formation data ($\approx$ free energy of formation) of different fullerenes calculated by Zhang et al. [15].

For arc experiments we can calculate the diameter distribution of tubes in the soot, not for the collarette, because the formation temperature range is only known for the soot (650 - 1100 °C) [3]. The other necessary parameter is the carbon content of the catalytic particle. The original composition of the anode material is usually about 99 at. % carbon and only 1 at. % metal. Then the carbon is transformed into the tubes and the metal accumulates in the particle until it is of almost pure metal. Therefore we set the parameter range for the carbon content to be between 0 to 1. The temperature and carbon content ranges were divided in 51 equidistant values each. Every carbon content value was combined with every temperature value. For the resulting 2601 combinations the diameters were (numerically) calculated with (1) - (6). In the laser experiments the lower temperature limit is set by the oven temperature. As the upper limit we also chose 1100°C as in the arc experiments, thought one should keep in mind that SWCNTs can be grown at slightly higher temperatures [6].

RESULTS

The detailed results for the pure metal catalysts Fe, Ni and Co in the arc are reported elsewhere [16]. For nickel and iron the calculated diameter distributions agree very well to the experimental data of Seraphin and Saito et al. [2,3]. In the case of cobalt the model overestimates the mean tube diameter (1.6 nm instead of 1.3 - 1.4 nm). May be this is a result of a weakness of the experimental data we use as the work of adhesion [11,17].

Fig. 2 shows the diameter distributions with the mixed metal catalyst Fe/Ni, which was calculated for the laser experiments for oven temperatures of 920°C. The corresponding tube diameters observed with TEM by Bandow et al. (Fig. 2 in [7]) are included in the Fig. 2 as grey squares. The original data was scaled up in order to compare the results with our calculations. The mean diameter as well as the width of the distributions are very similar to the experimentally observed ones.

Further mean diameters for other mixed metal laser experiments are shown in Fig. 3: The increase of the mean diameter with the formation temperature is correctly reproduced.

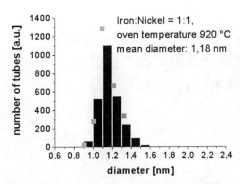

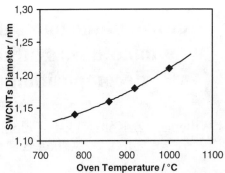

Figure 2. Calculated SWCNT diameter distribution of a laser experiment with an Fe/Ni - 1/1 mixture as the catalyst and an oven temperature of 920°C. The scaled abun-dances of the diameters observed by Bandow et al. (Fig. 2 in [7]) are included as grey squares for comparison.

Figure 3. Calculated mean diameters SWCNTs in the products of laser experiment with Fe/Ni - 1/1 mixtures as the cata-lysts and oven temperatures between 780 and 1000°C .

REFERENCES

1 Y. Saito, Carbon **33**, 979-988 (1995).
2 S. Seraphin, *J.Electrochem. Soc.* **142**, 290-297 (1995).
3 Y. Saito, T. Koyama, K. Kawabata, *Z. Phys. D* **40**, 421-424 (1997) .
4 M. Takizawa et al., *Chem. Phys. Lett.* **302**, 146-150 (1999).
5 T. Guo et al., *Chem. Phys. Lett.* **243**, 49-54 (1995)
6 A.G. Rinzler et al., *Appl. Phys. A* **76**, 29-37 (1998).
7 S. Bandow et al., *Phys. Rev. Lett.* **80**, 3779-3782 (1998). E. Anglaret, S. Rols, and J.-L. Sauvajol, *Phys. Rev. Lett.* **81**, 4780 (1998)
8 H. Kataura et al., *Jpn. J. Appl. Phys.* **37**, L616-L618 (1998).
9 F. Kokai et al., *J. Phys. Chem. B* **103**, 4346-4351 (1999).
10 A. Maiti, C. Brabec, J. Bernholc, *Phys. Rev. B* **55**, R6097-R6100 (1997).
11 H. Kanzow, A. Ding, *Phys. Rev. B* **60**, 11180-11186 (1999).
12 K. F. Peters, J. B. Cohen, and Y.-W. Chung, *Phys. Rev. B* **57**, 13430-13438 (1998).
13 H. Kanzow, submitted to *Chem. Phys. Phys. Chem.*, March (2000)
14 J. Thersoff, *Phys. Rev. B* **46**, 15546-15549 (1992).
15 B. L. Zhang, C. H. Xu, C. Z. Wang, C. T. Chan, and K. M. Ho, *Phys. Rev. B* **46**, 7333-7336 (1992).
16 H. Kanzow, C. Lenski, and A. Ding, to be published
17 W. Weisweiler, V. Mahadevan, *High Temp. - High Press.* **4** (1972) 27-34

Control of the outer diameter of thin carbon nanotubes synthesized by catalytic decomposition of hydrocarbons

I.Willems,[*a] Z. Kónya,[a] J.-F. Colomer,[a] G. Van Tendeloo,[b] N. Nagaraju,[c]
A. Fonseca[a] and J. B.Nagy[a]

[a]Laboratoire de Résonance Magnétique Nucléaire, Facultés Universitaires Notre-Dame de la Paix, 61
rue de Bruxelles, B-5000 Namur, Belgium
[b]EMAT, University of Antwerp (RUCA) Groenenborgerlaan 171, B-2020 Antwerp, Belgium
[c]St.Joseph's College, Residency Road, Bangalore 560-025, India

Abstract. Multi-wall carbon nanotubes were produced by catalytic decomposition of acetylene. Co-Mo, Co-V and Co-Fe mixtures supported either on zeolite or corundum alumina were used as catalysts. When Fe or V is added to Co, the amount of carbon deposit increases. The nanotubes were characterized by both low and high resolution TEM. From histograms representing the outer diameter distributions, it is made clear that the outer diameter of the nanotubes can be controlled, choosing the appropriate catalyst.

INTRODUCTION

Research in the field of carbon nanotubes has undergone an explosive growth since their discovery [1]. The catalytic decomposition of hydrocarbons in presence of different supported transition metal catalysts has already been investigated. In earlier studies [2,3], different synthesis parameters were investigated such as the time and the temperature of the reaction, the gas flows, the nature of the hydrocarbon used and the type of the catalyst. The best results were obtained with Fe and Co both of which gave high yield of multi-wall carbon nanotubes (MWNTs). The results of these experiments led us to investigate the influence of the mixtures of different transition metals such as Co/V, Co/Mo and Co/Fe as catalysts in the production of high yields of MWNTs. Our first results using mixtures of metals were the production of quasi-aligned MWNTs on Co-V catalysts and the production of single-wall carbon nanotubes (SWNTs) on Co-Fe-Ni catalysts [4]. In the present study, we report the influence of the metal contents and the nature of the support on the yield and quality of the nanotubes.

EXPERIMENTAL

Different transition metals (Co, V, Mo and Fe) and their mixtures were used as catalysts supported on zeolite (NaY) and corundum alumina (CA). The supported catalysts were prepared by impregnation of NaY/CA with a solution of the transition metal salts keeping the total metal amount at 5 wt%.

CP544, *Electronic Properties of Novel Materials—Molecular Nanostructures*, edited by H. Kuzmany, et al.
© 2000 American Institute of Physics 1-56396-973-4/00/$17.00

The syntheses of the carbon nanotubes were performed in a fixed-bed reactor. The acetylene was passed for 1h at a temperature of 700°C over the catalyst spread on a quartz boat. The flow of the carrier gas (nitrogen) and of the reacting gas (acetylene) were set to 300 ml/min and 30 ml/min, respectively. After the reaction, the carbon deposit was calculated as follows: carbon deposit (%) = $100(m_{tot.}-m_{cat.})/ m_{cat.}$
where $m_{cat.}$ is the initial amount of catalyst and m_{tot} is the total mass of the sample after reaction. The samples were characterized by transmission electron microscopy both at low (Philips CM 20) and high resolution (Jeol CX 200).

RESULTS

Carbon deposit

Earlier experiments in the field showed that cobalt is necessary for the formation of MWNTs of good quality, while iron is very active in the decomposition of hydrocarbons. A mixture of metals could combine these two advantages and led to a very high activity in the formation of MWNTs of good quality. The syntheses were first performed with catalysts containing only one supported metal. Later, the catalytic activity of Co/V, Co/Mo and Co/Fe catalysts supported either on zeolite or on corundum alumina were measured. The activities of the catalysts in the decomposition of acetylene, expressed as the percentage of carbon deposit, are given in Table 1.

TABLE 1. Carbon deposit (%) by acetylene decomposition over different catalysts.

Co(wt%)	0%	1%	2.5%	4%	5%
X*(wt%)	5%	4%	2.5%	1%	0%
Co-V/Zeolite	0	46	116	55	25
Co-V/Alumina	0	6	127	18	2
Co-Mo/ Zeolite	0	-	30	-	25
Co-Mo/ Alumina	0	-	25	-	2
Co-Fe/ Zeolite	48	38	120	144	25
Co-Fe/ Alumina	11	75	157	8	2

*X=V, Mo or Fe

At first it is observed that in general, mixtures of metals greatly improve the quantity of carbon deposition. The carbon deposits obtained with a single metal, i.e. iron or cobalt, are more than doubled when these metals are mixed together in equal amount or when cobalt is mixed with vanadium.

Molybdenum and vanadium alone are totally inactive in the production of nanotubes by the decomposition of acetylene. The addition of Mo to Co does not improve the activity of the catalyst on zeolite support, but it does on alumina support. On the other hand, iron alone on NaY is active in the synthesis of carbon nanotubes. But in this case, the quality of the nanotubes is poor [5]. The addition of Fe or V to Co improves the yield, but only for certain compositions of metals. However, a difference exists between the influence of iron and that of vanadium. As far as V is concerned, the increase of carbon deposit is important especially when Co-V is in 1:1 ratio. A smaller increase is recorded on zeolite for other ratios of Co and V i.e. 1%-4%, 4%-1%. As far as Fe is concerned, increases of carbon deposits are observed for particular metal contents, such as 2.5%-2.5% and 4%-1% on zeolite, and 2.5%-2.5% and 1%-4% on alumina. In conclusion, we can say that the most active supported catalysts for MWNTs synthesis contain Co mixed with V or Fe.

Mean outer diameter of nanotubes

To get a better idea of the influence of the catalyst on the thickness of the nanotubes, the outer diameters of the nanotubes have been measured from high resolution TEM pictures. The average outer diameters generally range from 7.5nm to 18nm. This is quite small in comparison with the nanotubes produced by arc-discharge [6] or laser ablation [7] methods.

TABLE 2. Mean outer diameter of nanotubes synthesized on different supported catalysts with equal metal contents.

	2.5%-2.5%
Co-V/NaY	7.5 nm
Co-V/CA	10.3 nm
Co-Mo/NaY	10.0 nm
Co-Mo/CA	10.0 nm
Co-Fe/NaY	17.8 nm
Co-Fe/CA	12.8 nm

From the data in Table 2, it is obvious that it is possible to synthesize nanotubes with a desired outer diameter distribution by choosing the catalyst to be used.

The nanotubes samples synthesized with Co-V/NaY contain tubes with smallest outer diameters, even if we consider the 1%-4% and 4%-1% metal compositions which give outer diameters of 9.2 nm and 8.1 nm, respectively. As already pointed out previously, this catalyst is very active, which makes it a very interesting candidate for the synthesis of narrow tubes. Explanation for its behavior may be found in the dispersion of the active particles at the surface of the support. Since the flow of acetylene is responsible for the thickness of the tubes, we can suppose that there is a better dispersion of the catalytic particles, in the case of Co-V mixtures, so that the quantity of acetylene, coming into contact with the active sites, is decreased. Besides, it is for mixtures of Co-V in equal amounts that we obtained not only the highest yields but also the thinnest tubes compared to the other compositions. In fact, if the amount of cobalt or vanadium is increased on the support, the outer diameters of nanotubes become larger.

For some applications of nanotubes, it can be interesting to have wider distributions of outer diameters. In this case, Co-V supported on CA and Co-Mo supported on CA or NaY (equal amounts of metals) could be used, knowing that the highest carbon deposit is obtained with Co-V supported on CA.

Finally, if we want to favor a distribution spread on a larger range of outer diameters, a mixture of Co and Fe supported on NaY or on CA can be chosen because the carbon deposit is high. However, the zeolite is preferred to corundum alumina because it is much easier to separate the nanotubes from zeolite than from alumina.

As far as the quality of the nanotubes is concerned, we can see, on high resolution pictures, that the best graphitization of nanotubes is observed in the samples synthesized with Co-Mo catalysts. The walls of nanotubes synthesized with Co-Fe or Co-V, are not very straight (Figure 1).

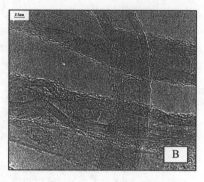

FIGURE 1. High magnification transmission electron microscope images of carbon nanotubes showing the graphitization of nanotubes synthesized on (A) Co-Mo/NaY and (B) Co-V/CA.

CONCLUSION

Depending on the catalyst used, it is possible to control some characteristics of the carbon nanotubes especially the outer diameter distributions and the carbon deposit. We can synthesize thin nanotubes, in the range of 7.5 nm to 9.2 nm, in high yields using a mixture of Co and V supported on zeolite NaY. These thin nanotubes are of special interest for their electronic and gas storage properties.

To favor the quality of the tubes, Co-Mo catalysts are suggested. Nevertheless, for applications requiring nanotubes of lower quality, Co-V catalysts will be prefered.

ACKNOWLEDGMENTS

I.W. acknowledges financial support from F.R.I.A., Belgium and K.Z. for Hungarian State Eotvos Fellowship. The authors thank the European Commission,TMR contract NAMITECH, ERBFMRX-CT96-0067 (DG12-MIHT) and the PAI initiated by the Belgian State, Prime Minister's Office of Science Policy Programming (4/10).

REFERENCES

1. Iijima S., *Nature* **354**, 56 (1991)
2. Yacamàn M.J., Yoshida M.M., Rendon L., Santiesteban J.G., *Appl. Phys. Lett.* **62**, 202 (1993)
3. Ivanov V., Fonseca A., B.Nagy J., Bernaerts D., Fudala A., Lucas A.A., *Zeolites* **17**, 416 (1995)
4. Colomer J.-F., Bister G., Willems I., Kónya Z., Fonseca A., Van Tendeloo G. and B.Nagy J., *J.Chem.Soc. Chem.Comm.* **14**, 1343 (1999)
5. Hernadi K., Fonseca A., B.Nagy J., Bernaerts D., Lucas A., *Carbon* **34**, 1249 (1996)
6. Journet C., Maser W.K., Bernier P., Loiseau A., Lamy de la Chapelle M., Lefrant S., Deniard P., LeeR., Fischer J.E., *Nature* **388**, 756 (1997)
7. Kanzow H., Schmalz A., Ding A., *Chem.Phys.lett.* **295**, 525 (1998)

Novel Purification Procedure and Derivatization Method of Single-Walled Carbon Nanotubes (SWNTs)

Michael Holzinger[a], Andreas Hirsch[a], Patrick Bernier[b],
Georg S. Duesberg[c,d], Marko Burghard[d]

[a] Institut für Organische Chemie, Henkestr.42, 91054 Erlangen, Germany.
[b] GDPC, Universite de Montpellier II, CC26, 34095 Montpellier Cedex 05, France.
[c] Physics Dept., Trinity College Dublin, College Green, Dublin 2, Ireland.
[d] Max-Planck-Institut für Festkörperforschung, Heisenbergstr. 1, 70569 Stuttgart, Germany.

Abstract. A new purification procedure is introduced, which uses the advantages of both, column-chromatography and vacuum-filtration. Potassium polyacrylate was used as a stationary phase. This method is based on the idea that the size of the existing cavities in the polymer increases during a swelling process in distilled water. The cavities are big enough to entrap nanoparticles, but allow for a free movement of nanotubes and bundles. The procedure starts with an oxidation step to remove part of catalyst and nanoparticles. In this step a chemical modification of the SWNTs occurs, namely the oxidation of cage carbon atoms to carboxylic groups as well as to hydroxyl- and carbonyl-groups.[1] In contrast to Haddon,[2] we use an alternative derivatziation of carboxylic acid groups in making amides in water. AFM images of the reaction products show clearly that the SWNTs have also been oxidized on their sidewalls.

INTRODUCTION

SWNT raw material prepared via arc discharge consists of tubes which are different in diameter and length. During the preparation of SWNTs, other carbon species such as fullerene polyhedra and other graphitized carbon structures, as well as amorphous carbon are formed simultaneously. There are also metal cluster impurities, sticking on the tips of the ropes and interconnecting the SWNTs. Therefore, effective purification methods are very important,[3] in order to explore the enormous potential of proposed applications. Further efforts concern the chemical behavior of SWNTs. For example, it is still unclear what really happens during the oxidation step. Another big obstacle is to obtain stable solutions in organic solvents. All these open questions constitute a serious challenge to synthetic chemists.

RESULTS AND DISCUSSION

The process involves three steps. First, raw material is treated in 65% nitric acid for 3h under reflux (typically 150 ml of acid *per* 100 mg of raw material). During this time a weight loss of about 20% takes place.

CP544, *Electronic Properties of Novel Materials—Molecular Nanostructures*, edited by H. Kuzmany, et al.
© 2000 American Institute of Physics 1-56396-973-4/00/$17.00

The next step is the treatment of the SWNTs with an ultrasonic tip for 1 min. This treatment reduces the size of the nanoparticles. Tubes and bundles are shortened as well, but to a lower extend. As a consequence, the size difference between SWNTs and degraded particles increases.

The final and most important step is a chromatographic separation using a stationary phase consisting of potassium polyacrylate swollen in distilled water. In contrast to common chromatographic procedures, the speed of this method can be increased by the application of vacuum. In this way, the swollen polymer particles are squeezed like a sponge until elution stops. SWNTs which are too big to be encapsulated by the cavities move through the space between the swollen polymer particles and elute in high quantity (more than 40 %mass) as the first fraction. The remaining material still contains a considerable amount of SWNTs which can be eluted in lower quality (Figure 2) by repeated swelling of the gel with distilled water and application of vacuum. Most of the degradation products remain within the cavities of the polymer.

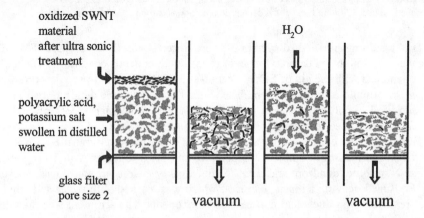

Figure 1:   Schematic representation of the filtering procedure using potassium polyacrylate swollen in distilled water.

Raman spectroscopy was performed to monitor the purification process. The D-line around 1350 cm^{-1} and the G-line, centered at 1580 cm^{-1}, is normally attributed to the carbon impurities and the SWNTs in a sample, respectively. Their relative intensities provide a measure for the purity of the sample.

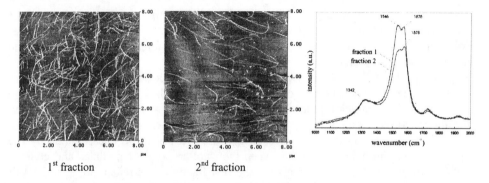

1st fraction   2nd fraction

Figure 1. AFM images and Raman spectra of the two collected fractions. The first fraction shows a large number of SWNT bundles and only a very small amount of impurities (single dots). These impurities mostly consist of amorphous carbon. On the right side, Raman spectra of the two collected fractions are shown. The different intensities of the G-line about 1560 cm^{-1} compared to the similar (scaled) intensities of the D-line at 1342 cm^{-1} reveal the decreased amorphous carbon content in the collected fractions.

The resulting purified dispersions were directly used to prepare SWNTs with chemically modified carboxylic acid groups under aqueous conditions. This goal could be achieved with the aid of EDC (1-ethyl-3-(3-dimethylamino-propyl)carbodiimide) as a water soluble coupling reagent. The addend is a Newkome dendron, that has already been used to solubilize C$_{60}$ in water.[4][5] A vast molecular excess of EDC and the dendron is needed to obtain a significant amount of product.

Th reaction was carried out by using a 0.5 mg/ml dispersion of purified SWNTs, which was stirred for 3 d at ambient temperature and pH 5 together with 10 mg/ml each of the Newkome dendron and EDC. The weakly acidic mixture was subsequently centrifuged, leaving a black sediment, which was washed with water and ethyl acetate to remove by-products and non-reacted components. The sediment was then dispersed in ethanol by ultrasonic agitation and centrifuged again. Finally, the supernatant solution, containing soluble tubes, was decanted.

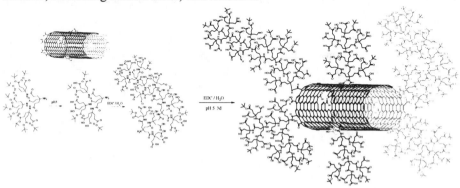

Figure 2. Amide coupling with dendritic amines in water.

In order to obtain solutions in organic solvents, the Newkome dendron was allowed to react in a weakly acidic aqueous dispersion. A very pleasant effect in this case is the partial polymerization of the dendron. Caused by the partial deprotection of the *tert*-butylic esters, generating free carboxylic acid groups, forming amide bonds with other dendra. Thus formed large dendra are eventually then attached to the carboxylic acid groups of the SWNTs both at the ends and the sidewalls.

As shown in Figure 4, the resulting dendritic polymer, which is covering the SWNTs, is sufficiently big to be detected by AFM.

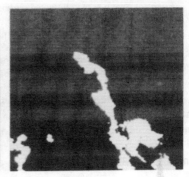

Figure 3. AFM images of modified, ethanol soluble SWNTs.

In conclusion, a straightforward purification procedure and a new method to functionalize SWNTs have been developed. As for the purification, there is no need for expensive filter systems which makes it possible to purify a large amount of raw material within hours. The resulting SWNTs show high purity in neutral aqueous environment. No surfactants or additives are required to obtain highly dispersed, stable suspensions. Amide linkage was successfully employed to decorate the dispersed and partially oxidized SWNTs with the Newkome dendra.

REFERENCES

[1]   G.S. Duesberg, J. Muster, S. Roth, M. Burghard, *Appl. Phys. A* **1999**, *69*, 269.

[2]   M.A. Hamon, J. Chen, H. Hu, Y. Chen, M. E. Itkis, A. M. Rao, P. C. Eklund, R. C. Haddon, *Adv. Mater* **1999**, *11*, 834.

[3]   M. Holzinger, A. Hirsch, P. Bernier, G. S. Duesberg, M. Burghard, *Appl. Phys. A,* **2000**, *70*, 1.

[4]   M. Brettreich, A. Hirsch, *Tetrahedron Lett.* **1998**, *39*, 273.

[5]   M. Brettreich, A.Hirsch, *Synlett* **1998**, 1396.

Growth of Elongated Structures in a Longitudinal Electrical Field

A.V.Eletskii

Russian Research Center "Kurchatov Institute". Kurchatov Square Moscow 123182 Russia

Abstract. Nano-sized elongated carbon structures have a trend to be formed in conditions of the arc discharge with graphite electrodes [1-3]. An effect of the applied electric field on both the growth rate and structural features of nanotubes formed was observed in [4]. In this work a simple model for the aggregation of elongated structures in presence of an applied electrical field is considered. The action of the field promotes the predominant alignment of a growing structure in the field direction and generates the formation of an elongated structure.

Attachment of particles in an electrical field. Digressing from the chemical nature of the aggregation process, consider an initial stage of growth of an elongated structure as a result of attraction of neutral spherical in a spatial inhomogeneous field. The interaction between two neutral particles at large distance R is determined by the induced dipole moments $\mathbf{D_1}$ and $\mathbf{D_2}$ and expressed by the relation

$$U = [\mathbf{D_1}\mathbf{D_2} - 3(\mathbf{D_1}\mathbf{n})(\mathbf{D_2}\mathbf{n})]/R^3 = -\alpha_1\alpha_2 E^2(3\cos^2\theta - 1)/R^3. \tag{1}$$

Here $\mathbf{n}$ is the unit vector directed along the straight line connecting the centers of dipoles, θ is the angle between the vectors $\mathbf{n}$ and $\mathbf{E}$; the dipole moment is expressed through the applied field strength $\mathbf{E}$ and the polarizability of the relevant particle via relations $\mathbf{D} = \alpha\mathbf{E}$. As is seen, at low angles $0 < \theta < \text{arc } \cos(1/\sqrt{3})$ particles attract to each other which causes the growth of an elongated structure. The normal F_n and tangential F_t components of the interaction force caused by the induced dipole moments are equal to

$$F_n = -\frac{\alpha_1\alpha_2 E^2}{R^4}(3\cos^2\theta - 1); \quad F_t = \frac{3\alpha_1\alpha_2 E^2}{R^4}\sin 2\theta. \tag{2}$$

Since the attractive potential has a notable value for only small angles θ, the tangential component of the interaction force between the particles can be neglected. In such a case the angle θ is not changed in result of a collision, and the time of approach of particles initially spaced at R is governed by the equation of motion for particles in a viscous substance:

$$\tau = \frac{2\pi\eta R^5 r_{ef}}{\alpha_1\alpha_2 E^2(3\cos^2\theta - 1)}, \tag{3}$$

where η is the viscosity of a substance, r_{ef} is the effective radius of a particle, characterizing viscous resistance to its motion in a substance. Averaging this relation over initial positions of particles results in:

$$\bar{\tau} = 2{,}82\eta r_{ef} / (\alpha_1\alpha_2 E^2 N^{5/3}), \tag{4}$$

CP544, *Electronic Properties of Novel Materials—Molecular Nanostructures*, edited by H. Kuzmany, et al.

where N is the number density of particles. Representing a particle as a conductive sphere of radius r its polarizability is estimated as $\alpha \approx r^3$. This reduces (4) to the form:

$$\tau_{ass} \approx \frac{\eta}{E^2}(r^3 N)^{-5/3} \approx \tau_0 (r^3 N)^{-5/3},$$ (5)

where $\tau_0 = \eta/E^2$ is the characteristic time of the task. For typical conditions in the arc discharge with graphite electrodes $\eta \sim 10^{-4}$ Pa·s, $E \sim 10$ V/cm, $(r^3 N) \sim 10^{-5}$, which provides $\tau_0 \sim 1$ s and $\tau \sim 10^7$ - 10^8 s. Such a magnitude appears too long, so the considered mechanism of association of atomic particles hardly plays a notable role in a real situation.

One more mechanism of the association of particles is related to diffusion approach of them to each other. The rate constant of this process is determined by the Einstein-Smoluchovski formula [5]

$$k_{dif} = 4\pi r_{ef}(D_1 + D_2),$$ (6)

where D_1, D_2 are the diffusion coefficients of particles in a gas, expressed through the Einstein relation

$$D = kT/6\pi r_{ef}\eta.$$

This provides estimation for the diffusion association time

$$\tau_{dif} = (Nk_{dif})^{-1} = 3\eta/NkT.$$ (7)

The ratio of magnitudes of characteristic association time determined by relations (7) and (5) is expressed in the form

$$\frac{\tau_7}{\tau_5} \approx \frac{E^2 (r^3 N)^{5/3}}{NkT}$$

This expression contains two small parameters: $(r^3 N) \sim 10^{-5}$ and $E^2/NkT \sim 10^{-9}$. Therefore it can be concluded that the diffusion mechanism of association of nanosize particles is the main one in typical conditions for arc discharge synthesis of nanotubes so that the electrical field does not play any notable role in the growth of a nanosize particle. The expression (7) provides an estimation for characteristic association time as $\tau_{dif} = 3\eta/NkT \sim 10^{-9}$ s. Apparently this estimation concerns to the origination of a structure.

Growth of an elongated cylinder in a longitudinal electrical field. The significance of the electrical field in the mechanism of the growth of a structure increases with rise in its length. This is caused by the effect of the amplification of electrical field in the vicinity of a long cylindrical tube. Angular dependence of the longitudinal electrical field promotes the preferential attachment of particles approaching from the edge side of a tube. This provides longitudinal growth of a cylindrical tube.

The electrical field applied to an elongated cylindrical aggregate induces total dipole moment

$$D_2 = \int_{-l}^{l} Cz^2 dz,$$

where axis z is directed along the axis of the aggregate, $\sigma(z) = Cz$ is the specific electrical charge accounted per unit of length of an elongated conductive cylindrical aggregate. The integration results in $C = 3D_2/(2l^3) = 3\alpha_\| E/(2l^3)$, where $\alpha_\| = -l^3/[3\ln(l/r)]$ is the longitudinal component of the polarizability of an aggregate and r is its radius.

The interaction potential for a particle, which is represented by an induced dipole D_1, and an elongated aggregate (D_2) is given by the expression

$$U(R) = \int_{-l}^{l} Czdz(\mathbf{D_1 n})/R^2 = \frac{3D_2}{2l^3} \int_{-l}^{l} zdz(\mathbf{D_1 n})/R^2 .$$

There was used the expression for the interaction potential of a charge e and a dipole D spaced at the distance R: $U(R) = e\mathbf{Dn}/R^2$ where the unit vector $\mathbf{n}$ is directed along the vector $\mathbf{R}$. Using the polar coordinated z and ρ and supposing the charge to be situated on the polar axis so that $R = \sqrt{(z-z')^2 + \rho^2}$, z' is the coordinate of the charge along the axis of an aggregate and z, ρ are the coordinates of a particle, one obtain

$$U(R) = \frac{3D_2}{2l^3} \int_{-l}^{l} zdz(\mathbf{D_1 n})/R^2 = \frac{3D_1 D_2}{2l^3} \int_{-l}^{l} z'(z'-z)dz'/R^3 . \qquad (8)$$

In the case of a long aggregate $l \gg R$ the main contribution into the integral (8) is determined by the region near the edge of a cylinder which allows to substitute the magnitude z' by l. This results in the following expression:

$$U(R) = -\frac{3D_1 D_2}{2l^2 R_0}, \qquad (9)$$

where $R_0 = \sqrt{z^2 + \rho^2}$ is the distance from a spherical particle to the end of the aggregate. The expression (9) is valuable under condition $z > l$. This is followed by the formula for the force acting between the cylindrical aggregate and a spherical particle in the longitudinal electrical field:

$$\mathbf{F} = -\frac{3D_1 D_2}{2l^2 R_0^2}\mathbf{n} = -\frac{3\alpha_1 \alpha_\| E^2}{2l^2 R_0^2}\mathbf{n} = -\frac{r^3 l E^2}{2R_0^2 \ln(l/r)}\mathbf{n} . \qquad (10)$$

Supposing the cylindrical aggregate to be fixed and representing the motion of a spherical particle under the action of that force through the Stokes formula

$$F = 6\pi r v \eta = -\frac{r^3 l E^2}{2R_0^2 \ln(l/r)},$$

one obtain the equation of motion for a particle:

$$v = \frac{dR_0}{dt} = \frac{F}{6\pi\eta r} = -\frac{r^3 l E^2}{2R_0^2 \ln(l/r)}.$$

Solving this equation results in the formula for the time of attachment of a particle to an aggregate:

$$\tau_E = \frac{4\pi\eta R_0^3 \ln(l/r)}{r^2 l E^2} \tag{11}$$

Averaging this magnitude with taking account the distribution of the distance R_0 between the nearest particle and the edge of an aggregate results in

$$\overline{\tau}_E = \frac{3\eta \ln(l/r)}{r^2 l E^2 N}, \tag{12}$$

where N is the number density of spherical particles. This expression is valuable with the proviso $Nl^3 >> 1$.

However the growth of an elongated structure is determined practically by the initial, most slow stage of its evolution. Indeed the averaged growth time of a cylindrical particle of length l is expressed by the relation

$$\tau_l = \frac{\eta[1 - \ln(l/r)r/l]}{3r^2 E^2 N} \tag{13}$$

As is seen, the time of growth of an elongated structure ($l >> r$) in an electrical field depends drastically on the initial size of the particle and does not depend practically on its final length l. The growth time decreases with rise in the electrical field strength E and vapor density N. Since in conditions which are typical for the arc discharge carbon nanotube production $\eta/E^2 \sim 1$ s, the growth time can be roughly estimated as $\tau_l \sim (3r^3 N)^{-1}$ s. Supposing $r \sim 0.1$ nm, $N \sim 10^{19}$ cm^{-3}, one obtain the characteristic time close to one day. This quantity seems to be overestimated and can be shortened considerably supposing that the carbon nanotube formation begins from not a single atom but from a fragment of the graphite layer, which size exceeds notably the above-adopted magnitude 0.1 nm.

ACKNOWLEDGMENTS

The work was supported by the Grant INTAS 97-11894.

REFERENCES

1. Iijima S., *Nature (London)* **354** 56 (1991).
2. Ebbesen T., Ajayan P.M. *Nature (London)* **358** 220 (1992).
3. Eletskii A.V. *Physics - Uspechi* **40** 899 (1997).
4. Srivastava A., Srivastava A.K., Srivastava O.N. *Appl. Phys. Lett.* **72** 1687 (1998).
5. Smirnov B.M. "*Physics of Fractal Clusters*". M., Nauka, 1991 (in Russian).

Characterization of carbon atoms chemical states in nanotubes containing soot materials and fullerene by XPS, XAES

A. Dementjev, A. Eletskii, V. Bezmelnitsyn, and K. Maslakov

RRC Kurchatov Institute, Kurchatov sq.1, Moscow 123182,Russia

Abstract. X-ray excited Auger electron spectroscopy (XAES) and XPS were used to studying chemical states of carbon atoms in various carbon compounds. There have been shown clear distinction in the XAES of different compounds, which permits to identify carbon containing contaminations (CCC) on a surface of nanotubes and fullerene.

INTRODUCTION

Understanding the carbon surface chemistry presents a serious challenge to surface scientists and surface sensitive methods. Auger electron spectroscopy has been considered as probing of the local electronic structure of a solid surface [1]. The interpretation of Auger line shapes is not simple because the Auger decay process is complicated by effects of screening, matrix elements and an interaction between two final states of holes [2]. The study of a carbon surface via Auger electron emission from (it is denoted as CKVV) is complicated by uncertainty in carbon containing contamination (CCC) of the surface. The information depth of Auger electron spectroscopy (E_k =280 eV) in investigation of a carbon surface is estimated as about 3 monolayers [3]. In this connection the question arises: what do we observe on a carbon surface: substrate or CCC?

It is well known that CCC is always observed on the oxide and metal surfaces prepared ex-situ. The atmosphere air, sample handling and/or contamination in the analysis chamber from vacuum pump oil may cause the CCC on a sample surface. The C1s XPS peak of CCC is usually used as a binding energy standard for studying insulators. The nature of CCC as base for charge reference in the XPS method has been investigated by many authors [4]. The aim of this paper is to show the potential of the X-ray excited Auger electron spectroscopy (XAES) combined with the X-ray photoelectron spectroscopy for the identification of chemical states of carbon atoms on surfaces of nanotubes containing substances.

EXPERIMENTAL

The XPS and the XAES data were obtained using the MK II VG Scientific spectrometer with oil pumps. Photoelectron and Auger processes were excited with an

CP544, *Electronic Properties of Novel Materials—Molecular Nanostructures*, edited by H. Kuzmany, et al.

Al K$_\alpha$ X-rays source. The vacuum in the analysis chamber was $1.5\cdot10^{-7}$ - $6\cdot10^{-8}$ Pa. The spectra were collected in the constant analyzer energy mode, with pass energy 10 and 20 eV for XPS, and 50 eV for XAES. The C1s and O1s XPS spectra were acquired with a 0.1 or 0.05 eV step size; and 0.1 or 0.25 eV step size was used for CKVV spectra. The C1s binding energy was used for an accurate calibration of the CKVV kinetic energy with respect to the Fermi level.

The nanotubes containing soot was produced through the conventional arc discharge method using the nickel-chromium alloy (weight ratio Ni:Cr = 80:20) as a catalyst. The water-cooled cylindrical chamber was 7 l in volume. The sharpened cylinder graphite rod of 6 mm in diameter was used as a cathode. The consumable anode was made of two similar rectangular graphite rods of 7x3.5 mm^2 in cross section with a thin tape of catalyst in between. The weight content of the catalyst in the anode material was amounted as 10%, which corresponds to atomic content about 2%. The arc discharge was burned at the He pressure 700 Torr, current 60 A, voltage 29-30 V and gap 4 mm. The discharge voltage and gap were automatically stabilized via moving the consumable anode. Therewith the rate of graphite gasification reached about 1 g/hour.

The soot formed as a result of thermal sputtering of graphite was deposited onto the side surface of the chamber. It had a cloth-like structure and could be easily separated as a whole from the wall forming a "sheet of paper". As evidenced by TEM observations the as-produced soot contain 20-25% of single walled nanotubes forming ropes of about 10 nm in diameter. The diameter of individual nanotubes ranges between 1.2 and 1.4 nm, as it follows from the Raman spectroscopy investigation of soot samples performed by E. Obraszova.

In order to purify the nanotubes containing soot from impurities it was subjected to the thermal treatment at about 300°C. To avoid the flammariation of soot particles the dosed air feeding at the level 0.5 cm^3/s was provided. The thermal treatment of the soot caused the 30-40% lowering in its mass, which is assigned to loss in amorphous graphite and carbon particles.

RESULTS AND DISCUSSION

Fig. 1 shows the comparison of the CKVV Auger spectra for polyethylene, methane and CCC. The CKVV spectrum of CCC coincides very closely in its structure with that for polyethylene. This demonstrates clearly the character of sp^3-bonds in carbon contamination. But there is a very important difference at the left part of these spectra which is due to the electron energy loss of Auger electrons as they escape from a surface [2]. The left part of the CKVV spectrum of CCC lies below than that for polyethylene but above than that for gaseous molecules of methane, neopentane and normal pentane [5]. The comparison of the CKVV Auger spectra in Fig.1 suggests the position of CCC molecules between the molecules and solid. The peculiarities I - IV in Fig. 1should be noted. These features can be used for identification of CCC on a carbon surface.

The observed Auger spectra of the soot under investigation are made up of various carbon compounds such as carbon nanotubes, fullerenes, amorphous graphite and carbon nanoparticles. The real composition of the soot can be evaluated through the simulation of the observed Auger spectrum combining it from those of constituents. For example, Fig. 2 shows summation of the CKVV Auger spectrum of fullerene with that for CCC with equal intensity. The summation shows decrease in the left part and change in the right part of the total spectrum. This qualitative data give an insight what happens with the CKVV Auger spectrum of the contaminated fullerene surface. The CKVV Auger spectra of all samples studied in this work show absence of the CCC.

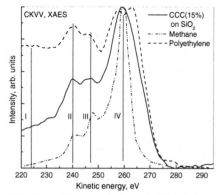

FIGURE 1. The comparison of CKVV Auger spectra for carbon containing contamination, methane and polyethylene.

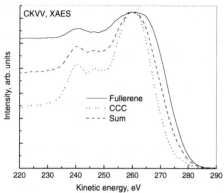

FIGURE 2. Simulation of the CKVV Auger spectrum of the fullerene with CCC by summing fullerene and CCC spectra with equal intensities.

Fig. 3 demonstrates typical C1s XPS spectra of nanotubes containing soot. During the treatment, we observed decrease in intensity of π-excitation with rise in the C-O intensity peak. The π-excitation has characteristic similar to that in fullerene [6]. Fig. 4 shows the change in CKVV spectra due to sample treatment. The left tail spectra have not notable inclination, which supposes low level of contamination of the samples.

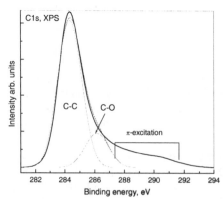

FIGURE 3. Typical C1s XPS spectra of nanotubes containing soot.

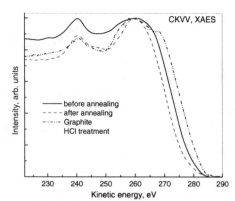

FIGURE 4. The comparison of nanoobes CKVV Auger spectra after different treatment.

The right part of the Auger spectrum for graphite is originated from a process involving two electrons in the p- band [7]. Differences in the right part spectra on Fig. 4 reflect differences in the p- band position for the samples. The composition and chemical interaction in the samples were determined by the intensity and binding energy of C1s, O1s, Ni2p and Cr2p XPS peaks and standard data [8]. The data for Ni2p and Cr2p show NiO and Cr_2O_3 states. The XPS data are summarized in table 1.

TABLE 1. Change in concentration of C, Ni, Cr, O in nanotubes containing samples as a result of thermal treatment (according C1s, Ni2p, Cr2p, O1s XPS data).

Sample treatment	C, at%	Ni, at%	Cr, at%	O, at%
Initial	97.5	2	0.5	0
As received	95.4	0.6	0.4	3.6
After annealing	46	16	4	34
After annealing, milling	82.5	2.5	1	14
Treatment HCl	96	0	0	4

CONCLUSIONS

The shape of N(E) CKVV Auger spectra provides an identification of the carbon containing contamination on the carbon surface prepared *ex-situ.* Obviously the Auger spectra should be used at beginning every analysis of carbon surface prepared *ex-situ*. This expands possibilities of surface sensitive methods. C1s XPS spectra show a partial oxidation of carbon atoms on a surface of the nanotubes containing soot. Seemly this oxidation state relates to soot particles, bacause the annealing does not change the Raman spectrum of nanotubes. The complicated chemical behavious of the nanotube contaning soot during annealing can be seen from the data of Table 1.

ACKNOWLEDGMENTS

The work was partially supported by the Grant INTAS 97-11894.

REFERENCES

1. Weissmann, R., and Muller, K., *Surface Sci. Rept.* **1**, 251 (1981).
2. Ramaker, D. E., *Critical Reviews in Solid State and Materials Science* **17**, 211 (1991).
3. Seah, M. P., and Dench, W.A., *Surf. Inter. Anal.* **1**, 2 (1979).
4. Barr, T. L., and Seal, S., *J. Vac. Sci. Tehnol.*, **A13**, 1239 (1995).
5. Rye, R. R., Jennison, D. R., and Hutson, J. E., *J. Chem. Phys.*, **73**, 4867 (1980).
6. Krummacher, S., Biermann, M., Neeb, M., Liebsch, A., and Eberhard, W., *Phys. Rev.*, **B48**, 8424 (1993).
7. Murday, J.S., Dunlop, B.I., Hutson, F.L., and Oelhafen, P., *Phys. Rev.*, **B24**, 4764 (1981).
8. Wagner, C.D., Riggs, W.M., Davis, L.E., Moulder, J.F., and Muilenberg, G.E., *Handbook of X-Rays Photoelectron Spectroscopy*, Minneapolis: Perkin-Elmer Corporation, 1979.

In-situ Raman and Vis-NIR Spectroelectrochemistry at Carbon Nanotubes

L. Kavan[*†], P. Rapta[*¶] and L. Dunsch[*]

[*]*Institute of Solid State and Materials Research, Helmholtztstr. 20, D - 01069 Dresden*
[†]*J. Heyrovský Institute of Physical Chemistry, Dolejškova 3, CZ-182 23 Prague 8*
[¶]*Slovak Technical University, Radlinského 9, SK-812 37 Bratislava*

Abstract. Electrochemical charging of SWCNT modifies the population of valence-band electronic states. Consequently, the resonance enhancement of Raman modes of SWCNT can be changed. SWCNT show reversible intensity/frequency *vs.* potential changes of the Raman-active tangential mode and RBM. The maximum intensity occurs at *ca.* -0.2 V *vs.* Ag/AgCl independent on the electrode material used as a support (Pt, Au, Hg). SWCNT are sensitive to photoanodic breakdown upon prolonged exposure to anodic bias and laser light. At positive potentials, SWCNT exhibit reversible bleaching of the first band-gap transition in semiconducting tubes (at 0.7 eV).

INTRODUCTION

Chemical doping of single-wall carbon nanotubes (SWCNT) by electron donors (alkali metals [1,2], anion radicals [3]) and electron acceptors (Br_2, I_2 [1-3]) was previously followed by electronic (Vis-NIR) [2,3] and Raman [1] spectra. The electronic structure of SWCNT was tuned, e.g. by reduction with molecules possessing different redox potentials [3]. SWCNT showed no distinct faradaic processes, but the effective charge-injection capacitances were, reportedly, as high as 283 F/g [4]. Carbon nanotubes exhibited enhanced electrocatalytic activity for the reduction of oxygen as compared to that of graphite electrodes [4]. Electrochemical oxidation of SWCNT in 18 M H_2SO_4 demonstrated that the Raman active tangential mode up-shifts by 320 cm^{-1} per hole per C-atom. The process was complicated by chemical oxidation at open circuit, and by irreversible overoxidation at anodic bias [5]. Here we report on *in-situ* spectroelectrochemistry at milder conditions, which allowed easy and reversible tuning of electronic properties of SWCNT.

EXPERIMENTAL SECTION

The purified SWCNT material was purchased from Tubes@Rice (sample No. P06049-6, suspended in toluene). It contained <1 wt% cobalt and nickel. Electrochemical experiments were carried out in aqueous 0.1 M KCl (saturated with nitrogen) by using HEKA IEEE-488 or EG&G PAR 273A potentiostats, with Pt auxiliary and Ag/AgCl reference electrodes. The working electrode was a thin film of SWCNT deposited on Pt or Au foils, ITO (indium-tin oxide conducting glass) or Hg. In the latter case a pool of Hg (diameter 8 mm) was positioned in a vertical glass tube and contacted electrically from the bottom side. A thin film of nanotubes was deposited by spraying of a freshly sonicated methanolic suspension of SWCNT on an electrode surface heated with hot air. The film mass on Au and Pt was typically between 0.03 to 0.06 mg/cm^2. For *in-situ* Raman measurements, the cell was equipped with a glass optical window. Raman spectra were excited by Ar$^+$ laser, ? = 514.5 nm (Innova 305, Coherent) and recorded on a T64000 spectrometer (Instruments, SA) interfaced to an Olympus BH2 microscope. The microscopic sample control was especially important in the case of mercury electrode, since the Hg-supported SWCNT film was mechanically unstable due to changes of surface tension. The ITO-supported SWCNT served for *in-situ* Vis-NIR spectroelectrochemistry. The spectra were recorded on a Shimadzu 3100 spectrometer. Due to limited stability of ITO at cathodic bias, the Vis-NIR spectra have been studied at positive potentials only.

RESULTS AND DISCUSSION

Electrochemistry

SWCNT show featureless cyclic voltammogram in aqueous 0.1 M KCl which points at charge injection over the whole potential range. This is in accord with the results reported for non-aqueous media [6], although, in our case, the voltammetric capacitances were lower by a factor of 2-3. In order to increase the electrochemical window to negative potentials, a mercury electrode was also used. In this case, the disturbing H$_2$ evolution is minimized, because of large hydrogen overvoltage on Hg.

In-situ Raman Spectroelectrochemistry

If a SWCNT electrode is contacted at open circuit with 0.1 M KCl, a small positive Raman-frequency shift is apparent in the region of the tangential mode. Analogous shift was reported also in 18 M H$_2$SO$_4$ [5]. The tangential mode further displays pronounced changes depending on the applied potential. The most striking effect is a gradual decrease in spectral intensity, if the applied potential sweeps both in negative and positive directions from certain optimum value (*ca*. -0.2 V *vs*. Ag/AgCl). The intensity drop was, to a great extent, reversible. Only at highly positive potentials the SWCNT exhibited irreversible changes due to breakdown (*vide infra*).

Figure 1 displays an overview of the intensity of the tangential mode at varying electrode potentials. This Raman band actually overlaps a multi-line feature (two E$_g$ and one A$_g$ modes), but its Lorentzian analysis confirms that all the individual components exhibit collective shifts without significant changes in their relative intensities [5]. This allows to plot simply the height of the peak envelope in Fig. 1. Apparently, the peak intensities were not much dependent on the material used as a support for SWCNT.

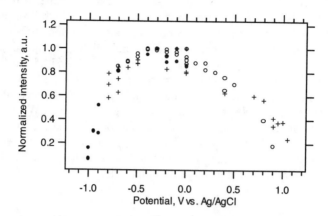

FIGURE 1. *Intensity of the main Raman band of tangential mode of SWCNT plotted vs. the applied potential in 0.1 M KCl (saturated with nitrogen). The nanotubes were deposited on platinum (+), gold (O) or mercury (●) electrodes. The intensity was normalized assuming the intensity 1 for the most intense peak during a potential excursion.*

Along with the drop of intensity, there was also a less pronounced reversible shift of the frequency of the tangential mode upon potential excursion (increased frequency at positive potentials and *vice versa*). A maximum positive shift was about 7 cm^{-1} at 0.8 V on SWCNT/Au electrode. We roughly estimate that, for a nanotube capacitance of 283 F/g [6] and potential difference of 1 V, the capacitive charge translates into *ca.* one electron per 30 carbon atoms. With reference to the work [5], we would expect a shift of *ca.* 10 cm^{-1} upon capacitive biasing to 1 V, which is not too far from our observation, although our double-layer capacitances were smaller.

The potential dependence of intensity/frequency is also apparent in the region of the A$_g$-symmetry radial breathing mode (RBM). Upon biasing, the RBM drops in overall intensity, which matches the results on chemical doping of SWCNT [1]. By scanning of the potential in both directions, the intensity of the higher-frequency RBM (at *ca.* 205 cm^{-1}) was less affected as compared to the intensity of the main peak at *ca.* 190 cm^{-1}. These features are demonstrated in Figure 2a.

Photoanodic Breakdown of SWCNT

Another characteristic issue was a pronounced sensitivity of SWCNT to photo-anodic breakdown. This manifested itself by "burning" of nanotubes after a prolonged exposure of the anodically polarized nanotubes to the laser light. At the illuminated site a cavity in the SWCNT film was clearly apparent under the microscope. A simultaneous application of the laser light and anodic polarization is essential for burning of SWCNT: no breakdown of nanotubes occurred (1) at the areas outside the laser spot of a polarized electrode and (2) by laser excitation of an electrode at less positive potentials or at open circuit. Also, in the absence of electrolyte solution, the SWCNT showed no signs of degradation by the laser light.

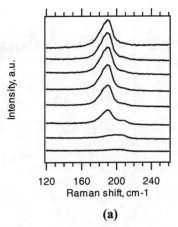 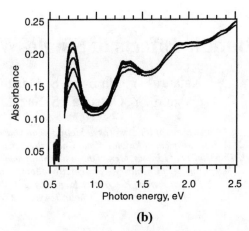

(a)   (b)

FIGURE 2. (a) Raman spectra of SWCNT/Au in aqueous 0.1 M KCl saturated with nitrogen. The applied potential was (from top to bottom in V *vs.* Ag/AgCl): -0.4, -0.2, 0, 0.2, 0.4, 0.6, 0.8. Spectra are offset for clarity. **(b)** Vis-NIR spectra of SWCNT/ITO in the same electrolyte solution. The applied potential was (from top to bottom in V *vs.* Ag/AgCl): 0, 0.2, 0.4, 0.6, 0.8.

In-situ Vis-NIR Spectroelectrochemistry

The Vis-NIR spectra of SWCNT (Figure 2b) show reversible bleaching of the first electronic transition at *ca.* 0.7 eV, which originates from semiconducting tubes [2,3]. The same effect occurs after both chemical oxidation and reduction of SWCNT [2,3]. The optical absorption at 514.5 nm is not significantly changed. Hence, the drop of Raman intensities (Figures 1, 2a) is not clearly understood, if we assume that they are controlled by resonance enhancement.

ACKNOWLEDGMENTS

Financial support from the DFG (L.K.) and Humboldt Foundation (P.R) is gratefully acknowledged.

REFERENCES

·1.   Rao, A. M., Eklund, P. C., Bandow, S., Thess, A. and Smalley, R. E. *Nature* **388**, 257-259 (1997).

2.   Kazaoui, S., Minami, N., Jacquemin, R., Kataura, H. and Achiba, Y. *Phys.Rev.B* **60**, 13339-13342 (1999).

3.   Petit, P., Mathis, C., Journet, C. and Bernier, P. *Chem.Phys.Lett.* **305**, 370-374 (1999).

4.   Britto, P. J., Santhanam, K. S. V., Rubio, A., Alonso, J. A. and Ajayan, P. M. *Adv.Mater.* **11**, 154-157 (1999).

5.   Sumanasekera, G. U., Allen, J. L., Fang, S. L., Loper, A. L., Rao, A. M. and Eklund, P. C. *J.Phys.Chem.B* **103**, 4292-4297 (1999).

6.   Liu, C., Bard, A. J., Wudl, F., Weitz, I. and Heath, J. R. *Electrochem.Solid-State Lett.* **2**, 577-578 (1999).

Bundle Effects of Single-Wall Carbon Nanotubes

H. Kataura[a], Y. Maniwa[a], S. Masubuchi[b], S. Kazama[b], X. Zhao[c],
Y. Ando[c], Y. Ohtsuka[d], S. Suzuki[d], Y. Achiba[d] and R. Saito[e]

[a] Department of Physics, Tokyo Metropolitan University, Hachiohji, Tokyo 192-0397, Japan
[b] Department of Physics, Chuo University, Kasuga Bunkyo-ku, Tokyo 112-8551, Japan
[c] Department of Physics, Meijo University, Shiogamaguchi Tempaku-ku, Nagoya 468-8502, Japan
[d] Department of Chemistry, Tokyo Metropolitan University, Hachiohji, Tokyo 192-0397, Japan
[e] Department of Electronics Engineering, University of Electro-Communications,
Chofu, Tokyo, 182-8585, Japan

Abstract. To see the bundle effects on the electronic and the vibrational properties of single-wall carbon nanotubes (SWNTs), we have measured the resonance Raman scattering of isolated SWNTs and thick bundles. For the measurements of the isolated SWNTs, we used an evacuated sample after bromine doping. A broad and asymmetric tangential mode band, which is a sign of the resonance of the metallic SWNTs and can be fitted by a Fano line shape, is not observed in the isolated SWNTs. This suggests the inter-tube interactions play an important role to the Fano interference. On the other hand, the purified sample shows very thick acquired bundles. We observed 4% higher breathing mode frequencies than in the pristine sample. Further, in the case of multi-wall nanotubes, we observed 5% higher breathing mode frequencies than the SWNTs. These results can be explained by the interlayer interactions.

INTRODUCTION

Pristine arc samples contain many isolated single-wall carbon nanotubes (SWNTs) as well as bundles. (1) Thus, in greater or less degree, measured optical responses are regarded as mixtures of signals from the isolated SWNTs and the bundles. If it is possible to wipe off the signal from the bundles, we can get information about optical properties of the isolated SWNTs. To realize this, we used bromine doped SWNTs. If the isolated SWNTs do not have any stable site for bromine molecules, the evacuation can remove the bromine molecules surrounding isolated SWNTs. Recently, Kazaoui *et al.* have shown that the bromine doped SWNTs do not show any absorption band below 1.8 eV. (2) If we measure the Raman spectra of the evacuated sample by excitations lower than 1.8 eV, we will obtain the Raman spectra only of the isolated SWNTs. Since the charge transfer suppresses the optical transitions in bundles, the Raman signals from the bundles can be wiped off.

Some theoretical works predicted the radial breathing mode (RBM) frequencies are strongly modified by the bundle effects.(3–5) We know that highly purified samples show very thick bundles. The bundle effects should be observed more clearly in the purified sample than in the pristine ones. Further, it is expected that the interlayer interactions in multi-wall nanotubes should be stronger than the inter-tube interactions in SWNT bundles. We will discuss about these results.

CP544, *Electronic Properties of Novel Materials—Molecular Nanostructures*, edited by H. Kuzmany, et al.

EXPERIMENTAL

Samples for a doping were prepared by a conventional electric arc method using NiY catalyst.(6) HRTEM revealed that a considerable amount of isolated SWNTs are existing in the sample. To avoid a doping inside SWNTs, pristine soot was installed into a quartz ampoule without any purification. Saturated bromine vapour was introduced to the ampoule at room temperature. After keeping one hour, the ampoule was evacuated by a rotary pump. Then the ampoule was sealed off.

Thick bundles of SWNTs were prepared by the purification of a SWNT sample fabricated by the laser ablation method using NiCo catalyst.(7) The purification was done by 15% H_2O_2 reflux for six hours.(8) Multi-wall nanotube sample was prepared by the carbon arc in hydrogen gas.

Raman spectra were measured using a JOBIN YVON U1000 double monochromator and HAMAMATSU PHOTONICS photon counting system interfaced to a personal computer. Ar^+, dye and Ti-sapphire lasers were used for the excitation.

RESULTS AND DISCUSSION

For 488 nm excitation, the fully bromine doped SWNTs show only one RBM band at 260 cm^{-1}, which is consistent with the result by Rao et al.(9) After the evacuation, we observe two RBM bands at 180 and 240 cm^{-1}. We confirmed that the two-band structure is stable up to 200 °C in vacuum. The band at 180 cm^{-1} is nearly the same to the RBM of the pristine sample and can be observed by any excitation energies in contrast with the 240 cm^{-1} band which has no intensity for excitations lower than 1.8

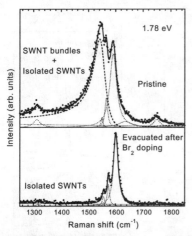

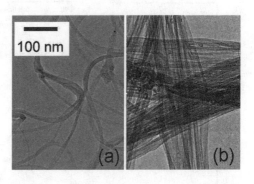

FIGURE 1. Near infrared Raman spectra of pristine and bromine doped SWNTs. Dashed curve indicates the fitted Fano line shape. Here, ω_0 = 1548 cm^{-1}, $1/q$ = −0.25 and Γ = 27 cm^{-1}.

FIGURE 2. TEM images of (a) pristine and (b) purified SWNTs fabricated by the laser ablation method using NiCo catalyst.

eV. This means that the sample consists of two kinds of SWNTs. One has stable bromine sites, and the other one does not have. We think the former should be the bundles and the latter the isolated SWNTs. We conclude that the 240 cm⁻¹ band is the RBM of the partially doped SWNT bundles and the 180 cm⁻¹ band is that of de-doped portions in the sample namely, the isolated SWNTs.

Now we can measure the Raman spectra of the isolated SWNTs by near infrared laser excitations. Figure 1 indicates the high-frequency Raman spectra of the pristine and the isolated SWNTs by the excitation in the metallic window of the sample. (10,11) In the case of the pristine sample, the spectrum can be fitted well by some Lorentzians and a Fano line shape, which is the sign of the resonance of metallic SWNTs. Surprisingly, however, the Fano line shape cannot be observed anymore in the case of the isolated SWNTs. The tangential modes can be fitted simply by three Lorentzians, instead. If the Fano line shape is originating from the metallic bundles and three Lorentzians are from the isolated SWNTs, the both Raman spectra can be explained simply. This result indicates that the Fano interference strongly connects with the inter-tube interactions. Indeed, the coupled phonon mode in the Fano line shape is 45 cm⁻¹ lower than the tangential mode of the isolated SWNTs. This frequency shift is the similar value to the inter-tube mode frequency. (3)

To see the bundle effect on the RBM frequencies we measured the resonance Raman spectra of isolated, pristine and purified SWNTs. Typical change in bundle diameter by the purification is indicated in Fig.2. TEM images show typical bundle diameters of the purified sample are larger than 50 nm. Figure 3 shows the low frequency spectra of each sample in the metallic window. Each sample shows similar resonance feature but RBM frequencies of the purified sample are 4% higher than the pristine one. This is consistent with the theoretical predictions of the bundle effect on the RBM frequencies.(3-5) Although the RBM frequencies of the isolated SWNTs are

(a)   (b)   (c)

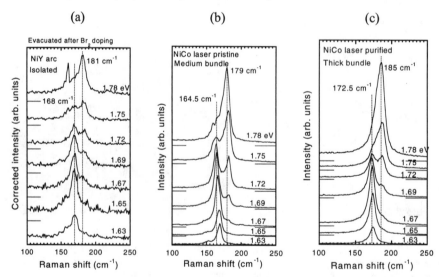

FIGURE 3. Resonance Raman spectra of (a) isolated SWNTs, (b) mixture of isolated SWNTs and medium bundles, and (c) thick bundles of SWNTs.

slightly higher than those of the pristine sample, this may be the effect of the bromine molecules remaining around the isolated SWNTs. Further, we found a interesting phenomenon. In the cases of the isolated SWNTs and the purified sample, RBM frequencies are independent of the excitation energy. In the case of pristine sample, however, the RBM peak position is strongly depending on the excitation energy. This strange result can be explained by the bundle effects by taking into account the fact that the pristine sample is a mixture of the isolated SWNTs and the bundles. The isolated SWNTs have lower RBM frequencies (3) and narrower electronic bandwidth than the bundles.(12) When the excitation energy is just the energy gap of a certain kind of SWNTs, isolated SWNTs are resonated predominantly. Thus, the RBM peak position locates the lowest frequency with the highest intensity. When the excitation is higher or lower than the energy gap, bundles are mainly resonated due to the broader band feature. Thus, RBM peak shows up-shift with decreasing the intensity. In the case of the purified sample, most of the SWNTs are inside of thick bundles. Consequently, the RBM frequency is stably high. Apparently, there is no cause to move the RBM peak of the isolated SWNTs.

Multi-wall carbon nanotubes produced by carbon arc in hydrogen gas show RBM peaks.(13,14) The peak frequencies are about 5% higher than the corresponding RBMs of SWNTs. Although there is no theoretical calculation about interlayer interaction in MWNTs, it is reasonable that the interlayer interaction in the MWNT is larger than inter-tube interactions in the SWNT bundle.

ACKNOWLEDGMENTS

The authors thank Mr. Misaki for taking HRTEM photographs of SWNT samples. This work was supported in part by Japan Society for Promotion of Science, Research for the Future Program. This work was supported in part by Grant-in-Aid for Scientific Research on the Priority Area "Fullerenes and Nanotubes" by the Ministry of Education, Science, and Culture of Japan.

REFERENCES

1. M. Yudasaka *et al.*, Chem. Phys. Lett. **312**, 155 (1999).
2. S. Kazaoui, N. Minami, R. Jacquemin, H. Kataura and Y. Achiba, Phys. Rev. B **60**, 13339 (1999).
3. L. Henrard, E. Hernández, P. Bernier and A. Rubio, Phys. Rev. B **60**, R8521 (1999).
4. D. Kahn and J. P. Lu, Phys. Rev B **60**, 6535 (1999).
5. U. D. Venkateswaran *et al.*, Phys. Rev. B **59**, 10928 (1999).
6. C. Journet *et al.*, Nature **388**, 756 (1997).
7. A. Thess *et al.*, Science **273**, 483 (1996).
8. K. Tohji, private communication.
9. A. M. Rao, P. C. Eklund, S. Bandow, A. Thess and R. E. Smalley, Nature **388**, 257 (1997).
10. H. Kataura, *et al.*, Synthetic Metals **103**, 2555 (1999).
11. H. Kataura *et al.*, Molecular Crystals and Liquid Crystals, in press.
12. Y. K. Kwon, S. Saito and D. Tománek, Phys. Rev. B **58**, R13314 (1998).
13. H. Kataura *et al.*, AIP Conference Proceedings **486**, 328 (1999).
14. H. Kataura *et al.*, Proceeding of Materials Research Society 1999 Fall meeting, in press.

Contribution to the characterization of nanotubes atomic structure : breathing mode of finite and infinite bundles

L. Henrard*, Ph. Lambin *, A. Rubio **

* *Laboratoire de Physique du Solide, Facultés Universitaires Notre-Dame de la Paix, Rue de Bruxelles,61. 5000 Namur. Belgium*
** *Departamento de Fisica Teorica, Universidad de Valladolid. E-47011 Valladolid. Spain*

Abstract.
The Raman spectroscopy is one of the most used tools for the experimental characterisation of the atomic structure of bundles of single-wall carbon nanotubes. More particularly, the low frequency range ($150 - 200\ cm^{-1}$) is dominated by the signature of the resonant fully radial (breathing) mode. A refined evaluation of the parameters that influence the excitation frequencies of these modes is then of primary importance for a reliable interpretation of experimental spectra. In the present contribution, we focus our attention on the effect of the inter-tube interaction. Indeed, we have developed a pair-potential approach for the evaluation of van der Waals interaction between carbon nanotubes in bundles. Starting from a continuum model, we show that the inter-tube modes range from $5\ cm^{-1}$ to $60\ cm^{-1}$, depending of the number of tubes in the bundle. We also found that the breathing mode frequency increase by 10 % if the nanotubes lie inside a infinite bundle as compared to the isolated tube. The radial-like modes and inter-tube modes of a finite bundles are considered and the transition from an isolated tube to an infinite array of tubes is detailed.

As proven by the present proceedings book, the research in carbon nanotubes (both single-wall nanotubes - SWNT - and multi-wall nanotube -MWNT -)is a very active field both because of their fascinating cylindrical structure and potential applications.

For what concerns SWNT, the control of the actual atomic structure during the production stage is still a major point of study. A first step is still the developement of reliable, fast and accurate characterization tools. For that purpose, the most used techniques are Raman spectroscopy, electron microscopy (HRTEM) and diffraction (ED), X-rays and neutron diffraction or scanning tunneling microscopy and spectroscopy (STM-STS). For each of those techniques, the first analysis of the experimental data have had to be re-read over by refined analysis. For example, first STM images have had to be re-analysed taking account of the cylindrical curvature of the tube [1]. And direct analysis of the X-rays (and neutron) diffraction

CP544, *Electronic Properties of Novel Materials—Molecular Nanostructures*, edited by H. Kuzmany, et al.

peak positions has led to an underestimation of the mean bundle lattice parameter due to tube form factor effect [2].

One of us (LH) has also performed a detailed and quantitative combined experimental/simulation study of the ED and HRTEM of individual single-wall nanotube bundle [3]. It was found, analysing the central line of diffraction, that single bundles are formed with nanotubes with a very narrow diameter distribution but that the mean diameter varies from one bundle to the other in a given 'macroscopic sample'. The analysis of the intensities related to the carbon-carbon bound orientation, then to the tube chiralities, has revealed a large number of tube symmetries whitin a single bundle.

We now turn to the Raman spectroscopy investigations. Besides the modes that have a direct counterpart in graphitic materials (C-C streching and shear), a fully symmetric A_1 mode (called radial breathing mode (RBM)) appears to be characteristic of closed shell carbon compounds and occurs at relatively low frequencies $(150 - 200 \ cm^{-1})$.

The first investigations [4] have led to a linear dependance of the excitation frequencies with the inverse of the tube diameter, $\omega = \frac{C}{d}$. This conclusion has been revisited by means of methods that account for the rehybridization of the C-C bound during the vibration and a slight chirality dependance of the linear law has been deducted [5]. For example, we found $C_{(n,n)} = 1301 \ cm^{-1}\text{Å}$ for armchair tubes and $C_{(n,0)} = 1282 \ cm^{-1}\text{Å}$ for zig-zag tubes [6]. Even if those numbers are surestimated when compared to ab-initio data due to the set of parameters we have chosen, the trends are adequately reproduced and we used the same nonorthogonal TB $\sigma + \pi$ scheme (see [7]) in the present paper.

Two main difficulties arise when the detailed interpretation of experimental Raman RBM excitation is considered. First, the Raman scattering is a resonant process due to the coincidence of the incident laser frequencies and the electronic van Hove singularities transitions. An analysis of the resonance effect on the RBM excitation have been recently provided [8].

The second difficulty in interpreting Raman data is the tube packing effect. We will not discuss the packing effect on the resonant activity (even if packing can modify transition energies (but not dramatically) and resonant process - that is still not fully understand for an isolated tube. The concern of the present work is the effect of the tube-tube interaction on the vibrational frequencies of the RBM of finite bundles.

For the tube-tube coupling, we used the long-range dispersion interaction between carbon-carbon Lenard-Jones (LJ) potential given by $V_{cc} = -\frac{C_6}{d^6} + \frac{C_{12}}{d^{12}}$, where d is the carbon-carbon distance and C_6 and C_{12} are parameters [9]. We have simplified further the model and we have considered carbon nanotubes as continuous cylinders. The van der Waals interaction energy per unit length between two carbon nanotubes, V_{TT}, can then be determined semi-analytically (see [6] for details) and the total energy of a bundle is obtained by summing V_{TT} over all tube-pairs.

First, we evaluated the energy of the bundle as a function of the number of tubes

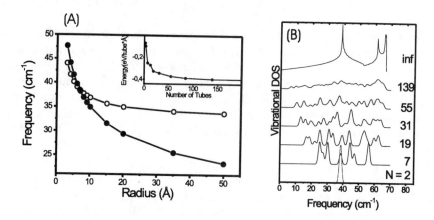

FIGURE 1. (a) Vibrational frequencies of SWNTs dimers as a function of the radius of the tube 1, R_1. The radius of tube 2, R_2 is $6.8\mathring{A}$ (open symbols) or equal to R_1. Inset : van der Waals bonding energies of bundles of SWNTs as a function of the number of tubes. (b) Inter-tube vibrational DOS with respect to the number of tubes in the bundle. An hexagonal symmetry is assumed and an broadening has been used for finite bundles.

(inset of figure 1(a)). Since the LJ tube-tube potential is short ranged, the bonding energy converges rapidly to the infinite array energy (-0.44 $eV/tube\mathring{A}$). However, a striking feature is that the energy goes down very sharply until $N = 50$ that is close to the average number of tubes in a bundle.

Figure 1(a) gives the frequencies of the vibrational modes of dimer of nanotubes. The solid symbols are for dimers made of tube with identical diameters and the open symbols give the vibrational frequencies of assymmetric dimers, one of the radii being kept fixed ($6.8\mathring{A}$). We note also that the nanotubes are not allowed to deform in this model.

For larger bundles, we deduced the inter-tube vibrational modes (eigen-values and eigen-vectors) from the diagonalisation of the dynamical matrix built up by the numerical evaluation of the second derivative of the total energy with respect to the tube center coordinates (see ref [6] for further details). On figure 2(b), we plotted the total inter-tube DOS for increasing number of tubes. The infinite bundle DOS was obtained from the computation of the dispersion relationship of a triangular lattice of tubes (1 tube per unit cell). The sound velocities have been found to be 1426 m/s and 806 m/s for the two phonon branches. These modes could be probed by inelastic neutron experiment or Brillouin scatterring and consequences on thermal and electrical transport properties of bundles of NTs have to be emphasized.

We now investigate the coupling between the inter-tube potential and the RBM. In ref [6], we have computed the RBM frequencies of infinite bundles of nanotubes.

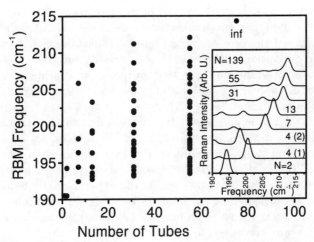

FIGURE 2. RB-based frequencies as a function of the number of tubes in the bundle. A hexagonal symmetry is assumed (except for $N = 1, 2$). Inset : Simulation of the non-resonant Raman spectra of finite size bundle (see text) for various number of tubes. An hexagonal symmetry is assumed except for $N = 2, 4$. $N = 4(1)$ is for a three-fold symmetry cluster (with one tube at the center) and $N = 4(2)$ is for a parallelepipe of tube.

We found (in a frozen phonons approach), an increase of the RBM by 10% when (10,10) tubes are bundled.

But, as we quoted before, bundles are finite. It is then of interest to know how the transition between isolated tubes and infinite bundles occurs in terms of RBM frequencies and what happens in the transition zone. First, we note that both for isolated tube and for infinite array, only one Raman active RBM exists. Coupling a finite number of tubes, more than one raman active RB-based modes will appear as a consequence of the translational symmetry breaking.

We dealt we finite bundle as follow. Using the same mixed procedure (TB + LJ) as before [6], we evaluated the second derivative of the total potential with respect to the tube radii variation and built the dynamical matrix, D. The diagonalisation of D gives the N eigenvectors and the N eigenfrequencies of the RB-based modes associated with the finite bundle made of N tubes. The eigenvectors described the relative breathing amplitudes of each tubes for a given RB-based mode. For example, the smallest bundle, the tube dimer, gives 2 RB-based modes. The lower frequency mode is characterized by an out-of-phase breathing of the tubes and the higher frequency mode by in phase breathing. For a (10,10) tube dimer, the frequencies are $\omega_1 = 190.5 \ cm^{-1}$ and $\omega_2 = 194.5 \ cm^{-1}$.

Figure 2 shows the RB-based mode frequencies as a function of the number of (10,10) tubes in a bundle. The isolated tube and the infinite bundle active mode are shown for comparison. As expected, RB-based modes for finite bundles range between the RBM of isolated tube and RBM of infinite array of tubes. Moreover, the higher frequency mode is characterized by an 'in-phase' breathing of all tubes and correspond the the Γ point Raman active mode in the limit of an infinite array.

We finally addressed the problem of the relative intensities of RB-based modes. We then need an evaluation of the bundle polarizability variation with respect to each normal modes of vibration. For that purpose, we considered all tubes as linear densities of dipoles with a polarizability per unit length that scaled like the square of the tube raddii and being 32.7Å^2 for a ground state (10,10) tube [10]. The polarisability of the bundle (and its variation due to the RB-based modes excitations) is then computed by letting the N linear densities of dipoles self-consistently interacting [11]. The results are given in the inset of figure 2 . The fact to be pointed out is that the higher-energy mode (in-phase mode) gives the essential of the Raman activity and, then, is the experimentally detected mode. The tube configuration (symmetry of the bundle) is also of considerable influence for small bundles but, as far as large bundles ($N > 50$) are concerned, RBM of finite bundles is already converged to infinite array of NTs case.

In conclusion, we have investigated inter-tube and RB-based modes of finite and infinite bundles of SWNTs. Inter-tube DOS have been given as a predictive result to be compared with inelastic neutron experiments. The influence of packing on RBM has also been emphasized. This effect has to be considered when comparison between experimental data and predicted RBM frequencies is made since produced SWNTs samples are made of bundle of tubes. The influence of a chirality and sharp diameter distribution in a single bundle [3] on the RBM linewidth has also to be taken into account (work in progress).

This work was supported by the belgian programs PAI/UAP 4/10. L.H. was supported by the FNRS.A.R. was supported by EU NAMITECH contract: ERBFMRX-CT96-0067 (DG12-MITH),Spanish DGES (PB98-0345) and by JCyL (VA28/99).

REFERENCES

1. L. Venema, V. Meunier, P. Lambin, and C. Dekker, Phys. Rev . B **61**, 2991 (2000).
2. S. Rols *et al.*, Eur. Phys. J. **10**, 263 (1999).
3. L. Henrard, A. Loiseau, C. Journet, and P. Bernier, Eur. Phys. J. **13**, 661 (2000).
4. R. Saito *et al.*, Phys. Rev. B **57**, 4145 (1998).
5. J. Kürti, G. Kresse, and H. Kuzmany, Phys. Rev. B **58**, R8869 (1998).
6. L. Henrard, E. Hernandez, P. Bernier, and A. Rubio, Phys. Rev. B **60**, R8521 (1999).
7. D. Porezag *et al.*, Phys. Rev. B **51**, 12947 (1995).
8. M. Milnera, J. Kurti, M. Hulman, and H. Kuzmany, Phys. Rev. Lett. **84**, 1324 (2000).
9. L.A. Girifalco and R.A. Lad, J. Chem. Phys. **25**, 693 (1956). The parameters of the Lenard-Jones potential for carbon are C_6=20 eV Å^6 and C_{12}=2.48 10^4 eVÅ^{12}.
10. L. Benedict, S. Louie, and M. Cohen, Phys. Rev. B **52**, 8541 (1995).
11. L. Henrard,. to be published.

Resonance excitation, intertube coupling and distribution of helicities in single-wall carbon nanotubes

M. Hulman[1], W. Plank[1], J. Kürti[2] and H. Kuzmany[1]

[1]Institut für Materialphysik der Universität Wien, Strudlhofg. 4, A-1090 Wien,
[2]Eötvös University, Department of Biological Physics, Budapest, Hungary

Abstract.
Photoselective resonance Raman scattering from laser ablation grown carbon nanotubes is presented. The resonance excitation was found to exhibit a quasi-oscillatory behaviour. Assuming a Gaussian distribution of diameters we exctracted values for a mean diameter and a width of the distribution within a simple model. In a more expensive calculation the contribution from all geometrically allowed tubes in a certain diameter range was considered. To match experiments and calculations, the frequencies obtained from the latter must be upshifted by ≈ 10 cm^{-1}. This is ascribed to the tube-tube interaction in the carbon nanotube bundles.

INTRODUCTION

Carbon nanotubes attract much interest in these years because of their unusual physical properties. Building of rolled up graphene sheet the single wall carbon nanotubes (SWCNT's) can be considered as one dimensional electronic systems. The ideal, defectless SWCNT thus exhibits typical 1D van Hove singularities in the electronic density of states (DOS).The singularities are symmetric with respect to the Fermi energy.Only chirality - the way how the SWCNT's are rolled up - determines whether they are metallic or semiconducting. The singularities in the DOS are very interesting from the point of view of optical spectroscopy as well. They can cause a resonant enhancement in Raman spectroscopy. If a sample is illuminated with laser light having an energy equal to the energy gap between corresponding van Hove singularities the resonant enhancement of some phonon lines in a Raman spectrum occurs. The Raman spectrum of SWCNT's has two important regions. The first one around 1580 cm^{-1} is very similar to the Raman mode in graphite. The second one between 140 and 220 cm^{-1} comes from a radial breathing mode (RBM) and is unique for the SWCNT's. Each nanotube has its own frequency of the RBM that scales inversely with its diameter. A typical sample of SWCNT's consists of a whole family of tubes with different diameters and chiralities. This leads to a broad band in the RBM region with a lot of fine structures in the Raman spectra. Identifying different modes in the spectrum we should be able to decide which SWCNT's are included in the sample.

CP544, *Electronic Properties of Novel Materials—Molecular Nanostructures*, edited by H. Kuzmany, et al.

We show in this work that one can take all geometrically allowed SWCNT's into account without loosing the fine structures in the Raman spectra. We estimated also a $\pi - \pi$ interaction energy γ_0, the basic parameter determining the electronic structure of SWCNT's.

EXPERIMENTAL

The preparation of SWCNT's was described elsewhere [1]. The as prepared material from the Rice University was heated in high vacuum of 10^{-7} mbar at the temperature 1000 K for 12 hours. Then the Raman spectra were taken using 27 different laser lines covering the whole visible region between 455 nm and 850 nm (1.46 - 2.73 eV). After that the raw material was purified by filtration from toluol and ethanol followed by the same temperature treatment as well as taking of Raman spectra. Measurements were performed by a triple spectrometer XY 500 (dilor, France) with a spectral resolution 2 cm^{-1}.

RESULTS AND DISCUSSION

All 27 Raman spectra taken are shown in the Fig.1.

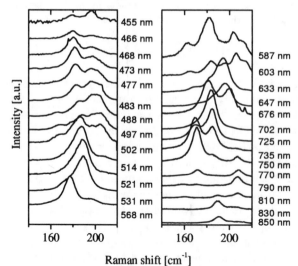

FIGURE 1. Raman line pattern for the RBM for excitation with 27 laser lines

On the first glance they seem to be rather chaotic but one can recognize a regular behaviour after a careful inspection. Positions of peaks depend on a laser excitation wavelength and exhibit an oscillating up and down-shift as the wavelength increases. Intensities of the modes behave similarly. In order to describe this behaviour rigorously we estimated the center of gravity (first moment) for each spectrum and plotted it as a function of the laser excitation energy in the Fig.2. We can see a quasi-oscillatory behaviour with a peak to peak distance approximately 0.6 eV. Amplitudes of peaks decreases with increasing laser energy.

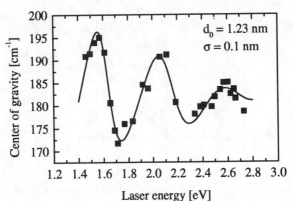

FIGURE 2. First moments for the Raman response of the RBM versus excitation energy (squares). The full drawn line is calculated from a simplified model

As it was already mentioned above the Raman spectrum is a sum of lines originating from different SWCNT's. In this sense the spectrum reflects a presence of different SWCNT's in the material. Since the frequency of the RBM scales as $1/d$ the distribution of diameters is one of the most important parameters characterising the material. In accordance with [1] we assume the distributions to be a Gaussian-like with the mean value 1.23 nm and a variance 0.1 nm.

In order to explain oscillations in Fig.2 we make use of the fact that energy gaps between van Hove singularities (transition energies)scales also as $\epsilon_i = A_i/d$ [2] where A_i chracterizes different pairs of the singularities. Our simple model is schematically drawn in Fig.3.

The radial arrows indicate the increase of transition energy (full lines) and the increase of RBM frequency (dashed line) with $1/d$. The dotted arrows connect the resonance energies between the van Hove singularities on the right side of the y-axis with the frequencies of the RBM on the left side of the y-axis. The hatched area represents the typical tube diameters. A laser with a certain energy ϵ excites a particular tube when ever this energy matches an electronic transition energy. The tubes with the corresponding frequencies $\nu(d)$ are observed in resonance. The center of gravity is then calculated as a weighted sum $\sum_i w_i \nu_i(d)/\sum_i w_i$ where w_i is the Gaussian function $(2\pi\sigma^2)^{-1/2} exp[-(d_i - d_0)^2/2\sigma^2]$ and i denotes different transition energies. Using expressions $d_i = A_i/\epsilon$ and $\nu_i = (C/A_i)\epsilon$ for the diameter and frequency, respectively and values $d_0 = 1.23$ nm and $\sigma = 0.1$ nm we can fit the experimental values in Fig.2 with a function depending on the energy ϵ only.

The result is drawn as a solid line in Fig.2 with A_i as fitting parameters,$i = $ 1-4 . Values for A_i obtained from the fit agree within 15% with theoretical values

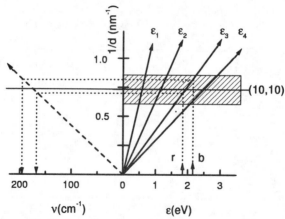

FIGURE 3. Schematic diagram for photoselective resonance scattering for SWCNT's

published in [2] for $\gamma_0 = 2.9$ eV.

As a next step we calculated Raman spectra using all geometrically allowed tubes in a diameter range 1.0 - 1.7 nm. A purpose was to prove that fine structures observed in the Raman spectra are retained and to obtain values for a $\pi - \pi$ interaction energy γ_0 and a width of electronic states α. Detail of calculations are described in [3]. We estimated center of gravity as a function of laser energy for various values of γ_0 and α and looked for a minimum of the standard deviation with respect to experimental values in Fig.2. The best result obtained is for $\gamma_0 = 3.025$ eV and $\alpha = 0.025$ eV. The values for d_0 and σ were set as before. As indicated in Fig.4 the fine structures in the spectra are still preserved.

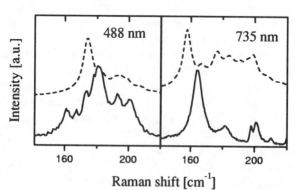

FIGURE 4. Observed (full line) and calculated (dashed line) pattern for the RBM for excitation with two different lasers

Results of calculations for the first moment are plotted in Fig.5. To get a better match between as calculated and as observed spectra we need an upshift of ≈ 10 cm^{-1} for the former. It has been recently calculated that an intertube coupling within carbon nanotube bundles increases the frequency of the RBM [4]. We see that the values for d_0, σ, γ_0 and α mentioned above lead to a very good agreement with experimental results.

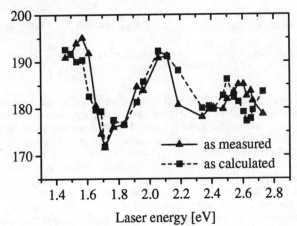

FIGURE 5. Center of gravity for the RBM versus energy for the exciting laser; calculated spectra were 10 cm^{-1} upshifted.

Summarizing we have provided a consistent explanation for details of our Raman experiments. Quasi-periodic oscillations are consequences of the fact that both the transition energies and the frequency of the RBM scale as $1/d$. Fine structures in the Raman spectra are preserved eventhough we take into account all allowed tubes from a certain diameter range. The discrepancy between as calculated and as measured results in Fig4. is ascribed to the intertube coupling that is in agreement with calculations.

This work was supported by the Fonds zur Föderung der wissenschaftlichen Forschung in Austria, Project Nr. P11943, and by Grants No. OTKA T022980 and No. OTKA T030435 in Hungary.

REFERENCES

1. A.G. Rinzler et al., Appl. Phys. **A 67**, 29 (1998).
2. R. Saito et al., Phys. Rev. **B 61**, 2981 (2000).
3. M. Milnera et al., Phys. Rev. Lett. **84**, 1324, (2000).
4. L. Henrard et al., Phys. Rev. **B 60**, R8521 (1999).

Raman Spectra of Single-Wall Carbon Nanotubes from 77 K to 1000 K

E.D. Obraztsova, S.V. Terekhov, A.V. Osadchy

Natural Sciences Center of General Physics Institute, Russian Academy of Sciences, 38 Vavilov street, 117942, Moscow, Russia, elobr@kapella.gpi.ru

Abstract. The Raman spectra of single-wall carbon nanotubes in a wide temperature range (77 K- 1000 K) have been measured during two different heating processes: in a vacuum furnace and in a spot of Ar^+-laser with a variable power density. A position of the tangential Raman mode was established to serve as a correct indicator of the material temperature because the mode downshifted from the position 1595 cm^{-1} (at 77 K) almost linearly with the temperature. The coefficient was -0.015 cm^{-1}/grad. The "breathing" Raman contour has shown two interesting peculiarities. At temperatures higher than 500 K the additional peaks, corresponding to small tubes (with diameters 0.1-0.2 nm less than those, contributed at room temperature) appeared and developed. The "Stokes to antiStokes ratio" has demonstrated an anomalous temperature behavior (up to a prevalence of the anti-Stokes signal under resonance conditions). These observations provided an additional information about the electronic structure of nanotubes.

INTRODUCTION

Nowadays the Raman scattering is one of the most effective and express techniques to characterize single-wall carbon nanotubes (SWNT)[1]. Usually the *tube diameter* is calculated from the position of the "breathing" mode [2]. The parameters of *electronic structure* of SWNT are estimated via monitoring the position of a dominating mode in the Raman "breathing" contour [3]. While the laser energy is tuned the different modes undergo a selective enhancement (Fig.1). Our earlier experiments [4] have shown that a similar effect may be induced by a variation of the laser power density only (without any change of the laser wavelength) (Fig.2). In case of excitation with a high laser power the additional "breathing" peaks corresponding to tubes with diameters smaller than those, observed at room temperature, appeared and developed. This phenomenon is supposed to have a thermal nature: heating broadens the spikes in one-electron DOS. This is equivalent to a narrowing of the mirror-band gap. As a result a resonance excitation becomes possible also for smaller tubes, which had at room temperature the inter-spike distance slightly exceeding the excitation energy value. To check if the laser really acts *just as a heater* we tried to get the same spectra transformation during a conventional heating process in a vacuum oven with a precise temperature control. We also monitored carefully the behavior of the "Stokes to antiStokes ratio" (I_{St}/I_{aSt}) at different temperatures and excitation energies to reveal the trends of its behavior in resonance conditions.

CP544, *Electronic Properties of Novel Materials—Molecular Nanostructures*, edited by H. Kuzmany, et al.
© 2000 American Institute of Physics 1-56396-973-4/00/$17.00

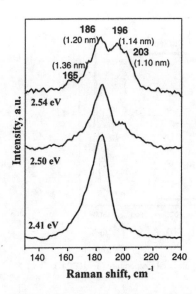

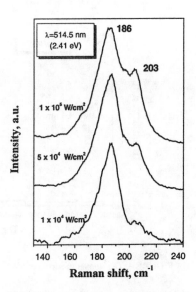

FIGURE 1. Transformation of the "breathing" contour in the Raman spectrum of SWNT induced by the excitation energy variation

FIGURE 2. Transformation of the "breathing" contour in the Raman spectrum of SWNT induced by the Ar+-laser power density variation.

EXPERIMENTAL

The experiments have been performed both with a bucky-paper and with a disperse SWNT produced by arc in *He* atmosphere with *Ni:Y₂O₃:C (1:1:2)* catalyst [4]. The Raman spectra in the vacuum thermostat were measured with a triple monochromator "Jobin-Yvon S-3000" in a micro-configuration. An Ar+-laser was used for the Raman spectra excitation and for the material heating.

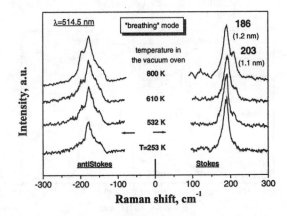

FIGURE 3. Termo-induction of the resonance Raman scattering for tubes with a small diameter during heating the SWNT material in the vacuum oven.

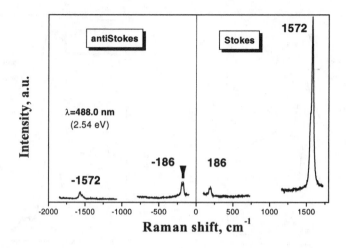

FIGURE 4. The Raman spectrum of SWNT in vacuum at elevated temperature (T≈ 1630 K).

RESULTS AND DISCUSSION

Raman spectra of SWNT heated in the vacuum oven is shown in Fig.3. It is seen that a conventional heating with temperatures higher than 500 K leads to the appearance of a high-frequency tail of a "breathing" Raman contour as in case of the laser heating (Fig.2). This confirms a thermal nature of the observed phenomenon.

The intensity ratio I_{St}/I_{aSt}, obeying the Boltzman law, is used usually as a measure of a sample temperature. This doesn't work for SWNT due to resonant scattering conditions. We have found another reliable temperature indicator in the Raman spectrum of SWNT. While the temperature increased the tangential mode downshifted from the position 1595 cm^{-1} (at 77 K) almost linearly with the temperature. The coefficient was -0.015 cm^{-1}/grad. The temperature values calculated with this coefficient from the

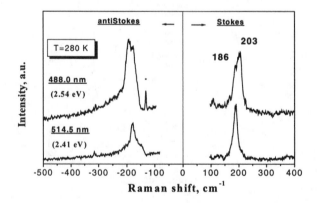

FIGURE 5. Dominance of the antiStokes component in the Raman spectrum of SWNT while the excitation laser energy is approaching the resonance value.

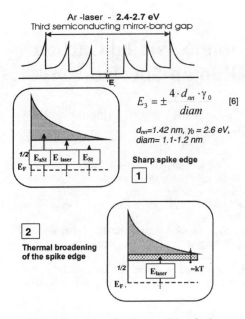

Ar -laser - 2.4-2.7 eV
Third semiconducting mirror-band gap

$$E_3 = \pm \frac{4 \cdot d_{nn} \cdot \gamma_0}{diam} \quad [6]$$

$d_{nn} = 1.42\ nm,\ \gamma_0 = 2.6\ eV,$
$diam = 1.1\text{-}1.2\ nm$

Sharp spike edge

1

2

Thermal broadening
of the spike edge

FIGURE 6. The scheme of resonant conditions for the Raman process in the individual nanotube:1- Stokes and antiStokes photons in case of a sharp spike edge; 2- in case of a thermal broadening of the spike edge.

positions of the tangential mode have coincided with the thermometer data. I_{St}/I_{aSt} ratio appeared to be very informative for characterization of the electronic DOS of SWNT. Changing the laser excitation energy we have observed the situation when the contribution of the antiStokes "breathing" mode became higher than the Stokes' one (Fig.4,5). The difference increased at elevated temperature. A linear dependence of the Raman intensity on the laser power density excluded a dramatic change in phonon occupation numbers, and, therefore, a stimulated Raman process. The observed effect may be ascribed only to a different extent of a resonance enhancement for photons with antiStokes and Stokes energies (Fig.6-1). Such asymmetry already has been observed for Stokes and antiStokes tangential modes in metallic nanotubes [5]. For them the energy difference is about 0.4 eV. But for the "breathing" mode this difference is much smaller – just 0.04 eV. The observation of the antiStokes dominance means that the spike in DOS has a really sharp low-energy edge. The excitation energy, corresponding to *the minimum value of the I_{St}/I_{aSt} ratio* gives a good estimation of the *third mirror-gap value* for a real SWNT material [6]. A specific shape of the antiStokes band (a long high-frequency tail and a sharp edge in a low-frequency range (Fig.5)) may serve as "a shape copy" of the third spike in the electronic DOS for semiconducting nanotubes. The measurements of a temperature dependence of I_{as}/I_s ratio and a heating-induced modification of the "breathing" contour shape allowed to estimate a thermal broadening of spikes (Fig.6-2) in electronic DOS of SWNT.

The work is supported by INTAS project 97-1700, by project 99023 of the Russian Federal Program "Fullerens and Atomic Clusters" and by ISSEP fellowships for S.V. Terekhov and A.V. Osadchy.

REFERENCES

1. Dresselhaus, M.S., Dresselhaus, G., Eklund, P.C., *Science of Fullerenes and Carbon Nanotubes,* Academic Press, New York, 1996
2. Fang, S.U., Rao, A.P., Eklund, P.C., et al., *J. Mater. Res.* **13** (1998) 2405.
3. Rao, A.M., Richter, E., Bandow, S., Chase, B. et al., *Science* **275** (1997) 187).
4. Obraztsova, E.D., Bonard, J.-M., Kuznetsov, V.L., et al., *NanoStruct. Mat.* **12** (1999) 567.
5. Brown, S.D.M., Corio, O., Marucci, A., et al., *Phys. Rev.B* **61** (2000) 37.
6. Charlier , J.-C., Lambin, Ph., *Phys. Rev. Lett.* **57** (1998) 15037.

Characterization of Single-Walled Carbon Nanotubes through Raman Spectroscopy

A. Marucci[a,§], S. D. M. Brown[a], P. Corio[a], M. A. Pimenta[b] and
M. S. Dresselhaus[a,c].

[a] Department of Physics, Massachusetts Institute of Technology, Cambridge, MA 02139, USA.
[b] Dep. de Fisica, Univers. Federal de Minas Gerais, C. P. 702, 30123-970 Belo Horizonte, Br.
[c] Department of Electr. Engineering and Computer Science, MIT, Cambridge, MA 02139, USA.
[§] present address: Lab. Phys. des Solides, UMR CNRS 8502, Univ. Paris-Sud, 91405 Orsay, France.

Abstract. The importance of Raman Spectroscopy in the characterization of single-walled carbon nanotubes (SWNTs) is emphasized. In particular, resonant Raman spectroscopy, performed using several laser excitation energy lines, explores both the vibrational and the electronic properties of the carbon nanotubes giving a broad picture of their characteristics. This work summarizes the main results obtained in the first-order, second-order and anti-Stokes Raman spectral regions.

INTRODUCTION

Raman spectroscopy has proven to be one of the most important tools in the characterization of SWNTs. The spectra collected through the resonant Raman scattering technique, performed using different laser excitation energies E_{laser}, are unique compared to other crystalline systems of carbon as a result of the presence of 1D singularities in the electronic density of states. Experimental results obtained through this technique confirm many theoretical predictions regarding 1D systems and identify many unique characteristics of the SWNTs. In particular, in the first-order and second-order Raman spectra, the bands correlated with the radial breathing mode (RBM) and the tangential mode (TM) are dependent on the diameter and the type of SWNTs in resonance (whether semiconducting or metallic), while the anti-Stokes spectra show an unusual asymmetry, due to the different SWNTs that are in electronic resonance in the Stokes and anti-Stokes spectral regions.

RESULTS AND DISCUSSION

According to resonant Raman theory, the signal from a species is enhanced when either E_{laser} or the scattered photon energy, $E_{scatt.phot.}$, approaches an allowed electronic transition between the valence and conduction bands of the species itself. This phenomenon becomes very unusual with SWNTs whose electronic density of states exhibits singularities at energy values which are dependent on the diameter as well as on the chirality of the SWNTs themselves [1]. Thus the transition energies for SWNTs (ΔE between the highest singularity in the valence band and the lowest

CP544, *Electronic Properties of Novel Materials—Molecular Nanostructures*, edited by H. Kuzmany, et al.
© 2000 American Institute of Physics 1-56396-973-4/00/$17.00

singularity in the conduction band), changes with the diameter and with the type of nanotube, metallic or semiconducting, as evidenced in the plot reported in ref. [2].

Figure 1 shows the spectra obtained in the first and second-order regions for a sample having a distribution of diameters 1.49±0.20 nm at E_{laser}=1.96 and 2.19 eV.

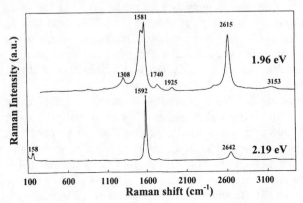

FIGURE 1. Wide range spectra of SWNTs with d_t= 1.49±0.2nm at E_{laser}=1.96 and 2.19 eV .

These spectra exhibits the RBM band at ~158 cm^{-1} (unfortunately not detected at E_{laser} = 1.96 eV), the TM band at ~1590 cm^{-1} plus an intense feature (especially for some values of E_{laser}) at 2615 ÷ 2640 cm^{-1} and a series of less intense peaks detectable with a different intensity scale. These weaker peaks have been assigned to the overtones and combination modes of the RBM and the TM and to peaks not related to these two modes, like the nonresonant feature at ~2420 cm^{-1} associated with the overtone of the K-point phonon in the Brillouin zone [3].

Table 1 contains a summary of the assignments of the peaks in the first and second-order region related to the RBM and the TM, collected at different E_{laser}.

TABLE 1 – Position (in cm^{-1}) of the components of some peaks in the first and second-order region.

E_{laser}	ω_{RBM}	$2\omega_{RBM}$	ω_{TM}	$2\omega_{TM}$	ω_{TM+RBM}	$\omega_{TM+2RBM}$
1.58 eV	150; 162	301; 330	1515; 1540; 1564; 1581; 1591; 1601		1737; 1762	1871
1.96 eV			1515; 1540; 1564; 1581; 1591; 1601	3082; 3122; 3153; 3178; 3203	1740; 1761	1925
2.19 eV	158; 171	311	1567; 1592; 1599	3119; 3150; 3172; 3201	1742; 1760	1930
2.41 eV	142; 155; 170; 178		1567; 1592; 1599	3141; 3166; 3195	1733; 1756	1990

As can be noted by analyzing Table 1, the second-harmonics of both the RBM and the TM, 2RBM and 2TM, as well as their combination modes TM+RBM and TM+2RBM, have positions and number of components dependent on E_{laser}. The variations of these second-order bands with E_{laser} are, however, in accordance with the variations of the corresponding first-order modes. It is, in fact, well known that the RBM frequency depends on the diameter of the SWNT (and thus changes shape and position with E_{laser} [1]) and that new downshifted components of the TM are present in the range of E_{laser} at which metallic nanotubes are in resonance [4]. The second-order bands exhibit the same trends and thus represent a confirmation of all these assumptions [3].

The strong peak at ~2630 cm^{-1} is the so-called G' band, the overtone of the D band at ~1315 cm^{-1}, which is found in all carbon materials exhibiting defects or imperfections. However, the G' band is present also in very pure materials like HOPG and it is thus an intrinsic feature of all carbon materials. As seen in Figure 1 both the D and the G' bands shift position with E_{laser}, a phenomenon that has recently been explained in terms of coupling between phonons and electrons around the K point in the Brillouin zone [6]. By analyzing the variation of the intensity of G' with E_{laser} for SWNTs, it can also be noted that the G' band is much stronger for values of E_{laser} for which the metallic SWNTs are in resonance. The greater electron-phonon coupling in metallic nanotubes that gives rise to this enhancement, has also been confirmed by anti-Stokes [6] and by SERS [7] measurements on SWNTs.

More details about resonance effects in SWNTs have been obtained through the analysis of anti-Stokes resonant Raman spectra [6]. As shown in Figure 2 the anti-Stokes spectra of the TM band exhibits an unusual asymmetry in shape relative to the Stokes TM band, as a result of the participation of the scattered photons in the resonant process.

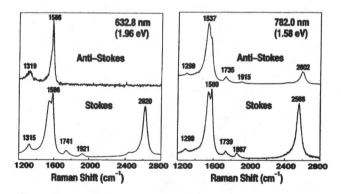

FIGURE 2. Stokes and anti-Stokes spectra at E_{laser}= 1.58 eV and 1.96 eV [6]. At 1.96 eV, the semiconducting tubes are in resonance in the anti-Stokes region while the metallic tubes are in resonance in the Stokes region.

The TM phonon is ~0.2eV and thus the Stokes and anti-Stokes $E_{scatt.phot.}$ differ by about ~0.4eV which is large enough to resonantly enhance the signal of different SWNTs in the two regions (for some values of E_{laser}, metallic SWNTs on one side and semiconducting SWNTs on the other).

The fact that the anti-Stokes process is selective of different nanotubes from the Stokes process implies also that the standard procedure of taking the ratio of the intensities of the Stokes and anti-Stokes signals as a measure of the local temperature of the sample cannot be used. Figure 2 also shows a higher intensity of the G' band when metallic SWNTs are in resonance.

CONCLUSIONS

The analysis of the Raman spectra of SWNTs performed using several E_{laser} over a wide range of frequencies (including the second-order and the anti-Stokes regions), permitted the exploration of many properties of this material. In particular the coupling between electrons, phonons and photons, during the resonant Raman process, causes a variation of the shape and position of the bands related to the RBMs and TMs in the first and second-order spectra and to the presence of an asymmetry in the Stokes and anti-Stokes signals. These unique phenomena are correlated with the dependence of the electronic transition energies of the SWNTs on their diameter and chirality and with the different values of the Stokes and anti-Stokes $E_{scatt.phot.}$.

AKNOWLEDGMENTS

This work has been supported by NSF grant DMR 98-04734. The measurements performed at the George R. Harrison Spectroscopy Laboratory at MIT were supported by the NIH grant P41-RR02594 and NSF grant CHE9708265.

REFERENCES

[1] R. Saito, G. Dresselhaus and M. S. Dresselhaus, *Physical Properties of Carbon Nanotubes*, Imperial College Press, London, 1998.
[2] H. Kataura, Y. Kumazawa, Y. Maniwa, I. Umezu, S. Suzuki, Y. Ohtsuka and Y. Achiba, Synth. Met. **103**, 2555 (1999).
[3] S. D. M. Brown, P. Corio, A. Marucci, M. A. Pimenta, M. S. Dresselhaus and G. Dresselhaus, Phys. Rev. B, **61**, 7734 (2000).
[4] M. A. Pimenta, A. Marucci, S. Empedocles, M. Bawendi, E. B. Hanlon, A. M. Rao, P. C. Eklund, R. E. Smalley, G. Dresselhaus and M. S. Dresselhaus, Phys. Rev. B **58**, R16012 (1998).
[5] M. J. Matthews, M. A. Pimenta, G. Dresselhaus, M. S. Dresselhaus and M. Endo, Phys. Rev. B **59**, R6585 (1999).
[6] S. D. M. Brown, P. Corio, A. Marucci, M. S. Dresselhaus, M. A. Pimenta and K. Kneipp, Phys. Rev. B Rapid Comm., **61**, R5137 (2000).
[7] P. Corio, S. D. M. Brown, A. Marucci, M. A. Pimenta, M. S. Dresselhaus, and K. Kneipp, in press Phys. Rev. B (2000).

Scanning Tunneling Microscopy and Spectroscopy of Multiwall Carbon Nanotubes

A. Hassanien[*1], M. Tokumoto[*], X. Zhao[†] and Y. Ando[†]

[*]*Electrotechnical Laboratory, 1-1-4 Umezono, Tsukuba Ibaraki 305-8568, Japan*
[†]*Department of Physics, Meijo University, Shiogamaguchi, Tempaku-ku, Nagoya 468-8502, Japan*

Abstract. We report on the structural analysis of multiwall carbon nanotubes (MWNTs), produced by DC arc discharge in hydrogen gas, using a scanning tunneling microscope operated at ambient conditions. On a microscopic scale the images show tubes condensed in ropes as well as individual tubes which are separated from each other. Individual nanotubes exhibit various diameters (2.5-6 nm) and chiralities (0-30°). For MWNTs rope, the outer portion is composed of highly oriented nanotubes with nearly uniform diameter (4-5 nm) and chirality. Strong correlation is found between the structural parameters and the electronic properties in which the MWNTs span the metallic-semiconductor regime. True atomic-resolution topographic STM images of the outer shell show hexagonal arrangements of carbon atoms that are equally visible by STM tip. This suggests that the stacking nature of MWNTs has no detectable effects on the electronic band structure of the tube shells. Unlike other MWNTs produced by arc discharge in helium gas, the length of the tubes are rather short (80-500 nm), which make it feasible to use them as a components for molecular electronic devices.

INTRODUCTION

Multiwall Carbon Nanotubes (MWNTs)[1] are of great interest because of their possible application as molecular electronic devices (field emitters[2], field effect transistors[3], ..etc). Combined with their chemical stability MWNTs show robust structures, which can survive severe strain and high currents. Moreover, as carbon nanotubes conduct electricity without heating, this would make them reliable for use as connectors in nano-electronic circuits. Furthermore, since the single shell electronic properties are very much dependent on its structural parameters[4-8] (tube diameter and chiralities), MWNT would offer a possibility of switching between semiconducing and metallic states within the same tube. This switching between high and low conduction could, however be of significant technological importance (for example, constructing nano-integrated devices). In an attempt to explore shell-shell interactions we present detailed investigations on the structural properties of the MWNTs outer shell. High quality topographic STM images reveal hexagonal atomic arrangement of the outer shell. Unlike graphite (e.g. HOPG), the atoms are equally visible by STM tip which suggests that the stacking nature of MWNTs shells (and therefore shell-shell interaction) has undetectable effects on the electronic band structure.

---

[1] Corresponding author, e. mail: hasanien@etl.go.jp

CP544, *Electronic Properties of Novel Materials—Molecular Nanostructures*, edited by H. Kuzmany, et al.
© 2000 American Institute of Physics 1-56396-973-4/00/$17.00

EXPERIMENTAL

The MWNTs samples used for this study were prepared by the usual arc discharge method with H_2 as exchange gas. The preparation method is described elsewhere[9]. A mat of the Pristine samples were collected and sonicated in ethyl alcohol for a few minutes prior to being cast on a highly oriented pyrolytic graphite (HOPG) substrate for STM measurements. We have carried out STM measurements using a Digital Instruments Nano-scope IIIa operated at room temperature in ambient conditions. High quality images of the nanostructure of MWNTs were obtained by recording the tip (Pt–Ir) height at constant current. Typical bias parameters are 400 pA tunnel current and 50 mV bias voltage. The images presented here have not been processed in any way. Scanning tunneling spectroscopy (STS) measurements were performed by interrupting the lateral scanning as well as the feed back loop and measuring the current (I) as a function of tip-sample voltage at a fixed tip-sample distance. A combination of STM and STS measurements on individual nanotubes allows us to investigate the structure and electronic properties, respectively.

RESULTS AND DISCUSION

We first summarize our results for atomic resolution STM images and scanning tunneling spectroscopy of MWNT, then we shall focus on the results relevant to equally visible hexagonal atomic arrangements of the outer shell that led us to role out the effect of shell-shell interaction in MWNTs.

In Fig.1 we show raw STM topographic image of MWNTs which are condensed in ropes. The background is the (001) surface of HOPG substrate. Although the diameter is fairly uniform (4-5 nm), the ropes have length between 80-500 nm.

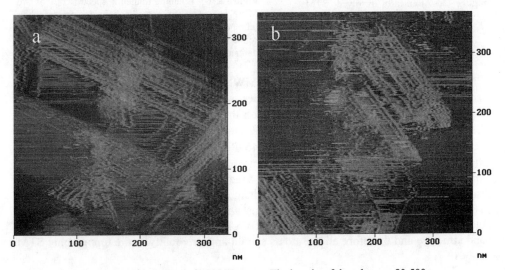

Figure 1. STM topographic images of MWNTs ropes. The lengths of the tubes are 80-500 nm

Figure 2 shows atomic resolution image of zigzag MWNTs with diameter 2.5 nm. The dark dots represent the centers of the carbon hexagons, which show a lattice on a cylindrical surface, the nanotube wall. The nearest neighbor distance of the dark spots is 0.25 nm which compares nicely with that of graphite. Geometrical distortions where the hexagon centers appear elongated along the tube circumference are due to tip-tube topology. As in our previous study of MWNTs produced by arc discharge in He gas[10], we have observed strong correlation between tube geometrical parameters and their electronic properties. Results are shown in Figure 3.

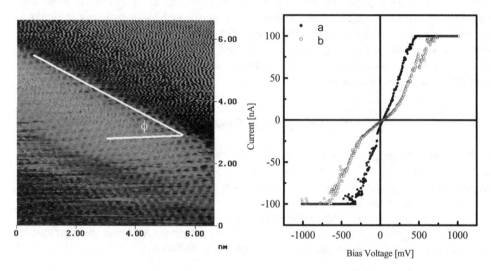

Figure 2. Atomically resolved STM topographic image of zigzag MWNT (ϕ =30° is the chiral angle). The dark dots are the centers of carbon hexagons.

Figure 3. I-V scanning tunneling spectroscopy of two MWNTs.; a) shows metallic behavior for armchair tube with diameter ~3.2 nm; curve b) indicates semiconducting behavior for zigzag tube with daimeter ~2.8 nm.

Next we address the stacking nature of MWNT. Figure 4 shows the highest atomic resolution ever obtained on carbon nanotubes, so far. The image allows us to identify clearly the exact location of carbon atoms on the outer shell within ±0.05 ⊕ accuracy. The distance a =2.42 ⊕ is a measure of the lattice constant of the MWNT which is in excellent agreement with graphite lattice, 2.46 ⊕. The hexagon highlights the underlying lattice of the outer shell. All hexagon corners (atomic position) have equal contrast, which means that the STM tip sense them equally. This is in contrast with the atomic resolution images of HOPG surface where triangular rather than hexagonal pattern is seen (STM tip sense preferentially surface atoms with no neighbors in the adjacent layer below). This suggests that the stacking of MWNTs shell does not modify the single shell band structure and therefore all the atoms should appear equally in the topographic STM images.

In summary, we have presented structural analysis of MWNTs produced by arc discharge in H_2 gas. Mesoscopic STM investigation shows short nanotubes (80-500 nm) which are condensed mainly in ropes with almost uniform diameters (4-5 nm). A strong relation is found between the structural parameters and the electronic properties, namely, MWNT can be metallic or semiconductors depending on the diameters and chiralities of the outer shell. Finally, the stacking of MWNTs shells has undetectable effects on the topographic images of the outer shell, which suggests that the shell-shell interaction doesn't affect the electronic properties of the single shells.

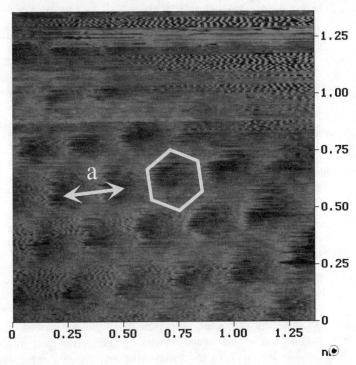

Figure 4. Atomically resolved STM image of zigzag MWNT. The hexagon highlights the outer shell structure.

REFERENCES

1. S. Iijima, Nature, **354**,56(1991).
2. Y. Saito, K. Hamaguchi, S. Uemura, K. Uchida, Y. Tasaka, F. Ikazaki, Y.Yumura, A. Kasuya, and Y. Nishina, Appl. Phys. A: Mater. Sci. Process. **67**,95 (1998).
3. S. J. Trans, A. R. M. Verschueren, and C. Dekker, Nature, **393**, 49 (1998).
4. N. Hamada, S. Sawada, and A. Oshiyama, Phys. Rev. Lett. **68**, 1579 (1992)
5. R. Saito, M. Fujita, G. Dresselhaus, and M. S. Dresselhaus, Appl. Phys.Lett. **60**, 2204 (1992).
6. J. W. G. Wildoer, L. C. Venema, A. G. Rinzler, R. E. Smalley, and C.Dekker, Nature, **391**,59 (1998).
7. T. Wang Odom, J.-L. Huang, P. Kim, and C. M. Lieber, Nature, **391**,62 (1998).
8. A. Hassanien, M. Tokumoto, Y. Kumazawa, H. Kataura, Y. Maniwa, S. Suzuki, and Y. Achiba, Appl. Phys. Lett. **73**, 3839 (1998).
9. X. Zhao, M. Ohkohchi, M. Wang, S. Iijima, T. Ichihashi and Y. Ando; *Carbon* **35**,*775 (1997)*.
10. A. Hassanien, M. Tokumoto, S. Ohshima, Y. Kuriki, F. Ikazaki, K. Uchida, and M. Yumura, Appl. Phys. Lett.**75**, 2755, (1999)

The spectroscopic investigation of the optical and electronic properties of SWCNT

X. Liu[1], H. Peisert[1], R. Friedlein[1], M. Knupfer[1], M. S. Golden[1], J. Fink[1], O. Jost,[2] A. A. Gorbunov,[2] W. Pompe,[2] and T. Pichler[1,3]

[1]Institute of Solid State and Materials Research (IFW) Dresden
P.O. Box 270016, D-01171 Dresden
[2]Institut für Werkstoffwissenschaft, TU-Dresden, Mommsenstr. 13, D-01062 Dresden
[3]Institut für Materialphysik, Universität Wien, Strudelhofgasse 4, A-1090 Wien

Abstract. The spectroscopic investigation of bulk samples of SWCNT is reviewed with examples taken from electron energy-loss spectroscopy in transmission, UV-vis and photoemission spectroscopy. It is shown that, given the insight obtained from EELS studies, UV-vis spectroscopy is a powerful express tool for the characterisation of the NT yield and diameter distribution in bulk SWCNT samples. In addition, using photoemission spectroscopy, apart from showing the use of annealing in UHV to remove many of the contaminants commonly found in purified SWCNT material and determining the work function of pristine SWCNT to be 4.65±0.1eV, we illustrate the similarity of the occupied electronic structure of SWCNT with that of graphite. Open questions, however, remain, such as the appropriate model for the changes of the electronic structure upon intercalation and the reasons for the absence of the singularities in the occupied density of states in the photoemission spectra.

INTRODUCTION

The investigation of the electronic properties of single-walled carbon nanotubes that has been carried out to date can, broadly speaking, be split into studies of individual tubes or bundles of NTs and those experiments giving information on the properties of the macroscopic SWCNT solid. The former class of experiments include STM/STS studies [1], transport measurements [2] (for example employing 'bias spectroscopy') or spatially resolved electron energy-loss spectroscopic (EELS) investigations [3]. The macroscopic probes employed to date include Raman spectroscopy [4], high resolution EELS in transmission [5,6], resonant inelastic x-ray scattering [7] as well as optical investigations [8-10].

In general, such experiments aim to clarify the fundamental optical and electronic properties of SWCNT, although it has been shown that, in some cases, they offer distinct advantages a characterisation tools of bulk SWCNT material itself [8,9]. In this paper, after a brief overview of the kind of information that can be obtained by studying the optical transitions between the occupied and unoccupied electronic levels in SWCNT, measured both by EELS in transmission and UV-vis absorption, we present some recent data from bulk films of SWCNT recorded using photoemission spectroscopy.

CP544, *Electronic Properties of Novel Materials—Molecular Nanostructures*, edited by H. Kuzmany, et al.

EXPERIMENTAL

The single-walled carbon nanotubes used for these studies were synthesised by the laser vapourisation of graphite, either from our own laser furnace [9] or from Tubes@Rice [11]. For the UV-vis characterisation of the soot, we applied the procedure introduced in Ref. [8] in which a well-sonicated soot-methanol mixture (1:100 by weight) was sprayed with an airbrush onto a quartz plate which was held at ~70°C. The optical absorption spectra were recorded using a commercial Shimadzu MPC-3100 instrument. The EELS in transmission measurements were carried out in a purpose-built high resolution spectrometer [12], with an energy and momentum resolution of 110 meV and 0.06 Å^{-1}, respectively. The photoemission measurements were carried out in a commercial PHI 5600ci system equipped with a helium discharge lamp (He I [hv=21.2] and He II [40.8eV]) and a monochromatic Al:Kα x-ray source (hv=1486.6eV). The UPS and XPS measurements were conducted at room temperature with an overall resolution of 100 and 350 meV, respectively. For both the EELS and photoemission measurements, thin films of SWCNT were fabricated by the spray-coating technique described above: for EELS onto KBr crystals (which were then subsequently dissolved to give free-standing NT films) and for the PES onto Pt foils.

RESULTS AND DISCUSSION

Optical properties and their use in the characterisation of bulk SWCNT material.

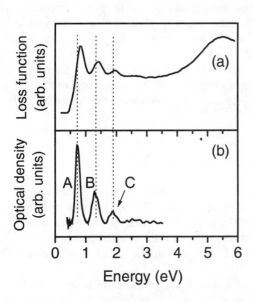

FIGURE 1. (a) The loss function of bulk SWCNT measured for q=0.08 Å^{-1} using EELS in transmission. (b) UV-vis spectrum of SWCNT after a background subtraction. The dotted lines indicate the energy positions of the main features A, B and C in the latter.

Fig. 1 shows, in the upper panel, a high resolution EELS in transmission measurement of the loss function of purified SWCNT, measured with q = 0.08 Å^{-1}. In the energy region shown, one can clearly see three low-energy features located at ~0.83, 1.4 and 1.96 eV, with indications for a fourth feature around 2.6 eV. These are collective excitations related to the transitions between the singularities in the SWCNT density of states (DOS), and have been discussed in detail elsewhere [5]. The larger structure located at ~5.5eV is the collective oscillation of the π-electron density of the NTs - the so-called π-plasmon - and is a characteristic of sp^2-hybridised conjugated carbon systems.

As can be seen from the lower panel of Fig. 1, the same interband transitions can be observed in a simple UV-vis experiment at 0.7, 1.3 and 1.9 eV [9,13]. There are two important points to make here, both resulting from the simplicity of the NT band structure (based upon that of a graphene sheet). Firstly, for the diameter distributions relevant here (1.0<Ø<1.4 nm), the energy separation of occupied and unoccupied van Hove singularities is characteristic for the semiconducting or metallic nature of the NTs (features A and B stem from wide gap semiconducting NTs and feature C from metallic NTs [5]). Secondly, the same quantity is also inversely proportional to the diameter of the NTs themselves. In this way, UV-vis spectroscopy of bulk SWCNT films offers an express method of determining the following: the NT yield (either calibrated against TEM, or relative, within a series of process parameter variations in the laser furnace synthesis); the type of NTs present and their diameter distribution in bulk samples [9].

The ease of the method means that in the optimisation of synthesis conditions or in investigations into the mechanism of SWCNT formation [14], UV-vis spectroscopy can feed back statistically significant (bulk) information into the synthetic activity on a short time-scale. As such it forms an excellent complement to more absolute, macroscopically insensitive, but sophisticated methods such as HRTEM. Furthermore, a detailed analysis of the UV-vis spectra of SWCNT soot as a function of the synthesis conditions has discovered periodic fine-structure in all three main absorption features, which occurs with a constant diameter increment of 0.07 nm [9]. One possible explanation for this apparent diameter grouping could be the preferred formation of SWCNT on the arm-chair side of the NT vector map [9]. This, of course, does not rule out the formation of smaller quantities of NTs on, or near to the zig-zag axis, which would then be 'picked out' in a resonant Raman experiment [15]. However, this is at present only a hypothesis, and more detailed theoretical treatment of the impact of, for example, excitonic effects, electron-phonon coupling and the consequences of bundle formation on the fine-structure in the optical spectra would be very valuable in this regard.

Photoemission spectroscopy of SWCNT

In the second part of this article we relate our first experiments in the investigation of SWCNT films using photoemission spectroscopy. The motivation for this is obvious: photoemission provides a direct probe of excitations from the occupied electronic states of the system in question and consequently has played a major role in the debate regarding the validity of the Luttinger liquid picture in the quasi-1D conductors [16].

Of course, as photoemission is surface sensitive, the preparation of clean surfaces that, in an ideal case, reflect the volume properties is of considerable importance. To

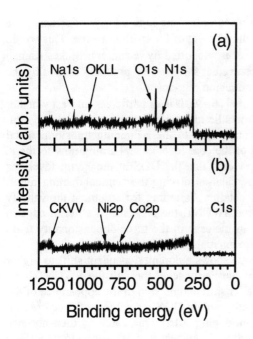

illustrate this, we show in Fig. 2 two overview core level spectra, which give an indication of the elemental composition of the SWCNT films, both as transferred into the spectrometer from the air, and after annealing to 650°C in ultra-high vacuum. It is clear that the 'purified' SWCNT contain both impurities from the preparation (e.g. Na from the NaOH neutralisation of the acid used in the purification, and Ni and Co catalyst) as well as adsorbed oxygen and nitrogen. The impact of adsorbed oxygen on the physical properties of SWCNT has been discussed recently [17].

As is clear from Fig. 2, the UHV anneal is enough to remove all the adsorbed gases, but does not remove the traces of Ni and Co catalyst which are still present on a 1% level.

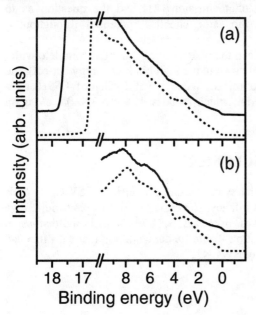

In Fig. 3 we show a summary of the valence band photoemission data both for pristine SWCNT (dotted lines) and for material which has been intercalated with K in UHV (solid lines), recorded both with He I (upper panel) and He II (lower panel) radiation.

Dealing first with the undoped material, for the higher energy photons the spectrum is quite similar to that of graphite, with a peak at 3 eV binding energy which can be assigned to a high density of states of the π-bands due to the flat band region near the M point of the Brillouin zone. This would imply a value of $\gamma_0 \sim 3$ eV, a conclusion that is supported by recent anti-Stokes Raman measurements [18]. A further peak is observed at $E_B = 8.0$ eV probably caused by the flat band region of the σ-bands at the same region of k-space.

In the light of the Luttinger liquid debate [16,19], it is of particular interest whether a Fermi edge can be observed from the metallic tubes. Unfortunately, the clear Fermi edge observed in the experiment is certainly due to the remaining transition metal catalyst particles mention in the context of the core level data. Turning now to the K intercalated material, it is immediately evident that the DOS at the Fermi level has grown significantly, compatible with intercalation driving the semiconducting tubes, which form the majority of the sample, metallic. Assuming for argument the validity of the Luttinger liquid scenario for pristine SWCNT, the presence of a Fermi edge in the intercalated material could point to an increase in the three-dimensionality upon intercalation, as is observed in the graphite intercalation compounds.

Although the C1s core level (not shown) shows a doping dependent shift to higher binding energy consistent with a shift of the Fermi level through the SWCNT band structure, the UPS data show a much smaller shift and no sign of the appearance of the graphite-like DOS minimum (the undoped graphite DOS resembles a 'V' centred on E_F) moving to higher binding energy upon doping. Also remarkable by their absence are the DOS singularities seen in STS studies. At present it is unclear as to why these are not observed in the UPS (we have also failed to observe them in spectra recorded with an energy resolution of 25 meV). Possible factors include the effects of electron-phonon coupling, which are known to be strong in the photoemission of solid fullerenes [20], the relatively short lifetime of the excited state which has been determined recently from two-photon photoemission [21] and the question as to whether all the SWCNT experience the same chemical potential throughout a disordered, low density film.

Finally, from energy difference between the photon energy and the complete width of the UPS spectra (from the Fermi level down to the high binding energy cut-off), the work function of the sample can be determined. We arrive at a value of 4.65±0.1 eV for the undoped (UHV-annealed) material, which reduces down to 3.9±0.1 eV upon 3% K intercalation.

CONCLUSION

Despite the world-wide interest in the synthesis and properties of SWCNT, their spectroscopic investigation is still in its infancy. Nevertheless, in this contribution we described, with the help of examples from EELS, UV-vis and photoemission spectroscopy, the wide range of information - from fundamental to practical - that can be gained from the spectroscopic study of SWCNT.

ACKNOWLEDGMENTS

This work was supported by the EU (IST Project *SATURN*), the Deutsche Forschungsgemeinschaft (FI439/8-1 and PO392/10-1) and and the SMWK (7531.50-03-823-98/5). T.P. thanks the ÖAW for funding an APART grant. We are grateful to H. Zöller for technical assistance.

REFERENCES

1. J.W.G. Wildoer, L.C. Venema, A.G. Rinzler, R.E. Smalley, and C. Dekker, Nature **391**, 59 (1998)
2. S. J. Tans, A. R. M. Verschueren, and C. Dekker, Nature **393**, 49 (1998); M. Bockrath, D.H. Cobden, P.L. McEuen , N.G. Chopra, A. Zettl, A. Thess, and R.E. Smalley, Science **275**, 28 (1997)
3. For example, see R. Kuzno, M. Terauchi, M. Tanaka, and Y. Saito, Jpn. J. Appl. Phys. **33**, L1316 (1994); and O. Stephan *et al.*, these proceedings
4. R. Saito, G. Dresselhaus, and M. S. Dresselhaus, Physical Properties of Carbon Nanotubes, (Inperial College Press, London 1998), Chapter 10
5. T. Pichler, M. Knupfer, M.S. Golden, J. Fink, A. Rinzler, and R.E. Smalley, Phys. Rev. Lett. **80**, 4729 (1998)
6. T. Pichler, M. Sing, M. Knupfer, M.S. Golden, and J. Fink, Solid State Commun. **109**, 721 (1999)
7. S. Eisenbitt, A. Karl, W. Eberhardt, J.E. Fischer, C. Sathe, A. Agni, and J. Nordgren, Appl. Phys. **A 7**, 89 (1998)
8. H. Kataura, Y. Kumazawa, Y. Maniwa, I. Umezu, S. Suzuki, Y. Ohtsuka, and Y. Achiba, Synthetic Metals **103**, 2555 (1999)
9. O. Jost, A. A. Gorbunov, W. Pompe, T. Pichler, R. Friedlein, M. Knupfer, M. Reibold, H.-D. Bauer, L. Dunsch, M.S. Golden, and J. Fink, Appl. Phys. Lett. **75**, 2217 (1999)
10. P. Petit, C. Mathis, C. Journet, and P. Bernier, Chem. Phys. Lett. **305**, 370 (1999)
11. Details can be found at http://cnst.rice.edu/tubes
12. J. Fink, Adv. Electr. Electron. Phys. **75**, 121 (1989)
13. As is usually the case, the energy position is slightly lower in the single-particle excitation (UV-vis) than in the collective excitation (ELS).
14. see, for example, O. Jost *et al*, these proceedings
15. M. Milnera, J. Kürti, M. Hulman, and H. Kuzmany, Phys. Rev. Lett. **84**, 1324 (2000)
16. see for example: M. Grioni, I. Vobornik, F. Zwick, and G. Margaritondo, J. Electron Spectrosc. Relat. Phenom. **100**, 313 (1999); J.Voit, these proceedings
17. P.G. Collins, K. Bradley, M. Ishigami, and A. Zettl, Science **287**, 1801 (2000)
18. S. D. M. Brown, P. Corio, A. Marucci, M. S. Dresselhaus, M. A. Pimenta, and K. Kniepp, Phys. Rev. **B61**, R5137 (2000)
19. M. Bockrath, D.H. Cobden, J. Lu, A.G. Rinzler, R.E. Smalley, T. Balents, and P.L. McEuen, Nature **397**, 598 (1999)
20. M. Knupfer, M. Merkel, M. S. Golden, J. Fink, O. Gunnarsson, and V. P. Antropov, Phys. Rev. **B47**, 13944 (1993); P. A. Brühwiler, A.J. Maxwell, P. Baltzer, S. Andersson, D. Arvanitis, L. Karlsson, and N. Mårtensson, Chem. Phys. Lett. **279**, 85 (1997)
21. T. Hertel and G. Moos, Chem. Phys. Lett. **320**, 359 (2000); T. Hertel *et al*, these proceedings

Rolling Nanotubes: Atomic Lattices as Gears and Contacts

M. R. Falvo[1,3], J. Steele[1,3], S. Paulson[1,3], R. M. Taylor II [2,3], S. Washburn[1,3], and R. Superfine[1,3]

[1]Dept. of Physics and Astronomy, The University of North Carolina, Chapel Hill, NC 27599-3255
[2]Dept. of Computer Science, The University of North Carolina, Chapel Hill, NC
3North Carolina Center for Nanoscale Materials, The University of North Carolina, Chapel Hill, NC

Abstract. The commensurate contact between lattices has profound affects on friction and on the electrical transport across the interface. We report on experiments in which multiwall carbon nanotubes (CNTs) are manipulated with AFM on a graphite (HOPG) substrate. We find certain discrete orientations in which the lateral force of manipulation dramatically increases as we rotate the CNT in the plane of the HOPG surface with the AFM tip. The three-fold symmetry of these discrete orientations indicates commensurate contact of the hexagonal graphene surfaces of the HOPG and CNT. As the CNT moves into commensurate contact, we observe the motion change from sliding/rotating in-plane to stick-roll motion. We have begun the electrical characterization of the nanotube/HOPG interface and find that the junction resistance is significantly lower when the lattices are in registry.

INTRODUCTION

The interaction between two bodies in contact is determined by the interaction between atoms. The arrangement of the atoms in two interacting surfaces has been shown to play a critical role in the energy loss that occurs when one body slides over a second both in experiment [1, 2], and simulation [3-5]. In particular, in the case of two contacting solid crystalline surfaces, the degree of commensurability has been shown to have a clear effect on friction [6-8]. Understanding the effect of these atomic interactions on energy loss [9-12], object motion and electrical transport is important for designing lubrication strategies and self-assembly processes, and will determine the forms of atomic-scale actuating devices[13].

In the present work, we describe experiments in which we are able to tune the commensurability between the two contacting atomically smooth crystalline surfaces. As a model system for such studies, CNTs and HOPG offer a well-defined geometry with atomically smooth surfaces that can remain relatively clean in ambient laboratory conditions. We show that the interlocking of the atomic lattices in the contact region of two bodies increases the force required to move the CNT, can determine whether the CNT slides or rolls, and has a dramatic effect on the contact resistance. In essence, the atomic lattice can act like a gear mechanism.

CP544, *Electronic Properties of Novel Materials—Molecular Nanostructures*, edited by H. Kuzmany, et al.

Experimental: Sliding and Conducting Nanotubes

The CNTs, prepared by the arc-discharge method [15]. A suspension was prepared by sonicating the CNT material in ethanol and then drip dispersing and evaporating onto HOPG. AFM [16] manipulations, performed in ambient conditions, employ an advanced operator interface called the nanoManipulator (nM) [17-19]. This system provides ability to perform complex manipulations, as well transparent switching between low force non-contact AFM for imaging and contact AFM for manipulation.

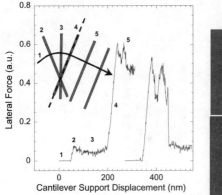
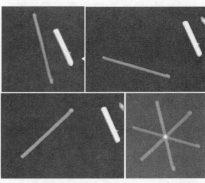

FIGURE 1. (Left) Lateral force trace as a CNT is rotated into (left trace) and out of (right trace) commensurate contact. The inset shows a top-view schematic the process for the left trace. (Right) Commensurate orientations of a CNT on graphite. The nanotube on the left is rotated in-plane into three commensurate orientations indicated by pronounced increase in lateral force as shown in (Left). Finally, the three images were overlain to emphasize the 3-fold symmetry of the commensurate orientations.

During each manipulation, the calibrated lateral force [20, 21] is monitored as a measure of the CNT substrate friction. As the AFM tip is pushed into contact with the CNT in a trajectory perpendicular to its axis, the CNT undergoes either a sliding in-plane rotation motion, or rolling with a constant in-plane rotational orientation[22] (Fig. 1 inset). The CNTs move as rigid bodies, which is expected for CNTs of this size (10-50nm diameter, 500-2000nm length) considering their high stiffness and the low friction of the graphite substrate [23]. For transport measurements, Si AFM tips were coated with gold over a chrome adhesion layer. The tip was placed on top of the nanotube[30], a bias voltage was applied and the current collected in the HOPG substrate.

Sliding and Rolling Nanotubes: When a CNT lying in an incommensurate state is manipulated, it slides smoothly and rotates in plane [22]. However, this motion is interrupted at discrete in-plane orientations in which the CNT "locks" into a low energy state [24, 25], indicated by an increase in the force required to move the CNT. Fig. 1 shows two lateral force traces illustrating the pronounced change in the force in going from the commensurate to incommensurate state and visa versa. The change in lateral force is roughly an order of magnitude, which is typical of our measurements. We resolve no gradual change in lateral force between the two states. The change of force as a function of in-plane rotation angle is discontinuous within our resolution (+/- 1 degree).

As the nanotube is rotated in-plane, several of these discrete commensurate orientations are observed, each separated by 60+/-1 degrees (Fig. 1). These orientations and the associated increase in lateral force are reproducible for a given tube. We believe this registered state corresponds to graphitic ABA stacking. Our hypothesis is supported by molecular statics calculations[26] of CNT on HOPG. These calculations show pronounced energy minima separated by 60-degree intervals as the CNT is rotated in-plane, corresponding to ABA stacking.

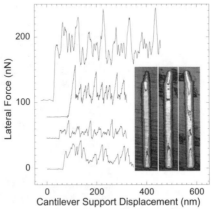

FIGURE 2. Lateral force traces obtained during the manipulation of a carbon nanotube on graphite. The signal has a periodicity of the nanotube circumference, and is different in the forward and reverse directions. The sequence of AFM images show the reorientation of the nanotube.

Fig. 1 (right) shows a second CNT lying in the same immediate area on the graphite substrate. While each CNT shows the complete set of commensurate locking behaviors described above, the two CNTs have lock-in orientations that differ by 11 degrees. In a series of manipulations (not shown), the tubes are rolled individually across the same region in order to verify that the difference in their orientations is not due to an inhomogeneity in the graphite substrate. If lock-in orientations are due to commensurate registry, the particular set of commensurate orientations is determined by the CNT chirality (the wrapping orientation of the outer graphene sheet of the CNT). Large multi-wall CNTs of different diameters are expected to have different chiralities [27] and should show different commensurate orientations[26].

Our evidence for rolling motion in the commensurate state is threefold [22]. First, when in commensurate contact, the nanotubes move without rotating in-plane. Second, for some nanotubes that have observable gross defects, we have observed changes in the topographical data that are consistent with rolling motion (Fig. 2 inset). Thirdly, our lateral force traces of AFM manipulation during rolling show periodicity equal to the nanotube circumference. Fig. 2 shows four such lateral force traces for four different nanotubes. In each case, a period consisting of several peaks is evident which is equal to the circumference of the nanotube within experimental uncertainty (<3%).

Transport through interlocking lattices: While the interlocking of lattices has been demonstrated to have profound effects on the mechanical properties of moving objects, it is also of great interest to measure the electrical properties of the contact region. First, the electronic characterization of the nanotube/substrate contact region may provide a direct, high-bandwidth insight into the changing contact region during the object motion. This hopefully can be correlated with the lateral force measurements. Second, the interlocking of atomic lattices is expected to have profound consequences in the electrical properties of the contact region, with general implications for nanotube/nanotube electrical contacts in devices [31]. Possible effects include (a) the changing of the atomic contacts from a phase-matched coherent array to a set which samples all phases, (b) the change in lattice plane spacing as the lattices slide to a new relaxed position, (c) a reorientation of the electronic wavefunctions at the Fermi level, which may have a profound effect in a semimetal such as graphite with isolated (in momentum space) pockets of conducting states.

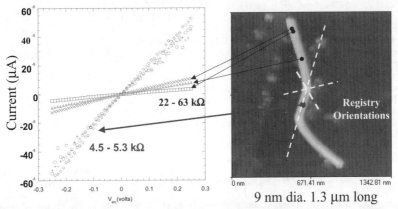

FIGURE 3. Electrical characterization of nanotube/HOPG interface. Inset Left is experimental geometry. I-V curves along the distorted nanotube show a distinct correlation of having a significantly lower resistance in registry versus out of registry.

We have begun such measurements by manipulating the nanotube on graphite with a metal-coated AFM tip. A bias voltage is applied to the tip and the current is measured to a contact pad applied to the top of the graphite substrate. The current runs from the AFM tip into the nanotube and into the graphite [30]. The measurements are performed either during the motion of the nanotube or after the nanotube has been stopped. In the former case the measurement is performed with the AFM tip on the side of the nanotube at a force determined by the force necessary to manipulate the tube, while in the latter case the AFM tip is applied to the top of the nanotube with a normal force determined by the feedback set point.

Our measurements show that the two-probe resistance of the nanotube/graphite interface is about ten times smaller when the lattices are interlocked as compared to when the lattices are out of registry. While most of our measurements have been performed on nanotubes that rotate in the plane without bending, an unusual example is shown in Figure 3. Here a nanotube that has a central section in registry with the end sections out-of-registry is measured in the two-probe geometry by moving the AFM tip along the ridge of the nanotube. We find the same correlation of contact resistance with lattice registry in a single nanotube.

ACKNOWLEDGMENTS

We thank Otto Zhou for providing the CNT material and the entire nanoManipulator team for their invaluable work. This work was supported by the National Science Foundation (HPCC, ECS), the Office of Naval Research (MURI), and the National Institutes of Health National Center for Research Resources.

REFERENCES

[1] Overney, R.M., H. Takano, M. Fujihira, W. Paulus, and H. Ringsdorf, Phys. Rev. Lett. 72 (1994) 3546-3549.

[2] Liley, M., D. Gourdon, D. Stamou, U. Meseth, T.M. Fischer, C. Lautz, H. Stahlberg, H. Vogel, N.A. Burnham, and C. Duschl, Science 280 (1998) 273-275.

[3] Sokoloff, J.B., Phys. Rev. B 42 (1990) 760-765.

[4] Harrison, J.A., C.T. White, R.J. Colton, and D.W. Brenner, Phys. Rev. B 46 (1992) 9700-9708.

[5] Sorensen, M.R., K.W. Jacobsen, and P. Stoltze, Phys. Rev. B 53 (1996) 2101-2112.

[6] Hirano, H., K. Shinjo, R. Kaneko, and Y. Murata, Phys. Rev. Lett. 67 (1991) 2642-2645.

[7] Sheehan, P.E. and C.M. Lieber, Science 272 (1996) 1158 - 1161.

[8] He, G., M.H. Muser, and M.O. Robbins, Science 284 (1999) 1650-1652.

[9] Tomanek, D., in *Scanning Tunneling Microscopy 3*, R. Wiesendanger and H.-J. Guntherodt, Editors(Springer-Verlag, Berlin, 1993) 269-292.

[10] Persson, B.N.J., *Sliding Friction: Physical Principles and Applications*, (Springer-Verlag, Berlin, 1998).

[11] Yoshizawa, H., Y.-L. Chen, and J. Israelachvili, J. Phys. Chem. 97 (1993) 4128 - 4140.

[12] Dowson, D., *History of Tribology*, (Longman Group Limited, London, 1979).

[13] Drexler, K.E., *Nanosystems*, (John Wiley and Sons, Inc., New York, 1992).

[14] Globus, A., W. Bauschlicher Jr., J. Han, R.L. Jaffe, C. Levit, and Srivastiava, Nanotechnology 9 (1998) 192-199.

[15] Ebbesen, T.W. and P.M. Ajayan, Nature 358 (1992) 16.

[16] Binnig, G. and H. Rohrer, Surf. Sci. 126 (1983) 236.

[17] Taylor, R.M., W. Robinett, V.L. Chi, F.P.J. Brooks, W.V. Wright, S. Williams, and E.J. Snyder, in: *Computer Graphics: Proceedings of SIGGRAPH '93* (Association for Computing Machinery SIGGRAPH, Anaheim, CA, 1993).

[18] Finch, M., V.L. Chi, R.M.I. Taylor, M. Falvo, S. Washburn, and R. Superfine, in: *Proceedings of the ACM Symposium on Interactive 3D Graphics* (Association for Computing Machinery SIGGRAPH, Monterey, CA, 1995) 13-18.

[19] Guthold, M., M. Falvo, W.G. Matthews, S. Paulson, A. Negishi, S. Washburn, R. Superfine, F.P. Brooks Jr., and R.M. Taylor II, J. Mol. Graphics Mod. 17 (1999) 187-197.

[20] Schwarz, U.D., P. Koster, and R. Wiesendanger, Rev. Sci. Instrum. 67 (1996) 2561-2567.

[21] Mate, M.C., G.M. McClelland, R. Erlandsson, and S. Chiang, Phys. Rev. Lett. 59 (1987) 1942-1945.

[22] Falvo, M.R., R.M. Taylor II, A. Helser, V. Chi, F.P. Brooks Jr. , S. Washburn, and R. Superfine, Nature 397 (1999) 236-238.

[23] Falvo, M.R., S. Washburn, R. Superfine, M. Finch, F.P.J. Brooks, V. Chi, and R.M.I. Taylor, Biophys. J. 72 (1997) 1396-1403.

[24] Falvo, M., Ph. D. thesis, University of North Carolina (1997).

[25] Liu, J., G. Rinzler, H. Dai, J.H. Hafner, R.K. Bradley, J.P. Boul, A. Lu, T. Iverson, K. Shelimov, C.B. Huffman, F. Rodriguez-Macias, Y.-S. Shon, R.L. Lee, D.T. Colbert, and R.E. Smalley, Science 280 (1998) 1253-1256.

[26] Buldum, A. and J.P. Lu, Phys. Rev. Lett. 83 (1999) 5050-5053.

[27] Dresselhaus, M.S., G. Dresselhaus, and P.C. Eklund, *Science of Fullerenes and Carbon Nanotubes*, (Academic Press, San Diego, 1996).

[28] Hirano, M., K. Shinjo, R. Kaneko, and Y. Murata, Phys. Rev. Lett. 78 (1997) 1448-1451.

[29] C. Dekker, Physics Today, vol. 52, pp. 22, 1999, and references therein.

[30] H. Dai, E. W. Wong, and C. M. Lieber, Science, vol. 272, pp. 523-526, 1996.

[31] M. S. Fuhrer, J. Nygård, L. Shih, M. Forero, Y.-G. Yoon, M. S. C. Mazzoni, H. J. Choi, J. Ihm, S. G. Louie, A. Zettl, and P. L. McEuen, Science, vol. 288, pp. 494-497, 2000.

Modeling of Nanotube as a Molecular Tool

István László

Department of Theoretical Physics, Institute of Physics
Technical University of Budapest, H-1521 Budapest, Hungary

Abstract. Molecular mechanics calculation was used in the study of a one wall nanotube under the action of a tip made of an other one wall nanotube. Neither a clean sliding nor a clean rolling was observed. The position of the lower part of the tip was determined with the tip-graphite and tip-tube interactions and its path was a complicated function of the upper tip position. These facts make difficulties in the possible application of a one wall nanotube as a molecular tool.

INTRODUCTION

The discovery of new crystalline and non-crystalline carbon structures made ways for several applications. Some of them are transistors [1], memory devices [2], nanometre-scale rolling and sliding of carbon nanotubes [3], and nanoprobes in scanning probe microscopy [4]. In scanning tunneling microscope (STM) [5] and atomic force microscope (AFM) [6] the crucial element is the probe tip. Usually the end of the tip is hundreds of angströms in diameter and in most of the cases the chemical composition and the shape are completely unknown. Dai *et al.* [4] attached individual nanotubes several micrometers in length to silicon cantilevers of conventional atomic force microscope.

In our computer experiment the tip was a 21 nm long (5,5) nanotube of 1680 carbon atoms and was closed by two halves of a C_{60} molecule at the ends. We studied the rolling and sliding of a 3.3 nm long (10,10) single wall nanotube of 520 atoms on a graphite layer of 474 atoms.

THE METHOD

The covalent carbon-carbon interaction was calculated with the help of a Brenner potential [7]. In the case of tube-layer, tip-tube and tip-layer interactions a 6-12 Lennard Jones potential was used with Girifalco's parametrization [8,9]. The energy minimization was based on conjugate gradient method.

CP544, *Electronic Properties of Novel Materials—Molecular Nanostructures*, edited by H. Kuzmany, et al.
© 2000 American Institute of Physics 1-56396-973-4/00/$17.00

The graphite layer was parallel with the x-y plane of our coordinate system and the z axis was directed upward and parallel with the tip. The (10,10) tube was placed along the x axis in symmetrical position over the origin. In each molecular mechanics run the atoms in the graphite layer and those in the top half C_{60} of the tip were kept in fixed positions. In a previous work [10] we studied the tip tube interaction by moving the tip vertically along the z axis just over the (10,10) tube. When the tip approached the tube surface there was a small attraction that changed to repulsion at smaller distances. There was a significant difference between the repulsion steepness of tip-graphite and tip-tube interaction.

RESULTS AND DISCUSSIONS

We studied four cases of tip manipulations. In each cases the upper half C_{60} of the tip moved horizontally on a line parallel and just over the y axis. The initial position was at $y = 20\mathring{A}$ and the motion was towards the negative y values. At the beginning of the simulations the (10,10) tube of radius $7.2\mathring{A}$ was just over the x axis. It had a radial deformation as it was described in references [10,11].

In case 1. the z coordinate of the upper C_{60} was $227\mathring{A}$ and the lower part of the tip was at $21\mathring{A}$ from the graphite layer. During the tip motion the tube position was unchanged. There was a small tube deformation, but it was not significant. (Figures 1.a. and 1.b.)

As it can be seen in Figures 2.a and 2.b., there was a significant change in the tube orientation for the second case, with $220\mathring{A}$ and $13\mathring{A}$ of initial tip endpoints. The motion of the tube was the following in the function of the y value of the upper half C_{60} position. From $20\mathring{A}$ to $8\mathring{A}$ there was a small deformation of the tube because of the tip-tube attraction. Figure 2.a. shows the $8\mathring{A}$ tip position. From tis value to $y = -10\mathring{A}$ we observed a small rolling in the direction of negativ y values. The final position of Figure 2.b. at $-46\mathring{A}$ was obtained by sliding and rotating.

The situation was the following in case 3., when the initial tip endpoits were $212\mathring{A}$ and $6\mathring{A}$. There were deformations by tip-tube attraction from $20\mathring{A}$ to $13\mathring{A}$ (Figure 3.a.) and those by tip-tube repulsion from $13\mathring{A}$ to $10\mathring{A}$ and then sliding to $3\mathring{A}$. The final motion was rolling from $3\mathring{A}$ to $-17\mathring{A}$ (Figure 3.b.). Figures 3.a and 3.b. show that there was not any rotation in this case.

b.   a.

FIGURE 1. Tip and tube position over the graphite layer in case 1. The position vector of the upper half C_{60} is (0, 13, 227) (a.) and (0, -15, 227) (b.) in $\mathring{A}$. The y and z axis are directed in order to the right and upward.

The final case was described with 205Å and 0Å initial tip endpoints. We observed sliding and rotating from 0Å to −17Å. These two positions are seen in Figures 4.a. and 4.b.

CONCLUSIONS

We have studied the rolling and sliding conditions of a one wall nanotube under the action of a tip made of an other one wall nanotube. We did not observed neither clean sliding nor clean rolling. In case 3. the dominant part of the motion was rolling and in cases 2. and 4. the dominant motion was sliding with a spectacular rotation in the x-y plane. The position of the tip was characterized by the position of the upper half C_{60}. The position of the lower part of the tip was determined with the tip-graphite and tip-tube interactions. Its path was a complicate function

FIGURE 2. Tip and tube position over the graphite layer in case 2. The position vector of the upper half C_{60} is (0, 8, 220) (a.) and (0, -46, 220) (b.) in Å. The y and z axis are directed in order to the right and upward.

FIGURE 3. Tip and tube position over the graphite layer in case 3. The position vector of the upper half C_{60} is (0, 13, 212) (a.) and (0, -14, 212) (b.) in Å. The y and z axis are directed in order to the right and upward.

FIGURE 4. Tip and tube position over the graphite layer in case 4. The position vector of the upper half C_{60} is (0, 0, 205) (a.) and (0, -17, 205) (b.) in Å. The y and z axis are directed in order to the right and upward.

of the upper tip position.

ACKNOWLEDGEMENTS

This work has been supported by AKP (Grant No.: AKP 98-30 2,2) and OTKA (Grant No.: T025017, T029813).

REFERENCES

1. S. J. Tans, A. R. M. Verschueren, and C. Dekker, *Nature (London)* **393**, 49 (1998).
2. Y.-K. Kwon, D. Tománek, and S. Iijima, *Phys. Rev. Lett.* **82**, 1470 (1999).
3. M. R. Falvo, R. M. Taylor II, A. Heiser, V. Chi, F. P. Brooks Jr, S.WAshburn, and R. Superfine, *Nature (London)* **397**, 236 (1999).
4. H. Dai, J. H. Hafner, A. G. Rinzler, D. T. Colbert, and R. Smalley, *Nature (London)* **384**, 147 (1996).
5. G. Binning, H. Rohrer, Ch. Gerber, and E. Weibel, *Phys. Rev. Lett.* **49**, 57 (1982).
6. G. Binning, C. F. Quate, and Ch. Gerber, *Phys. Rev. Lett.* **56**, 930 (1986).
7. D. W. Brenner, *Phys. Rev.* B **42**, 9458 (1990).
8. L. A. Girifalco, and R. A. Lad, *J. of Chem. Phys.* **25**, 693 (1956).
9. L. A. Girifalco, *J. of Phys. Chem.* **96**, 858 (1992).
10. I. László in *Electronic properties of novel materials- Progress in molecular nanostructures, XIII International Winterschool* , eds. H. Kuzmany, J. Fink, M. Mehring, S. Roth (AIP Conference Proceedings 486, Melville, New York, 1999. pages 355-358.).
11. T. Hertel, R. E. Walkup, and Ph. Avouris, *Phys. Rev.* B **58**, 13870 (1998).

Computer simulation of scanning tunneling spectroscopy of supported carbon nanotube aggregates

Géza I. Márk*,†, László P. Biró*,†, József Gyulai*, Paul A. Thiry†, and Philippe Lambin†

*Research Institute for Technical Physics and Materials Science, H-1525 Budapest
P.O.Box 49, Hungary, E-mail: mark@sunserv.kfki.hu
†Département de Physique, Facultés Universitaires Notre-Dame de la Paix
61, Rue de Bruxelles, B-5000 Namur, Belgium

Abstract. Among the various techniques to investigate carbon nanotubes scanning tunneling microscopy (STM) is one of the most promising since it gives access to both the atomic and electronic structures (STS). The interpretation of the experimental data remains delicate, however. We performed dynamical quantum-mechanical calculations aimed at understanding what controls the current-voltage characteristics in the STM investigation of different arrangements of nanotubes. The calculated STS spectra are directly comparable to experimental data. The theory allowed us to identify one component of pure geometrical origin responsible for an asymmetry of I-V curve, as often observed experimentally above a carbon nanotube. It is emphasized that this asymmetry strongly depends on the shape of the tip, and this may change during the experiment.

INTRODUCTION

A single wall carbon nanotube (SWCNT) consists of one cylindrical shell of graphene sheet with typical diameter of the order of $1\,nm$. A regular, multi-shell structure found experimentally is the "rope" or "raft" [1] of CNTs, which is built by placing side by side the CNTs at the van der Waals distance. To be investigated by STM the CNTs have to be supported on an atomically flat substrate, most frequently Au, or highly oriented pyrolitic graphite (HOPG). The complex structure of the system [2], through which the tunneling takes place makes the interpretation of the STS data delicate. The value of the tunneling gap [3] between the tip and the CNT does influence the STS results. If this gap is reduced below a certain limit switching from tunneling to point contact occurs [4]. Our wave packet dynamical code [2] is used to explore the way the current flows through a CNT in STM configuration.

CP544, *Electronic Properties of Novel Materials—Molecular Nanostructures*, edited by H. Kuzmany, et al.
© 2000 American Institute of Physics 1-56396-973-4/00/$17.00

TUNNEL CURRENT CALCULATION

The tunneling problem is studied in the framework of the two dimensional (2D) potential scattering theory [5]. The current density is determined by calculating the scattering of wave packets (WPs) incident on the barrier potential [2]. We study three tunneling situations. *i)* An STM tunnel junction with no CNT present, *ii)* one SWNT of 1 nm diameter in the STM junction, and *iii)* a raft modeled by three SWNTs. The effective surfaces [2] of the three model barriers are shown in *Fig. 1 inset.*

Fig. 1 shows WP transmission probabilities $P(\theta)$ for zero STM bias as the function of the incidence angle θ, the angle of the wave vector $\vec{k}_0 = (k_{x0}, k_{z0})$ of the initial WP relative to the normal direction. For each barrier the angular dependent transmission was calculated for WPs incident from the support ($P_+(\theta)$, tip positive) and for those incident from the tip ($P_-(\theta)$, tip negative). Incident WP energy is $E = E_F = 5\,eV$. For WPs incident from the support tunneling probability is decreasing with increasing angle because of the decreasing normal momentum of the WP. For WPs incident from the tip, the tunneling probability is influenced by the vortices [5] of the probability current density caused by multiple internal reflections inside the tip. This effect produces a plateau in the $P_-(\theta)$ functions with a shallow minimum around normal incidence. Hence the form of P_- is mainly determined by the particular tip geometry. The $P_+(\theta)$ functions for the raft model have a diffraction grating like characteristic with a strong peak around the normal incidence and smaller shoulders around $30 - 40°$ caused by the interference between the resonant states on the individual tubes.

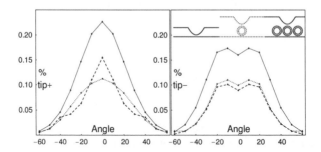

FIGURE 1. Angular dependent wave packet transmission probabilities for different numbers of nanotubes and for different tip polarities. Solid, dotted, and dashed line are for an STM tunnel junction with no nanotube, one nanotube, and a raft model of three nanotubes, respectively. Effective surfaces of the three model potentials are shown in the inset. 100% means the total transmission.

Fig. 2 shows the incidence energy dependence of the transmission probability of WPs with normal incidence through an STM tunnel junction with no CNT present

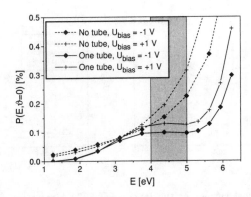

FIGURE 2. Energy dependent transmission of a wave packet incident from the normal direction for tip positive and negative $1\,V$ bias potential. Full (broken) lines are for one (zero) nanotube. The zero of the energy scale is always fixed to the band bottom of the *launching side* of the wave packet. On this energy scale always the states between $E = 4\,eV$ and $E = 5\,eV$ (shaded region on the figure) contribute to the tunnel current at zero temperature.

and through a CNT in the tunnel junction for positive and negative $1\,V$ biases. To model the non vanishing bias an electrostatic potential was added to the jellium potential. The tip, the CNT, and the support are assumed to be perfect conductors. The transmission for the STM tunnel junction with no CNT present follows an exponential like energy dependence characteristic of plane - plane tunneling. The presence of CNT, however, causes a plateau to appear between 3.8 and $5\,eV$. This plateau is a sign of resonant tunneling [2] caused by the two tunnel interfaces.

To estimate the tunnel current flowing through the real 3D junction we assumed that the the perpendicular-to-the-calculation-plane angle dependence of the transmission is like that of a plane-plane system and the tunneling channel is of cylindrical shape. $I(U_{bias})$ curves for the STM tunnel junction with no CNT present, for one CNT in the junction, and for three CNTs calculated with the assumption of a free electron like dispersion relation are shown in *Fig. 3 (a)*. All $I(U_{bias})$ curves of *Fig. 3 (a)* show some degree of asymmetry. These asymmetries are better displayed in the $I(U_{bias}) + I(-U_{bias})$ graphs of *Fig. 3 (b)*. The asymmetry of the tunnel gaps with CNTs are increasing with U_{bias} and has a much larger value than that of the STM junction with no CNT present. These asymmetries are of pure geometrical origin because of the free electron like DOS assumption. In our model the particular tip shape influences the structure of the probability current vortices produced in the tip and this effect is expected to influence the negative side of the STS curve when positive polarity means tunneling from sample to tip. In those STS measurements where larger tunneling current values were used during establishing the position of the STM tip before the feedback loop is switched off a point contact can occur between the CNT and its support. In this case unusual features are expected on the positive side of the STS spectrum while the negative side will

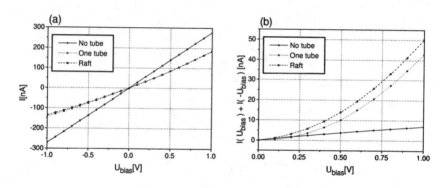

FIGURE 3. *a)* Tunnel current as the function of STM bias for an STM junction with no nanotube, one nanotube and for a nanotube raft. *b)* Tunnel current asymmetries for the curves in Fig. *(a)*.

not differ in shape from symmetric spectra but the magnitude of the tunneling current will increase significantly. These expectations are fulfilled by the experimental data [6]. The absolute values of our calculated currents are higher than those in STS measurements. This is mainly because in STM experiments the tunneling gap is determined in topographic mode. An $1\,nA$ current is expected at say $0.1\,V$ bias at gap of $0.4\,nm$. If U_{bias} is increased without modifying the gap value it would result a strong increase of the tunneling current. To avoid saturation of the tunneling current at bias values exceeding, say $1\,V$, when performing spectroscopy larger tunneling gap values are used than during topographic imaging.

Acknowledgments: This work was supported by the Belgian Federal OSTC PAI-IUPAP P4/10 program and the Hungarian OTKA Grant No T 30435. GIM and LPB gratefully acknowledge a grant from the Belgian Federal OSTC and the hospitality in FUNDP, Namur.

REFERENCES

1. L. P. Biró, S. Lazarescu, Ph. Lambin, P. A. Thirty, A. Fonseca, J. B. Nagy, and A. A. Lucas, Phys. Rev. B **56**, 12490 (1997).
2. G. I. Márk, L. P. Biró, and J. Gyulai, *Phys. Rev. B* **58**, 12645 (1998).
3. L. C. Venema, V. Meunier, Ph. Lambin, C. Dekker, *Phys. Rev. B* **61**, 2991 (1998).
4. G. I. Márk, L. P. Biró, J. Gyulai, P. A. Thiry, and Ph. Lambin, in *Electronic Properties of Novel Materials – Science & Technology of Molecular Nanostructures*, edited by H. Kuzmany et al, AIP Conference Proceedings 486, Melville, New York, 1999, p. 323.
5. A. A. Lucas, H. Morawitz, G. R. Henry, J.-P. Vigneron, Ph. Lambin, P. H. Cutler, and T. E. Feuchtwang, *Phys. Rev. B* **37**, 10708 (1988).
6. L. P. Biró, P. A. Thiry, Ph. Lambin, C. Journet, P. Bernier, and A. A. Lucas, Appl. Phys. Lett. **73**, 3680 (1998).

VI. NANOTUBES: TRANSPORT, DOPING, ELECTRONIC PROPERTIES

A brief introduction to Luttinger liquids

Johannes Voit

Theoretische Physik 1, Universität Bayreuth, D-95440 Bayreuth (Germany)

Abstract. I give a brief introduction to Luttinger liquids. Luttinger liquids are paramagnetic one-dimensional metals without Landau quasi-particle excitations. The elementary excitations are collective charge and spin modes, leading to charge-spin separation. Correlation functions exhibit power-law behavior. All physical properties can be calculated, e.g. by bosonization, and depend on three parameters only: the renormalized coupling constant K_ρ, and the charge and spin velocities. I also discuss the stability of Luttinger liquids with respect to temperature, interchain coupling, lattice effects and phonons, and list important open problems.

WHAT IS A LUTTINGER LIQUID ANYWAY?

Ordinary, three-dimensional metals are described by Fermi liquid theory. Fermi liquid theory is about the importance of electron-electron interactions in metals. It states that there is a 1:1-correspondence between the low-energy excitations of a free Fermi gas, and those of an interacting electron liquid which are termed "quasi-particles" [1]. Roughly speaking, the combination of the Pauli principle with low excitation energy (e.g. $T \ll E_F$) and the large phase space available in 3D, produces a *very dilute* gas of excitations where interactions are sufficiently harmless so as to preserve the correspondence to the free-electron excitations. Three key elements are: (i) The elementary excitations of the Fermi liquid are quasi-particles. They lead to a pole structure (with residue Z – the overlap of a Fermi surface electron with free electrons) in the electronic Green's function which can be – and has been – observed by photoemission spectroscopy [2]. (ii) Transport is described by the Boltzmann equation which, in favorable cases, can be quantitatively linked to the photoemission response [2]. (iii) The low-energy physics is parameterized by a set of Landau parameters $F^\ell_{s,a}$ which contain the residual interaction effects in the angular momentum charge and spin channels. *The correlations in the electron system are weak, although the interactions may be very strong.*

Fermi liquid theory breaks down for one-dimensional (1D) metals. Technically, this happens because some vertices Fermi liquid theory assumes finite (those involving a $2k_F$ momentum transfer) actually diverge because of the Peierls effect. An equivalent intuitive argument is that in 1D, perturbation theory never can work even for arbitrarily small but finite interactions: when degenerate perturbation

CP544, *Electronic Properties of Novel Materials—Molecular Nanostructures,* edited by H. Kuzmany, et al.

theory is applied to the coupling of the all-important electron states at the Fermi *points* $\pm k_F$, it will split them and therefore remove the entire Fermi surface! A free-electron-like metal will therefore not be stable in 1D. The underlying physical picture is that the coupling of quasi-particles to collective excitations is small in 3D but large in 1D, no matter how small the interaction. *Correlations are strong even for weak interactions!*

1D metals are described as Luttinger liquids [3,4]. *A Luttinger liquid is a paramagnetic one-dimensional metal without Landau quasi-particle excitations.* "Paramagnetic" and "metal" require that the spin and charge excitations are gapless, more precisely with dispersions $\omega_\nu \approx v_\nu|q|$ ($\nu = \rho, \sigma$ for charge and spin). Only when this requirement is fulfilled, a Luttinger liquid can form. The charge and spin modes (holons and spinons) possess different excitation energies $v_\rho \neq v_\sigma$ and are bosons. This leads to the separation of charge and spin of an electron (or hole) added to the Fermi sea, in space-time, or $q - \omega$-space. Charge-spin separation prohibits quasi-particles: The pole structure of the Green's function is changed to branch cuts, and therefore the quasi-particle residue Z is zero. Charge-spin separation in space-time can be nicely observed in computer simulations [5].

The bosonic nature of charge and spin excitations, together with the reduced dimensionality leads to a peculiar kind of short-range order at $T = 0$. The system is at a (quantum) critical point, with power-law correlations, and the scaling relations between the exponents of its correlation functions are parameterized by renormalized coupling constants K_ν. The individual exponents are non-universal, i.e. depend on the interactions. For Luttinger liquids, K_ν is the equivalent of the Landau parameters. As an example, the momentum distribution function $n(k) \sim (k_F - k)^\alpha$ for $k \approx k_F$ with $\alpha = (K_\rho + K_\rho^{-1} - 2)/4$. This directly illustrates the absence of quasi-particles: In a Fermi liquid, $n(k)$ has a jump at k_F with amplitude Z.

BOSONIZATION, OR HOW TO SOLVE THE 1D MANY-BODY PROBLEM BY HARMONIC OSCILLATORS

The appearance of charge and spin modes as stable, elementary excitations in 1D fermion systems can be rationalized from the spectrum of allowed particle-hole excitations. In 1D, low-energy particle-hole pairs with momenta between 0 and $2k_F$ are not allowed, and for $q \to 0$, the range of allowed excitations shrinks to a one-parameter spectrum $\omega_\nu \approx v_\nu|q|$, indicating stable particles (cf. Fig. 1). True bosons are then obtained as linear combinations of these particle-hole excitations with a definite momentum q. Most importantly, we now can rewrite any interacting fermion Hamiltonian, provided its charge and spin excitations are gapless, as a harmonic oscillator and find an operator identity allowing to express fermion operators as functions of these bosons. This is the complete bosonization program.

For free fermions, the Hamiltonian describing the *excitations* out of the ground state (the Fermi sea), a can be expressed as a bilinear in the bosons,

$$H = \sum_{\nu=\rho,\sigma} \sum_q v_\nu |q| \left(b^\dagger_{\nu,q} b_{\nu,q} + 1/2 \right) , \qquad (1)$$

with $v_\nu = v_F$, the Fermi velocity. Both the spectrum and the multiplicities of the states, i.e. the Hilbert space, of the fermion and boson forms are identical [3].

What happens in the presence of interactions? One possibility is that the interactions open a gap in the spin and/or charge excitation spectrum. The system then no longer is paramagnetic and/or metallic. With a charge gap, we have a 1D Mott insulator, with a spin gap a conducting system with strong charge density wave or superconducting correlations, and gaps in both channels imply a band insulator. Luttinger liquid theory cannot be applied anymore. In the other case, charge and spin excitations remain gapless: a Luttinger liquid is formed. Then, electron-electrons interactions will make $v_\sigma \neq v_\rho \neq v_F$, leading to charge-spin separation. Interactions will also renormalize the electronic compressibility and magnetic susceptibility, and the charge and spin stiffnesses, and by comparing the velocities measuring this renormalization to v_ν, the correlation exponents K_ν can be defined. The K_ν therefore only depend on the low-energy properties of the Hamiltonian. Two parameters per degree of freedom, K_ν and v_ν, completely describe the physics of a Luttinger liquid.

From model studies, e.g. on the 1D Hubbard model [6] and related models [4], the following picture emerges: (i) $K_\nu = 1$ describes free electrons, and $K_\sigma = 1$ is required by spin-rotation invariance. (ii) $K_\rho > 1$ for effectively attractive interactions, and $K_\rho < 1$ for repulsive interactions. (iii) For the 1D Hubbard model, K_ρ decreases from 1 to 1/2 as the electron repulsion U varies between $0\ldots\infty$. (iv) $K_\rho < 1$ decreases with increasing interaction range. For any finite range, there is a characteristic minimal K_ρ, which approaches zero, as the interaction range extends to infinity. (v) $v_\sigma \leq v_F$ for repulsive interactions. v_σ measures the magnetic exchange J. (vi) $v_\rho > v_F$ for repulsive interactions, and the more so the longer the interaction range. In the limit of unscreened Coulomb interaction, $v_\rho \to \infty$, and the charge fluctuations then become the 1D plasmons [7] with $\omega_\rho(q) \propto |q| \sqrt{|\ln|q||}$. (vii) Electron-phonon interaction decreases the v_ν, and most often also K_ρ. Interaction

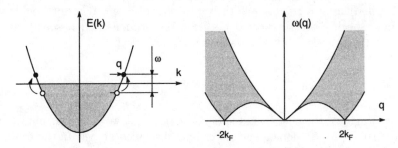

FIGURE 1. Particle-hole excitations in 1D (left). The spectrum of allowed excitation has no low-energy states with $0 \leq |q| \leq 2k_F$.

with high-frequency dispersionless molecular vibrations can enhance K_ρ and lead to superconductivity [8].

To complete our bosonization program, a local fermion operator must be expressed in terms of bosons. Exact operator identities are available for the Luttinger model [3,4] which can be summarized schematically as

$$\Psi_s(x) \sim \exp\left\{ i \sum_\nu \sum_p e^{ipx}(\ldots) \left(b_{\nu,p} + b_{\nu,-p}^\dagger \right) \right\} .$$ (2)

This fermion-boson transformation turns bosonization into a useful device: all correlation functions can be calculated as simple harmonic oscillator averages, repeatedly using the two important identities $e^A e^B = e^{A+B} e^{[A,B]/2}$ for $[A, B]$ a complex number, and $\langle e^A \rangle = \exp(\langle A^2 \rangle/2)$ valid for harmonic oscillator expectation values. As a consequence, *Luttinger liquid predictions for all physical properties can be produced.* Examples are given in the next section. [The behavior of the momentum distribution function $n(k)$ discussed above, has been obtained from the single-particle Green's function $\langle T\Psi(xt)\Psi^\dagger(00)\rangle$ in precisely this way].

Bosonization is an easy and transparent way to calculate the properties of Luttinger liquids. However, it is not the only method. More general, and more powerful is the direct application of conformal field theory to a microscopic model of interacting fermions. For Luttinger liquids, both methods become identical, and one might view bosonization as solid state physicist's way of doing conformal field theory. Also Green's functions methods have been used successfully.

PREDICTIONS FOR EXPERIMENTS

This section summarizes some important properties of Luttinger liquids in the form of experimental predictions. The underlying theoretical correlation functions can be found elsewhere [4]. We discuss a single-band Luttinger liquid. For multi-band systems, such as the metallic carbon nanotubes, the exponents differ from those give here, but can be calculated in the same way [9].

The thermodynamics is not qualitatively different from a Fermi liquid, with a linear-in-T specific heat (expected both for 1D fermions and bosons!), and T-independent Pauli susceptibility and electronic compressibility

$$C(T) = \frac{1}{2}\left(\frac{v_F}{v_\rho} + \frac{v_F}{v_\sigma} \right)\gamma_0 T , \quad \chi = \frac{2K_\sigma}{\pi v_\sigma} , \quad \kappa = \frac{2K_\rho}{\pi v_\rho} .$$ (3)

More interesting are the charge and spin correlations at wavenumber multiples of k_F which display the K_ρ-dependent power laws discussed above. In the electronic structure factor $S(k)$ and NMR spin-lattice relaxation rate T_1^{-1}, they translate into

$$S(k) \sim |k - 2k_F|^{K_\rho} + |k - 4k_F|^{4K_\rho - 1} , \quad T_1^{-1} \sim T + T^{K_\rho} .$$ (4)

The structure factor can be interpreted as showing fluctuations both of Peierls-type ($2k_F$) and of Wigner-crystal-type ($4k_F$) charge density waves, and the two terms in T_1^{-1} come from the $q \approx 0$ and $2k_F$ spin fluctuations. Evidence for such behavior has been found, e.g. in TTF-TCNQ [10] for $S(k)$, and $(TMTSF)_2ClO_4$ [11] for T_1^{-1}. Transport properties depend on the scattering mechanisms assumed. If we consider electron-electron scattering in a band with filling factor $1/n$, we obtain from the current-current correlations [12]

$$\rho(T) \sim T^{n^2 K_\rho - 3} , \quad \sigma(\omega) \sim \omega^{n^2 K_\rho - 5} . \tag{5}$$

The second law has apparently been observed in salts based on TMTSF [13]. These predictions ideally give information on the power-law behavior of correlations, and on the underlying value of K_ρ, which, of course, must be the same for different experiments in any specific material.

In order to see charge-spin separation, one must perform q- and ω-resolved spectroscopy (or time-of-flight measurements). Photoemission spectroscopy is the first choice because it directly probes single-particle excitations [14]. With some approximations, it measures the imaginary part of the electronic Green's function, and Luttinger liquid theory predicts, cf. Fig. 2 [15,16]

$$\rho(q,\omega) = \frac{-1}{\pi} \text{Im} G(q + k_F, \omega + E_F) \sim (\omega - v_\sigma q)^{\alpha - 1/2} |\omega - v_\rho q|^{(\alpha-1)/2} (\omega + v_\rho q)^{\alpha/2} . \tag{6}$$

One finds *two dispersing singularities* (with interaction dependent exponents; for α, cf. above) which demonstrates that the electron ejected from the material is composed out of two more elementary excitations. By q-integration, one can obtain the density of states, $N(\omega) \sim |\omega|^\alpha$, and by $\omega < 0$-integration $n(k)$. A practical comment: the easy part is the calculation of the Green's function in bosonization.

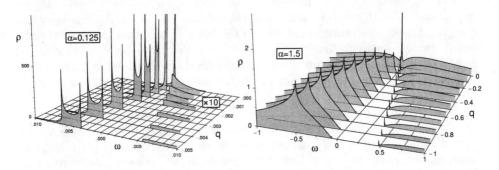

FIGURE 2. Spectral functions of a Luttinger liquid. The three signals represent the holon, the spinon, and the shadow bands (left to right). Left panel: weak/short-range interactions, $\alpha = 1/8$ ($K_\rho = 1/2$). Right panel: strong/long-range interactions, $\alpha = 1.5$ ($K_\rho = 1/8$).

The difficult part is the Fourier transformation if, e.g., the result must satisfy sum rules. Charge-spin separation is also visible, though with different exponents, in dynamical density and spin correlation at $q \approx 2k_F$ [16]. They can be measured, in principle, by EELS, inelastic neutron scattering, or Raman scattering.

STABILITY OF LUTTINGER LIQUIDS

Luttinger liquid theory crucially relies on one-dimensionality. Moreover, most of our discussion was for $T = 0$, and ignored phonons, lattice effects, impurities, etc. Are these factors detrimental to Luttinger liquids? In many cases, the answer will depend on the scales one considers.

Finite temperature is not a problem, and the correlation functions discussed above can be calculated for $T > 0$. Quite generally, however, divergences will be cut off by T whenever $T > \omega, v_\nu |q|, \ldots$. Also charge-spin separation will be masked in the spectral function when $(v_\rho - v_\sigma)q < T$ [17].

Interchain tunneling will introduce 3D effects. Depending on the on-chain interactions, it will either produce a crossover to a Fermi liquid (weak interactions), or to a long-range ordered 3D insulating or superconducting phase (strong interactions) [18]. In any event, *a Luttinger liquid is unstable towards 3D coupling at low enough temperature (scales)*. However, on high enough scales, it will be unaffected by 3D coupling, and coming from there, one will encounter a crossover temperature below which 3D correlations will build up, and 1D physics will be strongly modified. At still lower temperature, a phase transition may take place into a long-range ordered 3D state. When going to a Fermi liquid, the crossover is gradual, and Luttinger-like spectral functions can be observed somewhat off the Fermi energy [19].

Other sources of concern are phonons and lattice effects. Various studies of phonons coupled to Luttinger liquids have shown that depending on details of the electron-electron and electron-phonon interactions, a Luttinger liquid may remain stable, though renormalized, when phonons are added [8]. Alternatively, the electron-phonon interaction could lead to the opening of a spin gap, and thereby destabilize the Luttinger liquid. This situation is described by a different model due to Luther and Emery, but the correlation functions continue to carry certain remnants of Luttinger physics, like non-universal power laws stemming from the gapless charges (the system remains conducting), and charge-spin separation [20].

When the crystal lattice is important (commensurate band filling), the system may become insulating. For a 1D band insulator, Luttinger liquid physics is expected to be lost completely, although not much is known firmly [21]. More interesting is the case of a Mott insulator, brought about by electronic correlations. However even here, charge-spin separation still is seen, e.g. in photoemission both in theory [20] and in experiments on $SrCuO_2$ [22]. Moreover, far above the (charge or spin) gaps, they should no longer influence the physics, and genuine Luttinger liquid behavior is expected there.

HOW TO APPLY LUTTINGER LIQUID THEORY?

We discuss the example of the organic conductor TTF-TCNQ, starting from a recent photoemission study [23]. Photoemission shows a valence band signal whose dispersion is in qualitative agreement with a simple Hückel band structure, along the 1D chains, and no dispersion perpendicular. Two discrepancies between the data and a quasi-particle picture can be resolved in a Luttinger liquid picture: (i) the experimental dispersions are bigger than those expected from the density of states at the Fermi level within the Hückel band structure. (ii) The lineshapes are anomalous in that the signal on the TTF band has a tail reaching up to E_F at all k, while it has a low-energy shoulder with little dispersion on TCNQ. Both findings are consistent with Luttinger liquid spectral functions, with $K_\rho \ll 1/2$ on TTF, and with $1/2 < K_\rho < 1$ on TCNQ. Also recall that Luttinger liquids show more dispersion than Fermi liquids because of the upward renormalization of v_ρ by interactions.

Is this assignment consistent with other information? It is consistent for the TTF band. In fact, the magnetic susceptibility is rather independent of temperature [24], and diffuse X-ray scattering observes strong $4k_F$ density fluctuations at high temperatures [10]. It is not consistent, however, for the TCNQ band where the susceptibility is strongly T-dependent, with an activated shape. This can be taken as an indication of a spin gap, and suggests that the Luther-Emery model might be a better choice. The observation of $2k_F$ density fluctuations on TCNQ is consistent with both assignments, and the spectral function of the Luther-Emery model can at least qualitatively describe the data.

Evidence for or against such hypotheses must come from further experiments. Optics shows a far-IR pseudogap [25]. However, the consistency of the mid-IR conductivity with (5) should be checked. Also notice that [26] $\rho(T) \sim T$. One might look into the temperature dependence of the spin conductivity in view of theories discussing the manifestation of charge-spin separation in transport properties [27]. NMR could look for the T^{K_ρ}-term of Eq. (4), and Raman scattering could show if the values of v_ν measured through two-particle excitations superpose to the dispersions of the photoemission peaks. If successful, the Luttinger liquid theory will provide a consistent phenomenology for the low-energy properties of this material, and will have predictive power for future experiments.

ASPECTS OF MESOSCOPIC SYSTEMS

Due to the small sample size, boundary conditions become of importance, and may dominate the physics. As an example, for a quantum wire, the conductance is given by $G_n = 2nK_\rho e^2/h$ where n is the number of conducting channels [28]. When the wire is coupled to Fermi liquid leads, however, the interaction renormalization is absent [29], and $G_n = 2ne^2/h$ – a boundary effect!

The influence of isolated impurities on transport, or tunneling through quantum point contacts, is an important problem [28,30]. At higher temperatures (voltages), there will be corrections to the (differential) conductance $\delta G \sim T^{K_\rho-1}$, resp. $\delta(dI/dV) \sim V^{K_\rho-1}$. With repulsive interactions and at low energy scales, an impurity will cut the quantum wire into two segments with only a weak link between them. In this case, the conductance, resp. differential conductance, vary as $G(T) \sim T^{K_\rho^{-1}-1}$, resp. $dI/dV \sim V^{K_\rho^{-1}-1}$. The physical origin of this effect is the establishment of a strong Friedel oscillation around the impurity which will increasingly backscatter the electrons at lower energy scales.

An impurity can therefore be assimilated with open boundary conditions. This identifies the exponents just described as members of a larger class of *boundary critical exponents*. Quite generally, 1D interacting fermions with open boundaries and gapless excitations form a *bounded Luttinger liquid* state, rather similar to ordinary Luttinger liquids but with a different set of exponents and scaling relations [30]. The K_ν are properties of the Hamiltonian, and therefore independent of boundary conditions. The correlation functions, and their exponents, however depend on boundary conditions.

A particularly nice experiment demonstrating this relation, has been performed on carbon nanotubes [31]. With different preparations, it is possible to tunnel electrons from electrodes either into the end of nanotubes, or into their bulk. In the first case, conductance and differential conductance measure the power-laws just described for tunneling through a weak link, while for tunneling into the bulk, they measure the bulk density of states, described in the context of photoemission. The exponents differ slightly from those given here because of the peculiar band structure of the tubes and because the electrons tunnel from a Luttinger liquid into a normal metal [9]. The remarkable result of this work is that the various experiments can be described in terms of a *single coupling constant* $K_\rho \sim 0.28$.

OPEN QUESTIONS

The preceding discussion may suggest that one-dimensional fermions are completely understood, at least theoretically. However, many important questions remain open, both in theory and experiment. I now list a few of them.

One important problem relates to scales. While common folklore states that Luttinger liquids form on energy scales between the electronic bandwidth or the typical interaction energy, whichever is smaller, on the high-energy side, and the 3D crossover temperature on the low-energy side, it is not known with certainty if all predicted properties can indeed be observed in that range. Can both power laws and charge-spin separation be observed over the entire range? Some studies seem to suggest that, in the 1D Hubbard model, the Green's function power laws may be restricted to smaller scales [32]. Are these ranges the same for all correlations, or do they depend on the specific function considered? Do they depend on the specific Hamiltonian considered, e.g. on the interaction strength and range, and how?

Concerning mesoscopic systems, only Luttinger liquids with open boundaries are thoroughly characterized. It is conceivable that other boundary conditions (Fermi liquid leads, boundary fields or spins, superconductors) lead to new sets of critical exponents.

What is the spectral weight associated with Luttinger liquid physics in any given microscopic model, or in any given experimental system? Can one measure, in analogy to the quasi-particle residue Z in Fermi liquids, the weight of the coherent spin and charge modes, with respect to the incoherent contributions to the Green's function, or to any other correlation function? How sure can we be that this weight is sufficiently high, so that experiments (e.g. photoemission) actually see these excitations, and not just incoherent contributions or bare high-energy excitations? Is the high-energy physics, far from the Fermi surface, necessarily non-universal and strongly material- (model-) dependent, as is often claimed?

In the same way, the interpretation of some experiments, e.g. photoemission, rests crucially on the appropriateness of simple Hückel-type bandstructures. However, the materials investigated to date, are very complex, and there is no guarantee that these methods are appropriate. There are two ways out. (i) More sophisticated band structure methods become more performing as the computer power increases, and should attack the complex materials of interest here [33]. (ii) One might also look at novel structures where extremely simple 1D materials can be produced. One example for this direction are gold wires deposited on a vicinal Si(111)5x1 surface, where photoemission may have detected evidence for charge-spin separation and Luttinger liquid behavior [34]. (iii) In mesoscopic wires, both on semiconductor and tube base, we would love to have spectroscopic experiments made feasible which probe the dynamics of the elementary excitations beyond transport. As a first step, the study of "noise" might provide interesting insights [35].

ACKNOWLEDGEMENTS

I am a Heisenberg fellow of Deutsche Forschungsgemeinschaft, and received additional support from DFG under SFB 279-B4 and SPP 1073. Many of my contributions to this field are the fruit of collaborations with Diego Kienle, Marco Grioni, Anna Painelli, and Yupeng Wang. Much of my understanding of these matters is due to Heinz-Jürgen Schulz.

REFERENCES

1. P. Nozières, *Interacting Fermi Systems,* W. A. Benjamin Inc, New York (1964).
2. J. W. Allen, *et al.,* J. Phys. Chem. Sol. **56**, 1849 (1995).
3. F. D. M. Haldane, J. Phys. C **14**, 2585 (1981).
4. J. Voit, Rep. Prog. Phys. **58**, 977 (1995) contains a detailed introduction to Luttinger liquid theory, and examples for almost all statements made in this paper.
5. E. A. Jagla, K. Hallberg, and C. A. Balseiro, Phys. Rev. B **47**, 5849 (1993).

6. H. J. Schulz, Phys. Rev. Lett. **64**, 2831 (1990).

7. H. J. Schulz, Phys. Rev. Lett. **71**, 1864 (1993).

8. J. Voit and H. J. Schulz, Phys. Rev. B **37**, 10068 (1988); J. Voit, Phys. Rev. Lett. **64**, 323 (1990).

9. R. Egger and A. O. Gogolin, Phys. Rev. Lett. **79**, 5082 (1997); C. L. Kane, L. Balents, and M. P. A. Fisher, Phys. Rev. Lett. **79**, 5086 (1997).

10. J. P. Pouget, *et al.*, Phys. Rev. Lett. **37**, 437 (1976).

11. C. Bourbonnais, *et al.*, J. Phys. (Paris) Lett. **45**, L-755 (1984).

12. T. Giamarchi, Phys. Rev. B **44**, 2905 (1991).

13. A. Schwartz, *et al.*, Phys. Rev. B**58**, 1261 (1998).

14. M. Grioni and J. Voit, in "Electron spectroscopies applied to low-dimensional materials", edited by H. Stanberg and H. Hughes, Kluwer Academic Publ. (2000).

15. J. Voit, Phys. Rev. B **47**, 6740 (1993); V. Meden and K. Schönhammer, Phys. Rev. B **46**, 15753 (1992).

16. J. Voit, J. Phys. CM **5**, 8305 (1993) and Synth. Met. **70**, 1015 (1995); D. Kienle, Diploma thesis, Bayreuth (1997).

17. N. Nakamura and Y. Suzumura, Prog. Theor. Phys. **98**, 29 (1997).

18. C. Bourbonnais and L. G. Caron, Int. J. Mod. Phys. B **5**, 1033 (1991); E. Arrigoni, Phys. Rev. Lett. **83**, 128 (1999).

19. V. Meden, PhD thesis, Göttingen (1996).

20. J. Voit, Eur. Phys, J. **5**, 505 (1998).

21. V. Vescoli, *et al.*, Phys. Rev. Lett. **84**, 1272 (2000).

22. C. Kim, *et al.*, Phys. Rev. Lett. **77**, 4054 (1996)

23. F. Zwick, *et al.*, Phys. Rev. Lett. **81**, 2974 (1998).

24. T. Takahashi, *et al.*, J. Phys. C **17**, 3777 (1984).

25. H. Basista, *et al.*, Phys. Rev. B **42**, 4088 (1990).

26. J. R. Cooper, Phys. Rev. B **19**, 2404 (1979).

27. Q. Si, Phys. Rev, Lett. **78**, 1767 (1997); J. Voit, in *Proceedings of the Ninth International Conference on Recent Progress in Many-Body Physics,* Sydney (July 1997), ed. by D. Neilson, World Scientific, Singapore (1998), cond-mat/9711064.

28. C. L. Kane and M. P. A. Fisher, Phys. Rev. Lett. **68**, 1220 (1992).

29. I. Safi and H. J. Schulz, Phys. Rev. B **52**, 17040 (1995); D. L. Maslov and M. Stone, Phys. Rev. B **52**, 8666 (1995).

30. J. Voit, M. Grioni, and Y. Wang, Phys. Rev. B **61**, 7930 (2000); A. E. Mattsson, S. Eggert, and H. Johannesson, Phys. Rev. B **56**, 15615 (1997).

31. M. Bockrath, *et al.*, Nature **397**, 598 (1999).

32. M. Imada and Y. Hatsugai, J. Phys. Soc. Jpn. **58**, 3752 (1989); K. Schönhammer, *et al.*, Phys. Rev. B **61**, 4393 (2000).

33. Unpublished work by R. Claessen and P. Blaha (private communication) apparently suggests that the discrepancy between experimental and theoretical dispersions in TTF-TCNQ is not resolved by LDA.

34. P. Segovia, *et al.*, Nature **402**, 504 (1999).

35. S. R. Renn and D. P. Arovas, Phys. Rev. Lett. **78**, 4091 (1997).

Theory of Quantum Transport
in Carbon Nanotubes

Tsuneya Ando

Institute for Solid State Physics, University of Tokyo
7-22-1 Roppongi, Minato-ku, Tokyo 106-8666, Japan

A brief review is given of electronic and transport properties of carbon nanotubes mainly from a theoretical point of view. The topics include an effective-mass description of
electronic states, the absence of backward scattering except for scatterers with a potential
range smaller than the lattice constant, a conductance quantization in the presence of lattice
vacancies, and effects of phonon scattering.

1. INTRODUCTION

Graphite needles called carbon nanotubes (CN's) were discovered recently
[1,2] and have been a subject of an extensive study. A CN is a few concentric tubes
of two-dimensional (2D) graphite consisting of carbon-atom hexagons arranged
in a helical fashion about the axis. The diameter of CN's is usually between 20
and 300 Å and their length can exceed 1 μm. Single-wall nanotubes are produced
in a form of ropes [3,4]. The purpose of this paper is to give a brief review of
recent theoretical study on transport properties of carbon nanotubes.

Carbon nanotubes can be either a metal or semiconductor, depending on
their diameters and helical arrangement. The condition whether a CN is metallic
or semiconducting can be obtained based on the band structure of a 2D graphite
sheet and periodic boundary conditions along the circumference direction. This
result was first predicted by means of a tight-binding model.

These properties can be well reproduced in a k·p method or an effective-
mass approximation [5]. In fact, the method has been used successfully in the
study of wide varieties of electronic properties of CN. Some of such examples
are magnetic properties [6] including the Aharonov-Bohm effect on the band
gap, optical absorption spectra [7,8], exciton effects [9], lattice instabilities in
the absence [10] and presence of a magnetic field [11], magnetic properties of
ensembles of nanotubes [12], and effects of spin-orbit interaction [13].

Transport properties of CN's are interesting because of their unique topolog-
ical structure. There have been some reports on experimental study of transport
in CN bundles [14] and ropes [15,16]. Transport measurements became possible
for a single multi-wall nanotube [17-22] and a single single-wall nanotube [23-27].
In this paper we shall mainly discuss transport properties obtained theoretically.

2. ENERGY BANDS

Figure 1 shows the lattice structure and the first Brillouin zone of a 2D
graphite together with the coordinate systems. The unit cell contains two carbon
atoms denoted as A and B. A nanotube is specified by a chiral vector $\mathbf{L} = n_a\mathbf{a} + n_b\mathbf{b}$

CP544, *Electronic Properties of Novel Materials—Molecular Nanostructures*, edited by H. Kuzmany, et al.
© 2000 American Institute of Physics 1-56396-973-4/00/$17.00

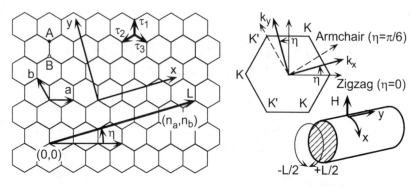

FIGURE 1. (a) Lattice structure of two-dimensional graphite sheet. η is the chiral angle. The coordinates are chosen in such a way that x is along the circumference of a nanotube and y is along the axis. (b) The first Brillouin zone and K and K' points. (c) The coordinates for a nanotube.

with integer n_a and n_b and basis vectors **a** and **b** ($|\mathbf{a}| = |\mathbf{b}| = a = 2.46$ Å). In the coordinate system fixed onto a graphite sheet, we have $\mathbf{a} = (a, 0)$ and $\mathbf{b} = (-a/2, \sqrt{3}a/2)$. For convenience we introduce another coordinate system where the x direction is along the circumference **L** and the y direction is along the axis of CN. The direction of **L** is denoted by the chiral angle η.

A graphite sheet is a zero-gap semiconductor in the sense that the conduction and valence bands consisting of π states cross at K and K' points of the Brillouin zone, whose wave vectors are given by $\mathbf{K} = (2\pi/a)(1/3, 1/\sqrt{3})$ and $\mathbf{K}' = (2\pi/a)(2/3, 0)$ [28]. Electronic states near a K point of 2D graphite are described by the **k·p** equation [29,5]:

$$\gamma(\sigma_x \hat{k}_x + \sigma_y \hat{k}_y)\mathbf{F}_K(\mathbf{r}) = \gamma(\vec{\sigma} \cdot \hat{\mathbf{k}})\mathbf{F}^K(\mathbf{r}) = \varepsilon \mathbf{F}^K(\mathbf{r}), \quad \mathbf{F}^K(\mathbf{r}) = \begin{pmatrix} F_A^K(\mathbf{r}) \\ F_B^K(\mathbf{r}) \end{pmatrix}, \quad (2.1)$$

where γ is the band parameter, $\hat{\mathbf{k}} = (\hat{k}_x, \hat{k}_y)$ is a wave-vector operator, ε is the energy, and σ_x, σ_y, and σ_z are the Pauli spin matrices. Equation (2.1) has the form of Weyl's equation for neutrinos.

The electronic states near the Fermi level can be obtained by imposing the periodic boundary condition in the circumference direction $\Psi(\mathbf{r}+\mathbf{L}) = \Psi(\mathbf{r})$ except for extremely thin CNs. The Bloch functions at a K point change their phase by $\exp(i\mathbf{K}\cdot\mathbf{L}) = \exp(2\pi i\nu/3)$, where ν is an integer defined by $n_a + n_b = 3M + \nu$ with integer M and can take 0 and ± 1. Because $\Psi(\mathbf{r})$ is written as a product of the Bloch function and the envelope function, this phase change should be canceled by that of the envelope functions and the boundary conditions for the envelope functions are given by $\mathbf{F}^K(\mathbf{r}+\mathbf{L}) = \mathbf{F}^K(\mathbf{r})\exp(-2\pi i\nu/3)$.

Energy levels in CN for the K point are obtained by putting $k_x = \kappa_\nu(n)$ with $\kappa_\nu(n) = (2\pi/L)[n - (\nu/3)]$ and $k_y = k$ in the above **k·p** equation as $\varepsilon_\nu^{(\pm)}(n, k) = \pm\gamma\sqrt{\kappa_\nu(n)^2 + k^2}$ [5], where $L = |\mathbf{L}|$, n is an integer, and the upper (+) and lower (−) signs represent the conduction and valence bands, respectively. Those for the K' point are obtained by replacing ν by $-\nu$. This shows that CN becomes metallic for $\nu = 0$ and semiconducting with gap $E_g = 4\pi\gamma/3L$ for $\nu = \pm 1$.

3. ABSENCE OF BACKWARD SCATTERING

In the presence of impurities, electronic states in the vicinity of K and K' points can be mixed with each other. Therefore, we should use a 4×4 Schrödinger equation

$$\mathcal{H}\mathbf{F} = \varepsilon\mathbf{F}, \quad \mathbf{F} = \begin{pmatrix} \mathbf{F}^K \\ \mathbf{F}^{K'} \end{pmatrix}, \quad \mathbf{F}^{K'} = \begin{pmatrix} F_A^{K'} \\ F_B^{K'} \end{pmatrix}, \quad \mathcal{H} = \mathcal{H}_0 + V. \quad (3.1)$$

The unperturbed Hamiltonian is given by $\gamma(\sigma_x k_x + \sigma_y k_y)$ for the K point and $\gamma(\sigma_x k_x - \sigma_y k_y)$ for the K' point. The effective potential of an impurity is written as [30]

$$V = \begin{pmatrix} u_A(\mathbf{r}) & 0 & e^{i\eta}u'_A(\mathbf{r}) & 0 \\ 0 & u_B(\mathbf{r}) & 0 & -\omega^{-1}e^{-i\eta}u'_B(\mathbf{r}) \\ e^{-i\eta}u'_A(\mathbf{r})^* & 0 & u_A(\mathbf{r}) & 0 \\ 0 & -\omega e^{i\eta}u'_B(\mathbf{r})^* & 0 & u_B(\mathbf{r}) \end{pmatrix}, \quad (3.2)$$

where $\omega = \exp(2\pi i/3)$. If we use a tight-binding model, we obtain the explicit expressions for the potentials as

$$u_A(\mathbf{r}) = \sum_{\mathbf{R}_A} g(\mathbf{r}-\mathbf{R}_A)u_A(\mathbf{R}_A), \quad u'_A(\mathbf{r}) = \sum_{\mathbf{R}_A} g(\mathbf{r}-\mathbf{R}_A)e^{i(\mathbf{K}'-\mathbf{K})\cdot\mathbf{R}_A}u_A(\mathbf{R}_A), \quad (3.3)$$

where $u_A(\mathbf{R}_A)$ is the local site energy at site $\mathbf{R}_A$ due to the impurity potential and $g(\mathbf{r})$ is a smoothing function having a range of the order of the lattice constant a and satisfying the normalization condition $\sum_{\mathbf{R}_A} g(\mathbf{r}-\mathbf{R}_A) = \sum_{\mathbf{R}_B} g(\mathbf{r}-\mathbf{R}_B) = 1$. The similar expressions can be obtained for $u_B(\mathbf{r})$ and $u'_B(\mathbf{r})$.

In the vicinity of $\varepsilon = 0$, we have two right-going channels $K+$ and $K'+$, and two left-going channels $K-$ and $K'-$. Figure 2 shows calculated scattering amplitude as a function of d/a for a Gaussian potential located at a B site and having the integrated intensity u in the absence of a magnetic field. The backward scattering probability decreases rapidly with d and becomes exponentially small for $d/a \gg 1$. The same is true of the intervalley scattering. This absence of the backward scattering for long-range scatterers disappears in the presence of magnetic fields although not shown explicitly.

It has been proved that the Born series for back-scattering vanish identically [30]. This has been ascribed to a spinor-type property of the wave function under a rotation in the wave vector space [31]. The absence of backward scattering has confirmed by numerical calculations in a tight binding model [32].

Because of the presence of large contact resistance between a nanotube and metallic electrode, the conductance usually exhibits a prominent effect of a single electron tunneling due to charging effects. An important information can be obtained on the effective mean free path and the amount of backward scattering in nanotubes [23-27]. In fact, the Coulomb oscillation in semiconducting nanotubes is quite irregular and can be explained only if nanotubes are divided into many separate spatial regions in contrast to that in metallic nanotubes [33]. This behavior is consistent with the presence of considerable amount of backward scattering leading to a strong localization of the wave function in semiconducting tubes. In metallic nanotubes, the wave function is extended throughout the whole region of a nanotube because of the absence of backward scattering.

321

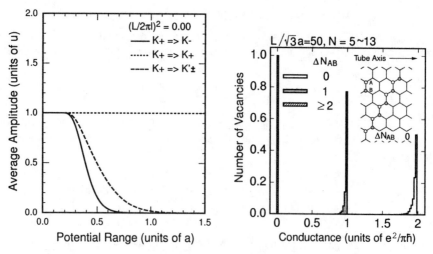

FIGURE 2. (left) Calculated effective scattering matrix elements versus the potential range at $\varepsilon = 0$ in the absence of a magnetic field. After Ref. [30].

FIGURE 3. (right) Calculated histogram of the conductance of nanotubes with a vacancy. After Ref. [35].

4. LATTICE VACANCIES

Effects of scattering by a vacancy in armchair nanotubes have been studied within a tight-binding model [34,35]. It has been shown that the conductance at $\varepsilon = 0$ in the absence of a magnetic field is quantized into zero, one, or two times of the conductance quantum $e^2/\pi\hbar$ for a vacancy consisting of three B carbon atoms around an A atom, of a single A atom, and of a pair of A and B atoms, respectively [35].

Numerical calculations were performed for about 1.5×10^5 vacancies and demonstrated that such quantization is quite general [36]. Figure 3 shows a histogram of the conductance for different values of ΔN_{AB}, where N_A and N_B are the number of removed atoms at A and B sublattice points, respectively, and $\Delta N_{AB} = N_A - N_B$. This rule was analytically derived in a k·p scheme later [37].

5. PHONON SCATTERING

A conductance quantization was observed in multi-wall nanotubes [38]. This quantization is likely to be related to the absence of backward scattering shown here, but much more works are necessary including effects of magnetic fields and problems related to contacts with metallic electrode before complete understanding of the experimental result. At room temperature, where the experiment was performed, phonon scattering is likely to play an important role [15,39,40].

Acoustic phonons important in the electron scattering are described well by a continuum model [40]. The potential-energy functional for displacement

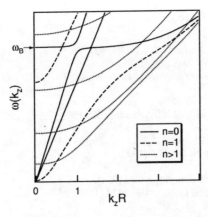 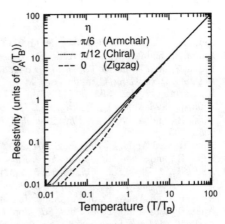

FIGURE 4. (left) Frequencies of phonons obtained in the continuum model.

FIGURE 5. (right) The resistivity of armchair (solid line) and zigzag (dotted line) nanotubes in units of $\rho_A(T_B)$ which is the resistivity of the armchair nanotube at $T = T_B$, and T_B denotes the temperature of the breathing mode, $T_B = \hbar \omega_B / k_B$.

$\mathbf{u} = (u_x, u_y, u_z)$ is written as

$$U[\mathbf{u}] = \int dx\,dy\, \frac{1}{2}\left(B(u_{xx}+u_{yy})^2 + \mu\left[(u_{xx}-u_{yy})^2 + 4u_{xy}^2\right]\right),$$
$$u_{xx} = \frac{\partial u_x}{\partial x} + \frac{u_z}{R},\quad u_{yy} = \frac{\partial u_y}{\partial y},\quad 2u_{xy} = \frac{\partial u_x}{\partial y} + \frac{\partial u_y}{\partial x}, \tag{5.1}$$

where the term u_z/R is due to the finite radius R of the nanotube. The parameters B and μ denote the bulk modulus and the shear modulus for a graphite sheet. Figure 4 shows phonon dispersions calculated in the continuum model (a small out-of-plane energy proportional to the curvature has been added although not important in the electron scattering).

The electron-phonon interaction is given for the K point by

$$V_{\text{el-ph}} = \begin{pmatrix} V_1 & V_2 \\ V_2^+ & V_1 \end{pmatrix},\quad V_1 = g_1(u_{xx}+u_{yy}),\quad V_2 = g_2 e^{3i\eta}(u_{xx}-u_{yy}+2iu_{xy}), \tag{5.2}$$

where g_1 is the deformation potential and g_2 describes the effective potential due to bond-length modification. The diagonal deformation-potential term does not contribute to the backward scattering as in the case of impurities and only the smaller off-diagonal term remains. Figure 5 shows calculated temperature dependence of the resistivity, where $\rho_A(T)$ is the resistivity of an armchair nanotube given by $g_2^2 k_B T / 2e^2 \hbar v_F^2 R\mu$ with $v_F = \gamma/\hbar$. The resistivity of an armchair CN is same as that obtained previously [39] except for a difference in g_2.

ACKNOWLEDGMENTS

The author acknowledges the collaboration with T. Nakanishi, H. Matsumura, R. Saito, H. Suzuura, and M. Igami. This work was supported in part by Grants-in-Aid for Scientific Research and Priority Area, Fullerene Network, from Ministry of Education, Science and Culture in Japan.

REFERENCES

1. Iijima, S., *Nature (London)* **354**, 56 (1991).
2. Iijima, S., Ichihashi, T., and Ando Y., *Nature (London)* **356**, 776 (1992).
3. Iijima, S., and Ichihashi T., *Nature (London)* **363**, 603 (1993).
4. Bethune, D.S., Kiang, C.H., de Vries, M.S., Gorman, G., Savoy, R., Vazquez, J., and Beyers, R., *Nature (London)* **363**, 605 (1993).
5. Ajiki, H., and Ando, T., *J. Phys. Soc. Jpn.* **62**, 1255 (1993).
6. Ajiki, H., and Ando, T., *J. Phys. Soc. Jpn.* **62**, 2470 (1993) [Errata, **63**, 4267 (1994)]; **65**, 505 (1996).
7. Ajiki, H., and Ando, T., *Physica B* **201**, 349 (1994); *Jpn. J. Appl. Phys. Suppl.* **34**, 107 (1995).
8. Ajiki, H., and Ando, T., *Jpn. J. Appl. Phys. Suppl.* **34**, 107 (1995).
9. Ando, T., *J. Phys. Soc. Jpn.* **66**, 1066 (1997).
10. Viet, N.A., Ajiki, H., and Ando, T., *J. Phys. Soc. Jpn.* **63**, 3036 (1994).
11. Ajiki, H., and Ando, T., *J. Phys. Soc. Jpn.* **64**, 260 (1995); **65**, 2976 (1996).
12. Ajiki, H., and Ando, T., *J. Phys. Soc. Jpn.* **64**, 4382 (1995).
13. Ando, T., *J. Phys. Soc. Jpn.* **69**, No. 6 (2000).
14. Song, S.N., Wang, X.K., Chang, R.P.H., and Ketterson, J.B., *Phys. Rev. Lett.* **72**, 697 (1994).
15. Fischer, J.E., Dai, H., Thess, A., Lee, R., Hanjani, N.M., Dehaas, D.D., and Smalley, R.E., *Phys. Rev. B* **55**, R4921 (1997).
16. Bockrath, M., Cobden, D.H., McEuen, P.L., Chopra, N.G., Zettl, A., Thess, A., and Smalley, R.E., *Science* **275**, 1922 (1997).
17. Langer, L., Bayot, V., Grive, E., Issi, J.-P., Heremans, J.P., Olk, C.H., Stockman, L., Van Haesendonck, C., and Brunseraede, Y., *Phys. Rev. Lett.* **76**, 479 (1996).
18. Kasumov, A.Yu., Khodos, I.I., Ajayan, P. M., and Colliex, C, *Europhys. Lett.* **34**, 429 (1996).
19. Katayama, F., *Master thesis* (Univ. Tokyo, 1996).
20. Ebbesen, T.W., Lezec, H.J., Hiura, H., Bennett, J.W., Ghaemi, H.F., and Thio, T., *Nature (London)* **382**, 54 (1996).
21. Dai, H., Wong, E.W., and Lieber, C.M., *Science* **272**, 523 (1996).
22. Kasumov, A.Yu., Bouchiat, H., Reulet, B., Stephan, O., Khodos, I.I., Gorbatov, Yu.B., and Colliex, C., *Europhys. Lett.* **43**, 89 (1998).
23. Tans, S.J., Devoret, M.H., Dai, H.-J., Thess, A., Smalley, R.E., Geerligs, L.J., and Dekker, C., *Nature (London)* **386**, 474 (1997).
24. Tans, S.J., Verschuren, R.M., and Dekker, C., *Nature (London)* **393**, 49 (1998).
25. Cobden, D.H., Bockrath, M., McEuen, P.L., Rinzler, A.G., and Smalley, R.E., *Phys. Rev. Lett.* **81**, 681 (1998).
26. Tans, S.J., Devoret, M.H., Groeneveld, R.J.A., and Dekker, C., *Nature (London)* **394**, 761 (1998).
27. Bezryadin, A., Verschueren, A.R.M., Tans, S.J., and Dekker, C., *Phys. Rev. Lett.* **80**, 4036 (1998).
28. Wallace, P.R., *Phys. Rev.* **71**, 622 (1947).
29. Slonczewski, J.C., and Weiss, P.R., *Phys. Rev.* **109**, 272 (1958).
30. Ando, T., and Nakanishi, T., *J. Phys. Soc. Jpn.* **67**, 1704 (1998).
31. Ando, T., Nakanishi, T., and Saito, R., *J. Phys. Soc. Jpn.* **67**, 2857 (1998).
32. Nakanishi, T., and Ando, T., *J. Phys. Soc. Jpn.* **68**, 561 (1999).
33. McEuen, P.L., Bockrath, M., Cobden, D.H., Yoon, Y.-G., and Louie, S.G., *Phys. Rev. Lett.* **83**, 5098 (1999).
34. Chico, L., Benedict, L.X., Louie, S.G., and Cohen, M.L., *Phys. Rev. B* **54**, 2600 (1996).
35. Igami, M., Nakanishi, T., and Ando, T., *J. Phys. Soc. Jpn.* **68**, 716 (1999).
36. Igami, M., Nakanishi, T., and Ando, T., *J. Phys. Soc. Jpn.* **68**, 3146 (1999).
37. Ando, T., Nakanishi, T., and Igami, M., *J. Phys. Soc. Jpn.* **68**, 3994 (1999).
38. Frank, S., Poncharal, P., Wang, Z.L., and de Heer, W.A., *Science* **280**, 1744 (1998).
39. Kane, C.L., Mele, E.J., Lee, R.S., Fischer, J.E., Petit, P., Dai, H., Thess, A., Smalley, R.E., Verschueren, A.R.M., Tans, S.J., and Dekker, C., *Europhys. Lett.* **41**, 683 (1998).
40. Suzuura, H., and Ando, T., *Mol. Cryst. Liq. Cryst.* (in press); *Physica E* (in press).

Conductance of Carbon Based Macro-Molecular Structures

S. Stafström , A. Hansson, and M. Paulsson

*Department of Physics and Measurement Technology, IFM,
Linköping University, SE-581 83, Linköping, Sweden*

Electron transport through metallic nanotubes and stacks of wide bandgap polyaromatic hydrocarbons (PAH) are studied theoretically using the Landauer formalism. These two systems constitute examples of different types of carbon based nanostructured materials of potential use in molecular electronics. The studies are carried out for structures with finite length that bridge two contact pads. In the case of perfect metallic nanotubes, the current is observed to increase stepwise with the applied voltage and the resistance is independent on the length of the tube. In the PAH stacks, the off resonance tunneling conductance decreases exponentially with the number of molecules in the stack and shows a near linear increase with the number of carbon atoms in each molecule.

INTRODUCTION

Molecular electronics has had an enormous development during the last decade [1]. At the microscopic level this has led to the possibility to measure the currents through individual molecules using scanning probe microscopy [1] or mechanically controlled break junctions [2]. In the case of carbon nanotubes (NT), it was shown that electrodes with a spacing at the sub-micrometer scale can be used to contact individual, single wall, carbon nanotubes (SWNT) [3]. Several other experiments dealing with the conductance of individual nanotubes have followed. [4,5].

Polyaromatic hydrocarbons (PAH) can by synthesized in many different sizes up to systems with 30-40 benzene units connected in the form of graphite flakes [6-8]. Derivatives of PAH with alkyl side chains replacing some of the hydrogens on the edge of the carbon flakes can form liquid crystals in which the planar molecules stack on top of each other while the alkyl chains isolate the π-electron systems of the individual stacks from each other. This situation is very interesting in the context of molecular- (or nano-) wires since electron transport is one-dimensional within the π-stacks. The liquid crystal phase could be considered as many parallel such wires over a wide area, with the possibility to be activated with specific phase differences that would allow for logical operations at the phase coherent quantum level.

The aim of this study is to investigate how electron transport along NT and stacks of PAH molecules depends on the size of these systems. In the case of NT we focus on the length of the tube whereas for the π-stacks the corresponding study deals with the number of molecules in the stack and the size of each molecule.

CP544, *Electronic Properties of Novel Materials—Molecular Nanostructures*, edited by H. Kuzmany, et al.
© 2000 American Institute of Physics 1-56396-973-4/00/$17.00

METHODOLOGY

The conductance is calculated using the many channel Landauer formula [9-11]:

$$G = \frac{2e^2}{\pi\hbar} Tr\left(\left|t(E)\right|^2\right)$$

(1)

where $t(E)$ represents the amplitudes for electron transmission between the two metallic contacts for an electron with energy E. In the results presented below we have excluded the prefactor, giving a dimensionless conductance, which is identical to the transmission probability ($T(E)$).

To calculate the conductance we have to specify the interactions in the molecular system as well as in the metallic contacts, and the coupling between these parts. All these interactions are here treated at the level of the tight-binding approximation. The metal contacts the carbon nanotube from the side. This setup resembles very much that used in conductance measurements [3,5]. In a similar way, the PAH-stacks are contacted from the bottom and top molecules. With the geometry and the interactions specified the calculations end up in a set of linear equations for the transmission and reflection amplitudes which was solved numerically using the Green function of the molecule. For the details of how the calculations are performed we refer to Paulsson [12].

RESULTS AND DISCUSSION

The transmission spectrum ($T(E)$) through the (5,5) NT is shown in Fig. 1 for a tube with 500 carbon. The parameters controlling the coupling between the contact and the tube are chosen to simulate the case of a high contact resistance [3]. As a consequence, resonances in the transmission probability appear almost exactly at the eigen energies of the tube. The dips in the peaks originate from the Green function of the NT, which is singular exactly at these energies. However, the dips are very sharp and do not affect the IV-curves shown below. Off resonance, the transport is an ordinary tunneling process, which drops exponentially away from the eigen energies.

As the length of the NT increases, the eigen energies and consequently also the peaks in the conductance spectrum come closer. The peak-to-peak spacing reduces on average linearly with increasing tube length. This will also increase the resonant contribution to the current linearly with the tube length. However, the off resonance regions, in which the conductance drops exponentially away from the center of the peak, will contribute less to the total current. By integrating the conductance spectrum over the energy we obtain the current-voltage (I-V) relation (see Fig. 2). The I-V curves are shown for two different tubes, with 500 and 700 atoms respectively. The steps in the curves in Fig. 2 occur as a result of resonance tunneling and correspond exactly to the peaks in the conductance spectrum. It is evident from these two curves

that the two effects discussed above exactly cancel and lead to a length independent
resistance in the case of perfect NT's (ballistic transport).

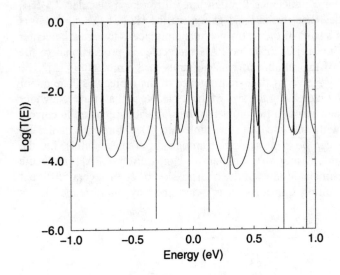

Fig. 1: Logarithm of the transmission coefficient for a (5,5) NT with 500 sites.

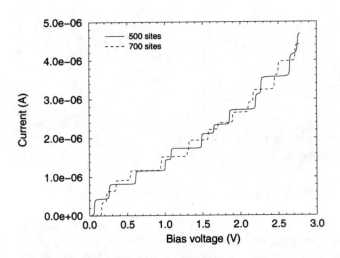

Fig. 2: I-V curve for (5,5) NT with 500 and 700 sites respectively.

A completely different conductance spectrum is obtained in the case of PAH (see Fig.
3) Here, the wide band gap causes a very small conductance in the gap region. The gap
size varies little with the number of PAH in the stack and this effect can be totally
neglected in comparison with the usual exponential conductance drop for increasing

tunneling distances and we observe a perfect fit to an exponential function for $T(E=0)$ as a function of L.

In Fig. 3 is shown the conductance through three different molecules, $C_{42}H_{18}$, $C_{60}H_{22}$ and $C_{114}H_{30}$. The decrease in the band gap when the size of the molecule is increased is clearly seen as a shift of the resonance tunneling peeks towards the center of the bandgap. Note that the bandgap in the PAH molecules is proportional to the inverse of the square root of the number of benzene units in the molecule [13]. The bandgap thus decreases very slowly with increasing size of the molecules. The major effect on the conductance by increasing the size of the molecule is that it gives a wider area for electron transmission through the molecule. In a classical picture the current would increase linearly with the area of the molecule. Quantum mechanically, however, the transmission can be seen as going through different channels, i.e. the eigenstates of the stack. Increasing the number of atoms in the PAH molecule increases the number of channels. Studies of the transmission through several different channels show that the resulting transmission is determined by interference effects between channels.

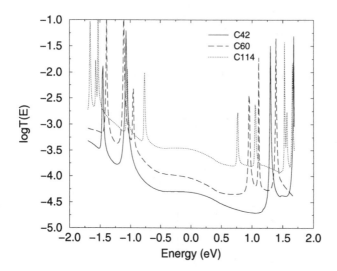

Fig. 3: Logarithm of the transmission coefficient for three different PAH molecules.

The phase change for an electron going through the molecular system depends on the electronic properties of the molecule together with the effective coupling of the molecule to the metallic leads, which in turn depends on the wave function of the channel in the metal. Thus, the phase differences between different paths will vary depending on the channel used in the metallic lead. In our case, many different channels (81 metallic bands at $E=0$.) are open in the metallic leads connecting the molecule to the electron reservoirs. The interference effects will therefore nearly average out. We find that the increase in the conductance is proportional to the number of atoms in the PAH molecule to the power of 1.28. The deviation from a linear behavior is a result of the band gap effect discussed above together with a small contribution from a net positive interference.

SUMMARY AND CONCLUSIONS

In this paper we have presented a method to calculate the IV characteristics of finite carbon NT and PAH contacted by two metal electrodes to the surface of the NT. The coupling between the contact and molecular systems is set to a small value. This results in a rather high contact resistance and a resonant tunneling type of conductance. For the NT, the conductance along the tube is independent on the tube length whereas in the PAH-stacks the conductance drops exponentially with the number of molecules in the stack. Finally, we found that the current increases almost linearly with the size of the molecules in the stack.

ACKNOWLEDGMENTS

Financial support from the Swedish Research Council for Engineering Science (TFR) and the Swedish Natural Science Research Council (NFR) are gratefully acknowledged.

REFERENCES

1. J. K. Gimzewski and C. Joachim, Science 283, 1683 (1999).
2. M. A. Reed et al., Science 278, 252 (1997).
3. S. J. Tans et al., Nature 386, 474 (1997).
4. S. Frank et al., Science 280, 1744 (1998).
5. T. W. Tombler et al., Nature 405, 769 (2000)
6. P. Herwig, C. W. Kayser, K. Müllen, and H. W. Spiess, Adv. Mater. 8, 510 (1996)
7. K. Müllen, and J. P. Rabe, Annu. New York Acad. Sci. 33, 205 (1998)
8. A. M. van de Craats et al., Adv. Mater. 10, 36 (1998).
9. R. Landauer, Philos. Mag. 21, 683 (1970).
10. M. Buttiker, Y. Imry, R. Landauer, and S. Pinhas, Phys. Rev. B 31, 6207 (1985).
11. P. F. Bagwell and T. P. Orlando, Phys. Rev. B 40, 1456 (1989).
12. M. Paulsson, Electron localization and conductance in conjugated systems, Licentiate Thesis No. 743, Linköping University, 1999.
13. J. Robertson and E. O'Reilly, Phys. Rev. B 35, 2946 (1987).

Band Structure, Conductivity and Thermoelectric Power of Carbon Nanotube Bundle

Viktor E. Kaminskii

Institute of Radio Engineering and Electronics, Russian Academy of Sciences, Mokhovaya 11, Moscow103907, Russia

Abstract. For electron transport in carbon nanotube bundle two-band model is proposed. A bundle is semiconductor with very narrow band gap. In effective mass approximation the Hamiltonian of kinetic energy is strongly non-parabolic. Both deformation potential and piezoelectric coupling with acoustic phonons determines carrier scattering rate. Piezoelectric coupling depends on bundle composition. Conductivity and thermoelectric power are calculated. Heavy and light holes are taken into consideration. Model parameters, which provide well agreement with available experimental data, are determined. It is shown that depending on both band and scattering rate parameter conductivity exhibits non-metallic or metallic behavior.

The resistivity measurements on the individual multiwall nanotubes give stronger or weaker but always non-metallic temperature dependence. For some samples of single-wall nanotube (SWNT) bundles the same temperature dependence is observed [1,2]. But for another there is a reversion to metallic temperature dependence [2,3] when temperature exceeds some value. In this paper the model of electron transport in a bundle of oriented carbon nanotubes is proposed. Since nanotube conductivity is non-zero at very low temperature we assume that bundle is the semiconductor with very narrow band gap. The tubular structure is the main feature of any bundle. It must have influence on the band structure. If the influence is significant the Hamiltonian of kinetic energy must have a universal form. Analysis of available experimental results shows that suitable for their description Hamiltonian can be expressed for conduction band as

$$H = E_e \ln\left(1 + \frac{H_e}{E_e}\right), \tag{1}$$

where $E_e = kT_e$ is the band parameter, H_e is the Hamiltonian for parabolic band in the effective mass approximation. We assume that the effective masses along axis of nanotube (m_z) and perpendicular to it (m) are different. For valence band the Hamiltonian has the same form with the band parameter $E_h = kT_h$. We take into

CP544, *Electronic Properties of Novel Materials—Molecular Nanostructures*, edited by H. Kuzmany, et al.

consideration both heavy (*hh*) and light (*lh*) holes. Since $m \neq m_z$ the conductivity is a tensor. For electron it is given by

$$\sigma_{xx} = \sigma_{yy} = \alpha D, \quad \sigma_{zz} = \alpha D \gamma^2, \tag{2}$$

where

$$\alpha = \frac{e^2}{2\pi^2} \left(\frac{2E_e}{\hbar^2} \right)^{3/2} \sqrt{m_z}, \quad \gamma = \sqrt{\frac{m}{m_z}}, \quad G = \exp\left(\frac{E}{E_e} \right),$$

$$D = \int_0^\infty (G-1)^{3/2} \left(-\frac{\partial F}{\partial E} \right) dE \int_{-1}^1 \frac{x^2 dx}{v(E,x)},$$

v is the momentum scattering rate. We consider the scattering by acoustic phonons and two coupling modes. Using equation (1) it can be shown that the scattering rate due to deformation potential and piezoelectric coupling are given respectively by

$$v_{DA} = \frac{1}{\tau_{DA}} G(G-1)^{1/2}, \quad v_{PA} = \frac{1}{\tau_{PA}} G(G-1)^{-1/2}(1+I), \tag{3}$$

where

$$I = (\gamma-1) \int_{-1}^1 |z-x| \left[(\gamma-1)^2 \left(z^2 + x^2 \right) + 2(\gamma-1)xz + 4\gamma \right]^{-1/2} dz .$$

Since carrier effective masses are unknown we assume that $\gamma^2 = m_e / m_{ze} = \gamma^2_{hh} = \gamma^2_{lh}$. In this case it follows from (2) that the ratio $\sigma_{zz} / \sigma_{xx} = \gamma^2$ and is independent of a temperature. For calculation band gap is taken equal to zero and value $\gamma^2 = 10$ is used. Figure 1 shows a measurement of $\rho(T)$ on several SWNT ropes in parallel [3]. Parameters for calculation are the following: $T_e = 1900$ K, $T_{lh} = 5000$ K, $T_{hh} = 2000$ K, $g_1 = m_{lh}/m_e = 0.5$, $g_2 = m_{hh}/m_e = 1.67$ and $p = \tau_{DA}/\tau_{PA} = 5 \cdot 10^{-5}$ for electron. In [3] ρ_{zz} measurements span the range 30-100 $\mu\Omega$cm. Using this result we estimate

$$\sigma_o = \alpha \tau_{DA} \frac{kT}{E_e} = (7\text{-}23) \times 10^4 \, \Omega^{-1} \text{m}^{-1}. \tag{4}$$

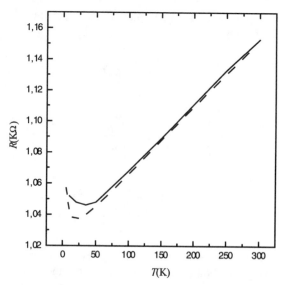

Figure 1. Experimental (solid line) and calculated (dashed line) resistivity of several SWNT ropes in parallel.

Figure 2(a) shows temperature dependence of resistivity, which is measured on SWNT bundle [1]. Calculated curve is fitted to it for the following parameters: $p=0.01$, $T_e=1400$ K, $T_{hh}=1100$ K, $T_{lh}=5000$ K, $g1=0.5$, and $g_2=1.67$. In [1] thermoelectric power of bundle is also measured. Figure 2(b) shows the experimentally determined and calculated TEP. For calculation the same parameters are used. Different temperature behavior of conductivity is the consequence of sample parameter difference. The parameter p is determined by bundle composition. If a bundle consists of identical nanotubes it is equal to zero. In this case it follows from equation (2) that for $T=0$ K

$$\rho_{xx} = \left\{ \frac{2}{3}\sigma_0 \left[\ln(1+\exp\eta) + \left(g_1 + g_2\right)\ln\left(1+\exp(-\eta)\right) \right] \right\}^{-1}, \qquad (5)$$

where $\eta = \lim_{T \to 0} \dfrac{E_f}{kT}$, E_f is the Fermi energy. Calculation shows that for $T>0$ K ρ_{xx} increases linearly with temperature. Such dependence is observed for one SWNT rope in [2]. In the same time another rope exhibits dependence that is similar to curve in Fig.2 (a). If p is not equal to zero its value determines mainly $\sigma(T)$ dependence at low temperature. The conductivity exhibits always the semiconducting behavior and the more is value of p the more is derivative $\partial\sigma/\partial T$. At high temperature T_e mainly determines $\sigma(T)$ dependence. If T_e is great (T_e>1400 K) an upturn to metallic conductivity dependence will be observed. Parameters p, T_e, g_1 and g_2 determine the crossover temperature. For different samples it can vary in wide range. If T_e is relatively low the conductivity will exhibit non-metallic behavior in the whole temperature range.

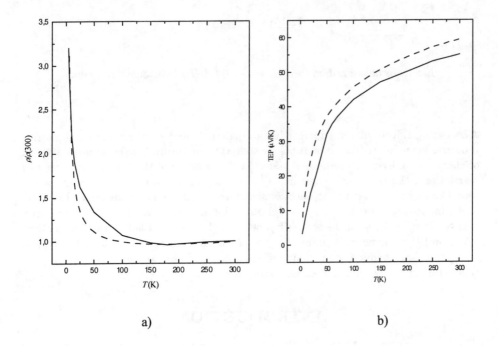

a)   b)

Figure 2. Temperature dependence of normalized resistivity (a) and thermoelectric power (b). Measurement on SWNT (solid line) and calculation (dashed line).

This work was supported by the Russian Foundation for Basic Research under Grant 98-02-17130.

REFERENCES

1. Hone, J., Ellwood, I., Muno, M. *et al.*, *Phys. Rev. Lett.* **80**, 1042-1045 (1998).
2. Lefebvre, J., Radosavljevec, M., Hone, J. *et al.*, in *XI11th Int. Winterschool on Electronic Properties of Novel Materials – Science and Technology of Molecular Nanostructures-1999*, edited by H. Kuzmany et al., AIP Conference Proceeding 486, New York, 1999, pp. 375-378.
3. Kane, C.L., Mele, E.J., Lee, R.S. *et al.*, *Europhys. Lett.* **41**, 683-688 (1998).

Impurity Scattering in Carbon Nanotubes with Superconducting Pair Correlations

Kikuo Harigaya

Electrotechnical Laboratory, Umezono 1-1-4, Tsukuba 305-8568, Japan

Abstract. Effects of the superconducting pair potential on the impurity scattering processes in metallic carbon nanotubes are studied theoretically. The backward scattering of electrons vanishes in the normal state. In the presence of the superconducting pair correlations, the backward scatterings of electron- and hole-like quasiparticles vanish, too. The impurity gives rise to backward scatterings of holes for incident electrons, and it also induces backward scatterings of electrons for incident holes. Negative and positive currents induced by such the scatterings between electrons and holes cancel each other. Therefore, the nonmagnetic impurity does not hinder the supercurrent in the regions where the superconducting proximity effects occur, and the carbon nanotube is a good conductor for Cooper pairs. Relations with experiments are discussed.

INTRODUCTION

Recent investigations (1,2) show that the superconducting proximity effect occurs when the carbon nanotubes contact with conventional superconducting metals and wires. The superconducting energy gap appears in the tunneling density of states below the critical temperature T_c. On the other hand, the recent theories discuss the nature of the exceptionally ballistic conduction (3) and the absence of backward scattering (4) in metallic carbon nanotubes with impurity potentials at the normal states.

In this paper, we study the effects of the superconducting pair potential on the impurity scattering processes in metallic carbon nanotubes, using the continuum $\mathbf{k} \cdot \mathbf{p}$ model for the electronic states. We find the absence of backward scatterings of electron- and hole-like quasiparticles in the presence of superconducting proximity effects, and the nonmagnetic impurity *does not hinder the supercurrent* in the regions where the superconducting proximity effects occur. Therefore, the carbon nanotube is a good conductor for Cooper pairs as well as in the normal state. This finding is interesting in view of the recent experimental progress of the superconducting proximity effects of carbon nanotubes (1,2).

CP544, *Electronic Properties of Novel Materials—Molecular Nanostructures*, edited by H. Kuzmany, et al.
© 2000 American Institute of Physics 1-56396-973-4/00/$17.00

IMPURITY SCATTERING
IN NORMAL NANOTUBES

We will study the metallic carbon nanotubes with the superconducting pair potential. The model is as follows:

$$H = H_{\text{tube}} + H_{\text{pair}}, \tag{1}$$

H_{tube} is the electronic states of the carbon nanotubes, and the model based on the $\boldsymbol{k} \cdot \boldsymbol{p}$ approximation (4,5) represents electronic systems on the continuum medium. The second term H_{pair} is the pair potential term owing to the proximity effect.

The hamiltonian of a graphite plane by the $\boldsymbol{k} \cdot \boldsymbol{p}$ approximation (4,5) in the secondly quantized representation has the following form:

$$H_{\text{tube}} = \sum_{\boldsymbol{k},\sigma} \Psi^{\dagger}_{\boldsymbol{k},\sigma} E_{\boldsymbol{k}} \Psi_{\boldsymbol{k},\sigma}, \tag{2}$$

where $E_{\boldsymbol{k}}$ is an energy matrix:

$$E_{\boldsymbol{k}} = \begin{pmatrix} 0 & \gamma(k_x - ik_y) & 0 & 0 \\ \gamma(k_x + ik_y) & 0 & 0 & 0 \\ 0 & 0 & 0 & \gamma(k_x + ik_y) \\ 0 & 0 & \gamma(k_x - ik_y) & 0 \end{pmatrix}, \tag{3}$$

$\boldsymbol{k} = (k_x, k_y)$, and $\Psi_{\boldsymbol{k},\sigma}$ is an annihilation operator with four components: $\Psi^{\dagger}_{\boldsymbol{k},\sigma} = (\psi^{(1)\dagger}_{\boldsymbol{k},\sigma}, \psi^{(2)\dagger}_{\boldsymbol{k},\sigma}, \psi^{(3)\dagger}_{\boldsymbol{k},\sigma}, \psi^{(4)\dagger}_{\boldsymbol{k},\sigma})$. Here, the fist and second elements indicate an electron at the A and B sublattice points around the Fermi point K of the graphite, respectively. The third and fourth elements are an electron at the A and B sublattices around the Fermi point K'. The quantity γ is defined as $\gamma \equiv (\sqrt{3}/2)a\gamma_0$, where a is the bond length of the graphite plane and γ_0 ($\simeq 2.7$ eV) is the resonance integral between neighboring carbon atoms. When the above matrix is diagonalized, we obtain the dispersion relation $E_{\pm} = \pm\gamma\sqrt{k_x^2 + \kappa_{\nu\phi}^2(n)}$, where k_x is parallel with the axis of the nanotube, $\kappa_{\nu\phi}(n) = (2\pi/L)(n + \phi - \nu/3)$, L is the circumference length of the nanotube, n ($= 0, \pm1, \pm2, ...$) is the index of bands, ϕ is the magnetic flux in units of the flux quantum, and ν ($= 0, 1, $ or 2) specifies the boundary condition in the y-direction. The metallic and semiconducting nanotubes are characterized by $\nu = 0$ and $\nu = 1$ (or 2), respectively. Hereafter, we consider the case $\phi = 0$ and the metallic nanotubes $\nu = 0$.

The second term in Eq. (1) is the pair potential:

$$H_{\text{pair}} = \Delta \sum_{\boldsymbol{k}} (\psi^{(1)\dagger}_{\boldsymbol{k},\uparrow}\psi^{(1)\dagger}_{-\boldsymbol{k},\downarrow} + \psi^{(2)\dagger}_{\boldsymbol{k},\uparrow}\psi^{(2)\dagger}_{-\boldsymbol{k},\downarrow} + \psi^{(3)\dagger}_{\boldsymbol{k},\uparrow}\psi^{(3)\dagger}_{-\boldsymbol{k},\downarrow} + \psi^{(4)\dagger}_{\boldsymbol{k},\uparrow}\psi^{(4)\dagger}_{-\boldsymbol{k},\downarrow} + \text{h.c.}) \tag{4}$$

where Δ is the strength of the superconducting pair correlation of an s-wave pairing. We assume that the spatial extent of the regions where the proximity effect occurs is as long as the superconducting coherence length.

Now, we consider the impurity scattering in the normal metallic nanotubes. We take into account of the single impurity potential located at the point $\boldsymbol{r}_0$:

$$H_{\text{imp}} = I \sum_{k,p,\sigma} e^{i(k-p)\cdot r_0} \Psi_{k,\sigma}^{\dagger} \Psi_{p,\sigma}, \tag{5}$$

where I is the impurity strength.

The scattering t-matrix at the K point is

$$t_K = I[1 - I \frac{2}{N_s} \sum_k G_K(k,\omega)]^{-1}, \tag{6}$$

where G_K is a propagator of a π-electron around the Fermi point K. The discussion about the t-matrix at the K' point is qualitatively the same, so we only look at the t-matrix at the K point. The sum for $k = (k,0)$, which takes account of the band index $n = 0$ only, is replaced with an integral:

$$\frac{2}{N_s} \sum_k G_K(k,\omega) = \rho \int d\epsilon \frac{1}{\omega^2 - \epsilon^2} \begin{pmatrix} \omega & \epsilon \\ \epsilon & \omega \end{pmatrix} \simeq -\rho\pi i \text{sgn}\omega \begin{pmatrix} 1 & 0 \\ 0 & 1 \end{pmatrix}, \tag{7}$$

where $\rho = a/2\pi L\gamma_0$ is the density of states at the Fermi energy. Therefore, we obtain

$$t_K = \frac{I}{1 + I\rho\pi i \text{sgn}\omega} \begin{pmatrix} 1 & 0 \\ 0 & 1 \end{pmatrix}. \tag{8}$$

The transformation into the energy-diagonal representation where the branches with $E = \pm\gamma|k|$ are diagonal has the same form of t_K.

The scattering matrix t_K in the energy-diagonal representation is diagonal, and the off-diagonal matrix elements vanish. This means that only the scattering processes from k to k and from $-k$ to $-k$ are effective. The scatterings from k to $-k$ and from $-k$ to k are cancelled. Such the absence of the backward scattering has been discussed recently (4).

IMPURITY SCATTERING WITH SUPERCONDUCTING PAIR POTENTIAL

We consider the single impurity scattering when the superconducting pair potential is present. In the Nambu representation, the scattering t-matrix at the K point is

$$\tilde{t}_K = \tilde{I}[1 - \frac{2}{N_s} \sum_k \tilde{G}_K(k,\omega)\tilde{I}]^{-1}, \tag{9}$$

where $\tilde{G}_K$ is the Nambu representation of G_K and

$$\tilde{I} = I \begin{pmatrix} 1 & 0 & 0 & 0 \\ 0 & 1 & 0 & 0 \\ 0 & 0 & -1 & 0 \\ 0 & 0 & 0 & -1 \end{pmatrix}. \tag{10}$$

The sign of the scattering potential for holes is reversed from that for electrons, so the minus sign appears at the third and fourth diagonal matrix elements.

The sum over k is performed as in the previous section, and we obtain the scattering t-matrix (with the same form in the energy-diagonal representation):

$$\tilde{t}_K = \frac{I}{1 + (I\rho\pi)^2} \begin{pmatrix} 1 + \alpha\omega & 0 & -\alpha\Delta & 0 \\ 0 & 1 + \alpha\omega & 0 & -\alpha\Delta \\ -\alpha\Delta & 0 & -1 + \alpha\omega & 0 \\ 0 & -\alpha\Delta & 0 & -1 + \alpha\omega \end{pmatrix} \tag{11}$$

where $\alpha = I\rho\pi i/\sqrt{\omega^2 - \Delta^2}$.

Hence, we find that the off-diagonal matrix elements become zero in the diagonal 2×2 submatrix. This implies that the backward scatterings of electron-line and hole-like quasiparticles vanish in the presence of the proximity effects, too. Off-diagonal 2×2 submatrix has the diagonal matrix elements whose magnitudes are proportional to Δ. The finite correlation gives rise to backward scatterings of the hole of the wavenumber $-k$ when the electron with k is incident. The back scatterings of the electrons with the wavenumber $-k$ occur for the incident holes with k, too. Negative and positive currents induced by such the two scattering processes cancel each other. Therefore, the nonmagnetic impurity *does not hinder the supercurrent* in the regions where the superconducting proximity effects occur. This effect is interesting in view of the recent experimental progress of the superconducting proximity effects (1,2).

SUMMARY

We have investigated the effects of the superconducting pair potential on the impurity scattering processes in metallic carbon nanotubes. The backward scattering of electrons vanishes in the normal state. In the presence of the superconducting pair correlations, the backward scatterings of electron- and hole-like quasiparticles vanish, too. The impurity gives rise to backward scatterings of holes for incident electrons, and it also induces backward scatterings of electrons for incident holes. Negative and positive currents induced by such the scatterings between electrons and holes cancel each other. Therefore, the carbon nanotube is a good conductor for the Cooper pairs coming from the proximity effects.

REFERENCES

1. A. Y. Kasumov et al, Science **284**, 1508 (1999).
2. A. F. Morpurgo, J. Kong, C. M. Marcus, and H. Dai, Science **286**, 263 (1999).
3. C. T. White and T. N. Todorov, Nature **393**, 240 (1998).
4. T. Ando and T. Nakanishi, J. Phys. Soc. Jpn. **67**, 1704 (1998).
5. H. Ajiki and T. Ando, J. Phys. Soc. Jpn. **62**, 1255 (1993).

Electrical and Thermal Transport in Carbon Nanotubes

David Tománek[1]

Department of Physics and Astronomy and Center for Fundamental Materials Research, Michigan State University, East Lansing, Michigan 48824-1116, USA

Abstract. Owing to their atomic-level perfection, carbon nanotubes exhibit unusually high electrical and thermal conductivity. Our electrical transport calculations, performed using a scattering technique based on the Landauer-Büttiker formalism, suggest that the conductance of inhomogeneous multi-wall nanotubes may show an unusual fractional quantization behavior, in agreement with recent experimental data. Our calculations also indicate that due to the combination of a large phonon mean free path, speed of sound and specific heat, the thermal conductivity of an isolated $(10, 10)$ carbon nanotube exceeds that of any known material, reaching the value $\lambda \approx 6,600$ W/m·K at room temperature.

Introduction

With the continually decreasing size of electronic and micromechanical devices, there is an increasing interest in materials that conduct electricity and heat efficiently. Carbon nanotubes [1], consisting of graphite layers wrapped into seamless cylinders, are now being produced routinely using carbon arc, laser vaporization of graphite, catalytic decomposition of carbon monoxide at high pressures, and chemical vapor deposition techniques [2]. These methods yield single-wall and multi-wall nanotubes that are up to a fraction of a millimeter long, yet only nanometers in diameter. Virtual absence of defects suggests that these molecular conductors should be ideal candidates for use as nano-wires that conduct electricity and heat efficiently.

The present study has been motivated by several open questions. The first relates to electron transport in nanotubes that is believed to be ballistic in nature, implying the absence of inelastic scattering. Recent conductance measurements of multi-wall carbon nanotubes [3] have raised a significant controversy due to the observation of unexpected conductance values in apparent disagreement with theoretical predictions. The second open question relates to the suitability of carbon

---

[1] In collaboration with Young-Kyun Kwon, Savas Berber, Stefano Sanvito, and Colin J. Lambert.

CP544, *Electronic Properties of Novel Materials—Molecular Nanostructures*, edited by H. Kuzmany, et al.
© 2000 American Institute of Physics 1-56396-973-4/00/$17.00

nanotubes to conduct heat efficiently in view of their atomically perfect structure and the stiffness of the interatomic bonds in self-supporting graphitic cylinders.

Electrical Conductance of Carbon Nanotubes

To address the conductance of multi-wall carbon nanotubes [4], we combined a linear combination of atomic orbitals (LCAO) Hamiltonian with a scattering technique developed recently for magnetic multilayers [5,6]. The parameterization of the LCAO matrix elements is based on *ab initio* results for simpler structures [7]. Our calculations can build on a number of published theoretical studies of the electronic structure of single-wall [8–10] and multi-wall carbon nanotubes [11–13]. Calculations for single-wall nanotube ropes [14,15] have shown that inter-wall coupling may induce pseudo-gaps near the Fermi level in these systems, with serious consequences for their conductance behavior.

Our scattering technique approach to determine the conductance of inhomogeneous multi-wall nanotubes [4] is based on the quantum-mechanical scattering matrix S of a phase-coherent "defective" region that is connected to "ideal" external reservoirs [5]. At zero temperature, the energy-dependent electrical conductance is given by the Landauer-Büttiker formula [16]

$$G(E) = \frac{2e^2}{h} T(E) , \qquad (1)$$

where $T(E)$ is the total transmission coefficient, evaluated at the Fermi energy E_F.

For a homogeneous system, $T(E)$ assumes integer values corresponding to the total number of open conduction channels at the energy E. For individual (n, n) "armchair" tubes, this integer is further predicted [17] to be an even multiple of the conductance quantum $G_0 = 2e^2/h \approx (12.9 \text{ k}\Omega)^{-1}$, with a conductance $G = 2G_0$ near the Fermi level. In the $(10, 10)@(15, 15)$ double-wall nanotube [13] and the $(5, 5)@(10, 10)@(15, 15)$ triple-wall nanotube, the inter-wall interaction significantly modifies the electronic states near the Fermi level and blocks some of the conduction channels close to E_F.

The experimental set-up of Ref. [3], shown schematically in Fig. 1(a), consists of a multi-wall nanotube that is attached to a gold tip of a Scanning Tunneling Microscope (STM) and used as an electrode. The STM allows the tube to be immersed at calibrated depth intervals into liquid mercury, acting as a counter-electrode. This arrangement allows precise conduction measurements to be performed on an isolated tube. The experimental data of Ref. [3] for the conductance G as a function of the immersion depth z of the tube, reproduced in Fig. 1(e), suggest that in a finite-length multi-wall nanotube, the conductance may achieve values as small as $0.5G_0$ or $1G_0$.

As nothing is known about the internal structure of the multi-wall nanotubes used or the nature of the contact between the tube and the Au and Hg electrodes, we have considered several scenarios and concluded that the experimental data can

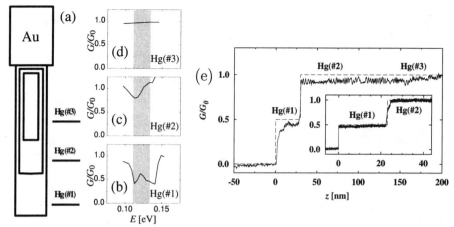

FIGURE 1. (a) Schematic geometry of a multi-wall nanotube that is being immersed into mercury up to different depths labeled Hg(#1), Hg(#2), and Hg(#3). Only the outermost tube is considered to be in contact with the gold STM tip on which it is suspended. The conductance of this system is given in (b) for the immersion depth Hg(#1), in (c) for Hg(#2), and in (d) for Hg(#3) as a function of the position of E_F. The Fermi level may shift with changing immersion depth within a narrow range indicated by the shaded region. (e) Conductance G of a multi-wall nanotube as a function of immersion depth z in mercury, given in units of the conductance quantum $G_0 = 2e^2/h \approx (12.9 \text{ k}\Omega)^{-1}$. Results predicted for the multi-wall nanotube, given by the dashed line, are superimposed on the experimental data of Ref. [3]. The main figure and the inset show data for two nanotube samples, which in our interpretation only differ in the length of the terminating single-wall segment. From Ref. [4], ©American Physical Society 2000.

only be explained by assuming that the current injection from the gold electrode occurs exclusively into the outermost tube wall, and that the chemical potential equals that of mercury, shifted by a contact potential, only within the submerged portion of the tube. In other words, the number of tube walls in contact with mercury depends on the immersion depth. The main origin of the anomalous conductance reduction from the theoretically expected integer multiple of $2G_0$ is the backscattering of carriers at the interface of two regions with different numbers of walls due to a discontinuous change of the conduction current distribution across the individual walls.

Thermal Conductivity of Carbon Nanotubes

To address the thermal conductivity of carbon nanotubes as a function of temperature [18], we made use of accurate carbon potentials [19] in equilibrium and non-equilibrium molecular dynamics simulations. The thermal conductivity λ of a solid along a particular direction, taken here as the z axis, is related to the heat flowing down a long rod with a temperature gradient dT/dz by

$$\frac{1}{A}\frac{dQ}{dt} = -\lambda\frac{dT}{dz} , \tag{2}$$

where dQ is the energy transmitted across the area A in the time interval dt. In systems where the phonon contribution to the heat conductance dominates, λ is proportional to Cvl, the product of the heat capacity per unit volume C, the speed of sound v, and the phonon mean free path l. Due to a virtual absence of atomic-scale defects, we expect l to be unusually large in carbon nanotubes. Also the heat capacity and speed of sound are expected to equal or even exceed those of diamond, which is known to have the highest measured thermal conductivity when isotopically pure. Hence, we suspect that isolated carbon nanotubes may be Nature's best heat conductors.

Precise measurements of thermal conductivity are very difficult, as witnessed by the reported thermal conductivity data in the basal plane of graphite [20] which show a scatter by nearly two orders of magnitude. Similar uncertainties have been associated with thermal conductivity measurements in "mats" of nanotubes [21].

Theoretical prediction of the thermal conductivity have proven equally challenging, albeit for different reasons. In a direct molecular dynamics simulation, construction of a periodic array of hot and cold regions along a nanotube introduces extra scattering centers that limit the phonon mean free path to below the size of the unit cell, thus significantly reducing the value of λ. Equilibrium molecular dynamics simulations based on the Green-Kubo formula, which relate λ to the integral over time t of the heat flux autocorrelation function, converge very slowly and require extensive ensemble averaging. We found that the most suitable approach to determine thermal conductivity in nanotubes combines the Green-Kubo formula with nonequilibrium thermodynamics [22,23].

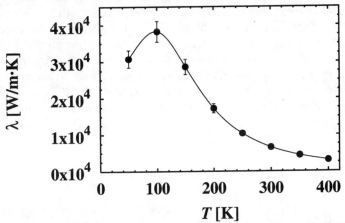

FIGURE 2. Temperature dependence of the thermal conductivity λ for a $(10, 10)$ carbon nanotube for temperatures below 400 K. From Ref. [18], ©American Physical Society 2000.

Results of nonequilibrium molecular dynamics simulations for the thermal conductance of an isolated $(10, 10)$ nanotube aligned along the z axis are presented in Fig. 2. We find that at low temperatures, when l is nearly constant, the temperature dependence of λ follows that of the specific heat. At high temperatures, where the specific heat is constant, λ decreases as the phonon mean free path becomes smaller due to umklapp processes. Our calculations suggest that at $T = 100$ K, carbon nanotubes show an unusually high thermal conductivity value of $37,000$ W/m·K. This value lies very close to the highest value observed in any solid, $\lambda = 41,000$ W/m·K, that has been reported [24] for a 99.9% pure ^{12}C crystal at 104 K. In spite of the decrease of λ above 100 K, the room temperature value of $6,600$ W/m·K is still very high, twice the reported thermal conductivity value of $3,320$ W/m·K for nearly isotopically pure diamond [25]. We also found this value to lie close to that of a hypothetical graphene monolayer. In graphite, we find that the inter-layer interaction reduces λ by one order of magnitude due to the reduced phonon mean free path. Similarly, we expect the high thermal conductivity value predicted for an isolated nanotube to decrease upon contact with a surrounding matrix, such as a nanotube "rope".

Summary and Conclusions

The calculations discussed above indicate that carbon nanotubes show unusual electrical and thermal conductance behavior. Results for the electrical transport indicate that the inter-wall interaction in multi-wall nanotubes not only blocks certain conduction channels, but also re-distributes the current non-uniformly across the walls. The puzzling observation of fractional quantum conductance in multi-wall nanotubes can be explained by back-scattering at the interfaces of regions with different numbers of walls. Sample-to-sample variations in the internal structure of the tubes offer a natural explanation for the observed variations of the conductance. Nonequilibrium molecular dynamics simulations suggest that carbon nanotubes may conduct heat exceptionally well, owing to a combination of a large phonon mean free path, high speed of sound and specific heat. The predicted thermal conductivity value $\lambda \approx 6,600$ W/m·K for an isolated $(10, 10)$ carbon nanotube at room temperature is twice that of isotopically pure diamond, Nature's best heat conductor.

Acknowledgment

The author gratefully acknowledges financial support by the organizers of the "International Winter School on Electronic Properties of Novel Materials: Molecular Nanostructures" (IWEPNM'2000) in Kirchberg (Austria), March 4-11, 2000, and by the Office of Naval Research and DARPA under Grant Number N00014-99-1-0252.

REFERENCES

1. S. Iijima, Nature **354**, 56 (1991).
2. M.S. Dresselhaus, G. Dresselhaus, and P.C. Eklund, *Science of Fullerenes and Carbon Nanotubes* (Academic Press, San Diego, 1996).
3. S. Frank, P. Poncharal, Z.L. Wang, and W.A. de Heer, Science **280**, 1744 (1998).
4. Stefano Sanvito, Young-Kyun Kwon, David Tománek, and Colin J. Lambert, Phys. Rev. Lett. **84**, 1974 (2000).
5. S. Sanvito, C.J. Lambert, J.H. Jefferson, and A.M. Bratkovsky, Phys. Rev. B **59**, 11936 (1999).
6. S. Sanvito C.J. Lambert, J.H. Jefferson, and A.M. Bratkovsky, J. Phys. C: Condens. Matter. **10**, L691 (1998).
7. D. Tománek and M.A. Schluter, Phys. Rev. Lett. **67**, 2331 (1991).
8. J.W. Mintmire, B.I. Dunlap, and C.T. White, Phys. Rev. Lett. **68**, 631 (1992).
9. R. Saito, M. Fujita, G. Dresselhaus, and M.S. Dresselhaus, Appl. Phys. Lett. **60**, 2204 (1992).
10. N. Hamada, S. Sawada, and A. Oshiyama, Phys. Rev. Lett. **68**, 1579 (1992).
11. R. Saito, G. Dresselhaus, and M.S. Dresselhaus, J. Appl. Phys. **73**, 494 (1993).
12. Ph. Lambin, L. Philippe, J.C. Charlier, and J.P. Michenaud, Comput. Mater. Sci. **2**, 350 (1994).
13. Y.-K. Kwon and D. Tománek, Phys. Rev. B **58**, R16001 (1998).
14. P. Delaney, H.J. Choi, J. Ihm, S.G. Louie, and M.L. Cohen, Nature **391**, 466 (1998).
15. Y.-K. Kwon, S. Saito, and D. Tománek, Phys. Rev. B **58**, R13314 (1998).
16. M. Büttiker, Y. Imry, R. Landauer, and S. Pinhas, Phys. Rev. B **31**, 6207 (1985).
17. L. Chico, L.X. Benedict, S.G. Louie and M.L. Cohen, Phys. Rev. B **54**, 2600 (1996), W. Tian and S. Datta, *ibid.* **49**, 5097 (1994), M.F. Lin and K.W.-K. Shung, *ibid.* **51**, 7592 (1995).
18. Savas Berber, Young-Kyun Kwon, and David Tománek, Phys. Rev. Lett. **84** (2000).
19. J. Tersoff, Phys. Rev. B **37**, 6991 (1988).
20. Citrad Uher, in *Landolt-Börnstein*, New Series, III **15c** (Springer-Verlag, Berlin, 1985), pp. 426–448.
21. J. Hone, M. Whitney, A. Zettl, Synthetic Metals **103**, 2498 (1999).
22. D.J. Evans, Phys. Lett. **91A**, 457 (1982).
23. D.P. Hansen and D.J. Evans, Molecular Physics **81**, 767 (1994).
24. Lanhua Wei, P.K. Kuo, R.L. Thomas, T.R. Anthony, and W.F. Banholzer, Phys. Rev. Lett. **70**, 3764 (1993).
25. T.R. Anthony, W.F. Banholzer, J.F.Fleischer, Lanhua Wei, P.K. Kuo, R.L. Thomas, and R.W. Pryor, Phys. Rev. B **42**, 1104 (1990).

Transport Properties of Single-walled Nanotube Mats under Hydrostatic Pressure

B. B. Liu [1,2], B. Sundqvist[1], O. Andersson[1], T.Wågberg[1], G. Zou[2]

[1]Department of Experimental Physics, Umeå University, 901 87 Umeå, Sweden
[2]National Lab of Superhard Materials, Jilin University, Changchun 130023,P.R.China

Abstract. We present electrical transport studies of single-walled carbon nanotube mats, synthesized by the arc discharge method with Ce/Ni as catalysts, under hydrostatic pressure up to 1.5GPa with a liquid pressure medium. These data were compared with the results at ambient pressure. The transport phenomena were described in terms of Mott's 2D variable range hopping (VRH) conduction up to 1.05GPa. An irreversible change is induced below 0.5GPa, and the resistance behaviour is reversible due to the strong interaction between tubes above 0.5 GPa. These results indicate that 2D VRH occurs within bundles.

INTRODUCTION

The study of electrical transport properties of bulk single-walled carbon nanotube (SWNT) materials attracts a lot of attention. However, the transport results differ from sample to sample[1]. Some results show semiconducting behavior with different laws and whether the behavior is dominated by on-tube or on-bundle effects or by interbundle contacts is still unknown. Therefore, further investigation is still meaningful. The effect of hydrostatic pressure(HP), which is expected to significantly change the interaction between tubes or bundles, gives us a further possibility to understand the characteristic properties. Under HP, Raman spectra and electric resistance show pronounced changes at 1.5GPa[2]. In this study, we systematically studied the electric resistance up to1.5 GPa, giving a different image of the transport behavior.

EXPERIMENTAL AND CHARACTERIZATION OF SAMPLES

SWNTs were synthesized by the arc discharge method with Ce/Ni as catalyst. As-grown mats were collected around the graphite cathode. The SEM image showed that the mat is formed as a whole network with the nanotube bundles entangled together. The bundle diameters are 10~30nm and the average length between entangled points is 100nm. Raman spectra showed lines typical for SWNTs including chiral tubes [3]. Mats were cut into pieces $5 \times 1 \times 0.2 mm^3$ for resistance R measurements. Silver paint and pressure contacts were applied in a four-probe configuration for R measurements under ambient and high hydrostatic pressure. The temperature T (2~300K) dependence of R and magnetoresistance MR (0~5 Tesla) were measured in an Oxford

CP544, *Electronic Properties of Novel Materials—Molecular Nanostructures*, edited by H. Kuzmany, et al.

2000 cryogenic system at ambient pressure. A 50/50 mixture of iso- and n-pentane was used as pressure medium, and temperaures down to 150 K were obtained by cooling with liquid N_2.

RESULTS AND DISCUSSION

The R(T) for an as-grown mat is shown in Fig.2(a) from 2 to 300K. Plotting lnR versus $T^{-1/3}$ results in a straight line, indicating that the resistivity $\rho(T)$ obeys

$$\rho(T)= \rho_0 exp(T_0/T)^{1/3} \tag{1}$$

typical for Mott's variable range hopping (VRH) in 2D. Here $T_0=13.8/\kappa_B g(E_F) \xi^2$, where ξ is the localization length and $g(E_F)$ is the density of states (DOS) at the Fermi level. ρ_0 depends on the phonon spectrum, scaling as v_{Ph}^{-1}. We note that straight line fits could not be obtained for lnR versus $T^{-1/4}$ or $T^{-1/2}$, as for 3D and 1D VRH.

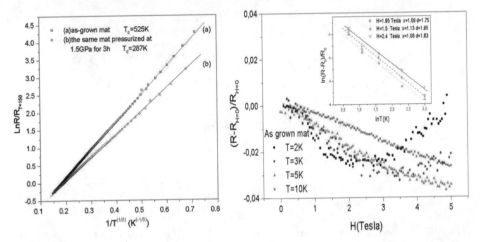

FIGURE 1. R (T) from 2 to 300K for (a) as-grown mat and (b) the mat pressurized at .5GPa for 3h.
FIGURE 2. (right) MR for as-grown mat. Inset shows $ln(R_H-R_0)/R_0$ vs lnT at fixed H.

The 2D VRH result is consistent with results reported by others [1(c)]. The values of T_0 for several as-grown mats were the same, approximately 525K, indicating that T_0 can have a characteristic value for mats. The resistance for a purified mat still obeyed 2D VRH with only a small change of T_0 (480K) [3], suggesting that 2D VRH is an intrinsic behaviour for nanotubes, consistent with the results found in Ref.[1(d)].

Fig.2 shows MR data from 2 to 10K. The overall MR is negative up to 200K. At low temperature, the MR is composed of negative and positive parts and shows a minimum. The negative MR has a T dependence in the weak field range,

$$[R(T,H)-R(T, 0)]/R(T,0) = -AH^\alpha T^{-x} \tag{2}$$

Where x=3/(d+1) and d is dimension. The $ln(R_H-R_0)/R_0$ versus lnT, extracted from fig.2, was shown in the inset. From the slope, the dimension was determined as about 1.7, very close to 2, indicating again that 2D VRH is the dominant behavior.

According to Mott's theory, hopping occurs between localized states near Fermi level, usually defects in disodered semiconductors. For mats disordered states could arise at bends, kinks, twists or dislocations due to different chirality and the high strain. Interbundle contacts could also create such states. Recently many studies show that different types of defects in real NTs locally alter the electronic properties. A VRH model based on defects thus seems very plausible. Since our system is 2D, contacts may have a little contribution to hopping states. 2D VRH thus mainly occur between localized states along bundles or tubes.

The application of HP results in a gradual decrease of the resistance observed up to 1.05GPa (Fig. 3). The R(T) at different P still obey 2DVRH with different T_0 as shown in Fig.4. The slope of $d\ln R/dP$ on decreasing P is the same for the whole range and almost the same as for the final part of the curve for increasing P.

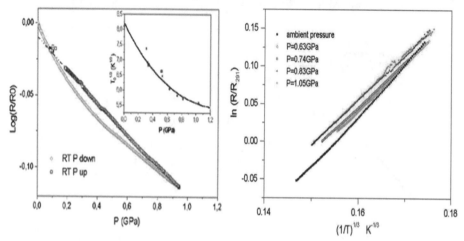

FIGURE 3. The pressure dependence of log (R/R $_{p=0}$). The T_0 $^{1/3}$ data in the inset are obtained from fig.4.

FIGURE 4. Resistance as a function of 1/T $^{1/3}$ at different pressures from 150K to RT.

R can be written $R=\rho/\,d\cdot Fc$, where ρ is the resistivity, d is the thickness and Fc is a geometrical correction factor. In the 2D VRH model,

$$d\ln R/dP=\underline{(1/T)T_0^{1/3}/dP}(a)-\underline{d\text{Ln}v_{Ph}/dP}(b)-\underline{d\ln(d\cdot Fc)/dP}(c). \qquad (4)$$

The average coefficient $d\ln R/dP$ ~-0.35 below 0.5GPa and -0.26~-0.18 up to 1.05 GPa at 293K. (a) can be estimated from the plot of $T_0^{1/3}$ versus P shown in the inset of Fig. 3. Data for T_0 were obtained from Fig.4 which shows lnR as a function of $T^{-1/3}$ at several pressures. These data were measured on several samples cut from the same mat. This term is -0.5 GPa^{-1} below 0.6 GPa and -0.15 ~-0.2GPa^{-1} above 0.6 GPa. (b) is related to the phonon spectrum. From Raman results [2(a)], $d\text{Ln}v_{radial}/dP= 0.05$GPa^{-1} and $d\text{Ln}v_{tangential}/dP=0.005$GPa^{-1} below 1.5 GPa. This term is 0.005-0.06GPa^{-1}, or approximately zero. (c) is related to the compressibility. It is large at low pressure and approaches -0.03GPa^{-1} (the c-axis compressibility of graphite) at HP. The zero-pressure thickness of the mat permanently decreased from 0.2 to 0.17mm after pressurized at 1.5GPa, with small change in the plane. So maximum (c) is -0.3 GPa^{-1}.

The changes in R with P are thus mainly due to the changes in $T_0^{1/3}$ and in the mechanical dimensions. Below 0.5GPa, both terms change significantly. Above 0.5GPa, the slope mainly arises from the change in T_0, indicating that the samples become much denser and that the interaction between tubes increases strongly.

T_0 decreases approximately exponentially with increasing P. Since it is inversely proportional to ξ^2 and $g(E_F)$, this decrease indicates that either ξ or $g(E_F)$ or both become larger under pressure. Pressure is expected to induce an increase of the DOS through buckling of the Fermi surface because of the interaction of the tubes. This effect was calculated by Kwon et al.[5], who found that the DOS at E_F increased by 7% in a rope compared with that in an individual tube. Alternatively, P might increase the number of localized states (defects). In contrast to non-hydrostatic pressure, hydrostatic pressure should only slightly improve the existing interbundle contacts and create relatively fewer new ones. Pressure might also induce an overlap of electron wave functions, decreasing the hopping energy and supplying more hopping paths, resulting in an increase of the localization length.

Pressure cycling showed that R is a reversible function of P above 0.5GPa, indicating that changes result mainly from interactions between tubes. However, irreversible effects were observed below 0.5 GPa by measuring R(T) at atmospheric pressure for the same mat pressurized at 1.5GPa for 3h (shown in Fig.1(b)) and obtaining T_0= 287K, which corresponds to about 0.5GPa in the inset of Fig.3. This results either because local deformation of tubes and bundles enhanced the density of "native defects" in bent or strained tubes or bundles or because of changes in the contact geometry and new intra-mat contacts. Raman studies under HP [2] also showed reversible changes up to 1.5GPa and the presence of irreversible change at low pressure. Since the resistance behavior follows 2D VRH mechanism both at ambient and high pressure, and T_0 largely decreases to 173 K at 1.05GPa from 525K at ambient pressure, indicating large changes in DOS and ξ under HP, we suggest that VRH occurs within bundles and not just along individual tubes.

CONCLUSION

In summary, electrical transport in SWNT mats is described in terms of Mott's 2D VRH conduction from ambient pressure to 1.5GPa. From the resistance under hydrostatic pressure, we conclude: (1) that the resistance decreases with increasing pressure, (2) that an irreversible change is induced below 0.5 GPa, and (3) that above 0.5GPa the resistance behavior is reversible due to the strong interaction of tubes. These results indicate that 2D VRH occurs within bundles.

ACKNOWLEDGMENTS

B.B. Liu thank Dr. E. McRae and Prof.D.Tománek for valuable discussions and the Wenner-Gren foundation for support. This work was financially supported by the Swedish Research Councils for Natural Sciences and TFR, and by the Chinese NSF.

REFERENCES

1. (a), Kaiser, et al., *Phys. Rev. B,* **57**, 1418 (1998). (b), Fisher, J. E., et al., ibid, 35, R4921 (1997). (c), Kim, G.T., et al., ibid, 58, 2529 (1998) (d), Yosida, Y., et al., J. Appl. Phys. 86, 999 (1999).
2. (a)Venkateswaran, U.D., et al., Phys. Rev. B, 59, 10928 (1999). (b), Gaál, R, et al., this volume.
3. (a)Liu, B.B, et al., Chem. Phys. Lett., 320 365 (2000). (b) Liu, B.B., et al., submitted
4. (a) Mott, N.F., Davis, E.A., Electronic Process in Noncrystalline Materials, Oxford University press, 1979. (b) Sivan, U., et al., Phys. Rev. Lett., 60, 1566 (1988).
5. Kwon, Y., Saito S., Tomanek D., Phys. Rev. B, 58 R13314 (1998).

Single Carbon Nanotube Electronic Devices

A.T. Johnson, J. Lefebvre, M. Radosavljevic, M. Llaguno, and J.F. Lynch

Dept. of Physics and Astronomy and Laboratory for Research on the Structure of Matter, University of Pennsylvania, Philadelphia, PA 19104

Abstract. We review recent progress towards the fabrication of engineered single nanotube circuits. Single wall carbon nanotubes are manipulated into circuits on a silicon dioxide surface using an AFM. Nanotubes can also be incorporated into an electron beam lithography resist system and used as "shadow masks" to create electrode pairs with sub-20 nm separation. Measurements of very short channel nanotube FETs indicate that the molecules may be highly doped due to exposure to the atmosphere, a fact not taken into account in earlier models of nanotube FETs.

INTRODUCTION

Single wall carbon nanotubes (SWNTs)[1] are a unique class of nanomaterials whose atomic geometry is directly coupled to the molecular electronic properties. For example, the precise wrapping of the underlying graphene sheet determines whether a nanotube is metallic or a semiconductor with a gap proportional to the reciprocal of the tube diameter. Nanotube circuit elements such as field effect transistors and quantum dots have been demonstrated.[2,3] Atomic scale defects, molecular adsorbates, and mechanical deformations are predicted to affect a tube's electronic properties[4-6] and could be used to tailor device operation. For example, we have used random deposition of nanotube circuits to demonstrate how impurity material can convert a nanotube transistor into a diode-like rectifier,[7] and others have observed similar behavior in a nanotube that was deformed, perhaps by an atomic-scale defect.[8] Controlled fabrication of single tube electronic circuits will let us explore a wide range of possible function, and here we discuss recent progress towards this goal. Our group has also developed scanning probe techniques to measure the *local* electronic properties of SWNT circuits with nanometer-scale resolution, as described in this volume by Freitag.[9] We conclude with recent measurements of nanotube FETs with channel lengths as low as 20 nm and discuss their implications for models of electron transport in such devices.

AFM MANIPULATION OF NANOTUBES

All SWNTs used in this work were made by the Smalley group (Rice) by laser ablation.[10] For AFM manipulation, the as-grown material is dispersed in dichloroethane and a few drops spun onto an oxidized degenerate Si substrate. AFM images show substrate covered with micron-length individual tubes and small diameter

CP544, *Electronic Properties of Novel Materials—Molecular Nanostructures*, edited by H. Kuzmany, et al.

bundles. Single tubes are manipulated with an AFM (DI Nanoscope III) using commercial Si tips (Nanosensors, Inc.). In contrast to other groups' success manipulating multi-wall nanotubes (MWNTs) with contact-mode AFM,[11] we find that SWNTs are disturbed by contact mode AFM, so we image and manipulate in tapping mode. SWNTs are manipulated by increasing the tip force over that used during imaging, and the scan rate is raised to 20 – 80 μm/sec. The high lateral speed partly disables the force feedback loop, allowing the tip to strongly interact with the SWNTs.

To controllably translate a SWNT, the tip is positioned at one end of the tube, the tip-sample force and tip scan rates are increased as above, and the tip moved 10 nm in the desired direction. Part of the SWNT is moved, but most of it remains in place due to substrate forces. The tip is moved backward 10 nm along the tube axis with the force reduced to the imaging level. The tip motion is repeated until the full tube is translated 10 nm, then the entire process repeated until the tube is translated the full desired distance. We can translate a 500 nm nanotube to within 10% of a desired micron-scale distance using a few thousand steps. Nanotubes are rotated by moving different segments of the SWNT by different amounts. As for translation, we rotate the tube by small increments until the full rotation is completed. Complex circuits can be created by AFM manipulation, as seen in Fig. 1(a)-(e). First experiments indicate that an SWNT bundle lying on top of another bundle can create a gate-tunable tunnel barrier in the lower bundle and a double quantum dot forms at low temperature.[3]

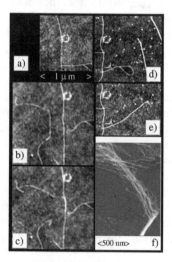

FIGURE 1. AFM manipulation of SWNTs into a "4-probe" configuration. (a) We begin with two tubes already placed into contact. (b) A second tube is brought in from the right. (c) The tube is placed on top of the main vertical tube. (d)-(e) The circuit is completed by opening the loop in the "lasso" tube and removing unwanted nanotubes. (f) Ropes cannot be manipulated successfully, but instead fall apart into their constituent nanotubes.

SWNT TEMPLATES FOR "NANOGAPS"

We use MWNTs, SWNTs, and SWNT bundles as templates for a nanometer-scale gap in a metal wire. SWNT segments have been contacted by these "nanogaps" resulting in very short nanotube quantum dots and field effect transistors.[12] Our method is a variation of the shadow mask technique pioneered by Dolan (Fig. 2).[13] In our process, the bridge is not made of resist but of a SWNT bundle or an individual MWNT. Carbon nanotubes are very anisotropic with nanometer-size diameters (1 – 2 nm for SWNTs and 5 – 100 nm for MWNTs and SWNT bundles) and micrometer lengths. They are resistant to processing because of their mechanical stiffness and chemical stability.

Fig. 2(b) illustrates the nanogap fabrication process after metal evaporation. Similar to a standard e-beam lithography process, two layers of resist are used but with a carbon nanotube layer incorporated between them. The fabrication steps are as follows: first, 80 nm of e-beam resist is coated onto an oxidized silicon substrate (300 nm SiO_2 on degenerate Si). Carbon nanotubes sonicated in isopropanol are deposited on top of the resist layer; and MWNTs or SWNT bundles located using an optical microscope or an AFM. Another 80 nm layer of e-beam resist is spun on top of this layer and baked at 190°C. The tri-layer resist is exposed in a 100 – 500 nm wide *continuous* line crossing the carbon nanotube. After development the trapped carbon nanotube is left free standing above the substrate (Fig. 2(b)). Finally, a 25 nm gold film is thermally evaporated with the carbon nanotube bridge preventing metal from reaching the area directly underneath it, and the unexposed resist is dissolved in acetone to remove unwanted metal and the carbon nanotubes.

SWNTs were contacted using the new nanogap fabrication method. Before making the nanogaps, raw SWNT material was sonicated in dichloroethane to break it into individual SWNTs and small bundles of 2 nm 5 diameter and then dispersed on a doped Si substrate. At room temperature, nanogaps without SWNTs show a resistance in excess of 100 G? , while those bridged by a SWNT or SWNT rope have resistance between 100 k? and 3 G? . Figure 2(c)-(d) show the two characteristic behaviors observed at 4.5 K for the differential conductance (dI/dV) of nanogaps contacting SWNTs as a function of bias and gate voltage.

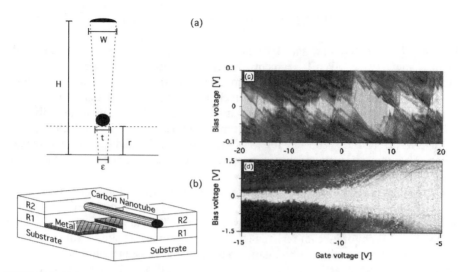

FIGURE 2. (a) Nanogap fabrication concept. The suspended nanotube acts as a shadow mask. (b) Schematic of sample after metal evaporation. (c) – (d) Gray scale plot of dI/dV for a metallic nanotube (c) and a semiconducting nanotube (d) taken at 4.5 K.

The first behavior (Fig. 2(c)) is typical of single electron charging in metallic SWNTs with Coulomb blockade diamonds of different sizes and has been attributed to charging in a multiple dot structure.[14] The maximum current at 0.1 V is 100 nA, corresponding to a resistance of 1 M? . From the largest diamonds, we extract a charging energy E_C = 80 meV and an energy level splitting of 20 meV. These are an order of magnitude larger than reported previously[14] and correspond to a dot size near 20 nm. Although E_C is larger than the room temperature thermal energy, no blockade is present at 300 K. At 77 K, the largest diamonds are still apparent. The smaller diamonds in the irregular spectra in Fig. 2(c) strongly suggest that longer parts of the SWNT under the electrodes also contribute to transport.

Figure 2(d) shows the 4.5 K transport spectrum of a semiconducting SWNT bridging a nanogap. At low bias voltage there is a gap structure that is discussed in the next paragraph. For a bias voltage larger than the gap (about 0.1 V), there is a strong field effect: the current at a bias voltage of V_b = 1 V increases by a factor of more than 10^4 as Vg is swept from -5 V to -15 V. The persistence of a very strong field effect as the electrode spacing is reduced to 20 nm may indicates that the energy band bending length is shorter than this (this could be caused by a doping level of 10^{-4} within the SWNT by exposure to atmosphere[15,16]) If the band bending length is very short, barriers to hole transport exist at the *ends* of the nanotube FET, not in the middle of the device as is commonly assumed.[2] Another possibility is that the carrier density, not a band-bending barrier, is the key factor in determining the circuit conductance, implying that transport in the circuit is predominantly diffusive.[15] With this assumption we find a rather low room temperature hole mobility μ_h = 1.0 m^2/V-s within the sample from dI/dVg = μ_h CV_b/L^2 and the measured values of the transconductance dI/dVg = 2 nA/V at a bias voltage V_b = 100 mV, the back-gate capacitance C = 0.1 aF, and the sample length L = 20 nm.

At room temperature, the gap in the I-V is reduced to zero at sufficiently negative gate voltage. In contrast, at 4.5 K the gap does not close even at a gate voltage as low as -15 V. We attribute this effect to the onset of single electron charging within the sample, as observed earlier in samples with an electrode spacing of 100 nm.[17] The 0.1 eV gap we observe (Fig. 2(d)) is close to that found for the metallic SWNT sample of Fig. 2(c), indicating that the smallest charging region formed within the nanotube is defined by the nanogap contacts and not induced by disorder.

CONCLUSIONS

Single wall carbon nanotubes have been manipulated into circuits on a silicon dioxide surface using an AFM. Nanotubes can also be incorporated into an electron beam lithography resist system and used as "shadow masks" to create electrode pairs with sub-20 nm separation. Measurements of very short channel nanotube FETs indicate that present models of their operation may need to be modified.

ACKNOWLEDGMENTS

We acknowledge useful discussions with J.E. Fischer, C.L. Kane, and E.J. Mele. Support was provided by Fonds FCAR (Quèbec) (J.L.), the U.S. NSF (grant number DMR98-02560), the David and Lucile Packard Foundation (A.T.J.) and the Alfred P. Sloan Foundation (A.T.J.).

REFERENCES

1. S. Iijima, Nature **354**, 56 – 59 (1991).
2. S. J. Tans, A. R. M. Verschueuren, and C. Dekker, Nature **393**, 49 – 52 (1998).
3. J. Lefebvre, J.F. Lynch, M. Llaguno, M. Radosavljevic, and A.T. Johnson, Appl. Phys. Lett. **75**, 3014 – 3016 (1999).
4. L. Chico, V. H. Crespi, L. X. Benedict, S. G. Louie, and M. L. Cohen, Phys. Rev. Lett. **76**, 971 – 974 (1996).
5. C. Kane and E. Mele, Phys. Rev. Lett. **78**, 1932 – 1935 (1997).
6. W. Clauss, D.J. Bergeron, M. Freitag, C.L. Kane, E.J. Mele, and A.T. Johnson, Europhys. Lett. **47**, 601-607 (1999).
7. R. D. Antonov and A. T. Johnson, Phys. Rev. Lett. **83**, 3274 – 3276 (1999).
8. Z. Yao, H. W. C. Postma, L. Balents, and C. Dekker, Nature **402**, 273 – 276 (1999).
9. M. Freitag and A.T. Johnson, this volume.
A. Thess, R. Lee, P. Nikolaev, H. Dai, P. Petit, J. Robert, C. Xu, Y. H. Lee, S. G. Kim, A. G. Rinzler, D. T. Colbert, G. E. Scuseria, D. Tomanek, J. E. Fischer, and R. E. Smalley, Science **273**, 483 – 487 (1996).
10. T. Hertel, R. Martel and P. Avouris, J. Phys. Chem. **B1998**, 910 – 915 (1998); M.R. Falvo, R.M. Taylor, A. Helser, V. Chi, F.P. Brooks, S. Washburn, and R. Superfine, Nature **397**, 236-238 (1999).
11. J. Lefebvre, M. Radosavljevic, and A.T. Johnson, submitted to Applied Physics Letters.

12. G. J. Dolan, Appl. Phys. Lett. **31**, 337 – 339 (1977).
13. S. J. Tans, M. H. Devoret, H. J. Dai, A. Thess, R. E. Smalley, L. J. Geerligs, and C. Dekker, Nature **386**, 474 – 476 (1997); M. Bockrath, D. H. Cobden, P. L. McEuen, N. G. Chopra, A. Zettl, A. Thess, and R. E. Smalley, Science **275**, 1922 – 1924 (1997); R.D. Antonov, Ph.D. Thesis, University of Pennsylvania, unpublished.
14. F. Leonard and J. Tersoff, Phys. Rev. Lett. **83**, 5174 – 5177 (1999).
15. R. Martel, T. Schmidt, H. R. Shea, T. Hertel, and P. Avouris, Appl. Phys. Lett. **73**, 2447 – 2449 (1998).
16. P.G. Collins, K. Bradley, M. Ishigami, and A. Zettl, Science **287**, 1801 – 1804 (2000).
17. P.L. McEuen, M. Bockrath, D.H. Cobden, Y. Yoon, and S.G. Louie, Phys. Rev. Lett. **83**, 5098 – 5101 (1999).

Locally Measured Electronic Properties of Single Wall Carbon Nanotubes

M. Freitag and A. T. Johnson

Dept. of Physics and Astronomy, University of Pennsylvania, Philadelphia, PA 19104

Abstract. We use a conducting-tip atomic force microscope (CT-AFM) to measure the local electronic properties of single wall carbon nanotube (SWNT) circuits. Tunnel current images of SWNTs in circuits can have 1 nm resolution, much better than that of standard AFM. By using the tip as a mobile electrode, we are able to measure variations of the local electrochemical potential within a SWNT bundle with millivolt resolution. We find that most of a bundle's resistance comes from hopping between the nanotubes in the bundle while the individual nanotubes are highly conductive. The conducting AFM tip can also be used as a local electrostatic gate of the SWNT circuit, and we make the observation that the gating of a SWNT bundle occurs at an isolated "hot spot".

INTRODUCTION

Single wall carbon nanotubes (SWNT)[1] constitute a class of molecules with unique electronic properties. A rich variety of devices might be constructed by either doping parts of a semiconducting nanotube or seamlessly joining different nanotubes through a pentagon-heptagon defect.[2,3,4] By contacting nanotubes with lithographic leads, the transport characteristics can be measured, but no local information about the circuit is obtained. Scanning tunneling spectroscopy on the other hand, can give information about the local density of states, but the current is passed perpendicular through the nanotubes into a homogeneously conducting substrate.[5,6] In this work we are merging the two techniques by using a scanning probe microscope with force feedback and a conducting tip[7,8] to determine local properties of SWNTs in active electronic circuits.

HIGH RESOLUTION CURRENT IMAGING

The resolution of AFM images is typically limited to 10 nm by the tip radius so that small features like single nanotubes in a bundle can not be resolved. In contrast, the exponential distance dependence of the tunnel current gives scanning tunneling microscopy (STM) much higher, even atomic, resolution.[9,10,11] The combination of AFM feedback with tunneling from a conducting tip allows current imaging on partly insulating substrates and thus enhanced resolution.

CP544, *Electronic Properties of Novel Materials—Molecular Nanostructures*, edited by H. Kuzmany, et al.

In contrast to conventional AFM, we use a quartz needle sensor[12] whose mechanical oscillations are detected electrically. The oscillation amplitudes are typically below 1nm,[7] a factor of 10-100 less than those of cantilever based non-contact AFM. This allows us to pass a tunnel current between the conducting tip and the sample. The oscillation amplitudes are large enough however, to prevent strong lateral forces that would disturb the SWNTs, as occurs in contact-mode AFM.

Figure 1 shows a nanotube bundle that is lying partly on an insulating SiO_2 surface and partly on a gold electrode. The high resolution current image Fig.1(c) is taken at the end of the small bundle where the substrate is insulating. Single nanotubes are clearly resolved and the feature to the right might be a small SWNT piece or some impurity (e.g. catalyst particle). Since there is no current from the tip into the non-conducting substrate, the image contrast is greatly enhanced.

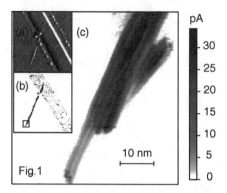

FIGURE 1. (a) Overview AFM error image showing the bundle partly on the gold electrode and partly on silicon-oxide. (b) Simultaneously acquired current image. (c) 50nm x 50nm current image of the end of the bundle with grounded electrode and a tip voltage of –1V. Individual nanotubes are clearly resolved.

PROBING LOCAL POTENTIALS WITH CT-AFM

The conducting AFM-tip can also be used as a mobile third electrode to pass currents into a biased nanotube bundle. Tunnel current images that are taken with a fixed tip voltage show a contrast inversion (the tunnel current changes its sign) along a zero-current line where the local electrochemical potential equals the tip potential.

Figure 2(a) shows a contacted SWNT bundle with an applied bias of 0.6 V. The current image Fig. 2(b) is taken in the small region indicated in Fig. 2(a). It shows that the zero-current line, where the local potential equals the tip potential of 0.442 V, is not strictly perpendicular to the direction of the bundle. Indeed it is parallel to this direction for a length of 50 nm, suggesting that adjacent nanotubes in the bundles can have different potentials. This is only possible if the conductance between the individual nanotubes is much weaker than the conductance along the length of a SWNT. Our findings support the view that bundles consist of highly conductive nanotubes that are only weakly coupled to each other. The resistance of the whole bundle results predominantly from the hopping between the tubes.

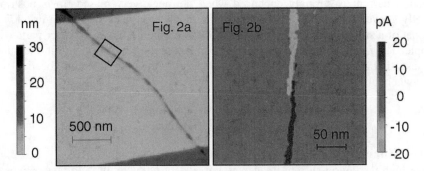

FIGURE 2. (a) Overview topography image showing a nanotube bundle connected to two electrical leads. The applied voltages are 0V (top electrode) and 0.6V (bottom electrode) respectively. (b) Current image in a zoomed-in region, taken with a tip-voltage of 0.442 V. In dark areas, the current flows from the sample into the tip, in bright areas, the current flows from the tip into the sample.

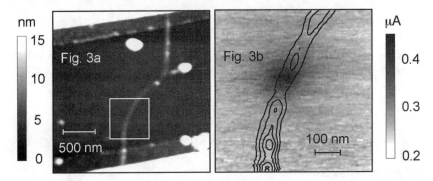

FIGURE 3. (a) Overview topography image showing a nanotube bundle and the electrical contacts. (b) Transport current as a function of the tip (gate) position in the region indicated by the rectangle in 3a. The tip voltage is -2 V and the applied bias between the leads is 0.2 V. The contour lines stem from a simultaneously acquired topography image. The image shows the "hot spot" located at a position with a bundle diameter of 1.7 nm, characteristic of a single nanotube.

LOCAL GATING WITH THE TIP

Standard transport measurements show field effect behavior in semiconducting nanotubes or bundles due to a voltage applied at a back gate. The three terminal technique can be used to resolve the gating behavior laterally.[13] In this mode, the AFM feedback forces are kept small so that no tunnel current flows between tip and sample. The voltage at the tip is solely used to locally gate the transport current.

Figure 3(a) shows the topography image of a nanotube bundle with a diameter varying between 1.5 nm and 7 nm. Figure 3(b) shows the transport current through the bundle as function of the tip position in a zoomed-in region. It reveals the surprising fact, that the gating is not uniform along the bundle, but instead highly localized. The transport current is strongly enhanced when the negatively biased tip is positioned directly above a "hot spot" where the bundle diameter is measured to be just 1.7 nm, i.e. that of a single tube, but nothing happens when the tip moves to other parts of the sample. The fact that the gating takes place at a bottleneck position within the bundle, suggests that a semiconducting nanotube is carrying the current in that location while everywhere else metallic nanotubes are present.

CONCLUSIONS

The ability to pass tunnel currents between tip and sample makes it possible to obtain high resolution current images of active electronic circuits. Voltages at specific points can be determined with nanometer resolution by nulling the tip-sample current. Finally, the tip can be used as a selective gate and the response of a SWNT bundle turns out to be localized in one single "hot spot".

ACKNOWLEDGMENTS

We acknowledge Marko Radosavljevic, Wilfried Clauss, David Bergeron, James Hone and Jacques Lefebvre for useful discussions. A. T. Johnson recognizes the support of the Packard Foundation, the NSF Grant number DMR-9802560 and an Alfred P. Sloan Research Fellowship.

REFERENCES

1. S. Iijima, Nature **354**, 56 (1991).
2. R. D. Antonov and A. T. Johnson, Phys. Rev. Lett. **83** (16), 3274 (1999).
3. L. Chico, V. H. Crespi, L. X. Benedict, S. G. Louie, and M. L. Cohen, Phys. Rev. Lett. **76** (6), 971 (1996).
4. Z. Yao, H. W. C. Postma, L. Balents, and C. Dekker, Nature **402**, 273 (1999).
5. J. W. G. Wildöer, L. C. Venema, A. G. Rinzler, R. E. Smalley, and C. Dekker, Nature **391**, 59 (1998).
6. T. W. Odom, J. L. Huang, P. Kim, and C. M. Lieber, Nature **391**, 62 (1998).
7. W. Clauss, J. Zhang, D. J. Bergeron, and A. T. Johnson, J. Vac. Sci. Technol. B **17** (4), 1309 (1999).
8. H. Dai, E. W. Wong, and C. M. Lieber, Science **272**, 523 (1996).
9. M. Ge and K. Sattler, Appl. Phys. Lett. **65**, 2284 (1994).
10. W. Clauss, D. J. Bergeron, and A. T. Johnson, Phys. Rev. B, **58** (8), 4266 (1998).
11. W. Clauss, D. J. Bergeron, M. Freitag, C. L. Kane, E. J. Mele, and A. T. Johnson, Europhys. Lett. **47** (5), 601 (1999).
12. Omicron Instruments, Taunusstein, Germany.
13. M. A. Eriksson, R. G. Beck, M. Topinka, J. A. Katine, and R. M. Westervelt, Appl. Phys. Lett. **69** (5), 671 (1996).

The Measurement of Work Function
of Carbon Nanotubes

Masashi Shiraishi, Koichiro Hinokuma and Masafumi Ata

Frontier Science Laboratories, SONY Corporation,
240-0036 Shin-Sakuragaoka 2-1-1, Hodogaya-ku, Yokohama, Japan

Abstract. The work function of multi-walled carbon nanotubes (MWNTs) is found to be about 5eV by the photo-electron emission method. For comparison, the work function of highly oriented pyrolytic graphite (HOPG) was also measured and it was determined that the work function of MWNT is larger than that of HOPG. MWNT has σ - π non-orthogonal valence state, while HOPG has an orthogonal state. Because of this non-orthogonality, HOMO level of MWNT is deeper, which gives a larger work function. To check the validity of this experimental result, the first principle *ab-initio* calculation was carried out and the result shows the validity of the experiment.

INTRODUCTION

Carbon nanotubes (CNTs) are a promising material for single electron transistors (SETs) [1], nanostructure diodes [2], hydrogen-storage material [3], for example. For an application to electronic devices, it is essential to find out the electric structure of the interface between CNT/metal and CNT/semiconductor, and the work function is one of the basic physical quantities to determine the interaction at the interface. There are various reports on the CNT work function [4]. The estimated value is, for example, 4.3eV using the UPS method and 1.3eV using the Fowler-Nordheim model in cold emitter experiment. The reported values are dispersed from 0eV to 8eV. Although, the same value as that of amorphous carbon is provisionally used as the work function of CNT, there is still a room for discussion. Here we show the CNT work function using photo-electron emission (PEE) method.

EXPERIMENTS

CNTs were synthesized using the arc discharge method without a catalytic metal. By TEM observation (not shown here), they are multi-walled carbon nanotubes of which diameter is about 10nm. Filtration process is used and most of graphites were removed. For the purification of CNTs, we applied heat at 450° in an oxygen atmosphere [5]. In this condition, only amorphous carbon is burned out and the CNTs suffer no damage and less surface contamination comparing with another purification methods using acids. PEE measurements were carried out by Riken-Keiki AC-1 [6]. In this measurement the photo electrons, which are emitted from the sample surface with D_2 lump excitation, are counted The first-principle *ab-initio* method *Q-Chem* ver.*1.2* was used to calculate the HOMO and LUMO levels of various sizes of graphene sheets and rings.

CP544, *Electronic Properties of Novel Materials—Molecular Nanostructures*, edited by H. Kuzmany, et al.
© 2000 American Institute of Physics 1-56396-973-4/00/$17.00

RESULTS AND DISCUSSION

The advantage of using PEE measurements is that it minimizes the disturbance caused by surface contamination. To measure the work function precisely, the purification process of CNTs is needed, and it means that the sample is exposed to the air before measurement and the sample surface is contaminated. Although UPS is another powerful method to measure the work function, such surface contamination prevents us from measuring the precise value. In this sense, unless the purification can be done in the ultra high vacuum, UPS is not a suitable method. On the other hand, PEE is not as sensitive to surface contamination as UPS measurements of the work function, because the penetration length of photons from D_2 lump is about 300Å and this fact makes us possible to measure the precise work function value of purified CNTs.

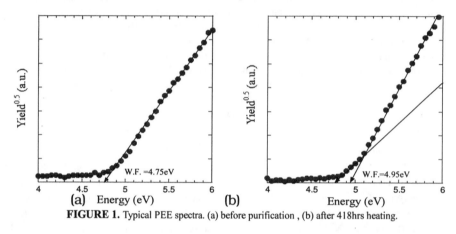

FIGURE 1. Typical PEE spectra. (a) before purification , (b) after 418hrs heating.

Figure 1 shows the typical PEE spectra from the before- and after-purified samples. The slopes are fitted by linear lines and the work function is calculated as an intercept of the fitted line and horizontal axis. In the before- purified sample, only one slope can be seen and the intercept is about 4.75eV. This signal originates from amorphous carbon. After 418hrs heating, although a small amount of amorphous carbon remains, the other slope can be observed. The interception is about 4.95eV. As noted above, only amorphous carbon is burned out and MWNT remains. Thus, the new signal is caused by the purification of CNT. As a result, the work function of MWNT is determined to be about 4.95eV, 0.2eV larger than that of amorphous carbon. To confirm the validity of this result, we measured highly oriented pyrolytic graphite (HOPG), C_{60} and C_{60}-polymer [7] using the same method, and found their work function to be 4.85, 6.5 and 5.5eV, respectively. Note that the work function of CNT is larger than that of HOPG. HOPG has σ - π orthogonal valence state, while C_{60} system and CNT do not. As a result of the mixing of σ and π valence states, their HOMO levels become deeper, so the measured work function using PEE is thought to be valid.

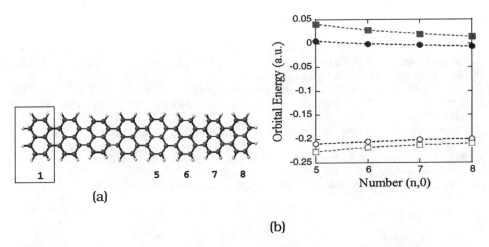

(a)

(b)

FIGURE 2. (a) Schematic for first principle *ab-initio* calculation. Carbon atoms form 6-membered rings and hydrogen atoms terminate the edges. The "sheet" structure is shown here. (b) The result of *ab-initio* calculation. 1 a.u. of the vertical axis is equal to 27eV. The open and closed circles are HOMO and LUMO levels of the "sheet" structure and the open and closed squares are HOMO and LUMO levels of the "ring" structure, respectively. The broken lines are to guide the eye.

First principle calculation by *ab-initio* method using 3-21G basis set supports our result. Figure 2(a) is the schematic of the calculated structure. Although hydrogen atoms terminate carbon atoms at the edge, this structure is very similar to a graphene sheet (we call this a "sheet" structure) and by rolling it up, a "ring" structure can be formed. The size of the "sheet" and the "ring" is changed by adding carbon hexagons and here the size is described by (n,m) index as in the CNT case. We calculated HOMO and LUMO levels in the "sheet" and "ring" structures whose indicies are (5,0), (6,0), (7,0) and (8,0). In Figure 2(b), the result is shown and it can be interpreted that the HOMO level in the "ring" structure is deeper than that of the "sheet" structure with the same index. This calculation has a good agreement with our experimental result and it supports the validity of the measured value. Besides, as the diameter of the "ring" beomes smaller, the HOMO-LUMO gap becomes larger and the HOMO level becomes deeper. The former result agrees with the previous work [8]. It is unclear why the HOMO-LUMO gap does not close in the (6,0) case, but it is probably because the length in the tubular axis is very short in our calculation. The latter result suggests that the work function of single-walled nanotubes (SWNTs) can be slightly larger, because the diameter of SWNTs is much smaller than that of our MWNTs.

REFERENCES

1. Ebbesen, T. W., Lezec, H. J., Hiura, H., Bennett, J. W., Ghaemi, H. F., and Thio, T., Nature **382**, 54-56(1996).
2. Yao, Z., Postma, H. W. Ch., Balents, L., and Dekker, C., Nature **402**, 273-275(1999).
3. Chen, P., Wu, X., Lin, J. and Tan, K. L., Science **285**, 91-93(1999).

4. Ago, H., Kugler, T., Cacialli, F., Salaneck, W. R., Shaffer, M. S. P., Windle, A. H., and Friend, R. H., J. Phys. Chem. B103, 8116-8121(1999).

Sinitsyn, N. I., Gulyaev, Y. V., Torgashov, G. V., Chernozatonskii, L. A., Kosakovskaya, Z. Ya., Zakharchenko, Y. F., Kiselev, N. A., Musatov, A. L., Zhbanov, A. I., Mevlyut, S. T., and Glukhova, O. E., Appl. Surf. Sci. 111, 145-150(1997).

Collins, P. G., and Zettl, A., Phys. Rev. B55, 9391-9399(1997).

Tian, M., Chen, L., Li, F., Wang, R., Mao, Z., Zhang, Y. and Sekine, H., J. Appl. Phys. 82, 3164-3166(1997).

5. Ochiai, Y., Ishikawa, S., and Mori, S., *Abstracts of 16th fullerene symposium(1998)* 1A-10, pp.129.(*in Japanese*)

6. Kido, J., Shionoya H., and Nagai, K., Appl. Phys. Lett. 67, 2281-2283(1995).

7. Ramm, M., and Ata, M., Appl. Phys. A. *will be published.*

8. Dresselhaus, M. S., Dresselhaus, G., and Saito, R., Sol. Stat. Comm. 84, 201-205(1992).

Electrical Properties of Singlewalled Carbon Nanotubes-PMMA Composites

C. Stéphan[1], T. P. Nguyen[1], S. Lefrant[1], L. Vaccarini[2] and P. Bernier[2]

[1]*Laboratoire de Physique Cristalline, IMN, Université de Nantes, France,*
[2]*Groupe de Dynamique des Phases Condensées, Université de Montpellier 2, France.*

Abstract. In this work, we report the results obtained from electrical measurements of poly(methyl methacrylate)-singlewalled carbon nanotubes composites thin films. These films were prepared by mixing the polymer with different nanotube concentrations. Electrical characteristics of the ITO-composite-metal structures were investigated by measuring the current versus the applied voltage at room temperature. Their evolution versus the nanotube concentration was studied in the order to determine the transport process in these materials. Such composites are promising for use as transport layers in multilayer diodes.

INTRODUCTION

Since their discovery in the early nineties[1], carbon nanotubes have generated much interest and have encouraged intensive investigation in view of their potential applications. Nanotubes offer a wide variety of properties which differ if we deal with multiwalled nanotubes (MWNT's) or singlewalled nanotubes (SWNT's) systems. Theoretical studies on the electronic structure of SWNT's have already been investigated, on one hand as a function of both their geometrical configuration (zigzag, armchair and chiral) and their diameter[2], and on the other hand when they are self-organized into ropes[3]. The combination of carbon nanotubes with polymers offers an attractive way to introduce new electronic properties. As an example, conjugated polymer-nanotubes composites can be used as active layers in organic light emitting diodes, but also polymer-nanotube composites can be used as a transport layer in multilayer diodes. Electrical measurements realized on composites based on poly(methyl methacrylate) (PMMA) and MWNT's[4] have shown a percolation behavior of the conductivity with a threshold of 0.5%. In the same manner, we have studied PMMA-SWNT's composites. In this work, we investigate the conductivity and the I-V characteristic behavior of the composites as a function of the SWNT's concentration in PMMA.

CP544, *Electronic Properties of Novel Materials—Molecular Nanostructures*, edited by H. Kuzmany, et al.
© 2000 American Institute of Physics 1-56396-973-4/00/$17.00

EXPERIMENTS

Carbon nanotubes were produced by the electric arc method using a mixture of 1 at.% Y and 4.2 at.% Ni as catalysts in the graphite rod[5]. They were then purified in a three step procedure: a nitric acid treatment, a cross flow filtration and a vacuum bake at 1600 °C. The nanotube powder and PMMA were mixed together in toluene with several concentrations specified as the weight percentage in the polymer. The solutions were then mixed in an ultrasonic bath during 8 hours. They were immediately deposited by spin coating onto glass substrates covered by an indium tin oxide (ITO) bottom electrode. The top electrode (MgAg), was then evaporated. Four samples of active area 2x2 mm^2 were obtained on the same substrate. The thickness of the films was determined by an Alphastep unit, ranging from several hundred nanometers to a few microns.

RESULTS AND DISCUSSION

The current-voltage characteristics of the composites were measured for several nanotube mass ratios in PMMA (fig. 1). The conductivity of pristine PMMA is 5×10^{-12} S/m. Up to 1%, we observe a dramatic increase to 6.4×10^{-3} S/m. For higher concentrations, the conductivity increases slowly and reaches 1.6×10^{-2} S/m. This behavior is characteristic of a percolation and can be described by an analytical model proposed by Fournier et $al.$ using the Fermi-Dirac distribution[6]:

$$\log \sigma_c = \log \sigma_n + (\log \sigma_p - \log \sigma_n)/[1 + \exp \{b(c - c_p)\}] \qquad (1)$$

where σ_c, σ_n, σ_p are the composite, nanotube and polymer conductivity respectively, c is the mass ratio, and b is an empirical parameter which describes the change in conductivity at the percolation threshold c_p. The plot in fig. 1 was best fitted by equation 1 with $c_p = 0.4$%.

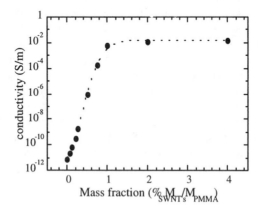

FIGURE 1. Semi-logarithmic plot of the composite conductivity for different nanotube concentrations in PMMA. The dotted line is a fit from equation (1).

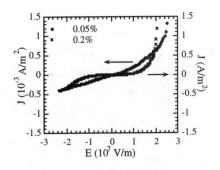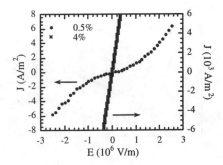

FIGURE 2. Current density-applied field characteristics of the ITO/composite/MgAg structures for the 0.05, 0.2%, 0.5% and 4% concentrations.

The current density-applied field (J-E) characteristics of diodes based on composites using four concentrations (0.05, 0.2, 0.5 and 4%) are shown on fig. 2. The characteristics of the 0.05 and 0.2% samples are asymmetric while those of the 0.5 and 4% are symmetric. Moreover, they become linear for concentrations higher than 2%. For low concentrations, the asymmetric behavior of the electrical curves can be explained by the asymmetric configuration of the sandwich structures (ITO/composite/MgAg) while in high concentration samples, charge carriers are hopping between the islands of nanotubes in the polymer matrix.

In fig. 3. a, we represent the log I-log V characteristics for several SWNT's concentrations. The low field region shows a linear characteristic with slope value of 1.4 for PMMA and 1 for composites. At high fields, quadratic characteristics are observed in pristine PMMA and weakly doped samples reflecting space charge effects.

The log I-$V^{1/2}$ curves (fig. 3. b) show linear characteristics at high fields. In this range, the current can be described by[7]:

$$I = I_0 \exp \left[\frac{\beta \left(V/d \right)^{1/2} - \Phi}{kT} \right] \qquad (2)$$

where V is the applied voltage, d is the film thickness, ϕ is the activation energy, and β is a constant depending on the conduction mechanism. In the Poole-Frenkel mechanism, β equals 2 $(e^3/4\pi\varepsilon\varepsilon_0)^{1/2}$ while it is $(e^3/4\pi\varepsilon\varepsilon_0)^{1/2}$ in the Schottky effect, where $\varepsilon\varepsilon_0$ represents the dielectric permittivity of PMMA. In order to distinguish between these effects, we compare the slope of the curves in fig. 3. b with the theoretical values of the coefficient β. The experimental value of β is close to that of the Poole-Frenkel effect for pristine PMMA and weakly doped samples (3.5×10^{-24} J $V^{-1/2}$ $m^{1/2}$) but increases in highly doped samples. The conduction mechanism in the high concentration PMMA-SWNT's composites could be modified by hopping process.

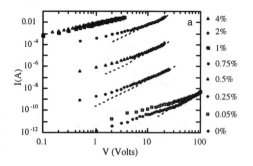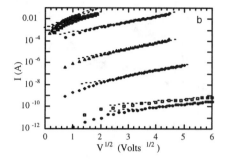

FIGURE 3. Current voltage plots in ITO/composite/MgAg structures: log I versus log V (a) and log I versus $V^{1/2}$ (b).

CONCLUSION

Electrical measurements on PMMA-SWNT's composites have shown a percolation behavior of the conductivity with a very low threshold (0.4%). The I-V characteristics are rectifying and asymmetric at low nanotube concentrations. For high concentrations, charge carrier hopping is likely to occur as evidenced by the symmetric and linear characteristics. The conduction mechanism at high field is limited by space charges and by an electrical field at lower applied field. These effects are significantly modified in high concentrated samples in which hopping of the charge carriers probably takes place.

ACKNOWLEDGMENTS

This work is supported by the European Community through its Training and Mobility of Researcher program under network contract: NAMITECH, ERBFMRX-CT96-0067(DG12-MIHT) and by the French CNRS program: ULTIMATECH.

REFERENCES

1. Iijima, S., *Nature* **354**, 56 (1991).
2. Wildoër, J. W. G., Venama, L. C., Rinzler, A. G., Smalley, R. E., and Dekker, C., *Nature* **391**, 59 (1998).
3. Delaney, P., Choi, H. J., Ihm, J., Louie, S. G., and Cohen, M. L., *Nature* **391**, 466 (1998).
4. Stéphan, C., Nguyen, T. P., Curran, S., and Lefrant, S., submitted.
5. Journet, C., Maser, W. K., Bernier, P., Loiseau, A., Lamy de la Chapelle, M., Lefrant, S., Deniard, P., Lee, R., and Fischer, J. E., *Nature* **388**, 756 (1998).
6. Fournier, J., Boiteax, G., Seytre, G., and Marichy, G, *Synth. Met.* **84**, 839 (1997).
7. Simmons, J. G., *J. Phys. D: Appl. Phys.* **4**, 613 (1971).

Role of the Metal in Contacting Single-Walled Carbon Nanotubes

V. Krstic[1*], G.T. Kim[2], J.G. Park[2], D.S. Suh[2], Y.W. Park[2], S. Roth[1,] and M. Burghard[1]

[1] *Max-Planck-Institut für Festkörperforschung, Heisenbergstr. 1, D-70569 Stuttgart, Germany*
[2] *Seoul National University, Seoul, Korea*

Abstract. Electrodes have been defined by e-beam lithography on top of Carbon Nanotubes in order to reduce the contact resistance between metal and Nanotube. Individual Single-Walled Carbon Nanotubes or individual thin bundles were contacted with different metals (Au, AuPd, Al, and Co). The electrode material has been varied in order to investigate the Nanotube/metal - contact. The adhesion properties of the metals and the Carbon Nanotubes on the substrates turned out to be crucial.

INTRODUCTION

Electrodes have been deposited on top of Carbon Nanotubes (CNTs) by several groups in order to minimize the contact resistance $R_{contact}$ between CNT and metal. Theoretical studies [1, 2] predict that the metal should have a minor influence on the contact properties, including $R_{contact}$. In contrast, a strong effect is expected from the extraordinary fast decay of the Fermi-level states of the CNTs perpendicular (z-direction) to its symmetry axis: $\psi \propto \exp\{-[2m_e/\hbar^2(V - E_F) + k^2]^{1/2} \ll z\}$, with potential in space V between CNT and metal, Fermi-level wave vector k and electron mass m_e [1].

In the present work, electrode arrays of different metals (Au, AuPd, Al and Co) have been defined by e-beam lithography on top of Single-Walled Carbon Nanotubes (SWNTs). Electronic transport was investigated in order to reveal the dependence of the contact properties on the metal.

EXPERIMENTAL

SWNTs are dispersed in aqueous surfactant solution and then purified by centrifugation [3]. As substrates Si wafers with a ~ 300 nm thick thermally grown oxide layer were used. To promote adsorption of the SWNTs, the wafer was treated with 3-(aminopropyl)-triethoxy-silane [3].

The electrode arrays were produced by covering the substrate by a two-layer poly(methyl methacrylate) resist system and using electron beam lithography to define the electrode arrays. Afterwards the desired metal was evaporated on the substrate and

CP544, *Electronic Properties of Novel Materials—Molecular Nanostructures*, edited by H. Kuzmany, et al.
© 2000 American Institute of Physics 1-56396-973-4/00/$17.00

finally the lift-off process was performed during which undesired metal is removed by mechanical stress. Electrical transport was measured under helium atmosphere.

RESULTS AND DISCUSSION

Gold-Palladium (AuPd)

In Fig. 1 a SFM-image of a AuPd electrode array on top of SWNTs and thin SWNT bundles is shown. The current/voltage (I/V) characteristics at room temperature (RT) and 4.2 K of the contacted CNTs (electrode pair (I)/(II)) are depicted in Fig. 2a) and 2b), respectively.

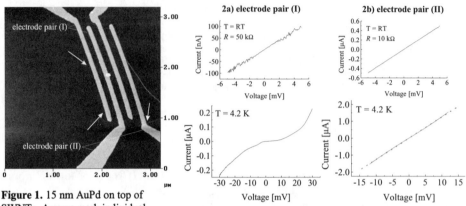

Figure 1. 15 nm AuPd on top of SWNTs. Arrows mark individual SWNTs and a thin bundle.

Figure 2. I/V-curves at RT and 4.2 K of pair (I) and (II). Dots are guide for the eye.

The contour of SWNTs and the SWNT bundle can be detected in the metal profile, indicating that their structural integrity is preserved during the evaporation process. Electrode pair (I) shows a RT resistance of about 50 kΩ > h/e^2. At 4.2 K step-like features are observed, characteristic of Coulomb-Blockade dominated transport. Pair (II) exhibits at RT only ≈ 10 kΩ < h/e^2, suggestive of ballistic transport. Compared to the I/V-curve of pair (I), no conductance fluctuations occur at higher voltages which are attributed to thermal instabilities of the contact region. At 4.2 K a power-law dependence $I \propto V^\gamma$ is observed ($\gamma \approx 0.9$) which could originate from, e.g., a non-constant density of states or Luttinger Liquid behaviour.

Gold (Au)

Fig. 3 shows a SFM-image of a Au electrode array contacting thin SWNT bundles. The RT I/V-characteristics are displayed in Fig. 4. Similar to AuPd, resistances smaller than h/e^2 are observed. However, the measured data yield $R^{(I)} \approx R^{(II)}$, $R^{(I)} + R^{(II)} \geq R^{(III)}$ and $\{R^{(I)}, R^{(II)}\} < R^{(III)}$. This indicates that ballistic transport is impeded along the SWNT bundle. Good contacts, even though not biased (as electrode Φ when measuring pair (III)), provide a finite probability for the travelling electron to propagate into. Therefore they represent a scattering centre perturbing ballistic transport.

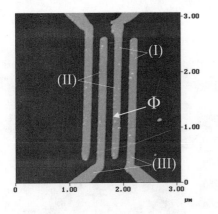 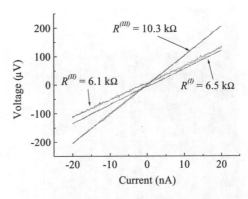

Figure 3. 20 nm Au on top of SWNTs.   **Figure 4.** I/V-curves of different electrode pairs at RT.

Occasionally, partially detached electrode lines are observed after lift-off, indicating that the adhesion of SWNTs to the surface-treated SiO_2 is of similar strength or stronger than the adhesion of the Au to this surfaces. It is noteworthy, that similar adhesion properties are observed for AuPd electrodes.

Aluminium (Al)

Despite the removal of residual H_2O by Ar-ion sputtering before Al evaporation at $p < 10^{-8}$ mbar and sample storage under Ar-atmosphere, no measurable current ($I > 0.5$ pA) could be observed for up to a few volts in any of the prepared samples. Interestingly, difficulties in contacting have been reported also for Al evaporated on Langmuir-Blodgett films consisting of aromatic molecules [4]. Moreover, the observed Al adhesion to the surface-treated substrate was much weaker compared to Au and AuPd.

Cobalt (Co)

Fig. 5 shows a SFM-image of a Co-electrode array. The resistance of both pairs (3 /16) and (3/4) exceeds 10 GΩ, whereas pair (15/16) shows $\approx$ 130 kΩ (corresponding to ~ 390 kΩ per bundle, as also found for other samples). The RT I/V-curve of pair (15/16) is depicted in Fig. 6a). The adhesion of Co to the substrate turned out to be weaker compared to Au and AuPd, but stronger than for Al.

Magnetoresistance of pair (15/16) (Fig. 6b)) changes significantly depending on the orientation of the magnetic field B relative to the substrate. This change could be due to weak-localization effects within the contacted SWNT bundles and or individual SWNTs. Aharanov-Bohm and weak-localization effects involving the electrodes are excluded, since the tunneling barriers between metal and SWNT are expected to randomize the phase of the electron wave-function. Further, no significant hysteresis could be observed in the up and down sweep of B, which would be characteristic of coherent spin transport [5].

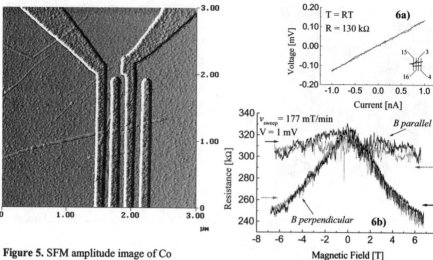

Figure 5. SFM amplitude image of Co (50 nm) on top of SWNTs.

Figure 6. I/V-curves at RT (6a)) and magneto-resistance at 4.2 K (6b)) of electrode pair (15/16). Arrows mark the sweep direction.

CONCLUSION

In contrast to SWNTs on top of electrodes [6], the resistance could be reduced below h/e^2 for Au and AuPd on top of tubes. As a consequence, non-biased electrodes may act as scattering centre and impede ballistic transport. The observed adhesion problems may lead to a mechanical increase of the SWNT/metal distance, $d_{SWNT/metal}$, during the lift-off process. Since the operator V of the potential in space V [1] is expected to be proportional to $d_{SWNT/metal}$, the transmission probability, which is proportional to $|<\psi| V |\psi_{metal}>|^2$, of an electron decreases, and hence $R_{contact}$ increases. For the larger resistances found for Co compared to Au and AuPd, this is a plausible explanation. The results obtained with Al demonstrate that metals can have a significant influence on the contact properties and the local electronic structure of SWNTs.

ACKNOWLEDGMENT

This work was partially supported by KISTEP (contract no. 98-I-01-04-A-026) Ministry of Science and Technology(MOST), Korea.

REFERENCES

[1] Tersoff, J., *Appl. Phys. Lett* **74**, 2122-2124 (1999).
[2] Delaney, P., Di Ventra, M., *Appl. Phys. Lett.* **75**, 4028-4029 (1999).
[3] Krstic, V., Duesberg, G. S., Muster, J., Burghard, M., Roth, S., *Chem. Mater.* **10**, 2338 (1998).
[4] Colliere, C.P., Wong, E.W., Belohradský, M., Raymo, F.M., Stoddart, J.F., Kuekes, P.J. Williams, R.S., Heath, J.R., *Science* **285**, 391-394 (1999).
[5] Tsukagoshi, K., Alphenaar, B.W., Ago, H., *Nature* **401**, 572-574 (1999).
[6] Krstic, V., Muster, J., Duesberg, G.S., Philipp, G., Burghard, M., Roth, S., *Synth. Met.* **110**, 245-249 (2000).

Effect of Pressure on the Electronic Properties of Carbon Nanotubes

Y.I. Prylutskyy[1], O.V. Ogloblya[1], E.V. Buzaneva[2],
S.O. Putselyk[2], P.C. Eklund[3], and P. Scharff[4]

Kyiv National Shevchenko University, Dept. of Physics[1] and Radiophysics[2],
Volodymyrska Str., 64, 01033 Kyiv, Ukraine (e-mail: prilut@office.ups.kiev.ua)
[3]Dept. of Physics, Penn State University, 104 Davey Hall, PA 16802-6300, USA
[4]Institute of Physics, TU Ilmenau, PF 100565, D-98684 Ilmenau, Germany

Abstract. The effect of pressure on the band gap for single-walled carbon nanotubes (SWCNT) has been calculated using the molecular dynamics approach. It was found that under the uniaxial homogeneous deformation of semiconductor SWCNT the gap varies linearly with pressure and strongly dependent on their chiral symmetry. An analytical expression for the determination of value of pressure characterizing the transition between semiconductor and metallic behavior was obtained and analysed in detail.

INTRODUCTION

Carbon nanotubes have been investigated intensively as a different form of one-dimensional (1D) material [1]. Each single-layer nanotube can be regarded as a rolled-up graphite sheet in the cylindrical form. The geometrical structure of this cylinder can be uniquely specified by two intergers (n,m). The fundamental symmetry considerations applied to the electronic band structure indicate that the (n,m) SWCNT are metallic or semiconductor, depending on whether or not its n-m is a multiple of 3. In this paper we calculated the geometrical structure and band gap of semiconductor SWCNT in dependence of external pressure using the proposed earlier molecular dynamics model [2]. We also study how the uniaxial homogeneous deformation influence on the band gap of SWCNT in dependence on their chirality n-m=3q±1, q is an integer. This seems of interest since such effect of pressure on the electronic properties of SWCNT has received little attention [3] in spite of papers to the predicted interesting mechanical properties [1-2, 4-7].

CALCULATION AND DISCUSSION

Let us examine the effect of uniaxial homogeneous deformation of semiconductor (n,m) SWCNT along their axes. In particular, we now calculate the

CP544, *Electronic Properties of Novel Materials—Molecular Nanostructures*, edited by H. Kuzmany, et al.
© 2000 American Institute of Physics 1-56396-973-4/00/$17.00

geometrical structure and band gap using the molecular dynamics model [2] adapted to take into account the modifications of two-center integrals with stress.

The equilibrium nanotube structure under the action of uniaxial compression P is determined by the minimum of the energy:

$$U = \frac{1}{2}\left[k_1 \sum_{i>j} d_{ij}^2(r,\varepsilon) + k_2 \sum_i \varphi_i^2(r,\varepsilon) \right] + P\pi R^2 L. \tag{1}$$

Here k_1 and k_2 are the force constants which describe the relative changing of the distance (d_{ij}) between the carbon atoms in hexagons and the changing of the value of angle (φ_i) in hexagons, accordingly; L is the length of tubule;

$$r = \frac{R - R_0}{R_0} = -\sigma\varepsilon, \tag{2}$$

where R is the nanotube radius at strain ε and σ is the Poisson ratio; $R_0 = \sqrt{n^2 + nm + m^2}\ \sqrt{3}\,d_0/2\pi$ for the (n,m) SWCNT, when the values of d_0=0.142 nm and φ_0=2.077 rad before deformation have been used for the C-C bond length and the angle in hexagons during the optimization of SWCNT structure.

The energy (1) was minimized over the parameters r and ε at constant pressure and under the condition of symmetry conservation of the nanotube. It should be emphasized that in our calculations we have used a non-orthogonal system of coordinates connected with the ith atom. Only three nearest neighbours of each ith atom were taken into account.

The numerical calculations show that the following values of parameters in the energy (1) k_1=0.504 kJ/m^2 and k_2=0.405×10^{-18} J/rad^2 [2] give the opportunity to obtaine the elastic constants (Young modulus and Poisson ratio), IR and Raman vibrational frequencies for considered below SWCNT close to the experimental ones in the absence of external pressure [8].

Since the deformation also modifies the different C-C bond lengths it is necessary to recalculate the overlap integrals between the different atoms. It was found that there are only two independent integrals β_x and β_y in dependence on the direction of X and Y axes in the graphite sheet

$$\beta_{x/y} = \beta_0 \frac{1}{1+x/y}, \tag{3}$$

where β_0=3.3 eV is the overlap integral between two π-electrons, located on the nearest carbon atoms before deformation [2];

$$x = \frac{3B(B-12A)}{54A - B}, \quad y = -\frac{63AB}{54A - B}. \tag{4}$$

Hear $A = \dfrac{k_2\varphi_0^2}{k_1 d_0^2} \approx 0.172$ and $B = \dfrac{9d_0}{8k_1\pi}\sqrt{n^2 + nm + m^2}\,P \approx \sqrt{n^2 + nm + m^2}\,P\times10^{-4}$

(in GPa^{-1}).

The value of electronic band gap of semiconductor (n,m) SWCNT as function of uniaxial compression P is determined by (if we neglect the effect of curvature [9])

$$E_g(P) = 2\left|\beta_x - 2\beta_y \cos\frac{q\pi}{n-m}\right| \approx$$

$$\begin{cases} E_g(0) + \beta_0\sqrt{n^2 + nm + m^2}\,P\times 10^{-4}, \text{ when } n-m = 3q+1 \\ E_g(0) - \beta_0\sqrt{n^2 + nm + m^2}\,P\times 10^{-4}, \text{ when } n-m = 3q-1, \end{cases} \tag{5}$$

where $E_g(0)$ is the gap before deformation [2]. Here we took into account that for considered below values of chirality (n,m) and pressure P: B/A<<1.

According to the expression (5) the value of pressure P_t (in GPa) characterizing the transition between semiconductor and metallic behavior of SWCNT is determined by the formular ($E_g(P_t)\equiv 0$):

$$P_t \approx \mp \frac{E_g(0)}{\beta_0\sqrt{n^2 + nm + m^2}}\times 10^4. \tag{6}$$

Let us discuss the obtained results. According to the data [10] the value of critical pressure may be found from the relation

$$P_c = 2E\left(\frac{h}{2R_0}\right)^3, \tag{7}$$

where E is the Young modulus and h=0.066 nm is wall thickness of SWCNT. For the SWCNT with diameters in the range of (1.0÷1.5) nm it was found that <E>=1.25 TPa, in the direction of the nanotube axis [6]. Thus, we can estimate the value of critical pressure for these nanotubes by the formular (7): $P_c \approx (0.2\div 0.7)$ GPa. For these ultimate values of pressure the band gap changes weakly with pressure (see the formular (5)). Finally, the carried out calculations are shown that for these types of semiconductor SWCNT: $P_t >> P_c$. Consequently, the concept of "transition pressure" [3] becomes senseless in this case.

Experimental measured energy distribution of density of states (DOS) is shown in Fig. 1. The value of energy gap for the samples of semiconductor SWCNT was

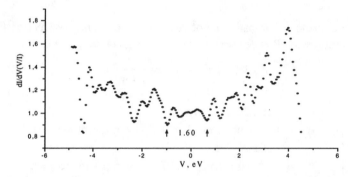

FIGURE 1. Experimental measured DOS versus energy for the samples of semiconductor SWCNT for that manifest itself the most reproducable (I-V) characteristics of Pt/Ir tip - carbon nanotubes layer - Pt/Ir tip structure.

estimated from the range with minimum of DOS near the Fermi level (in Fig. 1. near the point V≈0) using the scanning tunneling spectroscopy (STS). It is turned out that $E_g(0)≈1.6$ eV. From the computer simulation of both geometrical and electronic band structures of SWCNT we have obtained that this gap in the best way corresponds to the semiconductor (13,5) (R_0=0.63 nm and $E_g(0)$=1.55 eV) SWCNT or (13,6) (R_0=0.66 nm and $E_g(0)$=1.63 eV) one in the bundles.

The band gap as function of pressure is shown in Fig. 2. As we can see in the case of (13,5) SWCNT (n-m=3q-1) the gap decreases with pressure, whereas in the case of (13,6) SWCNT (n-m=3q+1) it increases. These different kinds of behavior can be understood in the 2D representation of tubule model by examining the effects of stress on the allowed lines in the Brillouin zones [3]. In the (n-m=3q+1) case the nearest corner allowed line moves away from the K point, whereas for the (n-m=3q-1) case it becomes closer.

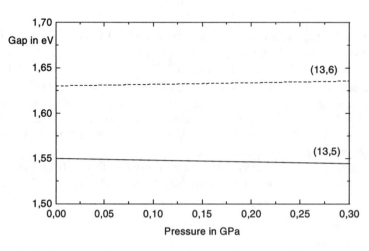

FIGURE 2. Calculated band gap versus pressure for the samples of semiconductor SWCNT.

ACKNOWLEDGMENTS

This work was supported by the CR NATO grant. The authors thank Dr. S. Fang (University of Kentucky) for kindly providing of SWCNT samples.

REFERENCES

1. Dresselhaus, M.S., Dresselhaus, G., and Eklund, P.C., *Science of Fullerenes and Carbon Nanotubes*, Academic Press, New York, 1996.
2. Prylutskyy, Yu.I., Durov, S.S., Ogloblya, O.V., Buzaneva, E.V., and Scharff, P., *Comput.Mat.Sci.* (2000) (accepted).
3. Heyd, R., Charlier, A., and Mc.Rae, E., *Phys.Rev.B* **55**, 6820 (1997).
4. Treacy, M.M.J., Ebbesen, T.W., and Gibon, J.M., *Nature* **381**, 678 (1996).
5. Dai, H., Wong, E.W., and Lieber, C.M., *Science* **272**, 523 (1996).
6. Krishnan, A., Dujardin, E., Ebbesen, T.W., Yianilos, P.N., and Treacy, M.M.J., *Phys.Rev. B* **58**, 14013 (1998).
7. Prylutskyy, Yu., Durov, S., Ogloblya, A., Eklund, P., and Grigorian, L., *Mol.Mat.* (2000) (accepted).
8. Prylutskyy, Yu.I., Ogloblya, O.V., Gubanov, V.O., Parnyuk, T.V., Buzaneva, E.V., Karlas A.Yu., Eklund, P.C., and Scharff, P., *Mol.Cryst.Liq.Cryst.* (to be published).
9. Mintmire, J.W., and White, C.T., *Carbon* **33**, 893 (1995).
10. Timoshenko, S., and Gere, J., *Theory of Elastic Stability,* McGraw-Hill, New York, 1988.

Interaction of Lithium atoms with graphitic walls

J. S. Arellano,[1] L.M. Molina, M. J. López, A. Rubio and J. A. Alonso

Departamento de Física Teórica, Universidad de Valladolid, 47011 Valladolid, Spain

Abstract. The interaction of a Li atom with a graphitic layer has been calculated using density functional theory. The presence of an heptagonal defect lowers the barrier for passage through the layer so much that it provides a plausible route for intercalation in multiwalled carbon nanotubes.

Intercalation of foreing atoms or molecules between the layers of graphite is a well known procedure of modifying the physical and chemical properties of this material [1]. Doping also leads to dramatic changes in the properties of the fullerite solid. For instance, alkali dopants take interstitial positions between the C_{60} molecules, and for an appropiate concentration the solid becomes superconducting [2]. Single-walled carbon nanotubes (SWCN) form ropes with a close-packed two-dimensional triangular lattice [3], and those ropes may be useful for intercalation and energy storage purposes. In particular carbon nanoropes have been investigated as possible candidates for anode materials in Li-ion battery applications [4]. Ab initio calculations [5] suggest that a high density of lithium can be intercalated in SWCN ropes since both the inside of the nanotubes as well as the interstitial channels between nanotubes are susceptible of intercalation. The multi-walled carbon nanotubes (MWCN) also hold promise for applications in lithium ion batteries since electrochemical intercalation of lithium between the concentric layers of MWCN has been achieved by several groups [6–8]. The experimentalists [7–9] propose that intercalation of alkali atoms may proceed through structural deffects in the graphitic walls, although the nature of those defects is unknown [10].

As a first step in the investigation of intercalation in nanotubes we study the interaction of lithium atoms with perfect and defective graphitic layers. For this purpose we have performed static calculations using the density functional formalism. We find a barrier for the passage of Li atoms through a perfect graphitic wall.

---

[1] Permanent address: Area de Física, División de Ciencias Básicas e Ingeniería, Universidad Autónoma Metropolitana Azcapotzalco, Av. San Pablo 180, 02200 México D.F., México. e-mail: jsap@hp9000a1.uam.mx

CP544, *Electronic Properties of Novel Materials—Molecular Nanostructures*, edited by H. Kuzmany, et al.
© 2000 American Institute of Physics 1-56396-973-4/00/$17.00

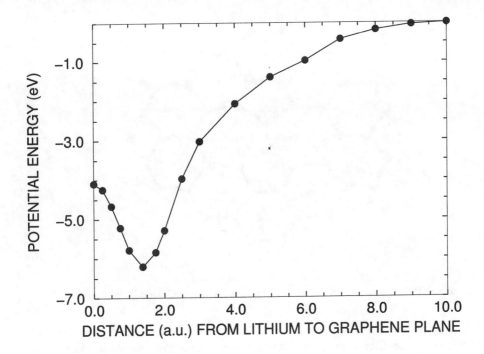

FIGURE 1. Potential energy of interaction between a Lithium atom on a graphene layer. The atom is in front of an hexagonal hole.

However that barrier is substantially lowered by the presence of an heptagonal defect.

The calculations employ the density functional formalism [11] with the local density approximation for exchange and correlation. Specifically, we have used the fhi96md code [12], which is based on a supercell geometry, and then allows for the use of a basis of plane waves. Nonlocal pseudopotentials describe the effect of the ion cores [13]. We have used previously the same method to study the interaction of H_2 with a graphene layer [14], and technical details can be found in that paper. We have calculated the energy of interaction of a lithium atom with a perfect graphene layer as a function of the distance d. The atom is always on a line perpendicular to the layer through the center of an hexagon of carbon atoms. The positions of the carbon atoms have been frozen in the calculation. Figure 1 shows the results. The two main features in the interaction potential are: (a) A minimum at an equilibrium distance $d_{eq} = 1.39$ a.u.; the binding energy, taking as zero the energy for separation d = 10 a.u., is E_b= 6.20 eV. (b) A value for the energy at d = 0 (that is with the Li atom in a configuration exactly on the graphene plane) which is lower than the energy for very large separation. On the other hand the difference

$$\Delta V = V(d = 0) - V(d_{eq}) \tag{1}$$

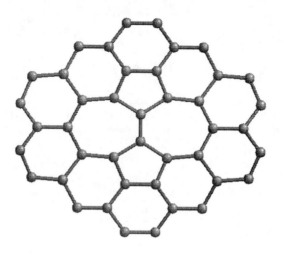

FIGURE 2. Relaxed defect obtained by rotating one C-C bond by 90 degrees.

equals 2.11 eV, which is a large value, but since $V(d = 10)$ - $V(d_{eq})$ is equal to 6.20 eV, the possibility of passing through the graphene layer seems to exist if the kinetic energy of the Li atom is not absorbed soon enough by the graphene layer. Dynamical simulations are required to explore this possibility.

The main role of a defect in the graphitic wall is to provide a more open site for the Li atom to pass accross. The presence of heptagonal defects has been indirectly established by electron microscopy on helicoidal and bent nanotubes [15,16]. Also Stone-Wales defects, where four hexagons are converted into two sets of pentagon-heptagon pairs, are known to exist in graphite structures [17]. We have investigated the heptagonal defect as the simplest possible defect that would perhaps favor Li intercalation. For this purpose we have first relaxed the structure of the graphene layer with the defect. This defect is formed when the bond between two neighbouring carbon atoms is rotated by 90 degrees. Two irregular pentagon-heptagon pairs are then formed. Relaxation makes the heptagons and pentagons more regular than the original ones. The relaxed defect is shown in Fig. 2. The structure of the defect was then maintained frozen when studying the interaction of the lithium atom with the defective graphene layer. With the atom above the center of an heptagon the distance d was varied. The interation potential has a similar shape to that of Fig. 1, although with crucial quantitative differences. The equilibrium position is now reached for $d_{eq} = 0.50$ a.u., that is, with the Li atom much closer to the graphene layer than in the case of the perfect graphene sheet. Furthermore the barrier ΔV of eq. (1) becomes 0.047 eV. That is, the existence of the heptagonal defect reduces the barrier by a factor of 45. It is then plausible that the heptagonal defects can provide an easy path for the intercalation of Li in multiwalled carbon nanotubes, as well as for the passage of Li atoms to the inner channel in single walled tubes, although dynamical simulations have to be performed to be confident.

ACKNOWLEDGEMENTS

Work supported by DGESIC(Grant PB98-0345), Junta de Castilla y León (Grant VA28/99) and European Community (TMR Contract ERBFMRX-CT96-0062-DG12-MIHT). L.M.M. is greatful to DGES for a Predoctoral Grant. M.J.L. acknowledges support from Universidad de Valladolid. J.S.A. wishes to thank the hospitality of Universidad de Valladolid during his sabbatical leave and grants given by Universidad Autónoma Metropolitana Azcapotzalco and by Instituto Politécnico Nacional (México).

REFERENCES

1. Dresselhaus M. S., and Dresselhaus G., *Adv. Phys.* **30**, 1399 (1981).
2. Hebard A. F., Rosseinsky M. J., Haddon R. C., Murphy D.W., Glarum S. N., Palstra T. T. M., Ramirez A. P., and Kortan A. R., *Nature* **350**, 600 (1991).
3. Thess A., Lee R., Nikdaev P., Dai H., Petit P., Robert J., Xu C., Lee Y. H., Kim S. G., Rinzler A. G., Colbert D. T., Scuseria G. E., Tomanek D., Fisher J. E., and Smalley R. E., *Science* **273**, 483 (1996).
4. Winter M., Besenhard J. O., Spahr M. E., and Novak P. *Adv. Mater.* **10**, 7259 (1998).
5. Zhao J., Buldum A., and Lu J.P., *Cond-Mat*/9910143 preprint.
6. Leroux F., Méténier K., Gautier S., Francowiak E., Bonnamy S., and Béguin F., *J. Power Sourc.* **81-82**, 317 (1999).
7. Marvin G., Bousquet Ch., Henn F., Bernier P., Almairac R., and Simon B., *Chem. Phys. Lett.* **312**, 14 (1999).
8. Gao B., Kleinhammer A., Tang X. P., Bower C., Fleming L., Wu Y., and Zhou O., *Chem. Phys. Lett.* **307**, 153 (1999).
9. Zhou O., Fleming R. M., Murphy D. W., Chen C. H., Haddon R. C., Ramirez A. P., and Glarum S. H., *Science* **263**, 1744 (1994).
10. Suzuki S., and Tomita M., *J. Appl. Phys.* **79**, 3739 (1996).
11. Kohn W., and Sham L. J., *Phys. Rev.* **140**, A1133 (1965).
12. Bockstedte M., Kley A., Neugebauer J., and Scheffler. M., *Comp. Phys. Commun.* **107**, 187 (1997).
13. Hamann D.R., *Phys. Rev. B* **40**, 2980 (1989).
14. Arellano J. S., Molina L. M., Rubio A., and Alonso J. A., *J. Chem. Phys.* (to be published).
15. Amelinckx S., *et al*, *Science* **265**, 635 (1994).
16. Terrones M., *et al*, *Philos. Trans. R. Soc. London A* **354**, 2055 (1996).
17. Terrones M. and Terrones H., *Fullerene Sci. Technol.* **4**, 517 (1996).

Electronic Structure of Doped Fullerenes and Single Wall Carbon Nanotubes

S. Eisebitt, A. Karl, A. Zimina, R. Scherer, M. Freiwald,
W. Eberhardt,
F. Hauke[1], A. Hirsch[1], Y. Achiba[2]

Institut für Festkörperforschung, Forschungszentrum Jülich, 52425 Jülich, Germany
(1) Institut füt organische Chemie, Universität Erlangen-Nürnberg, 91054 Erlangen, Germany
(2) Department of Chemistry, Tokyo Metropolitan University, Hachioji, Tokyo 192-0397, Japan

Abstract. Soft x-ray absorption and emission spectroscopy is being used to study and compare the electronic structure of (a) potassium doped single wall carbon nanotubes (*interstitial* dopant), (b) La@C_{82} (*endohedral* dopant) and (c) ($C_{59}N)_2$ (*on-cage* dopant).

The electronic structure of pristine fullerenes and fullerene like carbon cage systems [1] exhibit a very pronounced dependence on the geometric structure of the carbon cage. For example, single wall carbon nanotubes (SWNTs) can be both metallic or semiconducting with band gaps ranging from meV to eV, depending on the chirality vector of the tube. It is of course intriguing to further engineer the electronic properties of a given fullerene by doping - in a similar way as in "classical" semiconductor technology. For a solid formed from a fullerene, there are three different possibilities on how doping might be achieved: the dopant can be *interstitial* in between the individual fullerene units, it can be trapped within the carbon cage as an *endohedral* dopant, and it can be part of the carbon cage as an *on-cage* dopant. In this paper we want to give a comparative overview on the influence of doping in three selected fullerene systems, namely (a) potassium doped SWNTs (interstitial) [2], (b) La@C_{82} (endohedral) [3] and (c) ($C_{59}N)_2$ (on-cage) [4]. Details on the experimental setup and results for each system will be published elsewhere.

We have investigated both the unoccupied and the occupied electronic states in an atom selective fashion by soft x-ray spectroscopy. The unoccupied states are probed by soft x-ray absorption (SXA), where a core electron is promoted into empty electronic states. The occupied states are investigated using soft x-ray emission (SXE) spectroscopy, by monitoring the radiative decay of a core vacancy previously prepared by SXA. Here, a valence electron fills the core vacancy and the emitted photons are energy analyzed using a Rowland-type grating spectrometer. In first approximation, if electron correlation effects and wavevector effects can be neglected [5], both techniques probe the local partial density of electronic states (LPDOS). In the context of the study of dopants, it is important to point out that SXE is a bulk sensitive, atom selective and local technique. It is therefore possible to study the electronic structure *e.g.* exclusively in the vicinity of the N atom in ($C_{59}N)_2$. Furthermore, sample charging does not disturb the photon-in photon-out based measurement. The experiments were

CP544, *Electronic Properties of Novel Materials—Molecular Nanostructures*, edited by H. Kuzmany, et al.
© 2000 American Institute of Physics 1-56396-973-4/00/$17.00

performed using synchrotron radiation at beamline BW3 at HASYLAB/DESY and beamline 7 of the ALS/LBL.

Potassium doped SWNTs

We have doped purified SWNT material ("buckypaper") from the Smalley group [2] with K using a SAES getter source in UHV. The doping process was monitored by recording SXA spectra containing the C 1s and K 2p absorption edges (not shown). The electronic structure of the SWNT carbon cages has been monitored by the C 1s SXA and SXE. The absorption and emission energy scales are referred relative to each other with high accuracy by calibrating the Rowland spectrometer with the elastically scattered light from the synchrotron beamline. In Fig. 1, we can therefore plot both experiments on a common energy scale. In both the SXA and SXE spectra, differences due to the doping can be observed.

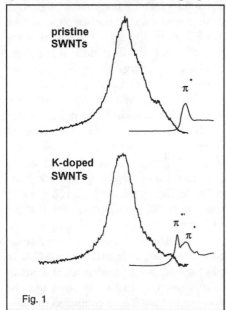

Fig. 1

For the pristine sample, the most prominent feature in the SXA spectrum in the vicinity of the absorption threshold is the π^* resonance, where the C 1s electron is promoted into unoccupied π states. In the doped sample, we observe two peaks π^* and $\pi^{*'}$. $\pi^{*'}$ is located at 1.25 eV lower absorption energy. During doping process, the π^* peak decreases continuously and the $\pi^{*'}$ peak emerges. We interpret these two peaks as being due to contributions of undoped (π^*) and doped ($\pi^{*'}$) material to the spectrum. (The energy shift between π^* and $\pi^{*'}$ is consistent with the different influence of screening on the C 1s core level and the core exciton formed after absorption into the π^* state. Similar shifts have been observed for C_{60} upon alkaline doping[6]) The smaller fwhm of $\pi^{*'}$ as compared to π^* suggests that the low energy states are being occupied by the doping electron, as expected. In the occupied states, the influence of doping can be observed by SXE without disturbing core excitonic effects, as there is no core hole present in the final state of the SXE process. The states close to the Fermi energy (E_F) which have been newly occupied due to the doping are directly visible in the SXE spectra as a peak at around 291 eV emission energy. We conclude that SWNTs can be doped by potassium in a way similar to the "traditional" doping in bulk semiconductors where the DOS at E_F is increased in the host material, in agreement with recent EELS results [7].

La@C82

In La@C_{82}, an important dispute regarding the electronic structure is the amount of the charge transfer from the La atom to the cage and the nature of the La-cage bond. It was demonstrated recently by resonant photoemission, that the La atom is not in the 3+ state but that some charge (~0.3 e) remains locally at the La atom and that the La-cage bonding involves hybridization and is thus not exclusively ionic [8]. SXE allows for a very direct check of this result, as it is an inherently local technique due to the required overlap with the core orbital in the dipole matrix element. In Fig. 2 we present an emission spectrum recorded after excitation with synchrotron radiation of 118.5 eV, which excites a La $4d^{-1}4f$ ^1P intermediate state.

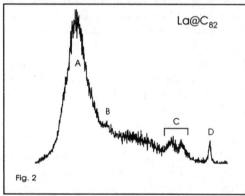

La@C_{82}

Fig. 2

Various peaks A-D can be observed in the radiative decay. Due to the presence of local excitations and strong electron correlation, the SXE spectrum does not represent the LPDOS in this case. The nature of the peaks A-C is well understood from excitation energy dependent emission studies on both La@C_{82} and LaF$_3$ as a reference compound and additional multiplett calculations on the La ion [9]. They are due to different $4d^{-1} \rightarrow 5p^{-1}$ core-to-core transitions, which we cannot discuss in this short overview article. Feature D, however, which we observe in La@C_{82} but not in LaF$_3$, is due to transitions involving valence band electrons localized at the La atom. Using the La 4d binding energy of 106.4 eV in La@C_{82} as measured by photoelectron spectroscopy [8,10] we can conclude that the valence band states giving rise to feature D are located at about 1 eV binding energy, in agreement with the resonant photoemission results [8]. Our result is therefore an independent and very direct proof of the incomplete charge transfer in La@C_{82}. In order to quantify the amount of charge located at the La atom, model calculations of the transition matrix element are needed. Finally we would like to mention that the relative spectral weight of feature A is much stronger in La@C_{82} as compared to LaF$_3$. This feature comes about from a SXE decay which takes place after the excited 4f electron has delocalized into the continuum and an electron from the ligand (F or C_{82}) has been captured onto the La ion during the 4d hole lifetime [9]. The observed difference in spectral weight reflects a faster electron transfer time in La@C_{82}, which is indicative of the partially covalent character of the bond.

$(C_{59}N)_2$

The azafullerene $(C_{59}N)_2$ can be conceived as a special C_{60} dimer, where one carbon atom on each buckyball has been replaced by a nitrogen atom. As the N atom has one electron more than a carbon atom, this situation is sometimes been referred to as "on-cage doping". However, it was shown recently by means of electronic structure calculations and EELS spectroscopy [11], that the extra electron is predominantly localized at the N atom. We have carried out a study of the local electronic structure at the N atoms in $(C_{59}N)_2$ as compared to the electronic structure on the carbon cage using SXA and SXE. For the unoccupied states (not shown), we observe no significant empty DOS at the N atom corresponding in energy to the low-energy carbon cage t_{1u} states, in agreement with the EELS results [11]. The C 1s and N 1s SXE spectra from $(C_{59}N)_2$ are presented in Fig. 3, allowing us to study the occupied electronic states atom selectively. The spectra are referred to a common binding energy scale by subtracting the respective core level binding energies from the SXE emission energies.

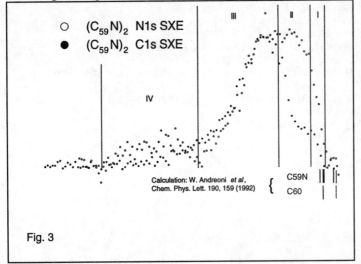

Fig. 3

The core level binding energies were determined by x-ray photoelectron spectroscopy (not shown) and are in agreement with the values reported in Ref.11. At the valence band maximum (region I), we observe a clear difference between the C1s and N1s spectra. The DOS locally at the N atom extends to lower binding energies than the C DOS. This small but significant difference is in good agreement with results from electronic structure calculations of $C_{59}N$ and C_{60} [12], which we have reproduced at the bottom of the figure (solid bars: occupied states, dashed bares: empty states). The calculations show that the N-induced highest occupied molecular orbital (HOMO) in $C_{59}N$ is 1.35 eV higher in energy than the HOMO in C_{60}. We are therefore directly observing the influence of the doping electron at the N site. Our observations are experimental proof for the statement that the electron is mainly localized at the N site. The "extra" electron introduced by the N atoms can therefore *not* act like a doping electron which can contribute significantly to the conductivity in the system. The intensity differences between the two spectra in regions II and IV are due to N 2p and distortion induced C 2s contributions, respectively, and will be discussed – along with other results on $(C_{59}N)_2$, La@C_{82} and K-doped SWNTs - in the forthcoming publications in more detail.

ACKNOWLEGMENTS

We thank the authors of Ref. 2 and in particular R.E. Smalley and J.E. Fischer for the supply with purified buckypaper. We also thank M. Waiblinger and A. Weidinger for the preparation of $(C_{59}N)_2$ films and D. Schondelmaier and S. Cramm for their help with the SWNT measurements.

REFERENCES

[1]For the ease of description, we adopt a nomenclature where SWNTs are also referred to as "fullerenes".
[2]A.G. Rinzler *et. al.* Appl. Phys. A **67**, 29 (1998)
[3]K. Kikuchi *et. al.*, Chem. Phys. Lett. **216**, 67 (1993)
[4]B. Nuber and A. Hirsch, Chem. Commun. **1996** 1421
[5]S. Eisebitt and W. Eberhardt, J. El. Spec. Rel. Phen. ≥**107** (2000)
[6]A.J. Maxwell *et. al.*, Chem. Phys. Lett **247**, 257 (1995)
[7]T.Pichler, M. Sing, M. Knupfer, M.S. Golden and J. Fink, Solid State Comm. **109**, 721 (1999)
[8]B. Kessler *et. al.*, Phys. Rev. Lett. **79**, 2289 (1997)
[9]Ph.D. thesis Jan Lüning, Reports of the FZ-Jülich Jül-3544, ISSN 0944-2952
[10]Barbara Kessler, private communication
[11]T. Pichler *et al.*, Phys. Rev. Lett. **78**, 4249 (1997)
[12]W.Andreoni, F. Gygi, M. Parinello, Chem. Phys. Lett. **190**, 159 (1992)

Multiple-Shell Diffusive Conduction in Multiwalled Carbon Nanotubes

P.G. Collins, R. Martel, K. Liu, P. Avouris

IBM T.J. Watson Research Center, Yorktown Heights, NY 10598, USA

Abstract. Multiprobe electrical transport measurements on individual multiwalled carbon nanotubes (MWNTs) are described. Data collected from many samples suggest that MWNTs are diffusive conductors with resistances dependent on both length and diameter. Furthermore, our measurements reveal room temperature nonlocal voltages, and combinations of semiconducting and metallic behaviors within a single MWNT. Both types of observations indicate conduction through more than one current path. A simple model in which two or more weakly coupled shells conduct in parallel is consistent with all of the observed behaviors.

Electrical transport in multiwalled carbon nanotubes (MWNTs) is a subject of significant current interest and debate. Unlike the simpler case of single-walled nanotubes (SWNTs), little consensus exists on the details of transport in MWNTs. On one hand, conductance quantization effects have been attributed to ballistic conduction in the single outer shell of a MWNT [1]. On the other hand, weak localization and Aharonov-Bohm oscillations in the low temperature magnetoresistance determine a coherence length for MWNTs indicative of diffusive transport [2], a result supported by recent scanning probe studies [3]. Analysis of the Aharonov-Bohm oscillations further suggests that MWNTs conduct only with their outer shell. On this point, the experimental results disagree with theoretical studies of MWNT conduction, which conclude that significant coupling between shells should exist and that more than one shell should contribute to transport [4,5].

Here, we use four-probe resistance measurements over a wide temperature range to explore the mechanism of MWNT transport. As shown below, these measurements support a diffusive mechanism for transport in MWNTs and indicate that more than one carbon shell participates in transport. In fact, our results provide evidence for participation by both semiconducting and metallic shells in the same tube.

The MWNTs used in this study were grown by standard arc discharge methods. The soot was ultrasonically dispersed in dichloroethane, centrifuged to separate the nanotube-containing portion, and then applied to Si wafers with prefabricated Au electrodes. Nanotube diameters were determined by atomic force microscopy (AFM).

Electrical resistance measurements were performed on individual MWNTs with a variety of contact configurations. Contact resistances of 1 - 10 kΩ were typically obtained, as determined by combinations of two-probe (2P) and four-probe (4P) measurements. The electrode spacings L were fabricated to range between 100 to 400

CP544, *Electronic Properties of Novel Materials—Molecular Nanostructures*, edited by H. Kuzmany, et al.
© 2000 American Institute of Physics 1-56396-973-4/00/$17.00

nm, allowing a true resistivity to be estimated from multiple measurements on the same nanotube. For example, three independent 4P resistances can be measured for a MWNT laying across six electrodes, and scaling these measurements by the different spacings L determines whether the nanotube has a constant resistance per unit length.

For a particular nanotube, the scaled resistances are generally in good agreement with each other. A small range in values can be attributed to uncertainties in L due to the electrode width and correspondingly ill-defined contact point. Notable exceptions to the rule also occur, especially when AFM images show contaminants on a particular nanotube segment.

Figure 1 depicts the MWNT 4P resistance per micrometer, as described above, plotted as a function of nanotube diameter D_t. The figure indicates a clear dependence on D_t, in contrast to the constant resistance expected for ballistic 1D conductors. The finite resistance of the MWNTs indicates that transport is not ballistic in our experiments, and this conclusion is amplified by the observed dependence of the resistance on both the length and diameter of the tubes.

The solid line in Fig. 1 shows a fit to the function $R = \rho L/(\pi D_t)$, with a resistivity $\rho = 700\,\Omega_\triangledown$. This fit suggests that the MWNT acts like a classical, two dimensional conductor with length L and width πD_t. However, given the size of the typical D_t, one would expect a MWNT to be further quantized into 1D electronic subbands having spacings ΔE greater than kT, even at room temperature. Analysis of the temperature dependence of MWNT resistivity further supports a quasi-1D interpretation [2]. In fact, 1D transport can also show the observed diameter dependence because the electronic density of states $N(E)$ at the Fermi level is proportional to D_t^{-1} in a carbon nanotube. A simple argument using Fermi's golden rule gives an elastic mean free path $l_e \propto N(E_F)^{-1} \propto D_t$. With a mean free path proportional to diameter, MWNT resistance should in fact be inversely proportional to diameter, as obseved in Fig. 1.

The origin of backscattering in MWNT conductors could arise from various sources, including the coupling between shells [4,5]. This intrinsic effect should affect most MWNTs but is difficult to experimentally confirm. Below, we describe experiments which indirectly suggest multiple shell conduction in MWNTs. The first is a measurement of nonlocal voltages in individual tubes. The second is a case of simultaneous semiconducting and metallic behaviors in a single MWNT.

Figure 2a shows a measurement of nonlocal voltages in a MWNT. The 2P I-V

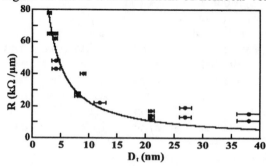

FIGURE 1. Four-probe resistance of various MWNTs plotted as a function of nanotube outer diameter D_t. Multiple points at the same diameter indicate resistances measured on different segments of the same MWNT and scaled by the segment length. The solid line is a fit to a $1/D_t$ dependence as described in the text.

characteristics are measured for segment BC, and independently but simultaneously voltages across other segments are also monitored (the voltage probes have input impedances > 100 MΩ and leak negligible currents). For a simple 1D conductor, the voltage drops in segments AB or DE must be zero. Instead, these segments exhibit large potential differences directly proportional to the total applied current. Although Fig. 2 depicts a specific measurement, the results are reproducible and consistent among all configuration possibilities for a particular sample. Many MWNTs measured have exhibited similar effects, though not always of the same magnitude.

Nonlocal voltages observed in multi-probe measurements can be generated by either classical or quantum effects. Quantum nonlocality arises in systems with long coherence lengths and is most often observed at very low temperatures [6]. Classical nonlocal voltages occur in systems possessing multiple current pathways and are particularly evident in anisotropic materials. The present measurements show minimal temperature dependence for 77K $<$ T $<$ 300K, so a classical model is appropriate.

The idealized MWNT, as a simple 1D conductor, seems an unlikely candidate to exhibit classical nonlocal effects, but electrical coupling between concentric shells can turn the conductor into a complex network. Fig. 2b depicts a schematic of the measurement and the simplest lumped element model of a multi-conductor MWNT: external electrodes contact the outermost shell of the MWNT with contact resistances Rc, and this outer shell electrically contacts an inner core (of at least one additional shell) via shell-shell coupling resistances Rss. With nonzero Rss values, the MWNT is changed from a simple linear conductor to a resistor network with multiple current paths. Furthermore, nonlocal voltages as seen in the experimental data can be easily generated by the sole requirement that the coupling Rss varies along the length of a MWNT. Sources of variation in the coupling could include enhancements near defects or at tube end caps, and in fact nonlocal voltages have been previously reported for MWNTs damaged with an ion beam [7]. For our undamaged tubes, the measured potentials indicate strong coupling at the nanotube ends: the voltage polarity

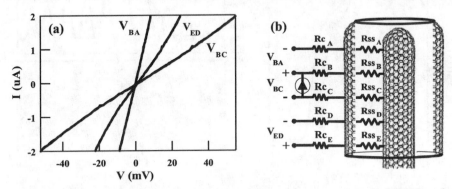

FIGURE 2. (a) Nonlocal voltages observed during a two-probe measurement of a MWNT are large and linear in the applied current. As indicated in the schematic (b), current flows only through electrodes b and c, but voltages are generated on all segments of the nanotube due to coupling between shells. The schematic indicates the contact resistances Rc between the electrodes and the MWNT outer shell, as well as resistors between shells to represent the coupling.

indicates current flow from electrode B towards A and from E towards D. Unfortunately, it is difficult to quantitatively determine the current fractions which flow in different segments without an independent measure of Rss.

Finally, we note that the MWNT in Fig. 2 exhibits linear I-V characteristics and "reasonable" resistance values both in 2P and 4P configurations. Therefore such characteristics by themselves are insufficient to determine whether a particular MWNT supports nonlocal voltages or multiple shell conduction. And while the measurement of nonlocal voltages strongly supports a multiple shell model, the absence of such voltages cannot be considered proof of single-shell conduction.

A second observable effect of multi-shell conduction should be the influence of different types of shells within a single MWNT. Theoretically, the electronic behavior of a single nanotube shell depends on the geometric diameter and chirality of the tube. A MWNT is comprised of many concentric graphene shells, each of which may be (semi)metallic or semiconducting. Therefore a MWNT with multiple conducting shells should show the distinctive signatures of each participating shell. For example, a clear manifestation of semiconducting behavior is the ability to modulate the MWNT conductivity under the electrostatic influence of a gate electrode. In our experiments, the underlying, heavily-doped silicon substrate is used as a back gate.

Figure 3 shows the low bias conductance of a 4 nm diameter MWNT at 4.2 K as a function of gate voltage. As observed in the case of SWNTs [8-10], the gate electrode very effectively modifies the MWNT. A gate potential of +2V depletes the MWNT of its hole carriers and shifts the Fermi level into the semiconducting gap. However, unlike in the case of SWNTs where the current modulation exceeds five orders of magnitude [8], the gate decreases the MWNT conductance by only two orders of magnitude. The large residual conductance of ~0.1 µS indicates that some conduction channel(s) remains open, even though the semiconducting tube has been depleted.

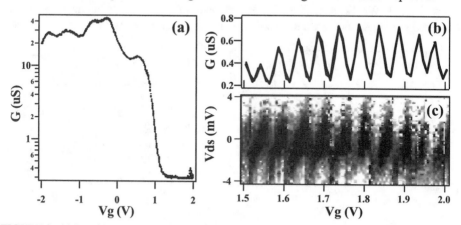

FIGURE 3. (a) Low bias conductance of a MWNT shows switching induced electrostatically by a gate bias Vg. The depletion of carriers at Vg > 1V indicates the MWNT is a p-type semiconductor. Higher resolution data of the same MWNT in the depletion region shows (b) conductance oscillations periodic in the gate bias and (c) a diamond-shaped greyscale plot of the conductance characteristic of Coulomb blockade in a quantum dot. Dark regions are conductance minima. All data are acquired at T = 4.2K.

Furthermore, closer inspection of the depletion region reveals a weak oscillatory structure in the conductance, an expanded view of which is shown in Fig. 3b. This oscillation corresponds to a periodically gapped behavior in the I-V characteristics. A series of such I-V characteristics, obtained as a function of gate voltage, are shown in the greyscale plot of Fig. 3c. Although surprising for a semiconducting nanotube, such oscillatory behavior is characteristic of transport through a metallic island in the Coulomb blockade regime, and is commonly observed in metallic SWNTs [9,10].

A simple model consisting of a semiconducting shell and a metallic shell in parallel quite easily explains the observed conductance behavior of this MWNT. In this model, the external electrodes contact and directly probe the semiconducting shell, while indirectly probing the inner metallic shell via shell-shell coupling Rss as described above. The residual conductance of ~0.1 µS measured when the semiconducting shell is depleted is then due to conduction through the metallic inner shell, limited by Rss. A large value of Rss is necessary to observe this effect, since it both allows the modulation of the semiconducting shell to be observable and also provides the tunnel barriers which define a quantum dot out of the metallic shell. Smaller values of Rss would allow metallic layers to short out any semiconducting shells. Since semiconducting MWNTs are rarely observed experimentally, small values of Rss might be quite common in MWNTs.

Fig. 3c indicates a maximum gap in the conductance of approximately 6 meV in the applied source-drain bias. In a Coulomb blockade system, this gap is the energy required to overcome the capacitative charging $\Delta E = e^2/2C$. By ignoring other coupling effects and focusing merely on the geometry of the tube, the capacitance and therefore the length L of the charging tube segment can be calculated from the 6 meV charging energy and the tube diameter of 4 nm. We obtain L ~ 200 nm, which is nearly equal to the 150 nm spacing between electrodes in this measurement. The agreement suggests that the entire region between the electrodes constitutes a coherent quantum dot, as reported for metallic SWNTs [9].

In conclusion, a simple, single shell model cannot account for the variety of behaviors observed in MWNTs. We conclude that conduction can involve multiple shells, and that this intershell coupling may contribute to the diffusive conduction.

REFERENCES

1. S. Frank, *et al., Science* **280** 1744 (1998).
2. A. Bachtold, *et al., Nature* **397** 673 (1999).
3. A. Bachtold, *et al.,*to be published. Preprint cond-mat/0002209.
4. Y. K. Kwon, D. Tomanek, *Phys. Rev. B* 58 16001 (1998).
5. S. Roche, F. Triozon, A. Rubio, D. Mayou, to be published.
6. S. Washborn, R. A. Webb, *Adv. In Phys.* 35 375 (1986).
7. T.W. Ebbesen, *et al., Nature* **382** 54 (1996).
8. S. J. Tans, *et al., Nature* **393** 6680 (1998); R. Martel, *et al., App. Phys. Lett.* **73** 2447 (1998).
9. S. J. Tans, *et al., Nature* **386** 474 (1997); A. Bezryadin, *et al., Phys. Rev. Lett.* **81** 4036 (1998).
10. M. Bockrath, *et al., Science* **275** 1922 (1997).

Superconductivity and normal state resistance of carbon nanotubes

A.Yu. Kasumov (1,2), R. Deblock (1), M. Kociak (1), B. Reulet (1), H. Bouchiat (1), S. Guéron (1), I.I. Khodos (2), Yu.B. Gorbatov (2), V.T. Volkov (2), C. Journet (3), P. Bernier (3), M. Burghard (4).

(1) Lab. de Phys. des Solides, Associe au CNRS, Bat 510, Universite Paris-Sud, 91405, Orsay, FR.
(2) Institute of Microelectronics Technology and High Purity Materials, Russian Academy of Sciences, Chernogolovka 142432 Moscow Region, Russia.
(3) GDPC University of Montpellier, France.
(4) Max Plank Institute for Solid State Research, Stuttgart, Germany.

Superconductivity in molecular systems (inorganic and organic) has been studied for 25 years on macroscopic samples consisting of a huge amount of molecules [1]. The discussions about the mechanism of superconductivity in these systems are still in progress. Discovery of a proximity effect in carbon nanotubes, *i.e.* the induction of superconductivity in a non-superconducting material that is connected to superconducting leads [2,3], has allowed to begin study of superconductivity in individual molecules. A single-walled carbon nanotube (SWNT) is a molecular wire with diameter about 1 nm and when connected its room temperature (RT) resistance is larger than the quantum of resistance $R_Q = h/(2e)^2 \approx 6.5$ kΩ (h - Plank constant, e - charge of an electron). There are theories stating that superconductivity is impossible in wires with normal resistance R_N larger than R_Q [4]. Recently these theories have received experimental confirmation [5]. In this paper we show data on single SWNTs which have $R_N \gg R_Q$ (up to $R_N \approx 10~R_Q$) and nevertheless undergo a proximity-induced superconducting transition.

The sample fabrication, described elsewhere [6], consists in the laser soldering of a suspended nanotube to superconducting electrodes, using Au or In. Because it is suspended between the edges of the slit in a membrane (fig.1), the nanotube can be imaged in a high resolution transmission electron microscope (HRTEM). The study in HRTEM shows that the nanotubes do not contain catalytic particles and drops of a melt with diameter ≥ 1 nm. Particles of smaller dimensions and individual atoms are outside detection range, and it is quite possible that the nanotubes contain impurity atoms of catalyst (Y, Ni) and solder (Au, In). Concerning atoms of catalyst, it was found that their average density is 1 atom every 4 nm of length of a tube, i.e. 1 atom out of 500 carbon atoms [7]. Outside detection range there are also impurity atoms (O, N, H) adsorbed on the surface of nanotubes, and nanotube structure defects (5-7 defects, vacancies). As a result of laser soldering the gas impurities can partially desorb, and part of natural defects can anneal. Thus a real nanotube is a nonhomogeneous conductor which contains defects and differs from the ideal model

CP544, *Electronic Properties of Novel Materials—Molecular Nanostructures*, edited by H. Kuzmany, et al.
© 2000 American Institute of Physics 1-56396-973-4/00/$17.00

(fig.1). Nevertheless the resistance of a real nanotube can be close to ideal value of 6.5 kΩ as the scattering on defects is strongly supressed [8]. Due to this fact ballistic transport in some tubes is possible even at room temperature [9]. We have also measured for some samples an almost ideal RT resistance value: $R \sim R_Q$ for individual SWNTs and MWNTs (multi-walled nanotubes), and $R \sim R_Q/N$ for ropes of SWNTs (here N is the number of tubes in a rope).

However the resistance of an overwhelming majority of samples largely exceeded the ideal value. Since 1994, we have measured the RT resistance of more than 3000 individual MWNTs, SWNTs, and ropes of SWNTs. Only about 30 samples had resistance close to ideal. It was possible to select nanotubes with low resistance quite fast due to our unique deposition technique [6]. Deposition of a nanotube, measurement of its resistance, removal if its resistance is too large and deposition of a new tube take about 5 minutes. Proximity induced superconductivity was observed [2] in selected nanotubes with low resistance.

We have also tried to create superconducting nanowires by deposition of thin layer (no more than 3 nm) of the metals such as In, Sn, Re, Mo-Ge on top of small ropes of SWNT. Surprisingly we always observed an increase of the resistance of the sample (SWNT rope plus metal) after deposition and we never detected any superconducting transition down to 1.5 K. Sometimes the deposition was destroying ropes. TEM observations of these samples always indicated a granular structure with crystallites in the nanometer range.

Here we present low temperature measurements of 3 individual SWNT mounted on Ta/Au/In superconducting contacts. There RT resistances are R_N=9 kΩ, 28 kΩ and 41 kΩ. For correct interpretation of transport measurements it is crucial to know what part of the tube resistance is due to the contacts: for a ballistic tube the entire resistance is due to the contacts, and in case of bad (low transmission) contacts the resistance can largely exceed R_Q [9]). In our case it is known for tubes SW1 and SW3. Tube SW1 was exposed to the electron beam of an electron microscope. The area of the contacts was shielded by a layer of membrane and metal opaque to electrons with an energy of 100 kV, so that only the part of the tube outside contacts (fig.1) was irradiated. The resistance of this sample after irradiation increased to 20 kΩ due to the radiation defects, and it is certain that this part of resistance (about 11 kΩ) is not due to contacts. The measurements presented here were obtained after this modification of the sample. Sample SW3 was exposed to the 30 kV Ga+ ion beam of a focused ion beam machine. As in the case of the electron beam, the ions did not reach the contact area, being "stuck" in the shielding membrane. The extremely small radiation dose: 10^{13} ions/cm^2 corresponded to 3 ions flown through a nanotube. The irradiation resulted in essential decrease of resistance of the sample down to 13 kΩ, apparently owing to the desorption of impurity atoms (in [10] all nanotubes were irradiated by Ga ions with a dose of $2 \cdot 10^{14}$ ions/cm^2, and their resistance remained very low, as low as 200 Ω). This definitely proves that the most part of resistance of SW3 before irradiation, 41 kΩ, is not due to contacts. Therefore the samples are non-homogeneous one-dimensional (1D) conductors, which resistance is determined by contacts and internal defects. For sample SW2 we have no direct measurement of the influence of defects on resistance.

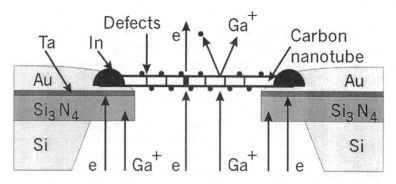

FIGURE 1. Schematic of a real nanotube with defects, modified by electron (e) and ion (Ga⁺) beams. Defects - adsorbed atoms and structural ones are shown as black dots and partitions correspondingly. On the picture are also schematically shown the desorption of and atom by Ga⁺ ion, the creation of a structure defect by e-beam, and shielding effect of Si_3N_4 membrane.

The measurement of temperature dependence of resistance shows that all three of the samples undergo a proximity-induced superconducting transition with critical temperature Tc ≈ 0.5 K, which is the Tc of the superconducting electrodes (fig. 2). This is the main result of this paper. Moreover besides the R(T) dependence, the magnetic field dependence of the resistance (fig.2), and the differential resistance Vs current curves (fig.3) definitely show the superconducting character of the transition. For samples SW1 (R(T) dependence for this sample was presented in [2]) and SW2 the transition is complete and the resistance drops from 26 kΩ down to less than 1 Ω, defined by the sensitivity of measurements (I = 1 nA, V = 1 nV). Supercurrents of the order of 100 nA are measured [2]. The resistance of sample SW3 does not go to zero, but saturates at a value of 1.5 kΩ.

This induced superconductivity in 1D conductors with $R_N \gg R_Q$ seems to be in contradiction with theoretical predictions. This could be due to the fact that the nanotubes cannot be considered as homogeneous wires. Also, the theory [4] predicts that superconductivity cannot exist in homogeneous diffusive wires of diameter less than 10 nm. However superconductivity in such wires built in porous media (glasses, zeolites and asbestoses) was investigated more than 30 years ago, and a superconducting transition was observed even for wires with diameter about 2 nm [11].

In addition, we note that superconducting single electron transistors, consisting of two Josephson junctions in series with $R_N \approx 50$ kΩ, are another example of systems through which supercurrents flow even though the normal state resistance of the junctions are larger than R_Q [12]. In the case of Josephson junctions, it is known that the electromagnetic environment of the junctions must be taken into account to describe the magnitude of the maximum supercurrent. It is therefore probable that in a similar manner, the normal state resistance of the nanotubes may not be the only parameter which determines whether or not a superconducting transition is possible, but that the circuit in which the nanotube is embedded probably plays a role as well.

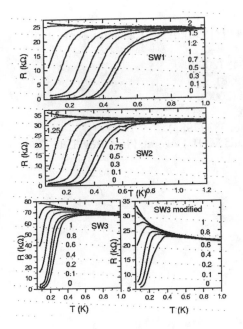

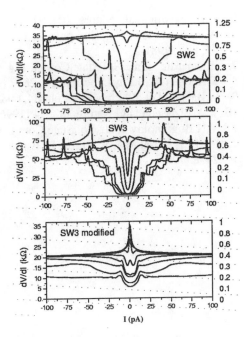

FIGURE 2. Resistance as a function of temperature for all samples, in various magnetic fields (fields are indicated in Teslas to the right of the curves). SW1 is 0.3 μm long and is mounted on Ta/Au contacts. RT resistance of SW1 was 9 kΩ after deposition and before TEM observation, and 20 kΩ afterwards. SW2 is 0.15 μm long and is mounted on Ta/Au/In contacts. SW3 is 0.2 μm long and is mounted on Ta/Cr/Au/In contacts.

FIGURE 3. Differential resistance as a function of current for samples SW2, SW3 and SW3m, at T=70 mK. (Fields are indicated in Teslas to the right of the curves).

REFERENCES

1. R.L.Greene, et.al., Phys.Rev.Lett. 34, 577 (1975); D.Jerome, et.al. J.Phys.(Paris)Lett., 41, L95 (1980).
2. A.Yu.Kasumov, et.al, Science 284, 1508 (1999).
3. A.F.Morpurgo, et.al., Science 286, 263 (1999).
4. G.Schon, Nature 404, 948 (2000); A.D.Zaikin et al.,Usp.Fiz.Nauk 168, 244 (1998); A.D.Zaikin et al., Phys.Rev.Lett. 78, 1552 (1997).

5. A.Bezryadin et al., Nature, 404, 971 (2000).
6. A.Yu.Kasumov et al., Europhys.Lett. 34, 429 (1996); A.Yu.Kasumov, et al., ibid. 43, 89 (1998).
7. E.Dujardin, et al., Sol.State.Comm. 114, 543 (2000).
8. T.Ando et al., J.Phys.Soc.Jpn. 67, 2857 (1998).
9. A.Bachtold et al., Phys.Rev.Lett. 84, 6082 (2000).
10. T.W.Ebessen et al., Nature 382, 54 (1996).
11. V.N.Bogomolov et al., Sol.State.Comm. 46, 383 (1983).
12. P.Joyez, P. Lafarge, A. Filipe, D. Esteve, and M. H. Devoret, Phys. Rev. Lett 72, 2458 (1994).

Tuning the Electronic Properties of Single Wall Carbon Nanotubes by Chemical Doping

P. Petit*, E. Jouguelet*, C. Mathis*, and P. Bernier[†]

*Institut Charles Sadron, 6, rue Boussingault, 67000 Strasbourg, France
[†]GDPC, Université Montpellier II, CC26, 34095 Montpellier Cedex 05, France

Abstract. The Fermi level carbon nanotubes can be tuned by redox reactions with solutions of radical-anions with Li^+ as a counter ion. This allows for selctively filling or depleting their density of states. The modifications of the electronic bands that occur upon charge transfer are probed by optical absorption spectroscopy performed on thin films. The bulk material can be reversibely doped by these redox reactions, and the induced shift of the Fermi level determines both the electrical conductiviy and the chemical composition of the material.

INTRODUCTION

High yield synthesis of carbon single wall nanotubes (SWNT) [1,2] has made it possible to investigate the electronic properties of macroscopic amounts of material [3,4] and to show the possibility to decrease its resistivity by more than one order of magnitude when exposed to potassium or bromine vapours [5]. By tuning the Fermi level of SWNT upon chemical doping, we show that it is possible to fill or deplete selectively the density of states (DOS) of the different kinds of tube (semiconducting and metallic) that constitute the material. The undergoing modifications of the electronic structure of SWNT are directly monitored by optical absorption spectroscopy performed on thin films [6]. The same redox reactions between radical ion salts of organic molecules and the bulk material also allows for controling the electronic properties of SWNT. The use of macroscopic amounts of SWNT allows to determine the number of electrons transferred onto the sample, leading to the relations between both conductivity and charge carrier concentration and the Fermi level of SWNT [7]. All the reported experiments and their interpretations lie on the fact that the redox potential and the Fermi level are identical, except for the reference level [8].

EXPERIMENT

Both SWNT synthesized by the electric arc and laser vaporization techniques [1, 2] were used in the present study.

CP544, *Electronic Properties of Novel Materials—Molecular Nanostructures*, edited by H. Kuzmany, et al.
© 2000 American Institute of Physics 1-56396-973-4/00/$17.00

To study the change in the electronic structure of SWNT upon doping, thin films were prepared by spraying a homogeneous suspension of SWNT in ethanol on a quartz substrate heated with hot air. The quarz plate supporting the SWNT thin film was inserted in a quartz optical cell connected to a U-shapped 'nss tube where the second branch consists of an ampoule containing neutral molecules in solution in pure THF and a piece of lithium. The whole apparatus was sealed under high vacuum. Electron transfer from lithium to organic molecules results in the radical-anion form of the molecule with Li+ as the counter ion. This experimental setup allows the doping of the sample by bringing the solution into contact with the film. Once the doping was achieved, the excess of doping solution was removed, the sample rinced by internal distillation of the solvent and then dried by cooling the ampoule at −70°C.

The material used for studying the electronic properties was purified by successive acidic treatments, cross flow filtrations and vacuum annealing, leading to negligible amounts of carbonaceous particles and metal catalysts with respect to SWNTs [9]. The samples used in the present study, typically of the order of 1 to 2 mg, were cut in the same bucky paper of about 15 mg resulting from the purification process. The experiments were performed in a glass apparatus sealed under high vacuum. The sample was fixed between copper jaws connected to airtight electrical feedthroughs allowing to check the redox reaction completion by measuring the change in the conductivity of the sample. A quartz optical cell connected to the glass apparatus was used to determine the change in the concentration of radical-anions induced by the redox reactions, e. g. the number of electrons transferred from the solution to the sample and thus the chemical composition LiC_x. As we always used an excess of reactive molecules compared to the number of carbon atoms, when a redox reaction is complete, the redox potential of the sample is that of the reactive molecules.

The molecules used for the present study were the radical-anions of naphthalene, benzophenone, fluorenone, anthraquinone and benzoquinone, with Li^+ as a counter ion, of respective redox potentials of 2.5, 1.8, 1.3, 0.85 and 0.4 eV.

RESULTS AND DISCUSSION

Doping Effect on the Electronic Structure of SWNT

The three sets of optical absorption bands observed in the optical spectrum of the pristine sample (figure 1, curve a) are identical to those reported by Kataura et al. [10]. The first two features at 0.68 and 1.2 eV originate from band gap transition in semiconducting tubes, whereas the third at 1.7 eV originates from metallic tubes.

All the features characteristic of the pristine material are removed after the sample is exposed to a solution of naphthalene-lithium (figure 1, curve e), providing evidence that, by setting the redox potential of SWNT to 2.5 eV, the initially empty states of both semiconducting and metallic tubes are filled, suppressing optical transition between mirror spikes in the DOS.

Doping the sample with fluorenone-lithium removes the absorption bands of energy lower than 1.3 eV. Thus, the effect of setting the redox potential of SWNT to 1.3 eV is to fill up the first and second peaks in the DOS of semiconducting tubes characterized by an energy gap lower than 1.3 eV. This experiment leads to an estimation of the redox potential of SWNT to 0.7 eV.

Exposure of the material to anthraquinone-lithium leads to the disappearance of the band at 0.68 eV, the other optical transitions being unaffected. A second estimation of the redox potential of SWNT of 0.5 eV is drawn from this experiment, close to the value reported for C_{60} [11].

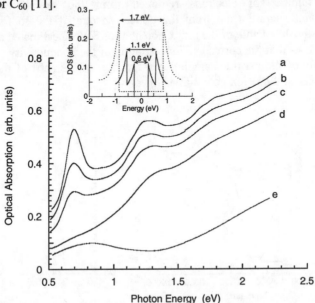

FIGURE 1. Evolution of the absorption spectrum of pristine SWNT (a) upon exposure to molecules of different redox potentials. Curve (b) and (c) show the progressive disappearence of the first optical absorption transition after exposure of the sample to anthraquinone-lithium for 1 sec. (b) and for 10 sec. (c). Curve (d) and (e) are the optical spectra of the sample after exposure to fluoreneone-lithium and naphthalene-lithium respectively. Inset: Schematic representation of the DOS of semiconducting tubes (solid line) and metallic tubes (dashed line).

These experimentsshow that , by choosing appropriate molecules, it is possible to modify selectively the Fermi level of SWNT and thus to modify the conducting nature of the individual tubes.

Doping Effects on the Electronic Properties of SWNT

The effect of doping SWNTs with the radical-anion of naphthalene on the resistance of a sample is shown in figure 2. As soon as the doping solution is brought into contact with the sample, its resistance drops, then slightly decreases and reaches a plateau when total charge transfer is achieved. The resistance of the doped sample is 1/15 that of the pristine sample and its chemical composition, determined from the change in the optical absorption spectrum of the doping solution before and after the reaction, corresponds to 1 lithium per 6 carbon atoms.

The reversibility of the doping using redox reactions is shown in figure 3. The sample previously doped with the radical-anion of naphthalene was successively exposed to solutions of neutral molecules of benzophenone, fluorenone and benzoquinone. For each "de-doping" reaction, we recorded the change in the resistance and the number of electrons removed from the sample was estimated at thermodynamic equilibrium from the optical spectra of the so formed radical-ion. Using the absolute value of the redox potential of SWNT estimated above, the inset in figure 2 shows that the corresponding variation of the conductivity (normalized to the conductivity of the pristine sample) as a function of the shift of the Fermi level (or equivalentely of the redox potential) is linear.

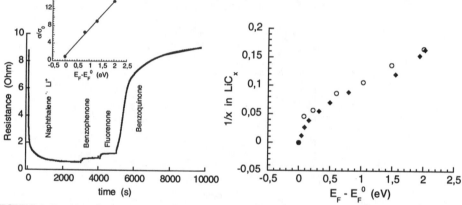

FIGURE 2. Doping and "de-doping" of SWNT by different organic species.

FIGURE 3. Energy dependence of the charge carrier concentration per carbon atom for two different samples (after [7]).

To establish a more accurate charge carrier concentration dependence with the redox potential, we exposed step by step a sample to a solution of naphthalene-lithium and recorded both the resistance and the charge transfer at each step at thermodynamical equilibrium. Using the linear dependence of the conductivity with the redox potential, we converted the conductivity axis into energy. The observed monotonous increase (figure 3) shows a one to one map between energy and charge concentration indicating that there is no stage of intercalation in SWNT. Furthermore, although this curve contains a discrete set of data points and taking into account the experimental errors, there is apparently no evidence of the expected features originated by the spikes in the DOS. This contradicts a priori theoretical calculations which have shown that van Hove singularities are present in the DOS of a rope of [10,10] tubes [12]. On the other hand, the ropes in the real material are made of tubes of different diameters and chiralities. Furthermore, X-ray diffraction and TEM experiments have shown that intercalation of alkali metals leads to the loss of the triangular lattice and that there is no long range ordering of the alkali metal decorating the tubes. Tube-tube interactions are then more complex than those used for calculations and if the tubes are not orderly decorated, the loss of long range ordering could explain the absence of van Hove singularities in figure 3 [13].

CONCLUSION

Single wall carbon nanotubes can be reversibly doped by redox reactions with no stages of intercalation. We established and quantified the one to one map between both Fermi level, charge carrier concentration and electrical conductivity and show the modifications undergone by the electronic bands upon doping.

ACKNOWLEDGMENTS

We are grateful to A. G. Rinzler and R. E. Smalley for supplying purified SWNT material.

REFERENCES

[1] Thess, A., Lee, R., Nikolaev, P., Dai, H., Petit, P., Robert, J.,Xu, C., Lee, Y. H., Kim, S. G., Rinzler, A. G., Colbert, D. T., Scuseria, G. E., Tomanek, D., Fischer, J. E., and Smalley, R. E., *Science* **273**, 483-, (1996).
[2] Journet, C., Maser, W. K., Bernier, P., Loiseau, A., Lamy de la Chapelle, M., Lefrant, S., Deniard, P., Lee, R., and Fischer, J. E., *Nature* **388**, 756-,(1997)
[3] Fischer, J. E., Dai, H., Thess, A., Lee, R., Hanjani, N. M., Dehaas, D. L., and Smalley, R. E., *Phys. Rev. B* **55**, R4921-,(1997).
[4] Petit, P., Jouguelet, E., Fischer, J. E., Rinzler, A. G., and Smalley, R. E., *Phys. Rev. B* **56**, 9275-, (1997).
[5] Lee, R. S., Kim, H. J., Fischer, J. E., Thess, A., and Smalley, R. E., *Nature* **388**, 255-, (1997).
[6] Petit, P., Mathis, C., Journet, and C., Bernier, P., *Chem. Phys. Lett.* **305**, 370-374, (1999).
[7] Jouguelet, E., Mathis, C., and Petit, P., *Chem. Phys. Lett.* **318**, 561-564, (2000).
[8] Reiss, H., *J. Phys. Chem.* **89**, 3783-,(1985).
[9] Rinzler, A. G., Liu, J., Dai, H., Nikolaev, P., Huffman, C. B., Rodriguez-Macias, F. J., Boul, P. J., Lu, A. H., Heymann, D., Colbert, D. T., Lee, R. S., Fischer, J. E., Rao, A. M., Ecklund, P. C., and Smalley, R. E., *Appl. Phys. A* **67**, 29-,(1998).
[10] Kataura, H., Kumawaza, Y., Maniwa, Y., Umezu, I., Susuki, S., Ohtsuka, Y., and Achiba, Y., *Synth. Met.* **103**, 2555-2558, (1999).
[11] Dubois, D., Moninot, G., Kutner, W., Jones, M. T., and Kadish, K. M., *J. Phys. Chem.* **96**, 7137-, (1992).
[12] Delanay, P., Choi, H. J., Ihm, J., Louie, S. G., and Cohen, M. L., *Nature* **391**, 466-, (1998).
[13] Kramert, B. , and MacKinnon, A., *Rep. Prog. Phys.* **56**, 1469-, (1993).

Chemical and Electrochemical Doping of Single Wall Carbon Nanotube Films Probed by Optical Absorption Spectroscopy

S. Kazaoui [a,*], N. Minami [a], H. Kataura [b] and Y. Achiba [b]

[a] National Institute of Materials and Chemical Research, 1-1 Higashi, Tsukuba, Ibaraki ,305-8565
[b] Faculty of Science, Tokyo Metropolitan University, Tokyo 192-0397, Japan

Abstract. The electronic properties of semiconducting (S) and metallic (M) single-wall carbon nanotube (SWCNT) films were modified either by chemical or electrochemical doping, while changes in their electronic structure were monitored *in-situ* by optical absorption spectroscopy. In both experiments the intensity of absorption features at 0.68, 1.2 and 1.8 eV were dramatically affected, demonstrating the possibility of tuning the Fermi level to specific bands in S- or M-SWCNTs.

INTRODUCTION

Owing to their unique structural and electronic properties, single-wall carbon nanotubes (SWCNT) constitute new class of materials that could contribute to the development of novel nano-scale electronic devices. Since the control of valence electrons is crucial in any of such device applications, their doping behavior is one of their most important solid-state properties that must be elucidated.

Indeed, the doping of SWCNT was originally studied through dc resistance and Raman measurements (1-3), but there is still paucity of information on the nature (semiconducting or metallic) and electronic states of SWCNTs that undergo charge-transfer.

In the present contribution, we have addressed this issue by the measurements of the change in optical absorption spectra of SWCNT thin films induced by both chemical and electrochemical doping. It should be stressed that, different from conductivity and Raman measurements, optical absorption can detect changes in specific electronic states (4-6).

EXPERIMENTS

SWCNTs were synthesized by the electric arc discharge method using Ni/Y catalyst (4). Films of SWCNT were prepared by dispersing the raw material in ethanol and then by spraying using an airbrush on top of a suitable substrate (4). Chemical doping of SWCNT films was carried out in gas phase under high vacuum with either electron

CP544, *Electronic Properties of Novel Materials—Molecular Nanostructures*, edited by H. Kuzmany, et al.
© 2000 American Institute of Physics 1-56396-973-4/00/$17.00

donors (Na, K, Rb, Cs, Ba) or electron acceptors (Br$_2$, I$_2$, TCNQ, TCNQF$_4$) with controlled stoichiometry, following the method already described in Ref. 6. Electrochemical experiments were performed on SWCNT films deposited on semitransparent Pt electrodes in 0.2 M LiClO$_4$ solution in acetonitril in the range -1.4 to 1.4 V vs. Ag (wire) with Pt as a counter electrode. Optical absorption spectra were recorded using a Shimazu 3100 spectrophotometer and electrochemical experiments were carried out using a Solartron SI 1260/1287 unit.

RESULTS AND DISCUSSIONS

Curves labeled x=0 in Fig. 1 represent typical optical absorption spectra of pristine SWCNT films. The main absorption features at 0.68 and 1.2 eV originate from band gap transition in semiconducting tubes, whereas the feature at 1.8 eV comes from metallic tubes, in good agreement with previous reports (6 and reference there in).

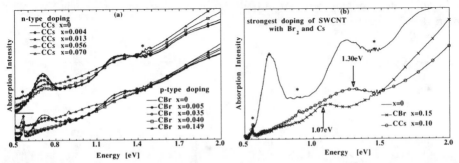

FIGURE 1. (a) Evolution of the optical absorption spectra of SWCNT films for various stoichiometry CM$_x$ (M=Br, Cs). The entire set of spectra for CCs$_x$ is offset for clarity with short line indicating the 0 level. (b) Absorption spectra for strongly doped samples. * indicates features coming from the quartz and the noise from the spectrometer.

Upon chemical doping with Br$_2$ and Cs dramatic changes of the optical absorption spectra of SWCNT films were observed (Fig. 1). A remarkable observation was the sizable decrease of the intensity of the absorption bands at 0.68 eV without a significant change of the features at 1.2 and 1.8 eV, at the initial stage of the doping (typically for x<0.005) irrespective of the nature of the dopants. Subsequent doping (typically for x<0.04) led to a dramatic depletion of both bands at 0.68 and 1.2 eV without a substantial alteration of the band at 1.8 eV. In this range of stoichiometry (0<x<0.04), we have observed neither a shift of the absorption bands nor an appreciable increase of the absorption background, again for all dopants. In this context, Fig. 2 specifically displays the variation of the normalized intensity of several absorption peaks (I$_x$/I$_{x=0}$) at 0.68, 1.2 and 1.8 eV versus stoichiometry CM$_x$ (M=Br, Cs). We note that essentially the same changes were observed for K, I$_2$ and TCNQ doped SWCNT. At high doping level (typically for x>0.04), only achieved with Cs, Br$_2$ and TCNQF$_4$, all the above-mentioned features including the one at 1.8 eV almost completely vanished. Moreover, two important changes were detected at this doping level: the gradual increase of the absorption background (due to free carrier absorption and Drude component) and the emergence of new bands at approximately 1.07 and 1.30 eV (their origin are not yet elucidated) as clearly evidenced for stoichiometry at $CBr_{0.15}$ and $CCs_{0.10}$, respectively (Fig. 1(b)). A striking observation was that SWCNT films exposed either to electron donors or electron-acceptors showed very similar changes in optical absorption spectra. The disappearance of several absorption peaks (in other word, inhibition of the electronic transitions) has been attributed to electron depletion from or filling to specific bands of the density of state (DOS) in the semiconducting or metallic SWCNTs. Furthermore, the simultaneous measurements of the change in absorption ($I_x/I_{x=0}$) and in dc resistance ($R_x/R_{x=0}$) for various dopants with controlled stoichiometry demonstrated their close correspondence, yielding the unambiguous interpretation about the doping effects (Fig. 2). Indeed, the resistivity decreased because the carrier density (either as electrons or hole) increased with the concentration of the dopants. These results are the direct experimental demonstration of the amphoteric doping behavior of both the semiconducting and metallic SWCNTs (more details were provided in Ref. 6).

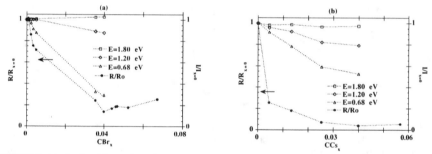

FIGURE 2. Normalized intensity of the absorption peaks ($I_x/I_{x=0}$) at 0.68, 1.2 and 1.8 eV and the normalized dc electrical resistivity ($R_x/R_{x=0}$) for various stoichiometry CM_x (M=Br, Cs). Dashed lines guide to the eye.

Upon electrochemical doping, the electrochemical potential (in other word, the Fermi level) of the system was finely tuned, while simultaneously and *in situ* recording the changes of the optical absorption spectra of SWCNT film (we used the double beam configuration to compensate as much as possible for the strong absorption features due to the electrolytic solution). Fig. 3(a) shows the absorption spectra of SWCNT film at several constant electrode potentials. A remarkable observation was that tuning the potential *reversibly* depleted all the absorption features. As the potential was increased from 0.2 V (vs. Ag), the peaks intensity decreased in a specific sequence: first 0.68, then 1.2 and finally 1.8 eV, in very good agreement with the results already obtained by chemical doping (6). Fig. 3(b) present the intensity of the peak at 1820 nm (or 0.68 eV) while scanning the potential (at 10 mV/sec). The peak intensity significantly decreased when the potential was lower than approximately 0.1

or higher than 0.9 volt, whereas the intensity was maximal in the range 0.1 to 0.8 volt (a window of 0.7 eV). The width of this window nicely agrees with the energy separation between the first pair of van Hove singularities of DOS of semiconducting tubes. Further works are in progress to fully elucidate the electrochemical behavior and to explore this avenue in order to modulate the electronic properties of SWCNT.

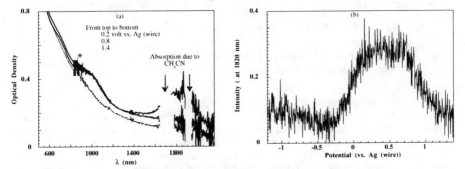

FIGURE 3. (a) Absorption spectra of SWCNT/Pt film at several electrode potential and (b) Intensity of the absorption peak at 1820 nm (0.68 eV) vs. electrochemical potential (vs. Ag (wire)).

In conclusion, tuning of the Fermi level of SWCNT by chemical or electrochemical doping allows to selectively fill or deplete their density of states, and thus allows to modify their electronic properties.

REFERENCES

* Corresponding author, e-mail kazaoui@nimc.go.jp

1. R.S. Lee *et al.*, *Nature* **388**, 255 (1997).
2. A.M. Rao *et al.*, *Nature* **388**, 257 (1997).
3. L. Grigorian *et al.*, *Phys. Rev. B* **58**, R4195 (1998).
4. H. Kataura *et al.*, *Synth. Met.* **103**, 2555 (1999).
5. P. Petit *et al.*, *Chem. Phys. Lett.* **305**, 370 (1999).
6. S. Kazaoui *et al.*, *Phys. Rev. B* **60**, 13339 (1999).

Pressure and Doping Dependence of Electronic Properties of Carbon Nanotube Ropes

R. Gaál[1], J.-P. Salvetat[1], J.-M. Bonard[1], L. Thien-Nga[1], S. Garaj[1], L. Forró[1], B. Ruzicka[2] and L. Degiorgi[2]

[1]Departement de Physique, EPFL, CH-1015 Lausanne, Switzerland
[2]Laboratorium für Festkörperphysik, ETH-Zürich, CH-8093, Switzerland

Abstract. We have performed DC transport and optical measurements on purified and doped thick films of single walled carbon nanotubes (SWNT) in the temperature range 4-300 K. The resistivity under hydrostatic pressure at room temperature shows nonmonotonic variation which is a manifestation of the symmetry change first seen in Raman scattering by Venkateswaran *et al* (Phys. Rev. B **59**, 10928 (1999)). Despite the non-metallic temperature dependence of the dc resistivity the optical conductivity shows Drude-response in agreement with the presence of metallic tubes in the mat. Measuring DC and optical conductivity on the same sample allowed us to separate the contribution of the intrinsic rope resistivity and that of the contact regions.

Charge transport studies on carbon nanotubes revealed up to now the most fascinating information about the electronic properties of these mesoscopic objects. Findings include weak localisation, Aharonov-Bohm oscillations [1] and indications of Luttinger-liquid behaviour [2]. These experiments have been carried out on either multiwalled nanotubes or ropes of single walled carbon nanotubes (SWNTs).

The SWNT is a very attractive system but it is known that torsions, twists, interaction with the substrate can all induce important local deformations which influence the electronic properties. These interactions seem to be partially absent in mats of nanotubes: one indication of this might be that in nanotube mats the temperature dependence of the resistivity is metallic above a characteristic temperature T^*. However the question rises why the resistivity turns to non-metallic at all.

Raman scattering has shown that SWNT ropes are sensitive not only to the above deformations but to hydrostatic pressure as well[3]. In the 1.5-1.7 GPa pressure range nanotubes deform from circular to hexagonal cross section. We carried out transport measurements under pressure to see the effect of it on the electronic properties.

Furthermore, carrier density in SWNT can be changed by doping as in the case of graphite. To have a clear picture of the effect of the charge transfer on the transport properties, it is necessary to separate the on-tube effect from the doping of the tube-tube contacts. In this respect, we have performed dc transport and reflectivity measurements on pristine and potassium doped buckypaper samples.

Buckypaper was made from purified SWNT suspension by filtration. Some samples were heated to 1200°C in order to eliminate traces of the nitric acid which acts as a dopant; these samples we call pristine. For the pressure dependent studies, the sample was placed in a piston-type pressure cell. Potassium doping was done on pristine samples in sealed tubes at 300 °C.

CP544, *Electronic Properties of Novel Materials—Molecular Nanostructures*, edited by H. Kuzmany, et al.

Fig.1 shows the temperature dependence of the resistivity of an 'as made', a pristine and a potassium doped buckypaper. Increasing doping level increases conductivity and pushes the resistivity minimum T^* to lower temperatures. The room temperature conductivity of the heat treated sample is about 20 $O^{-1}cm^{-1}$, potassium doping increases this by a factor of 12 and pushes T^* to 70 K. The as made sample is about halfway between the two. The systematic change of T^* and conductivity with the doping level can be explained if we note that the sample consists of metallic and semiconducting SWNTs in loose contact. In this picture the resistivity minimum is the result of a metallic on-tube conduction in series with an activated, hopping-like conduction of the contact regions. Doping will mostly enhance the conductivity of the semiconducting contact regions. Once the contact regions become better conducting, we pick up more from the temperature dependence of the intrinsic, metallic on-tube conductivity.

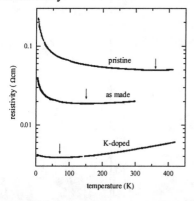
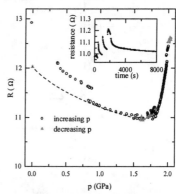

FIGURE 1. Temperature dependence of the resistivity of pristine, as made and potassium doped buckypaper. Arrows indicate the resistivity minimum temperature T^*.

FIGURE 2. Pressure dependence of the room temperature resistivity of a pristine buckypaper. The inset shows the long time scale relaxation of the resistivity seen in the 1.5-1.8 GPa pressure range.

Fig.2 shows the pressure dependence of the room temperature resistivity of a pristine paper. Upon increasing the pressure up to 1.5 GPa the resistance drops by about 10% and reaches a shallow minimum. Between 1.5-1.8 GPa we see long time scale relaxation of the resistivity indicating that the internal rearrangements are not instantaneous. Increasing the pressure further, the resistance rises sharply and keeps this tendency up to the highest pressure we could reach, 2 GPa. Upon lowering the pressure resistivity changes are reversible down to about 1.5 GPa.

We assign the initial decrease of the resistivity to compacting the buckypaper thus improving rope-rope contact. This change is partially irreversible: after the pressure cycle the sample resistance has decreased by about 8%. Above 1.8 GPa the resistivity increases steeply with pressure and a more striking feature is the reproducibility of this change upon decreasing pressure. This indicates that it is not due to rope contacts breaking up but rather some intrinsic change of the on-tube conductivity. We assign this change to the deformation of the tubes in the bundle as mentioned earlier. The change in the symmetry could open of a gap in the metallic tubes or increase the probability of Umklapp-processes thus leading to a rise of resistivity.

In order to further investigate the contribution of the contact regions to the total resistivity of a sample we carried out reflectivity measurements on both pristine and potassium doped buckypaper in the spectral range from 30 up to 3×10^4 cm^{-1}. The optical conductivity $\sigma_1(\omega)$ was obtained from Kramers Kronig (KK) transformations, standard extrapolations were used at high frequencies, while below the lowest measured frequency we performed Hagen-Rubens or constant value extrapolation for metallic or insulating behaviour respectively (see Ref. [4] for more technical details).

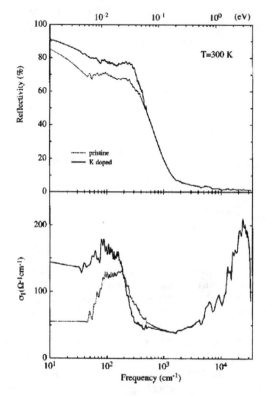

FIGURE 3. Optical reflectivity (a) and deduced optical conductivity (b) as a function of wave number (photon energy) for pristine and potassium doped SWNT sample.

Figure 3 shows the complete reflectivity spectra and the corresponding optical conductivity $\sigma_1(\omega)$ for the pristine and the potassium doped SWNT samples at T=300 K. Our spectra are perfectly reproduced by a Drude-Lorentz model with a Drude term and 3 harmonic oscillators at higher frequencies. The first oscillator describes the absorption at about 150-200 cm^{-1}; this one barely changes with doping and we assign it to a convolution of phonon modes. The other two oscillators are necessary to reproduce the peak at high frequency and they latter do not change with doping. The Drude term is

406

enhanced in the doped sample but even the undoped one has one: the parameters characterising the Drude term of the pristine sample are ω_{pD}=593 cm^{-1} for the plasma frequency and γ_D=115 cm^{-1} for the scattering rate, while for the doped sample we find ω_{pD}=984 cm^{-1}, with the same γ_D. In terms of conductivity, σ_{dc} =60 Ω^{-1} cm^{-1} for the undoped sample and 170 Ω^{-1} cm^{-1} for the doped one. The measured value of the dc conductivity of the pristine sample is 20 Ω^{-1}cm^{-1} at room temperature, much lower than the value deduced from optical conductivity. In our interpretation this is due to the fact that in pristine samples the dc transport is limited by the tube-tube (rope-rope) contacts, rather then the intrinsic conductivity of the metallic nanotubes. In the case of the K-doped sample $\sigma(\omega \to 0)$ increases by a factor of 3. The dc value changes much more (by a factor of 12-20), because tube-tube contact resistance changes dramatically upon doping. Once the passage from tube to tube has been improved, the percolating path through the sample picks up more from the intrinsic, high on-tube conductivity.

In conclusion, we have performed DC resistivity measurements on buckypaper under hydrostatic pressure in the 4-300 K temperature range. We have found non-monotonic variation of the room temperature conductivity in the pressure range where Raman spectra are shown to change drastically. We attribute this to a reversible deformation of the tubes at higher pressures. The temperature dependence of the conductivity indicates hopping conduction in the whole range of applied pressure. Optical measurements have revealed a Drude response in the optical conductivity despite the non-metallic behaviour in dc resistivity measurements. We attribute this feature to the non-metallic tube-tube contacts which govern the dc transport.

The work in Lausanne was supported by the NFP 36 program of the Swiss National Science Foundation. Work in Zürich was supported by the Swiss National Science Foundation.

REFERENCES

1. A. Bachtold *et al.*, *Nature* **397**, 673 (1999).
2. M. Bockrath *et al. Nature* **397**, 598 (1999).
3. U. D.Venkateswaran *et al.*, *Phys. Rev.* **B 59**, 10928 (1999).
4. A. Schwartz *et al.*, *Phys. Rev.* **B 58**, 1261 (1998).

Intercalation of Heavy Alkali Metals (K, Rb and Cs) in the Bundles of Single Wall Nanotubes

L. Duclaux, K. Méténier, P. Lauginie, J. P. Salvetat, S. Bonnamy, and F. Beguin

CRMD, CNRS-Université d'Orléans, 1B Rue de la Férollerie, 45071 Orléans cédex 2

Abstract. The electric-arc discharge carbon deposits (collaret) containing Single Wall Carbon Nanotubes (SWNTs) were heat treated at 1600°C during 2 days under N_2 flow in order to eliminate the Ni catalyst by sublimation, without modifications of the SWNTs ropes. Sorting this deposit by gravity enabled to obtain in the coarsest particles higher amount of SWNTs ropes than in other particle sizes. The coarser particles of the carbon deposits were reacted with the alkali metals vapour giving intercalated samples with a MC_8 composition. The intercalation led to an expansion of the 2D lattice of the SWNTs so that the alkali metals were intercalated in between the tubes within the bundles. Disordered lattices were observed after intercalation of Rb and Cs. The simulations of the X-ray diffractograms of SWNTs reacted with K, gave the best fit for three K ions occupying the inter-tubes triangular cavities. The investigations by EPR, and ^{13}C NMR, showed that doped carbon deposits are metallic.

INTRODUCTION

The electric-arc discharge method enables to produce carbon deposit on the cathode (called collaret) containing large amount of SWNT (Single Wall nanotube) bundles similar to those obtained by the laser pulsed technique [1]. The SWNTs are organized in a 2D triangular lattice and the nanotube diameter distribution is narrow close to the armchair (10, 10) diameter. Recent studies have shown that the SWNTs can be doped by electron donor improving their electronic conductivity [2]. After reaction with heavy alkali metals (cesium and potassium), the saturation composition of the doped SWNTs was found to be MC_8 [2-3]. Moreover, the authors have shown that the crystalline 2D structure of the nanotubes was destroyed and they have failed to prove clearly the intercalation of alkali metals in between the SWNTs. However, the observation of the 2D lattice after intercalation of alkali in between the tubes was previously reported [4-5]. Molecular dynamics calculations are not in agreement with all the experimental studies [6], as they predict for intercalated (10,10) armchair tubes a saturation composition KC_{16} with potassium atoms distributed all around the distorted tubes with a 2×2 lattice superimposed on the tube network. In this work, we have intercalated SWNTs by K, Rb, and Cs and studied their structure using X-ray diffraction. Annealed samples without ferromagnetic particles (Ni), were intercalated

CP544, *Electronic Properties of Novel Materials—Molecular Nanostructures*, edited by H. Kuzmany, et al.
© 2000 American Institute of Physics 1-56396-973-4/00/$17.00

with K and their electronic properties were studied using EPR and ^{13}C NMR.

EXPERIMENTAL

Purification treatment

The SWNTs were produced in the form of raw collaret deposit using the electric-arc discharge described elsewhere [1]. The pristine collaret carbon deposit is a mixture of graphitic carbon, amorphous carbon, SWNTs bundles and metallic particles of Ni and Y. The metallic particles (22.08 wt. % of Ni and 2.54 wt. % of Y) were eliminated by sublimation, using a long heat treatment (2 days) in nitrogen flow. The 2 days long heat treatment at 1650°C enables us to eliminate the Ni metallic particles as their X-ray peaks and ESR signals are not observed after heat treatment. After the thermal treatment some peaks due to yttrium nitride impurities appear on the diffractograms showing that YN is formed. Using the model previously reported by Thess al [7], the best aggreement of the simulated diffractogramm of the pristine SWNTs 2D structure with the experimental data, is obtained with a compact triangular network of nanotubes of 6.8 Å radius (a =16.7 Å).

Intercalation

The two bulb method was used for the vapor phase reaction with a small temperature difference between the alkali metal ($T_K=T_{Rb}=186$°C) and the raw collaret (T=202°C). Lower temperatures were used in the case of cesium reactions ($T_{Cs}=179$°C, $T_{collaret}=189$°C). After glove-box transfer, the samples were studied by X-ray diffraction on scelled capillaries using a curve position sensitive detector (INEL CPS 120). The composition was deduced from the weight uptake. EPR and ^{13}C NMR signals were recorded on the sample reacted with potassium.

RESULTS

Structure of the intercalated SWNT ropes

The composition of the purified "collaret" reacted with K and Rb was found to be MC_8 (M=alkali metal) in agreement with the composition observed by other authors [2-4] on bucky paper (of Rice University) reacted with K.

The *00l* lines due to the formation of first stage GIC with graphitic particles are also observed proving that the reaction of alkali metal has also occurred with the carbon deposit. After reaction with K, Rb or Cs, the 2D diffraction pattern of SWNTs ropes is clearly observed. Indeed, the *10* lines of the pristine SWNTs are shifted to higher distance values. As a result the a parameter increased from 1.67 nm for the pristine SWNTs to 1.85 nm, 1.87 nm, and 2 nm for the tubes intercalated by K, Rb and Cs respectively (Fig. 1). The expansion of the 2D lattice is attributed to the intercalation of the alkali metals in between the SWNTs inside the ropes.

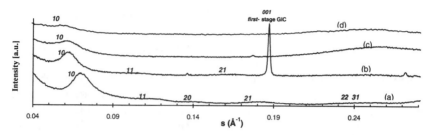

Figure 1. Diffractogram of SWNT ropes : pristine (a), intercalated by K (b), Rb (c), and Cs (d)

Assuming an ionic model where the alkali transfer a charge to the carbon nanotubes, the $K^{\delta+}$ and $Rb^{\delta+}$ ion can then occupy preferently the triangular voids because the inter-tube space between two adjacent tubes is not large enough to place an alkali ion (since the Van der Waals radius of the nanotubes is R+3.2/2 Å, R: radius of the tubes). The maximum occupancy of the triangular cavities is three alkali ion per cavity considering that the distance between first neighbor ions is close to the distance between two alkali atoms in a metallic state. Assuming that the distance between alkali atoms along the tube axis and in the normal section plane would be the same (ie d_{K-K}=4.9 Å), the global stoichiometry of the potassium intercalated sample would be KC_{13}. However it is clear that no periodic order of alkali atoms appear along the tube axis since no related diffraction lines are observed.

The single *10* peaks observed in the case of Rb and Cs intercalation, indicate that the 2D network is disordered. In some samples reacted with Cs, no diffraction lines were observed in agreement with the total disorder of the 2D array previously observed [2-3]. In our case, the use of a lower temperature of reaction has enabled us to obtain more ordered intercalated SWNTs ropes. Reaction with potassium enables us to obtain after reaction a less disordered 2D lattice since further *hk* peaks due to this structure appears on the diffractograms. Using a model of three alkali atoms per triangular cavities (d_{K-K}=4.9 Å), the best agreement of the simulation with the experimental data was obtained with in-plane MC_x composition (6<x<8).

EPR and ^{13}C NMR

Three EPR lines were observed on KC_8 intercalated carbon deposit, a very broad one (800 Gauss) peak to peak attributed to residual Nickel ferromagnetic impurities, a very thin one (1 Gauss) and a dysonian line attributed to the SWNTs (35 G), showing that SWNTs are metallic after doping.^{13}C NMR at 90 Mhz (8.5 T field) has been performed on heat treated collaret carbon deposit before and after K-intercalation. The pristine material spectrum looks quite similar to usual aromatic carbon materials ones i.e. a wide anisotropic powder quasi-axial distribution ranging from 0 to 220 ppm with main peak at 183 ppm (like graphene and frozen benzene). After intercalation with K, the main peak is diamagnetically highly shifted to 93-96 ppm and the sign of the anisotropy seems reversed, both effects were known from alkali graphite intercalation compounds (GICs). For example, the highfield shift for KC_8 GIC (90 ppm) attributed to the anisotropic dipolar interaction with conduction electrons and typical of a 90% charge transfer. Thus the present observations are well in agreement with an intercalation process in carbon material with a dominant sp^2 character.

CONCLUSION

Carbon deposit containing large amount of SWNTs can be reacted with alkali metal (M) up to MC_8 composition. The intercalation lead to an expansion of the 2D lattice of the SWNTs proving that the alkali metals are intercalated in between the tubes within the bundles. Disordered lattices were observed after intercalation of Rb and Cs. The simulations of the X-ray diffractograms of SWNTs reacted with K, give the best fit for three K ions occupying the inter-tubes triangular cavities. However, this model cannot explain the MC_8 composition suggesting that metalic potassium could be inserted in the mesopores of the material or inside the tubes. The investigations by EPR, and ^{13}C NMR, have shown that doped carbon deposits are metallic.

ACKNOWLEDGMENT

The authors thanks P. Bernier (GDPC, University of Montpellier, France) for supplying SWNTs collarets.

REFERENCES

[1] Journet C., Maser W.K., Bernier P., Loiseau A., Lamy de la Chapelle M., Lefrant S., Deniard P., Lee R. and Fischer J.E., *Nature*, **388**, 756 (1997).
[2] Lee R. S., Kim H. J., Fisher J. E., Thess A.and Smalley R. E., *Nature* 388 (1997).
[3] Suzuki S., Bower C. and Zhou O., *Chem. Phys. Letters* **285**, 230 (1998).
[4] Pichler T., Sing M., Knupfer M, Golden M. S. and Fink J., *Sol. State Com.,* 109,**721** (1999).
[5] F. Beguin, L. Duclaux, K. Metenier, E. Frackowiak , J. P. Salvetat, J. Conard, S. Bonnamy, P. Lauginie, in *Proceedings XIII IWEPNM-Science and Technology of Molecular Nanostructures*, ed. H. Kuzmany et al, pp 273-277 (1999)
[6] Gao G., Gagin T., and Goddard III W. A., *Phys Rev. Lett.* 80, 5556 (1998).
[7] Thess A. et al., Science 273, 483 (1996).

VII. OTHER MOLECULAR MATERIALS

On the Electronic Structure of Non Carbon Nanotubes

Gotthard Seifert

Theoretische Physik, Universität-GH Paderborn, 33098 Paderborn, Germany
E-mail:seifert@phys.uni-paderborn.de

Abstract. The electronic structure non-carbon nanotubes is discussed on the basis of general arguments by a conformal mapping of a two-dimensional layer structure onto the surface of a cylinder, similar to carbon nanotubes. The evolution of the band gap as a function of the tube diameter is characterized for phosphorus and metal chalcogenide nanotubes and analyzed in detail for MoS_2 nanotubes.

Carbon nanotubes became a challenging new field of research since their discovery [1]. Fullerene-like structures (IF) and nanotubes of WS_2 and MoS_2 where found also already in 1992 [2], but they were much less investigated experimentally and only recently the first results of theoretical calculations of such nanostructures were published [3]. Efficient methods for their synthesis have been developed by Tenne et al., methods, which can easily be scaled up [2] in contrast to the existing methods for carbon nanotubes and there are also many potential applications. The most obvious one is in the field of solid lubrication. Many data indicate that the IF materials of WS_2 and MoS_2 can serve as an improved solid lubricant [4]. Another field of application could be based on their rather well defined optical properties [3,5]. The applicability of WS_2 tips in scanning probe microscopy has been shown also recently [6]. But there is still a lack of systematic theoretical investigations to predict and understand the properties of these structures. There are up to now a few hints for other systems, as for example $B_xC_yN_z$ [7] and $NiCl_2$ [8]. Moreover, stable tubular structures of GaSe [9], GaN [10] and silsesquioxanes [11] have been predicted theoretically.

We will discuss here briefly some general considerations of the electronic band structure of nanotubes and describe the electronic properties of phosphorous and especially the abovementioned metalchalcogenide nanotubes.

As it was shown for carbon nanotubes [12,13] the electronic band structure of tubular structures can be related to the band structure of the corresponding layer material. In tubular structures periodic boundary conditions can only applied along the tube direction. Therefore the band structure of a tubular structure can be obtained just by folding the band structure of the corresponding 2D system

CP544, *Electronic Properties of Novel Materials—Molecular Nanostructures*, edited by H. Kuzmany, et al.

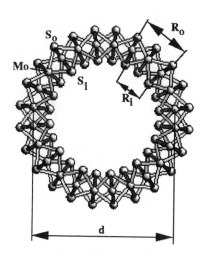

FIGURE 1. Top view along the tube direction of a (12,0) MoS_2 *nanotube*

onto the 1D Brillouin zone (BZ) of the tube. Since the systems considered here possess all a hexagonal layer structure, as graphene sheets, a very similar qualitative picture of the band structures results. In the case of (n,0) (zig-zag) tubes bands in the neighbourhood of the K-point of the layer structure are folded to the Γ point in the BZ of the tube. As in the case of the carbon nanotubes [12,13] band structures result with a direct gap at the Γ point, as it is seen for WS_2 [3], for example. Since the metal chalcogenides (MoS_2, WS_2) as well as black phosphorous (b-P) are semiconductors with a finite gap at the K point there will also be a nonvanishing gap for the tubes at the Γ point. I.e., a metallic behaviour for (3n,0) tubes, as proposed for carbon nanotubes, will not occur. Similarily, the (n,n) tubes will have an indirect gap see [14] for P-tubes and [3,15] for WS_2 and MoS_2 tubes, respectively. The folding arguments suggest also the same size of the indirect gap in the (n,n) tubes and the direct gap in the (n,0) tubes, which is indeed confirmed for the chalcogenide tubes [3,15]. In the case of P-tubes the gap size agrees also quantitatively quite well with that of the b-P monolayer [14]. Only for very narrow tubes ((3,3),(6,0) tubes) a clearly smaller gap was found, whereas already (15,0) and (10,10) tubes have already gaps as large as that of the monolayer [14]. However, for MoS_2 and WS_2 tubes the gap is considerably smaller than expected from folding the corresponding layer band structure. That means, the electronic band structure is considerably influenced by the curvature, when rolling the sheets. This can be understood considering the triple layer character (S-M-S) of the MS_2 systems. The rolling leads to a shrinking of the inner S layer (S_i) and a dilatation of the outer S layer (S_o) with respect to the hexagonal triple layer with a lattice constant $a = |\vec{a}| = |\vec{b}|$, $a \rightarrow R_i$, $a \rightarrow R_o$, respectively - see Fig.1.

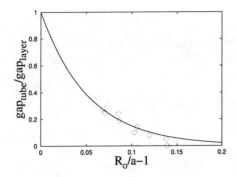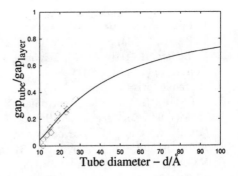

FIGURE 2. Gap size of MoS$_2$ nanotubes. In the **left** figure the gap size reduction by the increase of the S-S distance in the outer sulfur layer (R_o) - $x = R_o/a - 1$) is shown. In the **right** figure the evolution of the gap size as a function of the tube diameter (d) is shown.

The dilatation R_o/a is directly related to the tube diameter d:

$$R_o/a = 1 + \frac{c}{d}, \tag{1}$$

where c is the thickness of the triple layer. The dilatation of the outer layer of sulfur atoms (S_o) is the main reason for the band gap shrinking. Increasing the lattice constant of the 2D MS$_2$ triple layer the conduction band states at the K point are lowered in energy, hence the gap size is decreased. This downward shift is easily understandable, since these states are mainly antibonding Mo-d–S-p states. Correspondingly, the antibonding Mo-d–S-p states in the tubes are lowered in energy as the outer S layer is expanded. A plot of the calculated gap size for MoS$_2$ tubes g_{tube} [15] (divided by the gap size of the MoS$_2$ triple layer - g_{tube}/g_{layer}) - over $x = R_o/a - 1$ - see Fig.2 - suggests an exponential decrease of the gap size versus x. In combination with eq. (1) this allows an extrapolation of the gap size for tubes with large diameters - see Fig.2. Due to the strong similarity between MoS$_2$ and WS$_2$ nanotubes [3,15] the evolution of the gap size versus diameter is quite similar for WS$_2$ tubes. Moreover, qualitatively this result can certainly generalized for other not yet known non-carbon nanotubes, if their corresponding layer structures are semiconducting. A rather different behaviour is expected for metallic like nanotubes.

Acknowledgements.
The author thanks R. Tenne, E. Hernandez H. Terrones, Th. Frauenheim and M. Terrones for fruitful discussions, G. Jungnickel for computational assistance.

REFERENCES

1. S. Iijima, Nature **354** 56 (1991).
2. R. Tenne, L. Margulis, M. Genut, G. Hodes, Nature (London) **360**, 444 (1992);
3. G. Seifert, H. Terrones, M. Terrones, G. Jungnickel and T. Frauenheim, Sol. State Comm. **114**, 245 (2000).
4. R. Tenne, private communication.
5. G.L. Frey, S. Elani, M. Homyonfer, Y. Feldman and R. Tenne, Phys. Rev. B **57**, 6666 (1998). L. Margulis, G. Salitra, R. Tenne and M. Talianker, Nature **365**, 113 (1993); R. Tenne, M. Homyonfer and Y. Feldman, Chem. Mater. **10**, 3225 (1998).
6. A. Rothschild, S.R. Cohen and R. Tenne, Appl. Phys. Lett. **75**, 4025 (1999).
7. N.G. Chopra et al., Science **269**, 966 (1995); A. Loisseau et al., Phys. Rev. Lett. **76**, 4737 (1996); M. Terrones et al., Chem. Phys. Lett. **259**, 568 (1996); D. Goldberg et al., Appl. Phys. Lett. **69**, 2045 (1996).
8. Y.R. Hacohen, E. Grünbaum, R. Tenne, J. Sloand and J.L. Hutchinson, Nature (London) **395**, 336 (1998).
9. M. Cote, M.L. Cohen and D.J. Chadi, Phys. Rev. B **58**, R4277 (1998).
10. S.M. Lee, Y.H. Lee, Y.G. Hwang, J. Elsner, D. Porezag and Th. Frauenheim, Phys. Rev. B **60**, 7788 (1999).
11. D. Wichmann and K. Jug, J. Phys. Chem. **B103**, 10087 (1999).
12. N. Hamada, S. Sawada and A. Oshiyama, Phys. Rev. Lett. **68** 1579 (1992).
13. J.W. Mintmire and C.T. White, in Carbon nanotubes: preparation and properties (ed. T.W. Ebbesen), CRC Press Inc. (1997).
14. G. Seifert and E. Hernandez, Chem. Phys. Lett. **318**, 355 (2000).
15. G. Seifert, H. Terrones, M. Terrones, G. Jungnickel and T. Frauenheim, submitted for publication.

Self-Organized Growth of Indium-Tin-Oxide Nanowires

N. Roos[a], T.P. Sidiki[b], J. Seekamp, and C.M. Sotomayor Torres

University of Wuppertal, Institute of Material Science and Department of Electrical and Information Engineering, Gaußstr. 20, 42097 Wuppertal, Germany
[a] responsible author, email address : roos@uni-wuppertal.de
[b] present address : Philips Semiconductors Hamburg, Stresemannallee 101, 22529 Hamburg, Germany

Abstract. The observation of nanowire growth of thermally evaporated Indium-Tin-Oxide is reported. The influence of growth rate, layer thickness, substrate temperature, In:Sn ratio and oxygen partial pressure on the electrical, optical and morphological properties of the films is examined. By modifying these growth parameters conventional grainy ITO layers have been obtained. However, under certain conditions, dendrites grow over an initial grainy layer.
A range of parameters is established for dendritic growth. These dendrites have a length of hundreds of nanometers with diameters of tens of nanometers, depending on growth conditions. Despite this anisotropic growth, the resulting films exhibit good optical and electrical properties.

INTRODUCTION

Indium-Tin-Oxide (ITO) films offer the unique combination of very low resistivity, in the mΩcm range and high optical transparency, up to 95% in the visible part of the electromagnetic spectrum. Therefore, ITO is widely used as an electrical top contact for optical devices, eg. flat panel displays and solar cells. Different methods have been developed to grow these films, among them reactive thermal evaporation [1], magnetron sputtering [2] and laser ablation [3].

Although an optimum set of parameters to obtain low resistivity and high transparency films is well established, very little is known about the influence of growth conditions on the morphology of ITO. Here a study of thermal evaporation conditions leading to the growth of submicrometer-sized dendrites is presented. Due to their elongated shape and nanometer-sized diameters, they are referred to as nanowires in this work.

CP544, *Electronic Properties of Novel Materials—Molecular Nanostructures*, edited by H. Kuzmany, et al.
© 2000 American Institute of Physics 1-56396-973-4/00/$17.00

TABLE 1. List of ITO samples studied

Sample Nr.	$T_{Substr.}$ [°C]	Thickness [Å]	O_2 [mbar]	Pre-set rate [Å/s]	Measured rate [Å/s]	Pre-set In:Sn [%-wt]	ρ [mΩcm]	Transmission at 550 nm
4.1	400	516	5×10^{-5}	0.1	0.19	95:5	6.0	85%
4.3	400	515	1×10^{-4}	0.1	0.23	95:5	5.0	88%
4.5	400	513	2×10^{-4}	0.1	0.26	95:5	7.0	92%
4.7	400	529	5×10^{-5}	0.1	0.19	90:10	7.8	80%
4.9	400	533	1×10^{-4}	0.1	0.23	90:10	8.5	83%
4.11	400	534	2×10^{-4}	0.1	0.26	90:10	7.5	88%
4.13	200	531	1×10^{-4}	0.1	0.26	90:10	11	80%
4.15	100	535	1×10^{-4}	0.1	0.27	90:10	13	75%
4.17	400	559	1×10^{-4}	1.0	1.33	90:10		84%
4.19	400	521	0	0.1	0.09	90:10		
4.21	400	3055	1×10^{-4}	1.0	1.29	90:10		81%

EXPERIMENTAL DETAILS

The ITO films were deposited on silicon and glass substrates simultaneously, using a thermal evaporation chamber from the company "MBE Komponenten". The chamber is equipped with separate effusion cells for Indium and Tin in order to control the deposition rate and In:Sn ratio independently. Oxygen is admitted into the chamber through a needle valve and the O_2 content is controlled by its partial pressure to within ±10 %. The substrate can be heated from the backside with a radiation heater. The substrate temperature is measured 5 mm above the sample and the heater is controlled to within ±0.2 °C. The amount of deposited material is measured in situ by a quartz micro-balance. The microbalance was previously calibrated for Indium and Tin using data from an atomic force microscope and a surface profiler, respectively. The mass flow of metal is controlled to within ±6 % for Indium and ±14 % for Tin by setting the temperature of the effusion cells according to the calibration data. During the same process step, ITO films with the same parameters were evaporated on p-doped silicon with ρ=1-20 Ocm and on microscope coverslips. Immediately before use, silicon substrates were cleaned with acetone, immersed in HF and de-ionized water for 5 minutes each and then blown dry with Nitrogen. Glass substrates were cleaned with acetone. The samples studied here are listed in Table 1 with their corresponding growth parameters. The growth base pressure was about 5×10^{-6} mbar.

The samples evaporated on silicon substrates were cleaved and subsequently characterized with a Philips XL30 SFEG electron microscope and a Philips XL30 with an extension for energy dispersive x-ray analysis (EDX). Optical transmission data were obtained with a standard monochromator-based set-up. Sheet resistance measurements were performed on the films on glass substrate using a four-point probe set-up.

RESULTS AND DISCUSSION

The EDX data for sample 4.19 agree with the pre-set In:Sn ratio to within ±3 %-weight. The EDX data on oxygen content for samples 4.1 to 4.11 show no significant variation within the accuracy of our set-up. However, the increase in deposition rate with increasing O_2 partial pressure for samples 4.19 and 4.1 to 4.5 indicates that more Oxygen appears to be incorporated into samples grown at higher O_2 partial pressure.

Four-point-probe measurements show resistivities from 5.0 to 8.7 mΩcm, which is in good agreement with published results [2,4]. Optical transparency values ranged from 75 to 92% at 550nm, which also compare well with the literature [2,4].

Scanning electron micrographs of an Indium-Tin film without Oxygen (sample 4.19) show the bubble-like growth that is caused by the bad wetting of Indium and Tin on silicon and glass (see Fig. 1). Films of Indium-Tin-Oxide grown at deposition rates of approx. 0.2 Å/s cover the entire exposed substrate and show a grainy morphology with grain sizes of 20 to 50 nm (see Fig. 2). The Oxygen content and Tin doping have no influence on the shape of the deposited films as observed in scanning electron micrographs of samples 4.1 to 4.11. The films obtained in this range of parameters are suitable to act as transparent electrodes. Lowering the substrate temperature (samples 4.13 and 4.15) leads to poorer transmission and conductivity but has no visible effect on morphology.

Increasing the growth rate from 0.23 to 1.33 Å/s leads to a transition from grainy structures (see Fig. 2) to the formation of nanowires, which grow over an initial grainy layer (see Fig. 3). These nanowires typically have diameters of 10 to 20 nm and lengths of 40 to 50 nm for a given film thickness of 55.9 nm. The initial grainy layer is a common feature of all deposited oxides. At low deposition rates, it continues growing in a layer-by-layer-like fashion, while at higher deposition rates nanowires form. In Fig. 4, a sample grown at 1.29 Å/s is shown, which has a nominal thickness of 305.5 nm. It has the initial grainy layer with nanowires protruding from it, but in addition the voids between the wires have been partially filled with material, leading to a porous structure. With a growth rate between 0.2 and 1.3 Å/s it should be possible to obtain films that show only nanowires.

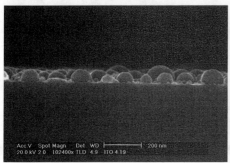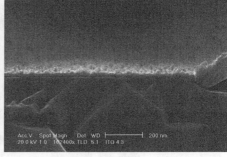

FIGURE 1. Indium-Tin film without Oxygen   **FIGURE 2.** 50 nm ITO film deposited at ~0.2 Å/s

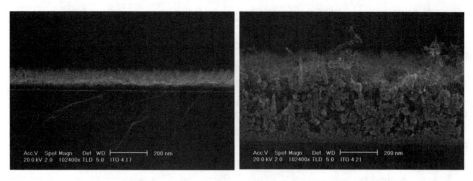

FIGURE 3. 50 nm ITO film deposited at ~1.3 Å/s   **FIGURE 4.** 300 nm (nominal thickness) ITO film deposited at ~1.3 Å/s

CONCLUSIONS

Using thermal evaporation, it is possible to grow ITO films under a wide range of different growth conditions. Nanowires have been observed to grow on top of an initial grainy layer at certain conditions. The resulting films still exhibit good electrical and optical properties. A set of conditions has been shown that favors the formation of nanowire. Further structural studies and measurements of transport are in progress.

REFERENCES

1. Thilakan, P., and Kumar, J., *Materials Science & Engineering B* **55**, 195-200 (1998)
2. Minami, T., Sonohara, H., Kakumu, T., and Takata, S., *Thin Solid Films* **270**, 37-42 (1995)
3. Kwok, H. S., Sun, X. W., and Kim, D. H., *Thin Solid Films* **335**, 299-302 (1998)
4. Baía, I., Quintela, M., Mendes, L., Nunes, P., and Martins, R., *Thin Solid Films* **337**, 171-175 (1999)

Synthesis of Carbon Nanofibers by Catalytic Pyrolysis of Hydrocarbons

M. Ritschel, C. Täschner, D. Selbmann,
A. Leonhardt, A. Graff, J. Fink

*Institute of Solid State and Materials Research Dresden, Helmholtzstraße 20,
D-01069 Dresden*

Abstract. Carbon nanofibers were produced by iron catalyzed pyrolysis of benzene in presence of hydrogen at temperature up to 1150 °C. The fiber diameter varied between 10 and 300 nm and is strongly depended on the total gas flow rate and the deposition time. A short detention time of the carbonfibers in the hot reaction zone is important for the production of nanofibers with a small diameter distribution. An acid treatment and annealing in an oxydizing or reducing atmosphere were performed for purification and etching of the nanofibers. The carbon nanofibers were characterised using x-ray analysis, scanning and transmission electron microscopy.

INTRODUCTION

Carbon nanofibers exhibit unique electronic and mechanical properties. Furthermore, it has been reported that carbon nanofibers can absorb considerable quantities of hydrogen. Therefore, a simple and effective process is desirable for mass production of such materials. The catalytic pyrolysis of hydrocarbons using ferrocene as precursor can satisfied the conditions for synthesis of carbon nanofibers.

This work has been studied the influence of synthesis parameter on the growth of fibers and the influence of chemical and physical treatments on the microstructure of nanofibers, respectively.

EXPERIMENTAL

A successful synthesis of carbon nanofibers by catalytic pyrolysis of hydrocarbons is depended on some important different process parameters as deposition temperature, gas flow rate, reaction time, or evaporation rate of the catalyst. The effect of these parameters was studied with regard to a high deposition rate and certain fiber diameters. The experiments were carried out with benzene as carbon source and ferrocene as catalyst. In benzene 1 vol% thiophene was dissolved. It has been reported, that sulphur containing in thiophene assisted the catalyst effect.

The used deposition equipment consist of a one meter long horizontal quartz hot wall reactor (inner diameter 40 mm). Benzene and ferrocene are separately introduced by argon or nitrogen as carrier gas in the reaction zone. The main hydrogen gas flow

CP544, *Electronic Properties of Novel Materials—Molecular Nanostructures*, edited by H. Kuzmany, et al.

coincides with the ferrocene gas before reaction zone. The ferrocene was reduced by hydrogen. The resulting Fe-atoms combine to clusters, which serve as seeds for the nanofiber growth. The saturation of the carrier gas (Ar, N_2) with the benzene/thiophene-mixture was taken placed in a bubbler at room temperature. The deposition was realized at normal atmosphere pressure. The optimal temperature range using benzene/ferrocene for the production of carbon nanofibers was 1100 °C up to 1150 °C. The evaporation temperature of ferrocene was hold at 110 °C. The deposition time was varied between 5 and 30 minutes.

RESULTS

The x-ray spectra of as grown material reveal the (002)-peak shifted to smaller theta(?)-values in contrast to pure graphite. The broadening of the (002)-peak and the weak intensity of (100), (101) and (004) peaks indicate carbon nanofibers with disordered crystallinity as deposition products.

The first formed fibers are blown out of the reaction zone by the gas flow. Then, the single nanofibers combine to bundles which deposited on cooler positions of the reactor wall and on the thermocouple tube in form of grey-black webs. If the deposition time was increased the web of fiberbundles becomes more dense and grow inside of the hot zone on and on. Secondly, more and more carbon was depositd on the fibers and lead to an increase of the fiber thickness to about 500 nm. Fibers with small diameters and a narrow distribution of diameters can be produced if the fibers are transported away the hot reactor zone by increasing of the gas flow as quickly as possible. In this way nanofibers can be produced with diameters of 10 to 20 nm. Fig. 1 illustrates the dependence of quantity and fiber diameter on the reaction time and the total gas flow rate. The inserted SEM-images show graphitic nanofibers. At a total flow rate of 180 sccm and a deposition time of 10 min fibers are obtained with a average diameter of 70 nm (Fig. 1a). With increasing flow rate the fiber diameter decreases (300 sccm to 25 nm – Fig. 1b), however the quantity is decreased, too. Besides, well-aligned nanofibers were synthesized at higher gas flow rates.

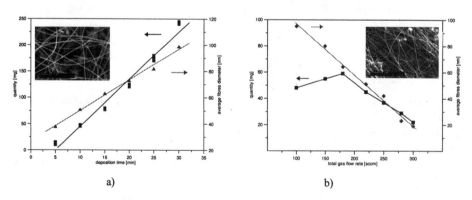

a)   b)

Figure 1. Quantity and diameter of nanofibers in dependence on a) deposition time and b) total gas flow rate

Results of TEM-studies are shown in Fig. 2. The carbon nanofibers consist of a core with an uniform diameter of 4 to 7 nm and a shell, which are formed from several graphitic carbon layers. The number of the carbon layers depend on the detention time of the fibers in the hot reaction zone. Furthermore, the TEM-studies shows, that very often each fiber core has an iron catalyst particle located mostly in the centre of one core end (Fig. 2a). Variable growth forms of fibers have been observed: Fibers with total closed graphitic shells, fibers with delaminated shells (Fig. 2b), fibers with two or three coexisted cores, and fibers which are branched (Fig. 2c). In the as grown fibers the graphite layers are oriented more or less parallel to the direction of the fiber axis.

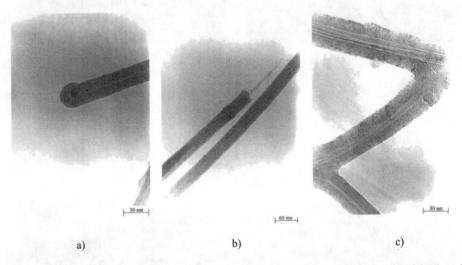

a)   b)   c)

Figure 2. TEM images of as grown fibers a) with a iron particle inside and b) with partly delaminated shell and c) branched fibers with several cores

The as grown material was purified by ultrasonic acid treatment and a following annealing in air atmosphere at temperatures between 490 and 550 °C for 2 to 3 hours or at 750 °C for 1 to 5 minutes in argon/oxygen gas mixtures. The microstructure of the fibers was examined after annealing higher 2000 °C, also.

Continued investigations of heat treatments for opening of the fiber ends are an important aim in the following.

Fig. 3 shows SEM-images of samples which were treated by acid and annealing in air flow. The samples show a reduced content of iron and carbon particles. The fibers are partly broken and the fiber ends are partly opened.

"As grown" material was heated at 750 °C for 1 - 5 minutes in a gas mixture of argon with 10 vol% oxygen. TEM-images show fibers annealing for 3 minutes, with partly etched shells. The fibers possess shells, which show a significantly dissolved feature of the outer carbon layers. At reaction time of more than 3 minutes the main parts of fibers were totaly dissolved. The selective dissolution of the fibers is attributed to the different reactivity between carbon-containing materials. For example the reactivity of the fiber core is lower than of graphitic carbon layers of the shells.

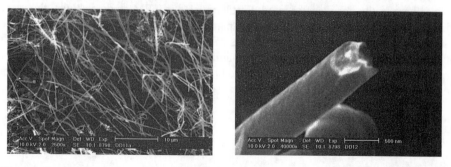

Figure 3. SEM-images show carbon nanofibers which purified by acid with following annealing in air

For the post-annealing at high temperatures the as-deposited material was put in an graphite crucible and heated up to 2500 °C for 15 minutes in argon atmosphere. HRTEM-images (Fig. 4) show ordered ranges of carbon fibers after the heat treatment. The graphite atomic planes of the fibers are arranged exactly parallel to the fiber core (Fig. 4a). Single ranges are observed which in the graphite atomic planes are tilted against the core by 45° (Fig. 4b).

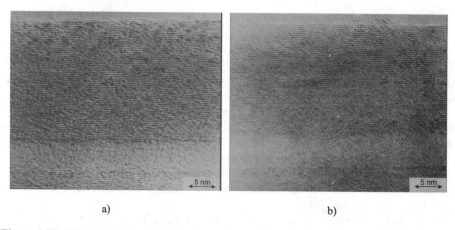

a)   b)

Figure 4. HRTEM-images of carbon nanofibers: a) graphite atomic planes arranged exactly parallel to the fiber core; b) graphite atomic planes arranged parallel to core and tilted against the core by 45° (right at the top) [TEM: IFW Dresden and MPI of Microstructure Physics Halle]

ACKNOWLEDGMENTS

This work is supported by BMBF, project 13N7575.

Nanotextured Carbon Films from Fluorinated Hydrocarbons: Kinetics and Mechanism of their Growth

Ladislav Kavan

J. Heyrovský Institute of Physical Chemistry, Dolejškova 3, CZ-182 23 Prague 8, Czech Republic

Abstract. Self-standing carbon films were prepared at 25°C by defluorination of gaseous perfluorohexane, perfluorocyclobutane, perfluorocyclohexane, perfluorodecalin, per-fluorocyclopentene, perfluorobenzene and perfluoronaphthalene with amalgams of Li and Na. The reaction allows growth of precise carbon films down to nm thickness. Some films contained C_{60} and carbon nanotubes. The film's thickness was evaluated from the UV-Vis reflectance spectra. The reaction with Li or Na was controlled by different kinetic equations, which contrasts to the reaction of alkali metal amalgams with solid fluoropolymers. A model elucidating these discrepancies has been proposed.

INTRODUCTION

Reduction of perfluorinated hydrocarbons (C_mF_n) with alkali metal (M = Li, Na, K) amalgams produces smoothly elemental carbon at room temperature [1]. If we neglect n-doping of carbon [1], the process is described by a general reaction:

$$C_mF_n + n M_{(Hg)} \rightarrow m\, C + n\, MF \qquad (1)$$

Depending on the precursor (C_mF_n), the reactions produce various graphitic and non-graphitic materials. Of particular interest is the formation of polyyne from poly(tetrafluoroethylene) (PTFE) [1,2], fullerene C_{60} and multi-walled carbon nanotubes [3] from perfluorocyclopentene, perfluorodecalin and perfluoronaphthalene. Carbonization of PTFE with $M_{(Hg)}$ gives a carbonaceous composite with interspersed nanocrystalline MF. The C-MF layer exhibits both electronic (σ_e) and cationic (M^+, σ_i) conductivities, hence, a short-circuited interfacial galvanic cell is formed:

$$M_{(Hg)} \mid C\text{-}MF \mid PTFE \qquad \text{(Scheme 1)}$$

A classical model [1] assumed that the reaction is governed by a self-discharge of the cell (Scheme 1). This easily leads to a kinetic equation for the thickness of C-MF layer (δ) [1]:

$$\delta = \sqrt{\frac{EaM\sigma_e\sigma_i t}{2F\rho(\sigma_e + \sigma_i)}} = k \cdot \sqrt{t} \qquad (2)$$

CP544, *Electronic Properties of Novel Materials—Molecular Nanostructures*, edited by H. Kuzmany, et al.
© 2000 American Institute of Physics 1-56396-973-4/00/$17.00

E is the cell voltage, $a = \delta/\delta_0$ (δ_0 is the thickness of the consumed PTFE), M is the molar mass (of CF_2 unit), t is the reaction time, F is Faraday constant and ρ is the density of PTFE. The rate constants, k (in $nm/s^{1/2}$ at 25°C) are: 57.9 (M=Li), 5.18 (M=Na) and 0.27 (M=K) [1]. Eq. (2) is valid also for numerous other fluoropolymers [1].

EXPERIMENTAL SECTION

Perfluorohexane, perfluorocyclobutane, perfluorocyclohexane, perfluorodecalin, perfluorocyclopentene, perfluorobenzene and perfluoronaphthalene were purchased from FLUOROCHEM, Ltd., ABCR, GmbH and Aldrich. Li and Na (from BDH) were dissolved in mercury to a concentration of 0.9-1.3 at%. The perfluorinated precursor (ca. 2-5 g) was first dried by shaking with ca. 10 ml of Li amalgam, and distilled. All operations were carried out in evacuated (<1mPa) all-glass apparatus. UV-Vis reflectance spectra of the growing film were measured at 25°C in-situ (Ocean Optics 2000 spectrometer, vacuum optical cell, Hellma). The film's thickness was evaluated from the reflectance spectra using the Nanocalc software.

RESULTS AND DISCUSSION

Immediately after a purified vapor of C_mF_n is exposed the surface of liquid $M_{(Hg)}$, an interfacial C-MF film starts to grow. Chemical analysis revealed a quantitative defluorination of C_mF_n for all the studied precursors (cf. Eq. 1). Carbonaceous films from C_5F_8, $C_{10}F_8$ and $C_{10}F_{18}$ contained about 1 % of C_{60} and multiwalled carbon nanotubes (for details see Ref. [3]). The prime thin films exhibited uniform interference color, which confirms that the film thickness is controlled in a sub-micron scale. The thickness was evaluated from the UV-Vis interferences in reflectance spectra (for $\delta v < \approx 2\mu m$, v is the refractive index). Since v is unknown for C-MF, the experimental δv values were used for kinetic analysis (supposing the refractive index to be independent of the layer thickness). Surprisingly, the carbonaceous film at the $C_mF_n(g)/Li_{(Hg)}$ interface grew linearly with time:

$$\delta v = K_{Li} \, t \qquad (3a)$$

(K_{Li} is a constant). On the other hand, the reaction of $C_mF_n(g)$ with Na-amalgam followed the "normal" kinetic behavior (cf. Eq. 2):

$$\delta v = K_{Na} \, t^{1/2} \qquad (3b)$$

Figure 1 displays an example of these two reactions for perfluorodecalin. There is a clear contrast with the reaction of $M_{(Hg)}$ with PTFE and other solid fluoropolymers [1], which are controlled by Eq. (2) independent of the alkali metal type (Li, Na, K). The actual values of rate constants, K_{Li} and K_{Na} (at 25 °C) are summarized in Table 1. Apparently, precise carbonaceous films (with thicknesses of the order of 10 to 10^2 nm) can simply be grown by adjusting the reaction time.

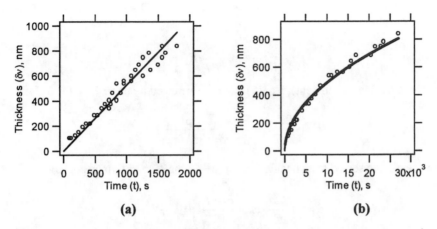

FIGURE 1. Growth of a film from gaseous perfluorodecalin and (a) Li-amalgam, (b) Na-amalgam

TABLE 1. Rate constants of the reaction of C_mF_n with Li/Na amalgams at 25 °C

Precursor	$K_{Li}\ [nm.s^{-1}]$	$K_{Na}\ [nm.s^{-1/2}]$
Perfluorohexane	0.03	-
Perfluorocyclobutane	0.2	-
Perfluorocyclohexane	0.002	-
Perfluorodecalin	0.5	5
Perfluorocyclopentene	7	9.5
Perfluorobenzene	0.06	-
Perfluoronaphthalene	0.15	18

To interpret the kinetic discrepancies, an upgraded model is proposed bellow. The interfacial reaction is still assumed to be governed by electrochemical mechanism (as in Scheme 1). Hence, the self-discharge current density, j follows the Faraday law:

$$j = \frac{nF\delta\rho}{aMt} \tag{4}$$

where n is the number of electrons (cf. Eq. 1). The cell voltage (E) is a sum of activation voltage (E_a) and ohmic drop (E_Ω):

$$E = E_a + E_\Omega \tag{5}$$

(Note that the classical model of the PTFE/$M_{(Hg)}$ interface (cf. Eq. 2 and Ref. [1]) considered only E_Ω but neglected E_a). The current density (j) is controlled by the total cell conductivity, σ:

$$j = \frac{E_\Omega\sigma}{\delta} \tag{6}$$

where $\sigma = \sigma_e\sigma_i/(\sigma_e+\sigma_i) \approx \sigma_i$; $(\sigma_e >> \sigma_i)$. However, the current density is also controlled by the activation voltage:

$$j = j_0 \cdot \exp\left(\frac{\alpha nF}{RT}E_a\right) \tag{7}$$

(α is charge transfer coefficient). Eq. (5) can be rearranged by using Eqs. (4, 6, 7):

$$E = \frac{RT}{\alpha nF} \cdot \ln\frac{\delta\rho nF}{j_0 tMa} + \frac{\delta^2\rho nF}{\sigma Ma} \tag{8}$$

Eq. (8) cannot be solved analytically, but for $\delta \to 0$, the second term is negligible:

$$\delta \approx \frac{j_0 tMa}{\rho nF} \cdot \exp\left(\frac{\alpha nFE}{RT}\right) = K_1 \cdot t \tag{9}$$

Analogously, for $\delta >> 0$ (or $E_a \to 0$), the first term is negligible:

$$\delta \approx \sqrt{\frac{E\sigma tMa}{\rho nF}} = K_2 \cdot \sqrt{t} \tag{10}$$

Eq. (10) is, apparently, identical to that, derived for a classical model (Eq. 2) [1].

To conclude, the reaction kinetics the $C_mF_n/M_{(Hg)}$ is rigorously described by Eq. (8), but (depending on the actual values of E_a and δ) it may approach two limiting situations described by Eqs. (9) and (10), respectively. For solid fluoropolymers, the second limiting scenario (Eq. 10) operates at all usual experimental conditions. The same holds for $C_mF_n(g)/Na_{(Hg)}$, where the growth rate is again limited by the film's Na^+-conductivity. (The corresponding rate constant for PTFE would be $K_{Na}=(5.18 \cdot v)$ nm/s$^{1/2}$, which is not very different from the rate constants K_{Na} collected in Table 1). Since the Li^+ conductivity is much higher, the reaction with $Li_{(Hg)}$ is controlled by the rate of the $C_mF_n(g)$ reduction at the film's surface (Eqs. 3a, 9). This is likely to be slower for gaseous reactants.

ACKNOWLEDGMENTS

This work was supported by the Grant Agency of the Czech Republic (contracts Nos. 203/98/1168 and 203/99/1015).

REFERENCES

1. Kavan, L. *Chem.Rev.* **97**, 3061-3082 (1997).
2. Kavan, L., Dousek, F. P., Janda, P. and Weber, J. *Chem.Mater.* **11**, 329-335 (1999).
3. Kavan, L. and Hlavaty, J. *Carbon* **37**, 1863-1865 (1999).

Synthesis of Silicon Nanowires and Novel Nano-Dendrite Structures

S. Sinha, B. Gao, and O. Zhou

Department of Physics & Astronomy and Curriculum of Applied and Materials Sciences
University of North Carolina, Chapel Hill, NC-27599, U.S.A.

Abstract. We report the synthesis and characterization of the silicon nanowires. It is observed that, by changing the growth conditions we can synthesize a branched structure which we name as Silicon Nano-dendrites (SiND). Similar structure has been observed before in metals but at a much larger scale. We propose a growth mechanism for this kind of novel structure based on the Vapor-Liquid –Solid (VLS) model. We also study the quantum confinement effects on the melting transition of the nanowires.

Nanonoscale materials have fascinated scientists from its very inception as it has made their dream of tailoring materials according to their requirement much closer to reality. Nanoscale materials can be 0-D [1] (e.g. quantum dots), 1D[2] (e.g. nanotubes and nanowires) and 2-D[3] (e.g. MBE grown films). Out of these three varieties the study and analysis of the 1-D nanoscale materials is of primary importance presently. We are reporting the synthesis of very interesting branched silicon nanostructures which we call Silicon Nano-Dendrites in this article. We will also discuss the dependence of various parameters in their synthesis and application possibilities along with a probable synthesis mechanism.

Silicon nanowires are expected to have different thermodynamic and electronic properties in comparison to bulk silicon because of its low dimension. Calculations[4] have predicted these dimensional effects (known as quantum confinement) to be observable only when the dimensions are lower than 25 nm for silicon. Recently there are reports[5,6] on the observation of photoluminescence on silicon nanowires and nanoparticles, an experimental evidence of the quantum confinement. We have also observed this effect on the melting point of the nano-particles and nano-wires. The synthesis of long wire like structures of silicon (diameter 10-100 micron), known as whiskers is well established[7]. Unfortunately the whiskers were too large to show any signs of quantum confinement and hence were not interesting to study. These silicon whiskers were synthesized by VLS (Vapor-Liquid-Solid) mechanism by flowing silane gas (SiH_4) over a heated crystalline silicon substrate containing metal (e.g. Fe, or Au) islands. The whiskers grew out from these metal islands. The whiskers grow when

CP544, *Electronic Properties of Novel Materials—Molecular Nanostructures*, edited by H. Kuzmany, et al.
© 2000 American Institute of Physics 1-56396-973-4/00/$17.00

excess silicon precipitates out from the liquid silicon/metal (eutectic) cluster as it cools. Hence the diameter of the liquid silicon/metal cluster decides the diameter of the whisker. However equilibrium thermodynamics puts a lower limit to the size of these clusters and hence the whiskers to 0.2μm. in diameter[8] produced by this VLS mechanism. Hence, non-equilibrium processes like Laser vaporization[9] or catalytic decomposition[10] is required to produce nanometer sized clusters. Other nanoscale materials like carbon, germanium, nitrides and carbides are also synthesized by the same technique[11].

Silicon nanowires have been synthesized[12] earlier by Laser ablation and we adopted the same method to study its growth and use that knowledge to synthesize this dendrite structure. A 532 nm Nd-YAG laser is used to ablate a $Si_{.9}Fe_{.1}$ target placed in a quartz tube (2" diameter and 4' long) which is placed within a three zone furnace. The central zone of the furnace is kept at 1175 C, while the outer zones are kept normally at 1150 C. The quartz tube is cooled by ice cold water on either sides and a constant flow of argon gas is maintained to prevent oxidation of the nanowires as well as maintaining a temperature gradient across the tube. A coated (non-reflecting) lens is used to focus the laser beam on the target with a spot size of $1mm^2$. The Laser spot is scanned both horizontally and vertically on the entire target. Silicon nanowires are formed in the back of the quartz tube and looked like a gray and mat-like material. In a typical setting the laser ablation produces 8mg of the mat-like material per hour. The Transmission Electron Microscope (TEM) is used to identify the silicon nanowires

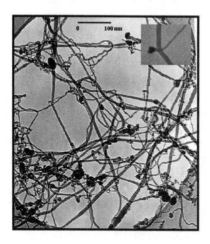

FIGURE 1. Typical TEM picture of SiNW. The long spaghetti like features are the nanowires and the black blobs are the nano -particles. The SiNWs are coated with a layer of SiO_2, which is measured (EDX) to be 15% of the weight . The inset shows a high resolution TEM picture of the nano-particles (Determined to be $FeSi_2$) They are approximately 10 nm in diameter.

from the mat (Figure-1) and we estimate that about 40%-50% of the mat are nanowires. The long and spaghetti-like silicon nanowires originate from the Si/Fe

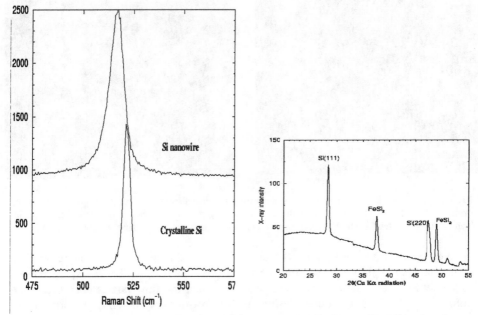

FIGURE 2 Left:- Raman spectra comparison of the SiNW and bulk silicon using Dilor micro-Raman spectrometer using 514 nm Laser light . We observe a shift of 5 cm^{-1} for the SiNW. Right:- Powder X-Ray spectra for SiNW taken in a Enraf Nonius FR590 system. It shows the silicon as well as the FeSi$_2$ peaks.

cluster (nanoparticles) and X--Ray as well as electron diffraction analysis determined it as FeSi$_2$. The X-Ray (Figure 2 –Right) also identified the main silicon peaks and aided in characterizing the sample.

After initially ablating the target for 3 to 4 hours when the argon gas flow rate is suddenly reduced, then some mat-like deposit is formed in the front portion of the quartz tube. The TEM picture (Figure 3b) of this material reveals a dendritic structure observed previously in metals[13] but in a much larger length scale. They typically comprise of a long primary backbone with a diameter of 15-75 nm with short and narrow secondary and in some cases tertiary branches emanating which are orthogonal to each other. Both the primary and secondary branches have a nanoparticle at their end. An intermediate energy density (175mJ/pulse) focussed over an area of 1mm^2 on the target, along with a sudden decrease of the argon flow-rate is essential for the formation of the Silicon Nano-Dendrites (SiND).

Similar branching and kinks but at a larger length scale have earlier been observed for silicon whiskers,[14] which were explained by a VLS growth mechanism. The sudden decrease of flow-rate, which produces momentary thermal instability is the key for synthesizing these structures. The Laser heats up the target instantaneously to form the eutectic Si/Fe cluster in liquid phase (liquid droplet).

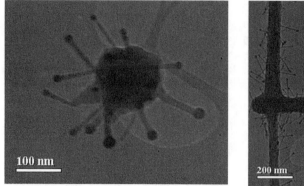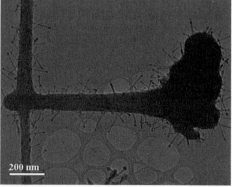

FIGURE 3. Left: SiNWs protruding out form the nanocluster. Some of these will grow to form the primary branches while the others will form secondary and tertiary branches (Dendrites). Right: Typical picture of the SiNDs. The SiNDs grow form the primary and secondary branches.

Normally (normal flow rate) the cluster is in a silicon rich atmosphere and the silicon condenses into liquid, from the vapor phase as it cools. However, the cluster spends much longer time at high temperatures when the flow rate is decreased. The volume of the liquid droplet during VLS growth depends on the surrounding temperature[15] as well as on the silicon availability in vapor state.

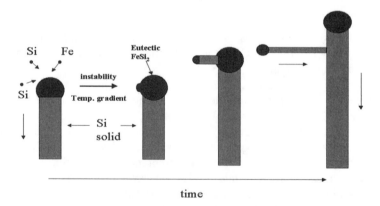

FIGURE 4. The figure shows the proposed growth of the SiND with time. The growth starts from a Si/Fe eutectic melt (nucleation site). The sudden instability of temperature gradient produced by the change in flow rate produces blimps on the melt surface. The SiNDs grow out from the melt but the primary branch grows faster in the direction shown by the arrow.

The sudden increase of temperature will cause a temporary instability in the droplet leading to an increase in volume[14]. This increase in volume will cause bulges on the surface of the liquid droplet (Figure 4). As this bulged liquid droplet starts cooling in a silicon rich vapor, the super-cooled silicon starts precipitating out forming nanowires in all directions of the droplet like a crown (Figure 3a). However, the system has only one general temperature gradient direction and also the volume of these bulges are much smaller compared to the parent droplet. The temperature gradient and the volume of the droplets favor growth in one specific direction and so it keeps on growing while the others stop. Hence they form a primary branch with a number of small branches (SiND).

We have also investigated some of the quantum confinement properties of the nano-silicon. We see a shift of 5 cm^{-1} in the Raman peak (Figure 3b) for the silicon nanowires compared to bulk silicon, a definite sign of quantum confinement[16]. We have observed the melting transition for the silicon nanowire at 1350 C compared to the 1423 C melting point for bulk silicon (Figure 5). There are different diameters of nanowire in the sample and hence the melting transition is broad. According to measurements on silicon nanocrystals[17] the lowering of melting point is inversely related to the diameter of the nano-particle. This surface – phonon instability model[18] predicts the melting point of a 10 nm diameter particle to be 1320 C. This is in good agreement with our data. The model assumes the melting to nucleate from an intrinsic surface defect. The melt nucleates on the solid particle surface and spreads, eventually forming a molten lobe, which finally envelopes the remaining solid. Hence the surface energy plays a key role in determining the melting temperature.

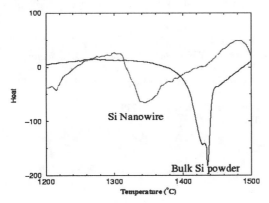

FIGURE 5. The comparison of the melting transition of SiNW and bulk silicon. 3~4 gm of the powder sample is heated in an argon atmosphere and the heat flow into the sample is measured as a function of the temperature. The melting point of SiNW is 1350 C, lower than bulk Si using a Perkin Elmer DTA-7 (Differential Thermal Analysis) calorimeter.

This SiND can have a number of important applications. The FeSi$_2$ particle can also be used as a catalyst for growing other nanoscale materials like carbon nanotubes. Then the silicon nanowire can be used as a template to grow nanotube-nanowire junctions in bulk. A better control on these SiNDs will aid us in non-lithographically patterning nano-junctions according to our requirement. This elegant method will speed up our endeavor in molecular electronics.

ACKNOWLEDGMENTS

We acknowledge Drs J. Lorentzen and L. McNeil for helping us with the Raman measurements. We also thank L. Fleming for help with the Laser ablation system. Finally we thank NASA and ONR for funding this research.

REFERENCES

1. Steigerwald, M.L.and Brus, L.E., *Annual Rev. Mater. Sci.* **19**, 471-495 (1989)
2. Ijima, S., *Nature*, **254**, 56-58 (1991)
3. Kapon,E., Hwang,D.M. and Bhat, R. *Phys. Rev. Lett.* **63**, 430 (1989).
4. Schuppler, S., Friedman, S.L., Marcus, M.A., Adler, D.L., Xie, Y.-H., Ross, H.F.M., Chabal, Y.J., Harris, T.D., Brus, L.E., Brown, W.L., Chaban, E.E., Szajowski, P.F., Christman, S.B., and. Citrin, P.H., *Phys. Rev. B* **52**,4910 (1995).
5. Geohegan, D.B., Puretzky, A.A., Duscher, G., and Pennycook, S.J., *Appl. Phys. Lett.* **73**, 438 (1998)
6. Lee, S.T., Wang, N., Zhang, Y.F., and Tang, Y.H., *MRS Bulletin* **24**, 36 (Aug. 1999)
7. Wagner, R.S., and Ellis, W.C., *Appl. Phys. Lett.* **4**, 89 (1964).
8. Hu, J., Odom, T.W., and Lieber, C.M., *Acc. Chem. Res.*, **32**, 435 (1999)
9. Thess, A., Lee, R., Nikolaev, P., Dai, H.,. Petit, P., Robert, J.,. Xu, C., Lee, Y.H., Kim, S.G., Rinzler,A.G., Colbert, D.T., Scuseria, G.E., Tomanek, G.E., Fischer, D., Smalley, J.E., *Science*, **273**, 483 (1996)
10. Kamins, T.I., Williams, R.S., Chen, Y., Chang, Y.-L., and Chang, Y.A., *Appl. Phys. Lett.* **76**, 562 (2000)
11. Duan, X., and Lieber, C.M., *J. Am. Chem. Soc.* **122**, 188 (2000)
12. Morales, A.M., and Lieber, C.M., *Science*, **279**, 208 (1998)
13. Porter, D.A., Easterling, K.E., and Reinhold, V.N., 1981, *Principles of Solidification by B. C. Chalmers*, Wiley, 1964 pp. 186- 259
14. Wagner, R.S., and Doherty, C.J., *J. Electrochem. Soc.: Solid State Science*, 93, (Jan. 1968)
15. Givargizov, E.I., *J. Cryst. Growth*, **20**, 217 (1973)
16. Li, B., Yu, D.P., and Zhang, S.-L., *Phys. Rev. B*, **59**, 3, 1645 (1999)
17. Goldstein, A.N., *Appl. Phys. A*, **62**, 33 (1996)
18. Wautelet, M., *J. Phys. D*, **24**, 343 (1991)

CdSe Nanoparticle Arrays Contacted on Electron Transparent Substrates

G. Philipp, M. Burghard, and S. Roth

Max-Planck-Institut für Festkörperforschung, Heisenbergstr. 1, D-70569 Stuttgart, Germany

Abstract. Arrays of CdSe clusters were obtained by deposition of the clusters (diameter $d \approx 4$ nm) on 20 nm thin, electron transparent Si_3N_4 substrates using a self-assembly process. The obtained cluster arrangements were characterized by transmission electron microscopy. To contact the cluster arrays, metal electrodes were defined by electron beam lithography on the Si_3N_4 substrates. A metallic layer evaporated onto the rear side of the membranes served as a backgate. The investigated junctions showed a very high resistance in spite of closely packed CdSe cluster monolayers.

INTRODUCTION

To perform electrical transport investigations on individual clusters it is necessary to contact these clusters to external sources and probes. Achieving electrical contact to one individual CdSe cluster [3] with diameter $d \approx 4$ nm in a reproducible way still constitutes a formidable technological challenge [1]. It is especially challenging to characterize the obtained cluster/electrode arrangement since transmission electron microscopy is required. In the present work, this is realized by using an electron transparent substrate. Suitable for this purpose are Si_3N_4 membranes that can be prepared using conventional silicon processing techniques [2]. The obtained Si_3N_4 membranes have a uniform thickness of 20 nm over an area of 200 μm x 200 μm with a roughness of less than 0.5 nm, can be easily handled using normal tweezers and are able to withstand several cooling cycles between room temperature and liquid helium temperature (T=4.2K).

DEPOSITION OF CDSE CLUSTERS ON SI$_3$N$_4$-SUBSTRATES

The nearly monodispersed 4 nm diameter CdSe nanocrystals [3] are stabilized by a ligand shell consisting of trioctyl phosphine oxide (TOPO) and thus are soluble in organic solvents. When nanocrystal solution is brought in contact with the unmodified surface of Si_3N_4-substrates, virtually no adsorption takes place.

CP544, *Electronic Properties of Novel Materials—Molecular Nanostructures*, edited by H. Kuzmany, et al.
© 2000 American Institute of Physics 1-56396-973-4/00/$17.00

To appropriately modify the silicon nitride surface prior to attaching CdSe nanocrystals, the substrate was first exposed to UV ($\lambda = 254$ nm) radiation in order to improve the wetting of the surface by aqueous media. The second step consisted in covering the substrate with 20 μl of a solution of 10 mg chitosane and 10 μl of 3–mercapto–propionic acid in 20 ml of water. After drying of the droplet, excess polymer was removed by immersing the substrate in hydrochloric acid (5 wt%) for 15 minutes, followed by thorough rinsing with pure water and drying in air. Subsequently, the substrate was placed for 2 hrs into a solution of CdSe nanocrystals in toluene, then rinsed with pure toluene, and finally dried in air. Fig. 1 shows a TEM micrograph of a CdSe nanoparticle assembly obtained on a

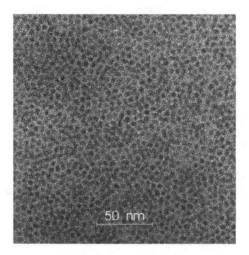

FIGURE 1. TEM-micrograph of an array of CdSe clusters deposited on a Si_3N_4 membrane with a chitosane salt pre-adsorbed as binding layer.

Si_3N_4 substrate. The resulting pattern is determined by the microscopic interparticle and particle/surface interaction. As the concentration at the interface increases, the nanoparticles first behave like a two-dimensional gas and finally form a densely packed monolayer on the membrane substrate with, however, no hexagonal close packing due to the relatively strong particle/surface interaction.

CONTACTING OF CDSE CLUSTERS BY METAL ELECTRODES

In order to contact these cluster arrangements to external sources and probes for electrical transport investigations, devices like the one shown as a schematical cross section in Fig. 2 have been fabricated.

The small gap metal electrodes were defined using a standard bilayer PMMA resist electron beam lithography process. The gap sizes obtained are in the order of 20 nm, but can be reduced to 5 nm by using a special shadow evaporation technique [4]. The CdSe clusters were deposited onto the substrates with metal

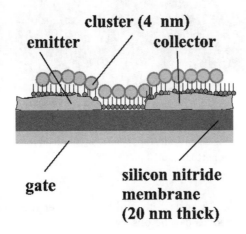

FIGURE 2. Schematic representation of a device consisting of CdSe clusters deposited into the gap between two Cr/AuPd electrodes using chitosane as a binding layer. The electrodes are defined by electron beam lithography on a Si_3N_4 membrane. The size of the gap can be as small as 5nm, if a special shadow evaporation technique is used.

electrodes as described in the previous section. A TEM micrograph of clusters deposited into the gaps between two metal electrodes is shown in Fig. 3. The size of the smaller gap in Fig. 3 is 20 nm whereas the size of the larger onbe is $\sim$ 50 nm. Electrical transport measurements of such devices revealed very high resistances in the $T\Omega$ range. This result is according to earlier findings [1], where only a very low yield of working devices was reported, and suggests that such metal electrodes are not the ideal means for directly contacting one *individual* cluster. Clusters with a few nm diameter are not accessible by lithographic techniques due to the limitation

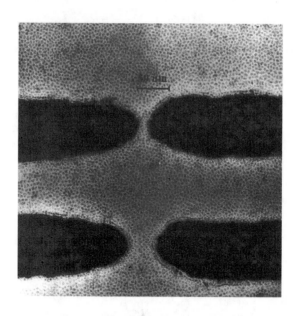

FIGURE 3. A TEM micrograph of CdSe clusters deposited into the gaps of two Cr/AuPd electrodes defined by electron beam lithography on a Si_3N_4 membrane. The size of the top gap around is 20 nm, that of the bottom gap 50 nm.

in resolution of the PMMA resist. In addition, cluster-like grains formed during evaporation of the electrodes can be observed in transmission electron microscopy. These grains might influence the measured current-voltage characteristics.

ACKNOWLEDGMENTS

The expert help of M. Kelsch with TEM specimen preparation and the supply of CdSe cluster material by A. Mews and Th. Basché (Universität Mainz, Germany) are gratefully acknowledged.

REFERENCES

1. D. L. Klein, *et. al.*, *Nature* **389**, 699 (1997).
2. J. W. M. Jacobs and J. F. C. M. Verhoeven, *J. Microscopy* **143**, 103–116 (1986).
3. C. B. Murray, *et. al.*, *J. Am. Chem. Soc.* **115**, 8706–8715 (1993).
4. G. Philipp, *et. al.*, *Microelectronic Engineering* **46**, 157–160 (1999).

Nanoparticles of CdCl$_2$ with Closed Cage Structures

R. Popovitz-Biro, A. Twersky, Y. Rosenfeld Hacohen, and R. Tenne

Department of Materials and Interfaces, Weizmann Institute, Rehovot 76100, Israel

Abstract. Nanoparticles of various layered compounds having a closed cage or nanotubular structure, designated also inorganic fullerene-like (*IF*) materials, have been reported in the past. In this work *IF*-CdCl$_2$ nanoparticles were synthesized by electron beam irradiation of the source powder leading to its recrystallization into closed nanoparticles with a nonhollow core. This process created polyhedral nanoparicles with hexagonal or elongated rectangular characters. The analysis also shows that, while the source (dried) powder is orthorhombic cadmium chloride monohydrate, the crystallized *IF* cage consists of the anhydrous 3R polytype which is not stable as bulk material in ambient atmosphere. Consistent with previous observations, this study shows that the seamless structure of the *IF* materials can stabilize phases, which are otherwise unstable in ambient conditions.

INTRODUCTION

Graphite nanoparticles were shown to be unstable in the planar form and undergo a spontaneous morphological transformation, which causes them to curl up into nanoballs (fullerenes) [1] or nanotubes [2]. The stimulation for this morphological change is the decrease in the size of one such layer, which increases the ratio between the dangling bonds at the periphery of the layer and the number of atoms within the layer (bulk). The larger the ratio, the greater the instability due to the high surface energy and therefore the propensity to form hollow cage structures. Deposing pentagons into the otherwise hexagonal network, imposes curvature into the layer. Thereby, the layer is able to close on itself and eliminate the dangling bonds. This curvature causes stress in the layer, due to the deviation of the sp^2 bonds from planarity. The elastic stress energy is nonetheless more than compensated by elimination of the dangling bonds. Therefore the total energy of the closed cage nanostructures is reduced relative to the planar form of graphite. However, large energy input is required in order to initiate this folding process, which explains the need for the high temperatures or other energy sources, in order to produce carbon fullerenes and nanotubes.

The research into C$_{60}$ and related structures, like nanotubes, paved the way for the discovery of such morphologies in other layered materials, such as MS$_2$ [3-5] (M=Mo,W, etc), other chalcogonides [6,7], BN [8], halides [9], oxides [10-12], and others. They were all found to create fullerene-like structures in a manner analogous to

CP544, *Electronic Properties of Novel Materials—Molecular Nanostructures*, edited by H. Kuzmany, et al.
© 2000 American Institute of Physics 1-56396-973-4/00/$17.00

carbon. The structures of the layered compounds are more complicated than the simple hexagonal structure of graphite sheets, and therefore the stress relief mechanism is more complicated, too. Although *IF* of suggested metal chalcagonides have been relatively thoroughly investigated [3-5], their stress relief structure is not yet fully understood. In this respect, it was proposed, that triangles and squares (rectangles, rhombi), rather than pentagons, will be preferred as the lower symmetry element, which is responsible for the curvature of the *IF* structures. This hypothesis was confirmed in recent studies of both MoS_2 and BN. Thus, MoS_2 polyhedra with octahedral structure were recently synthesized [13]. In another recent study, MoS_2 nanotubes with either three rectangles or one octagon and four rectangles (negative curvature) in their cup, were proposed to be stable structures[14]. *IF* structures of boronitride were shown [15,16] to contain B_2N_2 squares, rather than pentagons, which are typical of carbon fullerenes and nanotubes. MoS_2 crystallizes in a trigonal prismatic structure (2H), which is not likely to form octahedral polyhedra easily, but has a metastable tetragonal phase (1T) with Mo in octahedral coordination [17]. On the other hand, $CdCl_2$ crystallizes in a hexagonal lattice in which each Cd atom is surrounded by 6 Cl atoms in an octahedral coordination [18]. It was hypothesized that the octahedral environment of Cd in $CdCl_2$ will facilitate the formation of *IF* moieties with octahedral shape. In addition, *IF*-$CdCl_2$ can be conceptually made by pure evaporation/condensation process at 750°C, which was thought to be a good practice to obtain nanoparticles with closed cage structure. Previous work on the synthesis of closed cage (fullerene-like) structures of $NiCl_2$, revealed three typical topologies [9]: multilayer (onion) nanoparticles of a spherical shape; multilayer nanotubes, and close cages structures with hexagonal topology. The first two kinds of structures were observed before, while the hexagonal topology was unique to this compound. Since $NiCl_2$ crystallizes in the cadmium chloride (layered) structure, it was interesting to know if this latter topology is common to other materials with cadmium chloride structures, like $CdCl_2$ itself.

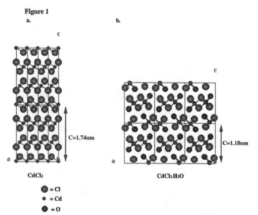

FIGURE 1. Schematic illustration of the crystal structure of (a.) $CdCl_2$ and (b.) $CdCl_2.H_2O$, both shown in the (010) projection.

The CdCl$_2$ structure is made of a staggered stacking of three Cl-Cd-Cl layers having interlayer distance (c/3) of 0.58nm and a rhombohedral unit cell (3R) (Space group *R3m*) [19], as shown in **Fig.1a**. The crystal of CdCl$_2$ is not stable in the ambient atmosphere. Like many other layered compounds, especially the more ionic ones like those containing halides, facile water intercalation into the van der Waals gaps occurs through the prismatic edges. It absorbs, first one water molecule and transforms readily into the monohydrate (CdCl$_2$.H$_2$O). The monohydrate is also a layered compound that packs in an orthorhombic space group (*Pnma*) [20] as shown in **Fig.1b** and layer to layer distance of 0.59nm (c/2). In this structure one water molecule replaces a chlorine atom and therefore each Cd atom is surrounded by 5 Cl atoms and one oxygen atom, with the hydrogen atom pointing out into the van der Waals gap, forming a hydrogen bond between two of the oxygen atoms of neighboring layers. Absorption of water through intercalation into the van der Waals gap of the layered monohydrate can lead to its exfoliation and eventually to dissolution of the salt by the self-absorbed water molecules. On the other hand, CdCl$_2$ containing more than one molecule of water, like CdCl$_2$.2.5H$_2$O, is known [21]. In this crystal, the cadmium atoms are octahedrally coordinated, and the octaheders are tightly bound together through hydrogen bonds. They are sensitive to the ambient atmosphere and therefore can not be studied in the air. Furthermore, absorption of water leads to the formation of higher hydrates and to the self dissolution of the salt in the absorbed water molecules. These chemical and structural changes explain the great difficulties encountered during the present study.

EXPERIMENTAL METHODS

CdCl$_2$ powder was dried at 200°C under Ar gas flow for 2 hr. X-ray powder diffraction confirmed that the dried powder consisted of the monohydrate. The powder was suspended in ethanol or CCl$_4$ and was dripped onto a Cu electron microscope grid and studied by transmission electron microscopy (TEM) (Philips CM120 operating at 120kV). Irradiation of the cadmium chloride monohydrate powder by the electron beam of the TEM for approximately 15 min led to a recrystallization and produced cage structures as reported below.

RESULTS AND DISCUSSION

Careful observation under the electron microscope was done on a preheated cadmium chloride material. According to XRD and electron diffraction (ED) analyses, the precursor consists mostly of CdCl$_2$.H$_2$O, which transforms *in situ* under the electron beam irradiation into *IF*-CdCl$_2$ nanoparticles, i.e. closed cage structures of the anhydrous 3R-CdCl$_2$. This chemical and structural transformation is demonstrated in the TEM micrographs shown in **Figs.2**. The low magnification image shows that under the e-beam irradiation, the macroscopic particle of the monohydrate breaks down and recrystallizes into new nanoparticles. The structural and chemical transformation is further elucidated using selected area electron diffraction from a 80x80 nm^2 region and high resolution imaging of this domain. Although a large number of nanocrystals contribute to the electron diffraction, the arc pattern with six-fold symmetry indicates that the crystals are arranged in a preferred orientation. The diffraction spots could be assigned and attributed to the anhydrous 3R-CdCl$_2$ polymorph. The diffraction pattern contains superposition of spots from both the {003} and {hk0} planes, which indicates that the two types of planes are parallel to the beam, i.e. the nanostructure constitutes a closed cage structure. Thus, the closed cage structures were formed by release of the water molecules from the monohydrate crystal and subsequent crystallization into the anhydrous form. Closer examination of the nanoparticles shows that most of the nanoparticles have a core-shell structure with closed cage CdCl$_2$ shell and an amorphous core of an unspecified chemical composition. This is not an uncommon situation and has been observed before, in e.g. the recrystallization of amorphous MoS$_3$ nanoparticles. After exposure to the e-beam irradiation [7] or pulsed electric signal [6], core-shell structures with MoS$_2$ shell and amorphous MoS$_3$ core have been formed. In both cases the core remains amorphous, since its recrystallization requires an out diffusion of molecules, through the already formed closed shell. In the present case, out diffusion of the water molecules through the IF-CdCl$_2$ shell is rather slow, once the first few layers of the shell are formed.

Figure 2

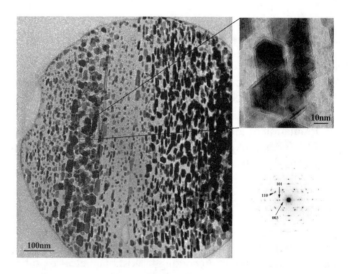

FIGURE 2. TEM micrograph showing CdCl$_2$ particles formed under e-beam irradiation. On the right hand side a magnified region and its electron diffraction pattern, exhibiting the hexagonal symmetry of the 3R-CdCl$_2$ polytype, are shown.

Figs.3 shows a few typical structures obtained after electron beam irradiation with 120keV for approximately 15 min. Thus **Fig.3a** and **d** show almost perfect hexagonal cage structures with sharp 120° inclinations. The distance between the fringes (0.58 nm) agrees with the c/3 distance between the $CdCl_2$ layers. Upon tilting such a hexagonal cage by 20°, four oppositely placed walls of the cage disappear, while two walls remained in focus. Such behaviour indicates that the walls which have disappeared are inclined in very sharp angles to the beam. This observation entails that the present cage structure of $CdCl_2$ is a rhomboid with some very sharp angles. **Fig.3** shows a few typical cage structures, which are mostly typified by 120° inclinations on one hand and 90° inclinations on the other hand. In many cases the rectangles possess a rather elongated shape. It is believed that the hexagonal (**Fig.3a**) and rectangular (**Fig.3b**) images are characteristic projections of similar cages taken at two perpendicular angles. Since the maximum tilting angle of the sample in the microscope is ±40°, it is not possible to tilt the same polyhedron from one projection into the other. Apparently, the cage structures are disposed on the substrate in two preferred orientations, either with their hexagonal or with their rectangular faces parallel to the substrate.

Figure 3

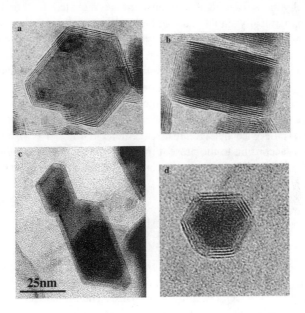

FIGURE 3. TEM micrographs showing an assortment of $CdCl_2$ cage structures with hexagonal or rectangular shapes. The distance between the $CdCl_2$ layers (fringes) is 0.58nm.

Hexagonal cage structures were previously observed in *IF*-NiCl$_2$, which crystallizes also in the cadmium chloride structure [9]. This fact lends support to the idea, that nanoparticles of layered compounds with octahedral coordination tend to form cage structures with hexagonal shape, rather than the initially hyphothyzed octahedral polyhedra.

It has been pointed out before [6,7,12] that phases, which are not stable in the bulk, can nonetheless form metastable *IF* structures. In that sense, the present case is not an exception. It is quite unexpected however, that while bulk 3R-CdCl$_2$ is hygroscopic, and therefore can not be studied in ambient conditions, *IF* structures of the same compound are fully stable in the ambient. This work reveals another important attribute of closed cage nanostructures. As discussed in the previous section, fullerene-like structures can be spontaneously formed by heating nanoparticles of layered materials. The energy input is required to start the folding process, which is stimulated by the introduction of pentagons, squares or triangles into the otherwise hexagonal array. Many of the compounds which form the closed structures, like CdCl$_2$, are unstable in ambient conditions. Therefore the formation of the closed *IF* nanostructures, can be considered as a means to stabilize high temperature phases. Obviously, mono (and higher) hydrates of CdCl$_2$ are unstable at elevated temperatures. Accordingly, the thermodynamically stable phase at high temperatures is the anhydrous CdCl$_2$. This high temperature phase can be made fully stable at ambient conditions, only in the seamless *IF* structure, which prevents water intercalation. This observation is, by no means, limited to this case, as has been shown for various other phases, like metal intercalated MoS$_2$ [6,7,22] or Tl$_2$O [12] and others.

The most important issue, which remains unresolved so far, is the molecular structure of the rhombi corners. According to the Euler rule, 6 squares are required to close the otherwise hexagonal arrangement of the Cd (Cl) atoms in the layer plane. However, according to the present model, the CdCl$_2$ cage consists of 12 corners, which could be formed by pentagons, only. Further work is necessary to elucidate this important issue.

ACKNOWLEDGMENT

This work was carried out with the support of Alfried Krupp von Bohlen and Halbach Stiftung (Germany), ACS-PRF (USA), AFIRST (France-Israel), Israel Academy of Sciences (Bikura), USA-Israel Binational Science Foundation and Israel Science Foundation.

REFERENCES

(1)Kroto, H. W.; Heath, J. R.; O'Brien, S. C.; Curl, R. F., Smalley, R. E. *Nature* **1985**, *318*, 162.

(2) Iijima, S. *Nature* **1991**, *354*, 56.

(3) Tenne, R.; Margulis, L.; Genut, M., Hodes, G. *Nature* **1992**, *360*, 444.

(4) Margulis, L.; Salitra, G.; Tenne, R., Talianker, M. *Nature* **1993**, *365*, 113.

(5) Feldman, Y.; Wasserman, E.; Srolovitz, D. A., Tenne, R. *Science* **1995**, *267*, 222.

(6) Homyonfer, M.; Alperson, B.; Rosenberg, Y.; Sapir, L.; Cohen, S. R.; Hodes, G., Tenne, R. *J.Am.Chem.Soc.* **1997**, *119*, 2693.

(7) Hershfinkel, M.; Gheber, L. A.; Volterra, V.; Hutchinson, J. L.; Margulis, L., Tenne, R. *J.Am.Chem.Soc.* **1994**, *116*, 1914.

(8) Chopra, N. G.; Luyken, J.; Cherry, K.; Crespi, V. H.; Cohen, M. L.; Louie, S. G., Zettl, A. *Science* **1995**, *269*, 966.

(9) Rosenfeld Hacohen, Y.; Grunbaum, E.; Tenne, R.; Sloan, J., Hutchinson, J. L. *Nature* **1998**, *395*, 336.

(10) Spahr, M. E.; Bitterli, P., Nesper, R. *Angew.Chem.Int.Ed.* **1998**, *37*, 1263.

(11) Avivi, S.; Mastai, Y.; Hodes, G., Gedanken, A. *J.Am.Chem.Soc.* **1999**, *121*, 4196.

(12) Avivi, S.; Mastai, Y., Gedanken, A. *J.Am.Chem.Soc.* **2000**, *122*, 4331.

(13) Parilla, P. A.; Dillon, A. C.; Jones, K. M.; Riker, G.; Schulz, D. L.; Ginley, D. S., Heben, M. J. *Nature* **1999**, *397*, 114.

(14) Seifert, G.; Terrones, H.; Terrones, M.; Jungnickel, G., Frauenheim, T. *Phys.Rev.Lett.* in press.

(15) Terrones, M.; Hsu, W. K.; Terrones, H.; Zhang, J. P.; Ramos, S.; Hare, J. P.; Castillo, R.; Prassides, K.; Cheetham, A. K.; Kroto, H. W., Walton, D. R. M. *Chem.Phys.Lett* **1996**, *259*, 568.

(16) Goldberg, D.; Bando, Y.; Stephan, O., Kurashima, K. *Appl.Phys.Lett.* **1998**, *73*, 2441.

(17) Qin, X. R.; Yang, D.; Frindt, R. F., Irwin, J. C. *Phys.Rev.B* **1991**, *44*, 3490.

(18) Pauling, L. *Proc of the Nat.Acad.of Sci.* **1929**, *15*, 709.

(19) Partin, D. E., O'Keeffe, O. *J.Solid State Chem.* **1991**, *95*, 176.

(20) Leligny, H., Monier, J. C. *Acta. Cryst.B* **1974**, *30*, 305.

(21) Leligny, H., Monier, J. C. *Acta.Cryst.B.* **1975**, *31*, 728.

(22) Ramskar, M.; Skraba, Z.; Stadelmann, P., Levy, F. *Adv.Mater.* **2000**, *12*, 814.

Structure and Dynamics in the Methylated Exopyridine Anthracene Rotaxane: ^{13}C, ^{1}H and ^{19}F Solid-State NMR studies

X. Bourdon [a], J. Leupold [a], M. Mehring [a], J. Thies [b], T. Kidd [b]
and T. Loontjens [b]

[a] 2. Physikalisches Institut, Universität Stuttgart, Pfaffenwaldring 57, 70550 Stuttgart, Germany
[b] DSM research, P.O. Box 18, 6160 MD Geleen, The Netherlands

Abstract. We report on the structural and dynamic characterization by solid state NMR of a new rotaxane consisting of a thread molecule enclosed by a macrocycle, which can eventually shuttle or rotate along/around the thread. Different kinds of slow and fast motions are detected in the methylated exopyridine anthracene rotaxane by ^{1}H, ^{13}C and ^{19}F relaxation time versus temperature measurements. Analysis of the proton decoupled and temperature dependent ^{19}F NMR spectra gives some indications of the possible motions of the macrocycle and breaking of the hydrogen bond in the solid-state at temperatures above 360K.

INTRODUCTION

The unique architectures and properties displayed by mechanically interlocked molecules [1-4] such as the simple hydrogen-bond assembled rotaxanes (consisting of a thread enclosed by a macrocycle) have become increasingly interesting to academic and industrialists alike. Indeed, these systems inherently offer a degree of control over the precise positioning of its substituents (which can be used to influence material properties) together with the possibility of switching their relative separation and orientation through the influence of external stimuli. Recently, it has been proved that the translational and rotational degrees of freedom available to these systems in the liquid state - as opposed to covalently connected functional groups - can be beneficially exploited to manipulate macroscopic properties at will by external photochemical and electrochemical stimuli [5-7]. Apart from the obvious basic interest of these novel systems, one crucial point in order to gain commercial applications concerns the recovery of the dynamic properties in the solid state.

In this contribution, we present the first solid state NMR study of a rotaxane, the methylated exopyridine anthracene rotaxane (PA-Rotaxane), as shown in Figure 1, which has an optically addressable peptide based molecular shuttle for possible laser writing applications. We have investigated the structural (no X-ray structure available) and mainly the dynamic properties in the PA-Rotaxane by ^{1}H, ^{13}C and ^{19}F NMR. Some structural information has been obtained and different types of slow and fast motions have been detected using different relaxation time measurements [8]. In this paper, we would like to emphasize the ^{19}F NMR. Indeed, the fluorine being only

CP544, *Electronic Properties of Novel Materials—Molecular Nanostructures*, edited by H. Kuzmany, et al.
© 2000 American Institute of Physics 1-56396-973-4/00/$17.00

present in the macrocycle and closely connected with it could be a suitable nucleus to detect motions of the macrocycle and indirectly a possible breaking of the hydrogen bond.

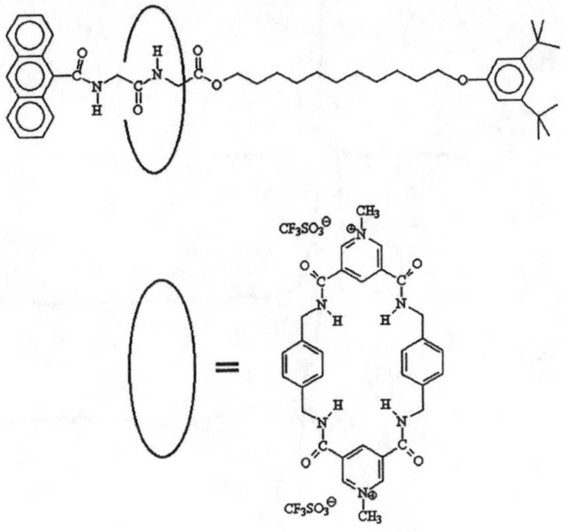

FIGURE 1. Molecular structure of the PA-Rotaxane.

RESULTS AND DISCUSSION

We have performed proton decoupled ^{19}F NMR experiments (^{19}F$_{dec}$) at various temperatures and some representative spectra can be seen in Figure 2.

The 80K spectrum is characteristic for a static CF$_3$ group (no rotation) with three anisotropic chemical shift tensors for each fluorine and three dipolar tensors (F-F dipolar couplings). According to our simulations (not shown) for a static CF$_3$ group, there still exist some rotations of this group at 80K but probably very slow. Therefore, the 80K spectrum can mainly be interpreted in terms of slow CF$_3$ group dynamics.

The 140K spectrum is characteristic for a rapidly rotating CF$_3$ group with one axial (chemical shift anisotropy) tensor and a single dipolar tensor. The simulation of this spectrum (not shown), taking in account a simple CF$_3$-group and its usual rotation around its C$_3$ axis, is in good agreement with the experiments.

The 293K spectrum can also be interpreted in terms of CF$_3$ rotation (simulation not shown) but there are some very interesting and noticeable changes compared to the 140K spectrum. Possible explanations of which are discussed later in this paper. One important point is that these changes can not be attributed to rotational or translational motions of the macrocycle at 293K. Indeed, in that case, the shape and the line width of the line would change dramatically when compared to the 140K spectrum.

The 410K spectrum has a completely different shape and line width compared with the 293K or 140K spectra. The line becomes narrower indicating rapid isotropic

449

rotation and the shape is completely different from the one expected for a classical rapidly rotating CF_3 group about a C_3 axis, implying that a more isotropic dynamic process occurs between 293K and 410K.

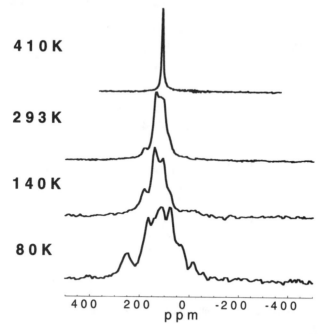

FIGURE 2. Comparison of static $^{19}F_{dec}$ NMR spectra at representative temperatures.

In order to go further trying to explain these results, we have calculated the second moment (M_2) of these spectra for all temperatures. This calculation is well known in solid state NMR and corresponds to the integration of the spectral function multiplied by $(v-v_0)^2$, where v_0 is the center frequency of the NMR line. This calculation corresponds more or less to a line width determination but is more precise by taking into account all the spectral features of the signal. The temperature dependence of $2M_2^{1/2}$ is represented in Figure 3. Three different dynamic regimes can be distinguished.

At low temperature up to about 110K, the $2M_2^{1/2}$ thermal evolution can be fully interpreted in terms of CF_3 rotation. We have estimated with a simulation of a completely static CF_3 group the second moment at 0K. According to this value, the CF_3 groups would become completely static below approximately 40K. We have calculated the activation energy associated to this first dynamic process and we obtained a value of 227meV (21.87kJ.mol^{-1}). Above 110K, the correlation of the dynamic process decreases when the temperature increases until about 270K after which a completely unexpected second dynamic process appears with an activation energy of 694meV (66.87kJ.mol^{-1}). As we have previously stated, at this temperature up to 360K, the assumption of macrocycle rotations can be excluded. This process can not be attributed to indirect effects coming from ^1H (because of the decoupling) or to ^{13}C because of its very weak natural abundance. Therefore, we assume that this second process could be attributed to flipping of the $CF_3SO_3^-$ group. Above 360K, a third dynamic process appears with an activation energy of 992meV (95.58kJ.mol^{-1}). This process is represented by a drastic line shape change. As mentioned previously, this could be consistent with some rotation or translation of the macrocycle. Currently, we do not fully understand the real nature of these hypothetical macrocycle motions (flipping, rotation or shuttling?). Moreover, we can not completely exclude the breaking of the ionic bond between the methyl pyridinium and the $CF_3SO_3^-$ groups, which could also explain this dynamic process. Nevertheless, we believe that the hydrogen bond is expected to be weaker in the solid state compared to the ionic bond. 2D ^{19}F NMR measurements and micro-calorimetry measurements are underway in order to confirm these assumptions.

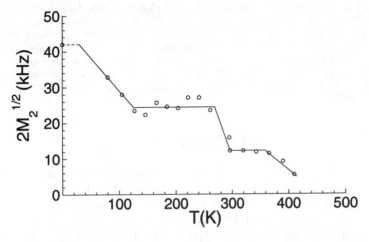

FIGURE 3. Second Moment thermal evolution of static $^{19}F_{dec}$ NMR spectra in the PA-Rotaxane

ACKNOWLEDGMENTS

We would like to acknowledge Y. Fagot-Revurat, C. Heppel, S. Krämer and K. Müller for valuable discussions and technical assistance. This work was supported by the DRUM European Community funded (Contract Number ERBFMRXCT970097) TMR network.

REFERENCES

1. G. Schill and H. Zollenkopf, *Nachr. Chem. Tech.*, **15** (1967) 149.
2. D. B. Amabilino and J. P. Sauvage, *Chem. Comm.*, (1996) 2441.
3. J. F. Stoddart et al., *Chem. Eur. J.*, **3** (1997) 1113.
4. D. A. Leigh et al., *J. Am. Chem. Soc.*, **118** (1996) 10662.
5. R. A. Bissel, Cordova, A. E. Kaifer and J. F. Stoddart, *Nature*, **369** (1997) 133.
6. J. P. Collin, P. Gavina and J. P. Sauvage, *New J. Chem.*, **21** (1997) 525.
7. M.Asakawa, G.Brancato, M.Fanti, D.Leigh, F. Zerbetto and S.Zhang, *J. Am. Chem. Soc.*,submitted.
8. X. Bourdon et al., *Ampere 2000 Proceeding*, submitted.

Direct Measurements of Electrical Transport Through DNA Molecules

D. Porath, A. Bezryadin, S. de Vries, and C. Dekker

Faculty of Applied Sciences, Delft University of Technology, 2628 CJ Delft, The Netherlands

Abstract. We present direct measurements of electrical charge transport through single DNA molecules. The molecules are electrostatically trapped between two metal nanoelectrodes separated by 8 nm and current flow through the DNA is measured upon application of a voltage between these electrodes. The measured current is negligible up to a threshold voltage followed by a sharp rise of the current. The conductance curves show a peak structure as a function of voltage, which suggests that the charge transport is mediated by the energy bands of the measured DNA. The existence of DNA between the electrodes is verified using DNase I enzyme control experiments.

INTRODUCTION

The question whether DNA is a good conductor was heavily debated over the past decade [1,2], spurred by fluorescence quenching optical measurements [3-11]. Direct electrical measurements on large number of molecules in films were performed as well, but these measurements were dominated by inter-chain hopping events [12,13]. Recently, Fink and Schönenberger have reported that a 600 nm long λ-DNA 'rope' can behave as a good linear conductor [14]. This result is in contrast with a previous experiment by Braun *et. al.* that measured no conductance across a similar λ-DNA molecule of 16 μm length [15].

Here we present direct electrical transport measurements through short and well-defined 10 nm long (30 base-pairs), double-stranded poly(G)-poly(C) DNA molecules that show semiconducting behavior down to cryogenic temperatures.

EXPERIMENTAL

In our experiments a DNA molecule is placed between two metal nanoelectrodes, that are 8 nm apart and current-voltage curves are measured. The DNA is well characterized by UV spectroscopy, mass spectrometry and gel electrophoresis. The electrodes are fabricated using electron beam patterning and etching processes. We create a 100 nm wide slit in the SiN layer with a local 30 nm narrow segment and underetch the SiO_2 layer to form two opposite freestanding SiN "fingers" that become the metallic nanoelectrodes after sputtering Pt through a Si mask (see Fig. 1). We then apply a voltage of up to 10 V to check that there is no current flowing between the electrodes (< 1 pA). The DNA is positioned between the electrodes by electrostatic

CP544, *Electronic Properties of Novel Materials—Molecular Nanostructures*, edited by H. Kuzmany, et al.
© 2000 American Institute of Physics 1-56396-973-4/00/$17.00

trapping [16-17] from a 1 μl droplet of dilute (<1 DNA molecule/(100 nm)^3) aqueous buffer solution [18]. A voltage is applied between the electrodes and a nearby DNA molecule is polarized and attracted by the field gradient in between the electrodes where it becomes trapped. When a current starts to flow through the molecule the voltage also drops on a large serial resistor (~2 GΩ), the field is reduced and trapping of additional molecules is prevented. The droplet is then dried by a flow of nitrogen and current-voltage measurements are performed.

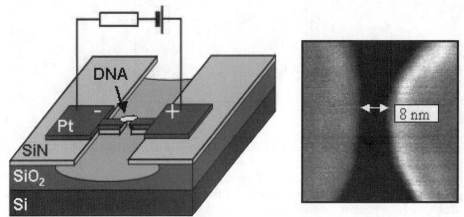

FIGURE 1. Schematic of a sample (left) and a scanning electron microscope (SEM) image (right) of an 8 nm gap (dark) and the electrodes (light).

RESULTS

The current-voltage curves measured between the electrodes after trapping the DNA show negligible current up to a threshold voltage followed by a sharp rise of the current (see Fig. 2). In most cases we measure up to a few nA. However, in some cases we have measured currents up to 150 nA. We have measured more than 20 samples showing similar behavior at room temperature in air (humidity 50%) and four samples at low temperatures down to 4 K. At room temperature the details of the curves do not reproduce well, although the general shape of the curve is conserved. At low temperatures the curves become more reproducible and details can be extracted from the derivatives of these curves. The current-voltage curves that we observe are asymmetric. This can be explained both by an asymmetry in the capacitance and resistance of the contacts or by an asymmetry in the transport behavior of the DNA molecule in opposite directions of the current flow along the molecule.

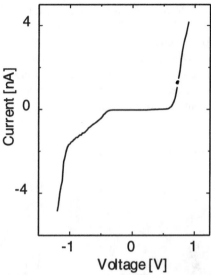

FIGURE 2. Typical current-voltage curve measured through DNA at 100 K.

These conductance curves of *dI/dV* versus *V* show a remarkable peak structure with peak spacing of 0.1-0.5 eV and a peak width of ~0.2 eV. This observation suggests that charge transport is mediated by the energy bands of the DNA [19]. The reproducibility and peak structure can be seen in Fig. 3a where we present three consecutive curves. When applying high voltages as in Fig. 3b the curves become noisier and less reproducible. The current values that we measure at a certain voltage can change considerably as can be seen by comparing the vertical scale of Fig 3 a and b. This can be related to structural changes in the molecule itself or to changes in the contacts.

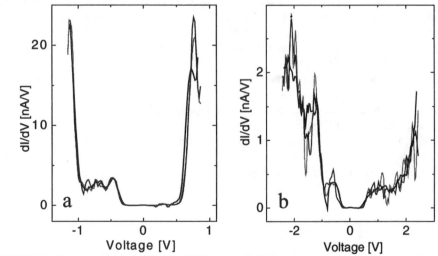

FIGURE 3. Peak structure observed in *dI/dV* versus *V*. Shown is a set of consecutive conductance curves measured at 100 K (a) and at 27 K up to higher bias voltages (b) through DNA.

CONTROL EXPERIMENTS

The existence of the DNA between the electrodes was verified by incubation of samples with trapped DNA in a solution containing DNase I enzyme [20] that specifically cuts DNA. After the incubation no current was observed anymore. This confirms that we indeed trap DNA and not any other (in)organic material. It shows also that the three-dimensional shape of the DNA was not damaged upon application of the voltage. This control experiment was repeated with similar enzyme solution but in the absence of Mg^{++} ions that activate the action of the enzyme (and in the presence of 10 mM EDTA that complexes Mg ions). Under such conditions the curves did not change, showing that in the original experiment it was indeed the enzyme that did the cutting of the DNA.

The length selectivity of the trapping procedure was checked in an attempt to trap a 10 nm long molecule in a 12 nm gap rather then in an 8 nm gap. These attempts were not successful, showing that measuring current through two or more molecules in series is unlikely because in these cases the 8 nm gap is not particularly favorable in comparison with a 12 nm gap. This also demonstrates that the DNA molecule is trapped with its long axis bridging the electrodes and not otherwise.

In an additional experiment we tried to trap from buffer solution without the DNA. This was not successful indicating that we do not measure ionic current through the solution or an electrochemical reaction. Moreover the measurements at low temperatures below −20 °C, where ionic conduction is suppressed, show that the currents that we measure are electronic in origin and that current flows indeed through the molecule itself and not through any traces of water or salt around the molecule.

SEM imaging showed a very small amount of material between the electrodes after trapping. This, as well as the very dilute DNA solution and the trapping procedure (large serial resistor) suggest that we measure current through only a single molecule, although we have no direct evidence for that.

DISCUSSION AND SUMMARY

Our measurements show that DNA molecules can transport high currents through the DNA with no significant damage to the molecule itself, as verified by the DNase I enzyme experiment. This current is electronic in origin and not ionic. The differential conductance can be taken as a measure for the density of states of the molecule if the contacts form tunneling barriers as is likely to assume in our experiment. Therefore, the peak structure that we observe in these curves suggests that current is mediated by the energy bands of the molecule. Coulomb blockade [21] phenomena probably contribute to the gap in the current-voltage curves that we measure. However, an object of size comparable to the 8 nm separation between the electrodes will contribute no more than a fraction of the observed gap. We also do not observe periodic steps in the current-voltage curves, as expected for Coulomb blockade steps. We therefore relate most of the observed gap to the energy difference between the Fermi level at the metal electrode and the onset of the energy band of the DNA.

ACKNOWLEDGMENTS

We thank L. Gurevich for assistance in the fabrication and measurements. We also thank E.W.J.M. van der Drift, A. van der Enden, L.E.M. de Groot, S.G. Lemay, A.K. Langen-Suurling, R.N. Schouten, Z. Yao, T. Zijlstra, M.R. Zuiddam, M.P. de Haas, J.M. Warman, L.D.A. Siebbeles, A.J. Storm, N.R.C. Kemeling and J. Jortner for assistance and discussions. We thank E. Kramer and E. Yildirim for the DNA characterization measurements. The work was supported by the Dutch Foundation for Fundamental Research on Matter (FOM).

REFERENCES

1. Taubes, G., Science **275**, 1420-1421 (1997).
2. Wilson, E. K., C&EN, July 2751-2754, (1998).
3. Arkin, M. R. et al., Science **273**, 475-480 (1996).
4. Lewis, F. D. et al. Science **277**, 673-676 (1997).
5. Barbara, P. F., Olson, E. J. C. Adv. Chem. Phys.1999, pp. 107, Chapter 13.
6. Meggers, E., Michel-Beyerle, M. E. & Giese B., J. Am. Chem. Soc. **120**, 12950-12955 (1998).
7. Beratan, D. N., Priyadarshy, S. & Risser, S. M., Chem. Biol. **4**, 3-8(1997).
8. Wan, C. et al., Proc. Natl. Acad. Sci USA **96**, 6014-6019 (1999).
9. Henderson, P. T., Jones. D., Hampikian, G., Kan, Y. & Shuster, B. G., Proc. Natl. Acad. Sci USA **96**, 8353-8358 (1999).
10. Jortner, J., Bixon. M., Langenbacher, T. & Michel-Beyerle, M. E., Proc. Natl. Acad. Sci USA **95**, 12759-12765 (1998).
11. Grozema, F. C., Berlin, Y. A. & Siebbeles, L. D. A. Int. J. Quant. Chem.Vol **75**, 1009-1016 (1999).
12. Elley, D. D. & Spivey D. I., Trans. Faraday Soc. **58**, 411-415 (1962).
13. Okahata, Y., Kobayashi, T., Tanaka, K., & Shimomura, M., J. Am. Chem. Soc. **120**, 6165-6166 (1998).
14. Fink, H. W. & Schönenberger, C., Nature **398**, 407-410 (1999).
15. Braun E. et al., Nature 391, 775-778 (1998).
16. Bezryadin, A. & Dekker, C., J. Vac. Sci. Technol. **B 15**(4), 793-799 (1997).
17. Bezryadin, A., Dekker, C. & Schmid, G., Appl. Phys. Lett. **71**, 1273-1275 (1997).
18. Buffer composition of 300 mM NaCl, 10 mM Na Citrate and 5 mM EDTA (Eurogentec Bel S.A.).
19. Porath, D., Bezryadin, A., de Vries, S. and Dekker, C., Nature **403**, 635-638 (2000).
20. Composition of the enzyme buffer solution: 10 mg ml^{-1} -5mM Tris-HCl, 5 mg MgCl$_2$, 10 mg ml$^-$1 (pH 7.5).
21. Grabert, H. & Devoret, M. H., *Single charge tunneling* (Plenum, New York, 1992).

Electrical Conduction through DNA Molecules

H.-W. Fink

Physik Institut der Universität Zürich, Winterthurerstrasse 190, H-8057 Zürich, Switzerland

Abstract. The first direct experimental evidence for effective charge transport though individual DNA molecules will be discussed. Visualization of the DNA molecules as well as mechanical and electrical manipulations is carried out in the Low Energy Electron Point Source Microscope whose principle of operation will be illustrated.

INTRODUCTION AND MOTIVATION

The DNA molecule, much longer known to us than carbon nanotubes, and extremely well understood by chemists and molecular biologists, has only recently been considered as a possible candidate for molecular wires. To a large extent this is due to the fact that DNA, compared to nanotubes, is a much less user-friendly object for physicists in as much as it can not be visualized by conventional electron microscopy without destroying it. While chemists have developed tools that allow them to know and control precisely what they are doing to an ensemble of molecules, experiments on just a single species require new approaches.

Once these limitations are overcome, the prospects of employing DNA as molecular wires are truly exciting. A great number of catalytic reactions, carried out by enzymes, are well known. They allow for multiplying, cutting and joining DNA molecules. Length determinations on a large ensemble of molecules can be done by electrophoresis techniques. Filtering techniques are at hand to achieve mono-disperse molecules at almost any desired length from a few nucleotides to macroscopic length scales. Last, but not least, the DNA is soluble in water and clever methods have been developed in recent years to stretch out the molecules [1] on various surfaces, including those of silicon wafers.

Charge Transfer versus Charge Transport Experiments

Charge transfer processes and their distance and time dependence are important fundamental issues in reaching an understanding about the functions of biological systems. The question is, how effectively an electric charge, an electron or a missing electron, can be transferred from one part of the molecule to another one, a certain distance away. The charge can be introduced into the biological system by a particular designed molecule attached to a specific site of the system under study and detected by another molecule at some distant position that absorbs this charge. Chemists have developed ingenious methods to study those effects on large ensembles of molecules. The pioneering work of Barton et al. [2] has provided a detailed understanding of

CP544, *Electronic Properties of Novel Materials—Molecular Nanostructures*, edited by H. Kuzmany, et al.

electron transfer in the DNA double helix. Electron donor respectively acceptor molecules are intercalated into DNA molecules and short time laser pulses are used to monitor the electron emission and subsequent absorption processes.

These charge transfer processes are relevant for the biological function of the DNA and possibly related to such important processes like damage repair. However, they can not directly address the question of electrical conductivity since the absolute rate at which of those transfer processes occur are not accessible from these data. Besides, the injected charges are "hot" electrons and the ability of a system to transfer the charge to some distant place does not necessarily make it a conductor in a physical sense. A piece of matter is electrically conducting if it incorporates charge carriers that are able to move freely under ordinary thermal conditions, even in the absence of an external force like an applied chemical potential. The conventional way of probing the presence of mobile charges is done by arranging for a small disturbance of the system by applying a small electrical potential difference. In a conducting material, the response of the system to the applied potential will be an electrical current following the direction of the potential gradient.

The main problem of doing this with DNA molecules is to identify them and to attach leads to the molecule. The tools and methods we used to achieve this are described below.

PROBING DNA CONDUCTIVITY

The LEEPS Microscope

With the invention of the Low Energy Electron Point Source (LEEPS) Microscope [3], it has become possible to image individual freestanding DNA molecules [4]. The low energy of the electrons, between 20 and 300 eV, provides for a high contrast in imaging unstained molecules and avoids radiation damage. The principle setup of this lens-less electron microscope is illustrated in Figure 1.

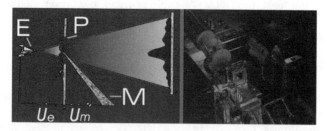

FIGURE 1. The LEEPS microscope principle and a view of the instrument.

A point source for electrons (E) is placed in close proximity to the sample (P). The atomic sized emission area of the point-source tip provides a coherent spherical wave of high brightness at low emission voltages (Ue) between 20 and 300 Volts. The coherence of the electrons makes the projection image at a distant detector to a Gabor type in-line hologram. It is brought about by the interference of the scattered wave from the object with the unscattered reference wave. The reconstruction of the hologram obtained with low energy electrons reveals the shape of the object. In this way, we could image single λ-DNA molecules. In order to not just visualize the nanometer sized objects, but to also be able to manipulate them during observation, a manipulation-tip (M) is placed between the sample and detector.

Manipulating Nanometer Sized Objects

The action of the manipulating-tip is illustrated in Figure 2 showing the mechanical manipulation of a single carbon nanotube that has been placed onto a 7x7 µm carbon grit from Quantifoil® GmbH. Once the manipulating-tip is maneuvered into the sample plane, contact to the nanotube can be established and the tube is forced to follow the motion of the tip. The left image in Figure 2 provides an overview of one grit element while the subsequent images are taken with the electron source closer to the sample to obtain higher magnified images.

FIGURE 2. Manipulation of a carbon nanotube on a 7x7 µm square holes Quantifoil® grit.

Once an object is in contact with the manipulating-tip it is just a matter of applying an electrical potential between the manipulating-tip and the grounded sample holder to probe the conductivity of the observed object. We have used this technique to address the question of electrical conductivity in DNA molecules.

Current versus Voltage Characteristics of DNA Molecules

Depositing carbon nanotubes onto a micro-machined grit as illustrated in Figure 2, is essentially just a matter of choosing the proper concentration in the liquid. The deposition of DNA molecules appeared to be somewhat more demanding since they do apparently have a tendency of being entangled to form robes. The very left image in Figure 3 shows such a situation with entangled DNA molecules. The sample preparation is done in the following steps. First, a drop of buffer solution, containing λ-DNA molecules, was placed onto a Quantifoil® sample holder that contained a regular array of 2 µm diameter holes. In order to stretch the molecules over the holes, we used blotting paper to create a flow parallel to the sample surface. After transfer of the sample holder into the vacuum chamber, LEEPS images as shown at the very left of Figure 3 revealed. While over some of the holes no DNA molecules could be found, others showed a network of DNA ropes. We attributed this to poor binding of the DNA molecules to the carbon substrate. Measurements on DNA ropes of the order of one µm length revealed resistance values ranging from 25MΩ up to values as high as

1GΩ despite the fact that most of the ~17μm long λ-DNA molecules should be in contact with the substrate surface. The situation could be improved by depositing gold onto the carbon sample holder. This resulted in less rope formation as evident from the second image displayed in Figure 3 where the manipulation-tip is in contact with two DNA ropes.

FIGURE 3. Stretching λ-DNA molecules over 2 μm diameter holes to probe their electrical conductivity by contact with a manipulating-tip. See text for explanations.

The gold film probably offered more, albeit unspecific, binding sites to the DNA. Thus, the tendency of the molecules to interact with each other rather than with the surface of the sample holder appeared to be reduced. This understanding of the situation was supported by the fact that the measured resistances were lower. A better binding to the gold surface should come along with a lower contact resistance between the molecules and the metal. The second contact to the molecules is done via the apex of the manipulation-tip that was initially not prepared in any special way. Since the electrochemically etched tungsten tip was most likely covered by an oxide layer, the tip was heated in vacuum above 2000K and subsequently covered with a gold layer. With this measure, we ended up at resistance values for a 600nm long DNA rope (see image three of Figure 3) to which contact was made with a gold covered tungsten tip, of 2.5 MΩ. This corresponds to a resistivity of the order of 1mΩcm [5]. Since we have no a priori knowledge about the potential drops at the contacts the quoted resistivity value need to be viewed as an upper limit for the DNA molecule alone.

COMPARISON WITH OTHER STUDIES

Following the above described first direct measurements of contacting DNA molecules and probing their ability to transport charges, experiments on small synthetic DNA molecules have been reported more recently by the Delft group [6]. They use an electric field to trap the molecules between two electrodes. Electrophoretic forces are known to act equivalently to a hydrodynamic flow [7], which implies that an electric field is able to stretch the molecules. However, the Delft group reported that trapping occurred only with molecules that matched the gap exactly, slighter shorter ones would not be stretched and trapped. This indicates that the behavior of the DNA under those conditions is different. After all, the electric fields employed in the trapping experiments, reported by the Delft group, are about 6 orders of magnitude larger than typical fields for electrophoresis experiments. It might also be conceivable that the absence of buffer ions, that needed to be replaced by pure water to avoid unwanted ion currents during trapping, creates a situation which changes the DNA conformation in a not yet understood fashion. A direct comparison to the experiments described here seems therefore not sensible and the lack of compatibility of the results is not all that surprising.

Even more recently, experiments by Gruner et al [8] have been presented. They employ microwave techniques to address the question of DNA conductivity. This is a macroscopic experiment performed on an ensemble of λ-DNA molecules that are arranged inside a cavity. Their findings suggest that a 600nm long DNA molecule would exhibit a resistance of about 300MΩ, roughly 2 orders of magnitude larger than our findings. While the mechanism of charge transport through DNA is not yet understood in any detail, they offer to consider a model which takes the flexibility of the DNA into account rather then a static model. Intramolecular vibrations could lead to a larger overlap of the π-electrons of the base stacking for a given point in time and space and thus facilitate electron transport. In the Gruner experiments [8] a persistence length of 60nm is assumed which is the result of the interplay between Brownian agitation by the solvent molecules and the desire of the polymer to be in a straight configuration. A 600nm long DNA molecule would thus exhibit on average about 10 kinks, respectively sites where the directionality of the molecule changes. Those kinks are most likely places where the overlap of the π-electrons is less compared to the straight parts. Accepting this considerations and the fact that the DNA molecules used in our experiments are prepared in a straight configuration in contrast to the Gruner study, a resistance which is lower by a factor 100 appears not too surprising.

I have to admit that the above discussed model is based on pure speculation at this point in time. However, it seems certain that the high flexibility of the DNA, its ability to undergo conformational changes driven by external chemical and physical forces, will have to be taken into account in efforts to model any specific experiment. This poses challenges for the experimentalists, in particular when individual molecules are the subject of interest, in designing experiments with a high degree of confidence in understanding and controlling the conditions that define the structure and the shape of the DNA under observation.

REFERENCES

1. Austin, R. H., Brody, J. P., Cox, E. C., Duke, K., and Volkmuth, W., *Physics Today*, 32-38 (February 1997).
2. Barton, J.K., "The DNA Double Helix: A Medium for Long Range Charge Transport" in *Yearbook of Science & Technology 1999*, McGraw-Hill 1999, p.124.
3. Fink, H.-W., Stocker, W., and Schmid, H., *Phys. Rev. Letters* **65**, 1204-1206 (1990)
4. Fink, H.-W., Schmid, H., Ermantraut, E., and Schulz, T., *J. Opt. Soc. Am.* **14**, No. 9, 2168-2172 (1997)
5. Fink, H.-W., and Schönenberger, Ch, *Nature* **398**, 407-410 (1999)
6. Porath, D., Bezryadin, A., de Vries, S., and Dekker, C., *Nature* **403**, 635-638 (2000)
7. Bakajin, O. B., Duke, T. A. J., Chou, C. F., Chan, S. S., Austin, R. H., and Cox, E. C., *Phys. Rev Letters* **80**, 2737 (1998)
8. Gruner, G. *this issue*

Conductivity and Dielectric Constant of the DNA Double Helix

G. Gruner

Department of Physics, University of California Los Angeles, Los Angeles CA 90095

Abstract. We have measured the conductivity and dielectric constant along the lambda phage DNA (λ-DNA) double helix in a buffer and in a dry environment at microwave frequencies. Our experiments give evidence for charge delocalization and thermally driven charge transport. We discuss how the structural features of the DNA duplex are related to the conductivity and dielectric constant.

INTRODUCTION

The electronic properties of the DNA double helix have attracted ample interest recently, and the question whether charges are delocalized along the DNA duplex have been investigated by a range of studies of chemical nature, by monitoring charges injected through photoreceptors (1).

Early attempts to measure the conductivity of the DNA double helix were performed on pressed pellets (2), and the results are certainly influenced by charge transport between the DNA strands in close proximity to each other. Under such circumstances the conductivity as evaluated may not be related to the motion of charges along the double helix. Experiments on oriented film (3) also suffer from this shortcoming, but, at the same time give evidence for significant conductivity along the DNA duplex. Recent measurements (4) of direct current (dc) current-voltage (I-V) characteristic across individual DNA segments (using two probe contact configurations) lead to low resistance, and to an electrical conductivity of the other of 1000 (Ohms cm)$^{-1}$. However a different group has found insulating behavior (5).

Here we report on conductivity and dielectric constant measurements in the micro and millimeter wave spectral range. We find that both σ_1 and ε_1 can be accounted for by assuming that the electron states are localized along the DNA double helix with a localization length extending many base pair distances and temperature driven hopping between the localized states leading to finite conduction. We also discuss the relation between transport and structural features of the DNA duplex.

CP544, *Electronic Properties of Novel Materials—Molecular Nanostructures*, edited by H. Kuzmany, et al.
© 2000 American Institute of Physics 1-56396-973-4/00/$17.00

EXPERIMENTS

We have used a configuration which does not require contacts to be attached to the specimen under study. The conductivity was evaluated from the measured loss of highly sensitive resonant cavities, loaded with the material, and operating at micro and millimeter wave frequencies. The technique and the analysis which leads to the evaluation of the conductivity from the measured losses is well established (6), and such cavity configurations were used earlier by us to measure the electrical conductivity of various linear chain compounds displaying large electronic or ionic conductivity. The material is placed in the high electric field region of the cavity and the resulting change in the quality factor, Q of the cavity is measured. Q is inversely proportional to the loss W, and the loss due to the specimen is evaluated from the change (decrease) of Q upon the sample being inserted in the cavity. We treat the DNA strands as thin wires, (of the diameter of 2 nm), and we define the conductivity σ as j/E where j is the electric current density induced along the helix axis. For randomly coiled DNA strands the loss due to motion of electric charges W is, to a good approximation, (6) given by

$$W = \frac{1}{3}V\sigma_1 E_0^{\ 2} \tag{1}$$

where V is the volume of the conducting medium (see below), E_0 is time averaged applied ac field at the position of the sample, the factor of 1/3 results from a geometrical average of random orientations of the DNA with respect of the applied uniform electric field and σ refers to the real part of the complex conductivity. This then leads to a change of the quality factor, given as

$$\frac{1}{Q} = \frac{G}{3}V\sigma_1 \tag{2a}$$

where G is a known geometrical factor. Similarly, the dielectric constant is related to the change of the resonant frequency f, and the appropriate relation is

$$\frac{\Delta f}{f} = \frac{G}{3}V\varepsilon_1 \tag{2b}$$

DNA specimens used in this study are lambda phage DNA, (λ-DNA) extracted from E. coli and purchased from Sigma and BioLabs (7). For DNA lyophilized in a buffer (referred to later as "DNA in buffer"), obtained from Sigma the amount of DNA is established through the measurement of the intensity of the 260 nm absorption band, and the total weight including the DNA and the buffer material is also measured. We find that 11% of the measured weight is due to the DNA strands, 22% due to the solvent and 67% is water. A lyophilized buffer, identical to that of used for the preparation of the DNA was also prepared by us and we have evaluated to loss the buffer alone. Although this loss and frequency shift depends somewhat of the lyophilization process we have found that in all cases both are independent of temperature, in contrast to the strongly temperature dependent loss and frequency shift which we associate with the DNA helix. We have also purified λ-DNA from solutions using well-established purification procedures, and removed most of the water environment. We refer to the duplex "dry DNA". A comparison of the weight of DNA as evaluated from the intensity of the 260nm absorption band and by direct weight measurements indicates that $85 \pm 5\%$ of the weight is due to the DNA double helix. Consequently, there are approximately 3 water molecules for each base pair for this dehydrated, dry state. DNA in a water reach environment assumes the B form while in a dry environment the A form is found, and in a dry environment the base pair arrangement is disordered.

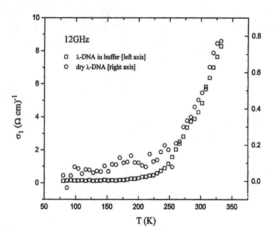

FIGURE 1. Conductivity of λ-DNA in buffer and dry λ-DNA versus inverse temperature as measured at 12GHz.

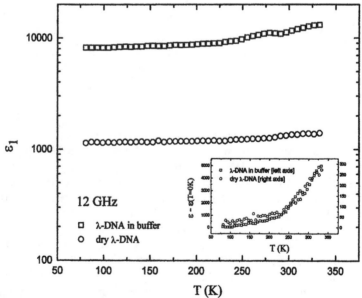

FIGURE 2. Dielectric constant of purified λ DNA in buffer and dry λ DNA in buffer versus temperature. The insert shows the temperature dependent part of the dielectric constant in detail.

In Fig. 1 and Fig. 2. we display the temperature dependence of the conductivity and dielectric constant, both measured at 12GHz as the function of temperature. Several features are of importance. We find a strongly temperature dependent conductivity which disappears at low temperatures. The dielectric constant is large, even at T=0 and also increases with increasing temperature. We have found by conducting experiments at several frequencies, that the conductivity is only weakly frequency dependent in the micro and millimeter wave spectral range.

ANALYSIS AND DISCUSSION

In the light of the complex nature of the DNA when in a complex, multicomponent environment, there are several contributions to the charge response to applied ac electromagnetic fields. The charged counter ions surrounding the duplex contribute to the conductivity at low frequencies and also because of the large mass the spectral weight of this contribution is small. The water in which the DNA duplex is embedded leads to a significant dielectric loss, and this is mainly responsible to the (temperature independent) buffer loss we measured. There may be a contribution to the loss coming from the response of the water molecules in the hydration layer around the DNA. However due to the interaction between the counterions and the water dipoles dipole orientation at low frequencies is not likely. We believe therefore that the dielectric constant and the conductivity we have measured reflect the response of electronic charges along the DNA duplex.

Several theories have been proposed to account for charge transport along the DNA double helix. Long range electron tunneling has been proposed (8), together with a scenario involving both tunneling and hopping between base pairs (9), polaronic transport (10) and variable range hopping between localized states (11), with localization introduced by disorder, together with more exotic mechanisms (12,13). These issues have not been connected to, but are undoubtedly related to issues involving the nature of electronic states in one dimension (1D). Such "1D effects" are well known and explored, they include the importance of electron-electron interactions (leading to a new for of quantum liquid called the Luttinger Liquid (LL)), the importance of fluctuation effects, and also the theorem that all electron states are localized for whatever small amount of disorder. There are also firm theoretical predictions on the nature of electrical conductivity, both in the presence of disorder and in the presence of electron-electron interactions (see for example Ref 14 and references therein).

At this stage only a few comments can be made. First, the strong temperature dependence of the conductivity argues against simple tunneling scenarios, and gives evidence that temperature driven processes, which may involve also the reorientation of base pairs, (15,16) play an important role. Second, the large dielectric constant argues for long-range charge polarization in the presence of an applied electric field. Third, the difference between the DNA in and without buffer can be understood as in a dry environment the duplex assumes a disordered form (17). Both the dielectric constant and the conductivity is larger when the duplex is more ordered, giving evidence that charge delocalization and transport is strongly influenced by disorder induced localization.

The large dielectric constant and the conductivity both argue for a finite, and large localization length, and the temperature dependence gives evidence for temperature driven charge delocalization along the DNA duplex. The dielectric constant for a metallic 1D strand of finite length ℓ is given by (18)

$$\varepsilon_1 \approx 1 + \left(\ell q_0\right)^2 \tag{3}$$

where q_0 is the screening length, comparable to the inverse lattice constant for typical metallic densities. One possibility is that the polarization buildup is determined by the tertiary structure, by the coiling features of the duplex. In a simple minded picture a coiled DNA may be represented as straight DNA segments, separated by short regions (extending over few base pairs) over which the coiling takes place. Assuming that the relevant length scale is the so-called persistence length (the length scale over which the DNA can be thought as a more or less rigid and straight section), approximately 600A we arrive at a dielectric constant of 10^5, about one order of magnitude larger that measured. This indicates that the charges involved are less (per base pair) than what would correspond to metallic density. Because of base pair misalignment, the short regions act as barriers between the straight sections. The temperature dependence of both ε_1 and σ_1 can be accounted for by hopping over barriers, this hopping being mediated by temperature. Given the random coiling features of the duplex, the hopping between segments is expected to be random, and can be represented by a random distribution of barrier heights. This problem in one dimension has been considered (19), and the solution is a temperature dependent conductivity and dielectric constant at finite frequencies, the two having similar temperature dependences, as indeed observed here. Increased disorder leads to decreased length scale and thus to decreased conductivity and dielectric constant, again in agreement with our observations.

CONCLUSIONS

There are significant unresolved issues, many of those related to the fact that the electronic structure of DNA is dependent on different factors and can be significantly varied by changing the environment, such as the water content. At the same time is expected that base pair sequences also influence the charge propagation.

Clearly much remains to be done before the intricacies of charge transfer and charge transport are understood, and the relation to biological processes is firmly established.

REFERENCES

1. Dandliker, P. J.., Holmlin, R.E., and Barton, J. K., *Science*. **275**, 1465 - 68 (1997)

Lewis, F.D., Wu, T., Zhang, Y., Letsinger, R.L., Greenfield, S.R., and Michael R. Wasielewski, M.R., *Science* **277**, 673-676. (1997)

Hall, D. B., Holmlin, R. E., Barton, J.K., *Nature* .**382**, 731-735 (1996)

2. Pohl, H., and Kauzman, W., *Rev. Mod. Phys.* **36**, 721 (1968)

3. Okahata,, Y., Kobayashi, T., Tanaka, K., and Shimomura, M., *Journal of the American Chemical Society* **120**, 6165-6166. (1998)

4. Fink, H.W., and Schonenberger, C., *Nature*. **398**, 407 (1999)

5. Braun, E., Eichen, Y.,Sivan, U.,Ben-Yoseph, G., *Nature* **391**, 775-778 (1998)

6. Grüner, G., "Waveguide Configuration Optical Spectroscopy," in *Millimeter and Submillimeter Wave Spectroscopy of Solids*, edited by G. Grüner, Series: Topics in Applied Physics, Vol. No. **74**, Ch. 4., Springer Verlag, Berlin 1998, pp.111-166

7. Lyophilized from a solution in 1mM Tris-HCl, pH 7.5, 1mM NaCl, 1mM EDTA and 330 gr/ml DNA, product No. D9768. Lyophilization led to 0.5mg of DNA (as measured by the intensity of the absorption at 260nm) in total weight of 4.9 mG. Approximately 80% of the DNA have the full length and approximately 20% are shared, shorter segments.

8. Harriman, A., *Angew. Chem. Int. Ed. Engl* **38**, 996 (1999)

9. Jortner, J., Bixon, M., Langenbacher, T., Michel-Beyerle, M. E., *Proceedings of the National Academy of Sciences of the United States of America*, **95**,12759-12765 (1998)

10. Henderson, P. T., Jones, D., Hampikian, G., Kan, Y., Schuster, G. B., *Proceedings of the National Academy of Sciences of the United States of America*, **96**, 8353-8358 (1999)

11. Giese, B. ,. Wessely, S , Spormann, M., Lindemann, U. , Meggers, E. , Michel-Beyerle, M. E. , *Angew. Chem. Int. Ed. Engl.* **38**, 996 (1999)

12. D.L. Cox, R. Sing and S.K. Pati (to be published)

13. A. Castro-Neto and A. Bishop (to be published)

14. Vescoli, V., Degiorgi, L., Henderson, W., Grüner, G., Starkey, K.P., and Montgomery, L.K. , *Science*, **281**, 1181-1184 (1998)

15. Wan, C., Fiebig, T., Kelley, S. O.., Treadway, C. R.., Barton, J. K., Zewail, A. H., *Proceedings of the National Academy of Sciences of the United States of America* **96**, 6014-6019.(1999)

16. Bruinsma, R., Rudnick, J., and Gruner, G., (to be published)

17. Falk, M. , Kaufman, K.A. , and Lord, R.C. , *J. Am. Chem. Soc.* **85**, 3843 (1962)

18. Rice, M.J., and Bernasconi, J., *J. Phys. F. Metal Phys.* **3**, 55 (1973)

Holczer, K., Grüner, G. , Mihály, G. , Jánossy, A. , *Solid State Comm.* **31**, 145 (1979)

19. Alexander, S., Bernasconi, J., Schneider, W.R., Orbach, R., *Reviews of Modern Physics* .**53**, 175-98. (1981)

Alexander, S., Bernasconi, J., Schneider, W.R., Biller, R., Clark, W.G., Gruner, G., Orbach, R., Zettl, A., *Physical Review B (Condensed Matter)* **24**, 7474-7. (1981)

The Electronic Properties for DNA/CNT Chip on Si

E. V. Buzaneva, A. Yu. Karlash, S. O. Putselyk, Y. V. Shtogun, K. O. Gorchinskyy, Y. I. Prylutskyy, S. V. Prylutska, O. P. Matyshevska[*], and P. Scharff[**].

[*]National Taras Shevchenko University, Vladimirskaya str. 64,0033, Kyiv, Ukraine,
[**]Institute of Physics TU Ilmenau, PF 100565, D-98684 Ilmenau, Germany.

Abstract. We have studied the effects of the electronic structure self-formation and charge transfer for the planar system of layers consisting of DNA unwrapped double helix, DNA double helix, DNA unwrapped double helix with carbon nanotubes (CNT) in the separated sections with common boundaries on the silicon substrate – DNA chip on Si. The electronic structure of (DNA/CNT) layer is formed as a result of changes of electron levels in adenine, cytozine, guanine and thymine which were directly revealed by UV, IR spectroscopy. The transport properties of such a system are determined by the charge transfer through the barriers in the DNA double helix/unwrapped double helix and in the DNA unwrapped double helix/CNT heterostructures that was proved by tunneling spectroscopy investigations.

INTRODUCTION

Our hypothesis about possibility of the new electronic material construction by self-assembling of DNA/CNT layers based on theoretically predicted [1,2] and experimentally confirmed [3,4] facts that single DNA molecule and carbon nanotube (CNT) are able to conduct electrical charges. Electronic properties investigation of DNA molecules and CNT in such layers has great significance in view of their using in DNA nanotechnology for the construction of "lab on a chip" electron systems. The results of these investigations could be used as basis for further development of the engineering principles of the molecular computer, for example [5].

In this work some experimental results which supported the proposed model for engineering of DNA chip were obtained. In our investigation we used the DNA/CNT and the DNA layers adsorbed from DNA in NaOH buffer solution with and without CNT on Si substrate. On the Si substrate in the DNA/NaOH buffer solution with and without irradiation by visible light the regulated system of hydrogen bonds is transformed/destroyed and the DNA molecule structure turns from double helix into chaotic ball structure. This change has a co-operative character. The self-organization of the DNA/CNT layer under the adsorption on the substrate could be stimulated by this process. The UV, IR spectroscopy as well as tunneling spectroscopy were carried out for study the electronic structure of DNA/CNT, DNA double helix, DNA unwrapped double helix layers and charge transfer along these layers and through the barriers between them.

CP544, *Electronic Properties of Novel Materials—Molecular Nanostructures*, edited by H. Kuzmany, et al.
© 2000 American Institute of Physics 1-56396-973-4/00/$17.00

EXPERIMENTAL

The DNA of ukrainian production with nucleic acids such as adenine A-29.0, guanine G-21.2, cytozine C-21.2, thymine T-28.5 mol.% were used. Single-walled carbon nanotube bundles (from Dr. S. Fang, University of Kentucky) were mixed with DNA/NaOH buffer solution. The upper layer of this mixture and DNA/NaOH buffer solution as single droplets were deposited on Si (100), quartz and KRS substrates and dried with and without 2 h irradiation by visible light ($\lambda > 500$ nm) for the layer formation in the separated sections with common bounds on the Si substrate - for DNA chip on Si formation - as well as for the single layer formation on quartz and KRS substrates.

UV absorbtion spectra of single layers on quartz substrate between 200 and 850 nm were recorded on Hitachi 850 Spectrafluorimeter. IR transmittance and reflectance spectra of single layers on a Si and KRS substrates were recorded between 400 and 4000 cm^{-1} on Specord M-80 Care Zeiss Jena spectrometer at T=293 K. I –V characteristics of Pt/Ir tip – DNA (or DNA/CNT) layer – Pt/Ir tip structures were registered by Tunneling Spectrometer constructed in University Laboratory.

RESULTS AND DISCUSSION

The structure of DNA and CNT in DNA, DNA/CNT layers.

In the DNA/nanotubes layer the absorption peak wave length decreases on 5.8 nm in comparison with that one in DNA/NaOH layer, which corresponds to increased on 0.01 eV electron transiton enegry in DNA molecule. This may be caused by changing the hydrogen bonds under DNA/nanotube interaction.

In typical optical absorption spectra of the DNA layer the wide peaks were observed at around 259 nm associated with the $\pi-\pi^*$ transition of nucleic acids and they are the same as for DNA in aqueous buffer solutions in our experiment and in [6]. These results directly indicate that the DNA structure is not changed in the dried layer. In these spectra we also observed absorption increase in DNA layer adsorbed from DNA/NaOH solution in comparison with DNA layer adsorbed from DNA/Tris-HCl solution (at around 259 nm optical density increases from 0.38 to 0.47*10^{-4} M^{-1}cm^{-1}). This is known as hyperchromatic effect that testified the DNA unwrapping of double helix in chaotic ball structure in this buffer solution and DNA has the same structure in the layer. It was also found that the optical density at 259 nm increases from 0,4*10^{-4} M^{-1}см$^{-1}$ in the DNA/NaOH layer to 0,5*10^{-4} M^{-1}см$^{-1}$ in the DNA/NaOH/nanotubes layer that corresponds to partial unwrapping of the DNA double helix caused by DNA/nanotube interaction. The peculiarity of the typical optical absorption spectrum of (DNA/CNT) layer consists of changing of absorption peak wave length which we relate to known electron transition in the DNA molecule from 285.4 nm in the DNA/NaOH layer to 279.6 nm in DNA/CNT layer. We evaluated the electron transition energy to be increased on 0.01 eV. These results directly confirmed that the electronic structure of (DNA/CNT) layer is formed as a result of changes of electronic levels in nucleotide bases of DNA molecule under DNA/nanotube interaction.

In the typical absorption spectrum of (DNA/CNT) layer recorded between 300 and 850 nm we observed peaks with wavelengths corresponding to photon energies: 1.67, 1.86, 2.1, 2.56, 2.8, 3.34 eV. They correspond to reported ones for thin films of single-walled CNT prepared by both laser evaporation and electric arc methods in [7]. Using this fact we could suppose that tube's structure does not change in (DNA/CNT) layer.

From typical IR reflectance/transmittance spectra of (CNT/DNA) layer/Si structures (Fig.1) using their comparison with spectra of Si substrate, DNA layer/KRS and Si/(CNT/DNA) layer structures we found:

(i) the infrared-active modes at 845, 870 cm^{-1} and 1512, 1583, 1596, cm^{-1} as the narrow peaks in the reflectance spectra. These frequencies differ from corresponding ones of graphite 868 and 1590 cm^{-1}[7], and the frequency 870 cm^{-1} is the same as predicted by recent calculations [8] for armchair tubes, another one (845 cm^{-1}) is smaller than predicted by the same calculations, 850 cm^{-1} for chiral tubes and frequencies 1512, 1583, 1596 cm^{-1} correspond to also predicted by the calculations for armchair tubes (1580, 1510cm^{-1}) and for chiral tubes (1575 cm^{-1}) and as for zigzag tubes (1510cm^{-1}) (Fig.1a and 1b).

(ii) the features in the increasing of reflection in range 682-721 cm^{-1} as well as increasing of adsorption at 1458 cm^{-1} which originates in the spectrum of (CNT/DNA)layer/Si structure testify that structure of DNA is double helix unwrapping in (CNT/DNA)layer and presence of CNT with IR active phonons calculated for zigzag tubes (770, 650 cm^{-1}).

(iii) strong line at 1374 cm^{-1} which characterizes E_{1u} mode for armchair tubes [7].

(iv) several weak and broad peaks were observed between two lines 1374 and 720 cm^{-1} but we identified only the lines at 870, 845. IR active modes at 1234 cm^{-1} also correspond for calculated mode (1230cm^{-1}) in armchair tubes.

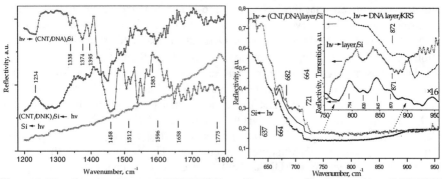

Figure 1. IR reflectance spectra of (CNT/DNA)layer/Si structure, Si substrate (a) and (CNT/DNA) layer/Si, Si/(CNT/DNA)layer structure and Si substrate (b). The spectral regions are represented where the IR active phonons are expected for armchair, zigzag and chiral tubes from the force field calculation [8]. The reflection angle is 75^0 to the normal (Brewster angle, p-polarised radiatiation). Inset: IR reflectance spectrum of (CNT/DNA)layer/Si and IR transmittance spectra of (CNT/DNA)layer/Si and DNA layer/KRS. Absorption peaks of Si substrate are 630, 664 cm^{-1}.

(ivv) detected lines at 1458,1512,1583,1596,1658,1775cm⁻¹ correspond to IR active phonons that expected for armchair, zigzag and chiral (1775 cm⁻¹) tubes from the force field calculation [8]. Presented identification is a continuation of the discussion concerning to the question how many different types of tubes are in a single wall CNT bundle.

The electronic structure of the layers and charge transport in layers, heterostructures

For the identification of conductivity type of CNT we used the values of optical transitions in CNT layer found from UV spectra (300-850nm) and calculation results of the band structure of the single wall CNT obtained by zone-folding method [7]. The features at 1.67, 2.56, 2.8 eV we associated with electronic transitions between pairs of singularities in semiconducting tubes.

This fact and our results obtained for the layer conductivity of structures of DNA double helix and unwrapped double helix with and without CNT on Si were used for the formation of the planar system of these layers in the separated sections with common bounds on the Si substrate – DNA chip on Si. We created the system of

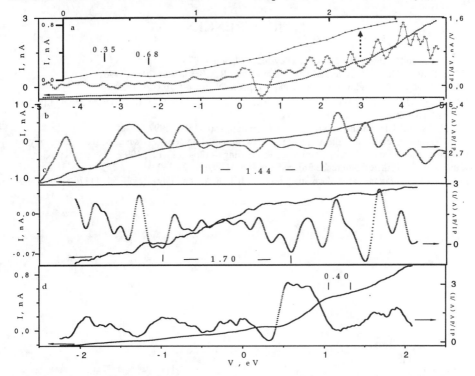

Figure 2. Typical I-V, (dI/dV)-V and dI/dV(V/I)-V characteristics of DNA unwrapped double helix/CNT heterostructure(a), CNT layer(b), DNA unwrapped double helix layer (c) and DNA unwrapped double helix/DNA double helix heterostructures (d). dI/dV(V/I) corresponds to the normalized differential conductance, that represents the shape of density states. The characteristics don't change principally at the increase of distances between tips in the range of 50-10² μm

heterostructures as DNA unwrapped double helix/CNT, DNA unwrapped double helix/ DNA double helix and formed Pt/Ir tip–DNA (or DNA/ CNT) – Pt/Ir tip anostructures. I-V characteristics of nanostructures (Fig.2a,d) as well as layers from CNT (Fig.2b) and from DNA unwrapped double helix with CNT (Fig.2c) have the following features: a non-linearity; diode behavior; negative differential resistance of heterostructure (Fig.2a) with dependence on the voltage that is typical for resonance tunneling through double barrier; the irregularity in increasing of a current with the steps that are typical for Coulomb blockade (Fig.2a-c) and for jumping transport of charge carriers (Fig.2d). The part of these features can be conditioned by the electronic structure of these layers and their interfaces. To study the electronic structure we investigated the shape of the normalized differential conductance, $(dI/dV)/(I/V)$, that represents the shape of density of states (DOS) of the layer. It can be seen from the normalized differential conductance versus voltage curves that there is a voltage gap at the low applied bias (1.44 and 1.70 eV for nanostructures on CNT and DNA with CNT, respectively). The voltage dependence of the differential conductance as well as the normalized conductance exhibits a clear peak structure that is typical in the DOS for 1D systems - CNT (Fig.2b).

REFERENCES

[1]   D.Dee and M.Baur, The J.of Chem.Phys.**60**,.541-560, (1974)
[2]   R.Saito, G.Dresselhaus, and M.Dresselhaus, Phys.Rev.B **53**,.2044-2050,(1996).
[3]   D.Porath, A.Bezryadin, S. de Vries and C.Dekker, Nature **403**, 635-637,(2000).
[4]   S.Tans, M.Devoret, H.Dai, A.Thess, R.Smalley, L.Geerligs, and C.Dekker, Nature **386**,474,(1997).
[5]   J.M.Tour et al., Science **278**, 252-254, (1997).
[6]   N.Higashi, T.Inoue and M.Niwa, Chem Commun.,i50/-1508,(1997)
[7]   C.Thomsen, et al, in "Electronic properties of novel materials – science and technology of molecular nanostructures", AIR Conference Proceedings 442pp. 123-127, 1998.
Kataura H., et al., AIR Conference Proceedings 486, 1999, pp.328-332.
[8]   Kürti J. et al. in the same place. pp.278-283.; Jishi R.A. et al. Phys.Rev.B. **51,** 11176 (1995)

Resonance Raman Scattering from Carbyne Materials

T. Danno[1*], K. Murakami[1], M. Krause[2], and H. Kuzmany[2]

1 Department of Lifestyle Design, Kochi Prefectural University
5-15 Eikokuji-Cho, Kochi 780-8515 JAPAN
2 Institut für Materialphysik der Universität Wien
Strudlhofgasse 4, A-1090 Wien, AUSTRIA
*tdanno@cc.kochi-wu.ac.jp

Abstract. Resonance Raman scattering from carbyne films obtained by dehydrocholorination of poly (vinylidene chloride) was studied under various laser excitation energies at room temperature. Four dominant Raman lines at around 1150, 1290,1530 and 2150 cm^{-1} were observed. Each line showed clear dispersion effect between 10 and 40 cm^{-1} / eV . The scattering line at 1150cm^{-1} showed a slightly shift of 5 cm^{-1} towards the higher wave number when the specimen was exposed to ambient atmosphere. The intensity ratio of line at 2150 cm^{-1}, which was referred to as the polyyne structure, to the one at 1530 cm^{-1} referred as the polyene structure increased with increasing of the energy of exciting laser. The averaged number of the conjugated atoms consisting of the polyyne structure was estimated to be about 10 from the comparison of the calculation of Σg mode frequency.

INTRODUCTION

Carbyne has attracted the particular interest of researchers from the point of view of the simplest linear chains of carbon atoms and the next allotrope of carbon[1]. We have obtained the carbyne films by dehydrochlorination of solid poly(vinylidene chloride) using 1,8-diazabicyclo[5,4,0]undec-7-ene (DBU), which is the strongest organic base so far known, in polar solvents[2]. The film shape was kept by the two step treatments, where the amorphous and the crystalline region is dehydrochlorinated stepwise.

Raman measurement is powerful tool to investigate the structure of carbon materials. Resonance Raman spectroscopy, especially, gives data about the electronic structure, which might be interpreted in terms of the molecular and crystallographic structure of carbonaceous materials, including carbyne. Akagi et al. measured Resonance Raman spectra for the carbyne material obtained from chlorinated polyacetylene by dehydrochlorination[3]. Raman spectra show the three dominant lines at 2157($\nu_{C\equiv C}$), 1556 ($\nu_{C=C}$),1145 cm^{-1} by 458 nm laser excitation. The number of conjugated sp carbon was estimated to be 12 from the dispersion effect. Kijima et al. obtained similar spectra with two dominant lines at 2150 and 1550 cm^{-1} for the carbyne material electropolymerized from diiodoacetylen[4]. Kastner et al. obtained the carbyne-like films by electrochemical reductive preparation from PTFE[5]. They estimated the

CP544, Electronic Properties of Novel Materials—Molecular Nanostructures, edited by H. Kuzmany, et al.
© 2000 American Institute of Physics 1-56396-973-4/00/$17.00

conjugation length of C≡C as 8-14 C atoms from the resonance Raman measurements incorporated with a quantum calculation of Σ_g mode for olygoynes. Also the anomalous large intensity ratio of Raman line at 2150 cm^{-1} to 1550 cm^{-1} was reported.

In this paper we measured resonance Raman scattering from carbyne material obtained from PVDC by chemical dehydrochlorination. After the spectrum was compared with the one for previous papers, the conjugation length of sp carbon was estimated using the results of quantum calculations.

EXPERIMENT

Carbyne was obtained from PVDC film spin-coated on the KBr cleavage as described previously[2]. The specimen was stored under N_2 atmosphere also during the Raman measurements. Raman spectra were recorded at room temperature with an argon ion laser and a krypton ion laser excitations using a line focus of 0.05mm x 2.0mm. Another conditions for Raman measurements were described elsewhere in detail[6].

RESULTS AND DISCUSSION

Figure 1 shows the Raman spectrum of carbyne film with 458 nm excitation. This spectrum has characteristic features compared with those of previous papers as following:

1. Strong luminescence was observed as a background.

2. Four dominant Raman lines at ca. 2150, 1540, 1300, 1150 cm^{-1} were observed.

3. The line width of each dominant line was narrower than those of previous researches.

4. The intensity of a broad scattering around 1300 cm^{-1} was relatively weak.

5. An intensity ratio of the line at 2150 cm^{-1} to the one at 1540 cm^{-1} was relatively strong.

The line at 2150 cm^{-1} was assigned to the stretching mode of sp-bonded carbon ($v_{C≡C}$). The line at 1540 cm^{-1} is common for all forms of polycrystalline graphite

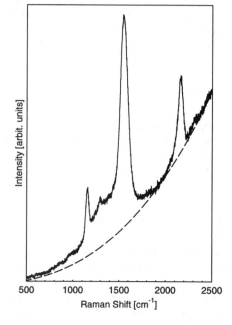

FIGURE 1. Raman spectrum of the carbyne film with 458 nm excitation. The background was assumed as a third order polynominal (broken line).

or „amorphous carbon" with a broad distribution of sp^3- and sp^2-bonded carbon atoms. The assignments of the lines at 1300 and 1150 cm^{-1} are still not clear but they might be assigned to v_{C-C} (1300 cm^{-1}) and $v_{C=C}$ (1150 cm^{-1}) of sp^2-bonded carbon analogous to the vibrations in *trans* conjugated polyenes and *trans*-polyacetylen[3,7].

Each of Raman lines showed the apparent dispersion effects. Table I shows the summary of the dispersion effects focused on the stretching mode of sp-bonded carbon and the one for the amorphous carbon. The dispersion of the Raman shift of $v_{C\equiv C}$ for present work is smaller than those of previous works[8]. The linear relation between laser energy and vibrational frequency was interpreted by the conjugation length model, which is determined by the relation between the conjugation length and gap energy or vibrational frequency of the system, respectively [9]. For the oligoynes, quantum calculations of the HOMO-LUMO transition energy and the vibrational frequency has been carried out as a function of the number of conjugated sp-bonded carbons[5]. From these results following relation was obtained,

Table I. Dispersion effects of the peaks and the intensity ratio for the Raman lines from carbyne film.

Laser wave length [nm]	Laser energy [eV]	Peak and intensity ratio		
		$v_{C\equiv C}$ (cm^{-1})	$v_{C=C}$ (cm^{-1})	$Iv_{C\equiv C}$ / $Iv_{C=C}$
647	1.92	2135	1508	0.14
568	2.18	2149	1507	0.23
514	2.41	2154	1522	0.22
(after air exposure :		2157	1520	0.18)
458	2.71	2158	1543	0.35
Dispersion [cm^{-1} / eV]		28	46	

$$v(N) = 1750 + \frac{3980}{N} \quad [cm^{-1}]$$

where N is the number of carbon atoms, $v(N)$ is the frequency of $v_{C\equiv C}$[5]. Using above relation, the average number of conjugated carbon in the present material was estimated to be 10, which means 5 C≡C. Similar results were obtained by the previous works[3,4,5].

The intensity ratio of the line at 2150 cm^{-1} to the one at 1540 cm^{-1} determines the carbyne content in the sample. Present result is very similar to those of other materials which were obtained by various method except the results for the material obtained by the electrochemical reduction of PTFE, which showed anomalously strong intensity ratio over 1.0. The intensity ratio showed also the clear dispersion. This results is also similar to the one previously observed[5].

Effects of the air exposure (40 days) is shown also in Table I. Both the peaks shifted some extent upwards and the intensity ratio decreased about 18%, respectively, after air exposure. These results suggest the further cross-linking of sp-bonded carbons to graphite like materials. The present results, however, are much different from those of carbyne obtained by the electrochemical reduction of PTFE, where the Raman line of v_C

$\equiv$C shifted about 20 cm^{-1} upwards after the same period of aging[5] and the intensity ratio decreased about 30%[8]. Carbyne molecules are embedded in the precursor (PVDC in this case), which protect carbyne from the oxygen or water attacking, whereas the carbyne molecules are formed on the film surface of PTFE by the electrochemical reduction method.

ACKNOWLEDGMENT

Support of this work by the Project for International Research Exchange of the Kochi Prefectural University is acknowledged.

REFERENCES

1. Kudryavtsev,Yu.P., Heimann,R.B., in *Carbyne and Carbynoid Structures*, eds. R.B.Heimann, S.E.Evsyukov, L.Kavan, Dordrecht/Boston/London, KLUWER ACADEMIC PUBLISHERS, 1999, Ch. 1, pp. 1-7.
2. Danno, T., Murakami, K., and Ishikawa, R., in *Electronic Properties of Novel Materials - Science and Technology of Molecular Nanostructures* (AIP Conference Proceedings 486), eds. H.Kuzmany, J.Fink, M.Mehring, and S.Roth, Melville, New York, American Institute of Physics, 1999, pp213-216.
3. Akagi,K., Nishiguchi,M., Shirakawa,H., *Synthetic Metals*, **17**, 557-562(1978).
4. Kijima,M., Sakai,Y., Shirakawa,H., *Synthetic Metals*, **71**, 1837-1840(1995).
5. Kastner,J., Kuzmany,H., Kavan,L., Dousek,F.P., Kürti,J., *Macromolecules*, 28(1), 344-353(1995).
6. Krause M., Hulman, M., Kuzmany, H., Dennis, T.J.S., Inakuma, M., and Shinohara, H., in *Electronic Properties of Novel Materials -Science and Technology of Molecular Nanostructures* (AIP Conference Proceedings 486), eds. H.Kuzmany, J.Fink, M.Mehring, and S.Roth, Melville, New York, American Institute of Physics, 1999, pp136-139.
7. J.C.W.Chien, *POLYACETYLENE Chemistry, Physics and Material Science*, Academic Press, Orland, 1984, pp213
8. Kavan,L., Kastner,J., in *Carbyne and Carbynoid Structures*, eds R.B.Heimann, S.E.Evsyukov, L.Kavan, Dordrecht/Boston/London, KLUWER ACADEMIC PUBLISHERS, 1999, Ch. 6, pp. 343-356.
9. Kuzmany, H., *Phys. Stat. Sol(b)*, **97**, 521(1980)

A Near Field EEL Spectroscopy Study of Surface Modes in Nanotubes and Onions from Various Layered Materials

O. Stéphan*, M. Kociak*, L. Henrard†, K. Suenaga‡, E. Sandré*, and C. Colliex*

*Laboratoire de Physique des Solides, Bâtiment 510, Université Paris-Sud, 91405 Orsay Cédex, France
†Laboratoire de Physique du Solide, Facultés Universitaires Notre-Dame de la Paix, 61 rue de Bruxelles, 5000 Namur, Belgium
‡JST-ICORP, Department of Physics, Meijo University, Nagoya 468-8502, Japan

Abstract. We present a Near Field Electron Energy Loss Spectroscopy study of the low-loss energy region (1 to 50 eV), recorded on multishell nanotubes and hyperfullerenes from various layered materials. We concentrate on the study of surface modes excited in a near-field geometry where the coupling distance between the electron beam and the surface of the nano-objects is accurately monitored. Similarities between surface collective excitations in the different layered nanostructures are pointed out. Relying on a classical continuum dielectric model taking fully into account the anisotropic character and the hollow geometry of the nanoparticles, we show that the detected modes are directly related to the in-plane and out-of-plane components of the dielectric tensor of the associated planar bulk material. When the wall thickness decreases, we show that the excitation spectrum is dominated by the « in plane modes ». We discuss the validity of this model for nano-objects with vanishing wall thickness.

INTRODUCTION

From a structural point of view, nanotubes and onions can be described as the counterparts of layered planar materials within a spherical or cylindrical geometry. Thanks to their well defined geometry and a large range of number of layers (from 1 to 100 typically), they are perfect test objects to investigate macroscopic concepts like the dielectric constant at a nanometer scale.

EXPERIMENTAL

In order to investigate their dielectric properties, we performed measurements on a STEM VG HB501 field emission gun operating at 100 keV and fitted with a Gatan 666 parallel-EELS spectrometer and a couple of high angle annular detectors for simultaneous dark field mode imaging. Following the spectrum image approach [1], collections of EELS spectra were acquired by scanning under digital control a 5Å electron probe across a chosen isolated nano-object whose morphology (schematically described by an inner radius r and an outer radius R) was first identified by means of

CP544, *Electronic Properties of Novel Materials—Molecular Nanostructures*, edited by H. Kuzmany, et al.

bright field images. In order to obtain a high accuracy in the definition of the successive impact parameters (defined as the distance between the beam and the center of the particle), the annular dark field (ADF) profile was simultaneously acquired (see figure 1a). A typical sequence of spectra obtained by scanning the probe every 3 Å perpendicularly to the axis of a multishell carbon nanotube is displayed on Fig. 1.

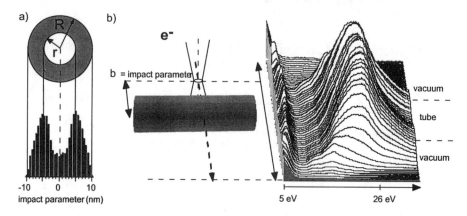

FIGURE 1. a) Definition of outer and inner radii (R and r respectively) for a nanoparticle. The experimental ADF profile is shown as the way to accurately determine them. b) Line spectrum across a multishell carbon nanotube of 25 nm outer diameter with corresponding scattering geometry.

The line scan mode allows a study of the excitation probability of the different modes as a function of the impact parameter. In particular, a striking observation from fig. 1 is the detection of electronic excitation modes for a non-penetrating beam and for a coupling distance as large as 10 nm. In this paper, we concentrate on data acquired for such non-penetrating configuration where the near-field decaying from the particle out into vacuum is probed.

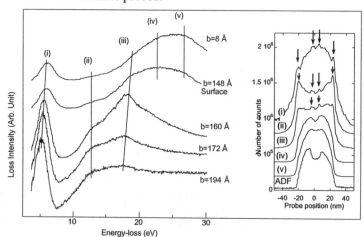

FIGURE 2. Selection of EEL spectra extracted from the line-scan displayed on Fig. 1 for different impact parameters. Inset: intensity profiles of the modes labelled from (i) to (v) and ADF profile.

A selection of spectra extracted from the previous line scan is displayed in figure 2. Three energy features, labelled (i), (ii) and (iii), are detected for non-penetrating geometry at respectively 6 eV, 13 eV and 17 eV, the latest mode shifting to higher energy at smaller impact parameters. Three extra features at 6.5 eV, 23 eV and 27 eV occur when the beam intersects the nano-objects. These bulk modes are similar to those detected in graphite and are commonly interpreted as collective excitations of π electrons (mode (i)) and $\pi+\sigma$ electrons (modes (iv) and (v)) [2]. The intensity profiles displayed in inset provide a direct representation of the spatial dependence of the modes excitation probabilities and therefore valuable information about the nature of the excitations. By comparison with ADF profiles, it can be seen that bulk modes (i), (iv) and (v) are localised within the nano-object. The situation is different for modes occurring in a non-penetrating geometry ((ii) and (iii)). The excitation probability is non zero outside the particle and decays exponentially into vacuum. Each mode shows a maximal probability on both the inner and outer surfaces of the tube, leading to the identification of these modes as surface modes. Mode (i) (not resolved from the bulk mode) is the surface (and volume) mode associated with π bulk mode while modes (ii) and (iii) involve π and σ electrons.

We note that very similar results were obtained from C onions of similar size [3].

THE DIELECTRIC MODEL

In order to understand the nature of these surface excitaions, we relied on a dielectric model, taking fully into account the anisotropy of the particles, their geometry and describing the Coulombic coupling between the beam and the nano-object. This approach is based on the description of the 'spherical' or 'cylindrical' anisotropy by a dielectric tensor $\bar{\varepsilon}(\omega)$, locally diagonal in spherical or cylindrical coordinates [3] and directly transferred from that of the bulk planar material. Here, we present simulation results for hollow spherical particles for which the EEL spectrum calculation can be completely analytically driven. The knowledge of $\bar{\varepsilon}(\omega)$ leads to an analytic expression for the dynamic multipolar polarizability $\alpha_l(\omega)$ by resolving the non-retarded Maxwell's equations and matching the boundary conditions. $\alpha_l(\omega)$ is defined as the response function of the molecule at a multipolar field of order l [4].

The EEL spectrum is then computed for a given impact parameter outside the particle within the semi-classical approximation: the coulomb field of an electron moving near the surface of a medium polarizes this medium; this induced electromagnetic field acts back on the electron, which suffers an energy loss. The induced electromagnetic (EM) field is related to the known quantity $\alpha_l(\omega)$. Thus, assuming that the STEM electron probe is made up of classical electrons moving in a straight line, it is possible to compute the work of an electron along its trajectory as:

$$W(b) = -q \int_{-\infty}^{+\infty} dt \, (\vec{v}.\vec{E}^{ind}(\vec{r}_e(t))) = \int_0^{\infty} \hbar\omega \, P(\omega,b) d(\hbar\omega) \qquad (1)$$

where W is the total energy loss, b the impact parameter (see fig 1 a), $\vec{E}^{ind}$ the induced field, $\vec{r}_e$ the position of the electron and $\vec{v}$ is the classical velocity of the electron, and the probability $P(\omega, b)$ of an energy-loss $\hbar\omega$ is a function of $Im\alpha_l(\omega)$ [4]. Therefore, one of the main inputs to our model is the dielectric tensor component of the bulk planar material. We used tabulated values extracted from optical measurements [5] or *ab initio* DFT LDA calculation.

DISCUSSION

Fig. 3 shows, the simulation of the EELS spectra of onion-like particles (r=1.8 Å, R=150 Å, E_0=100 keV) for various impact parameters when the beam is not intersecting the onion. The 6 eV (i) and 17-18 eV (iii) resonances are associated with the in plane component ($\varepsilon_\perp(\omega)$) of the dielectric constant of graphite, they are attributed to surface excitations of π electrons associated with π-π^* transitions and of σ electrons associated with σ-σ^* transition respectively. The weak peak at 14 eV (ii) involve both π-σ^* and σ-π^* transitions associated with the out of plane component ($\varepsilon_{//}(\omega)$) of the dielectric constant [6]. As shown in the inset of Fig 3 when a lorentzian model with a vanishing small damping (0.1 eV) is used, it appears that the major resonances observed in the simulation based on a tabulated dielectric constant, result from a multimodes structure. This behavior is characteristic of an anisotropic nano-object and is different to the isotropic hollow spherical shell case where two surface modes (called tangential and radial resonances) are clearly separated.

The change in the multipolar decomposition of the electron field explains the shift for the in-plane mode as a consequence of an increase in the contribution of high momentum modes with decreasing impact parameter [3].

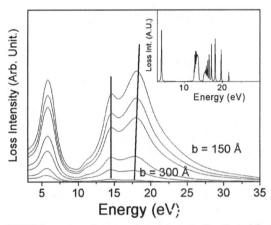

FIGURE 3. Simulated NFEEL spectrum from a carbon onion as a function of the impact parameter.

This description seems to extend to all anisotropic particles of cylindrical and spherical geometry whose valence electrons are in sp^2 configuration. Indeed, a similar near-field behaviour of 5 shell BN nanotubes was observed. A π surface mode was detected at 5.5 eV while in-plane and out-of-plane surface modes involving π and σ electrons were also observed at 17 eV and 12 eV respectively, the highest energy mode, shifting to higher energies with decreasing impact parameters [3].

Then the question arises of the limit of validity of such description of the dielectric response of a nano-object to a coulombic external field when using the dielectric tensor of the associated planar bulk material as an input data. In fig.4, we show the near field EEL spectrum recorded from a WS_2 nanotubes made of one layer of WS_2 together with a simulation using the dielectric model. Surprisingly for such thin nano-object, a good agreement between experiment and simulation is observed. Also worth nothing is the predominance of the in plane modes (at low and intermediate energy) at the expense of out of planes modes around 20 eV (as compared to WS$_2$ nanotubes with higher number of layers (not shown here). For a clear identification of the contribution of every component of the dielectric tensor in the EELS spectrum, two curves corresponding to artificial dielectric tensors where $\varepsilon_\perp$ and $\varepsilon_{//}$ are respectively set to one (respectively dotted and dashed lines) are also shown in fig. 4b).

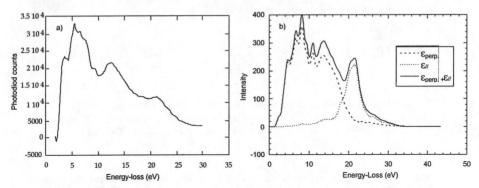

FIGURE 4. Experimental (a) and simulated (b) NFEELSpectra from a monolayer WS$_2$ nanotube.

CONCLUSION

We have performed an extensive Near Field EELS study of the surface excitation modes in spherical and cylindrical particles made of various layered compounds with different electronic properties (from semi-metal graphite, semi-conducting WS$_2$ to insulating BN). We show a clear sensitivity of these near field experiments to the anisotropic character and the geometry and size of the particles. Similarities in the interpretation of the electron excitations in all anisotropic nano-objects have been stressed. More specifically, we have shown that the dielectric response of such nano-object is correctly reproduced when directly transferring the in plane and out of plane components of the dielectric tensor of the planar bulk materials. This approach seems to remain valid in the limit of objects with a few number of layers. In this limit, we show that the modes associated with the in plane component of the dielectric constant predominate at the expense of the out of plane modes.

ACKNOWLEGMENTS

L.H. is supported by the Belgian FNRS.

REFERENCES

1. C. Colliex, M. Tencé, E. Lefèvre, C. Mory, H. Gu, D. Bouchet, and C. Jeanguillaume, *Mikrochim. Acta* **114/115**, 71 (1994).
2. J. Daniels, C.V. Festenberg, H. Raether, and K. Zeppenfeld, in "Optical constants of solids by electron spectroscopy", Springer tracts in modern physics, Vol. 54 (Springer, Berlin - 1970) pp 126-130.
3. M. Kociak et al. Phys. Rev. B, *in press*, to be published 15 May 2000.
4. A.A. Lucas, L. Henrard, Ph. Lambin. *Phys.Rev. B* **49**, 2888 (1994).
5. B.T. Draine, *Ap.J.* **333**, 648 (1988); B.T. Draine, *Princeton Observatory Report*. Princeton 1987.
6. E. Tossati and F. Bassani, *Nuovo Cimento B* **65**, 161 (1970).

Mechanical Properties of Individual Microtubules Measured Using Atomic Force Microscopy

N.H. Thomson[1], A. Kis[1], S. Kasas[2], G. Dietler[3], A. J. Kulik[1] and L. Forró[1].

[1]*Institut Genie Atomique, Département de Physique, EPFL, Lausanne, CH-1015.*
[2]*Insitut de Biologie Céllulaire et de Morphologie, Université de Lausanne, CH-1005.*
[3]*Département de Physique, Université de Lausanne, Lausanne, CH-1015*

Abstract. Atomic force microscopy has been used to investigate the mechanical properties of microtubules. Method of Salvetat et al. [1], originally developed for measuring the Young's modulus of carbon nanotubes, has been extended to a biological system of similar geometry. Preliminary measurements of Young's modulus of microtubules in this configuration give a value of 74 MPa. Compression measurements on the same system give a value of 22 MPa, which is of the same order of magnitude, while values previously reported in the literature have a dispersion of 3 orders in magnitude, from 1MPa to 1GPa. Further improvements of our method will enable comparison between the properties of different types of microtubules, for example, between those from healthy and diseased tissue.

INTRODUCTION/MOTIVATION

Microtubules are protein polymers, microns long, that are a central structural element of the internal skeleton (cytoskeleton) of the living cell. They provide mechanical stiffness for the cell, shape its spatial development, act as a transport network (like railways for other proteins) and they are involved in complex cell processes, such as the separation of chromosomes during cell division.

Differences in the mechanical properties of cells from healthy tissue and those from diseased states, such as cancer or Alzheimer's, demand not only descriptive, but a quantitative characterization at the molecular level. One of the important parameters could be the Young's modulus, determined from the mechanical response. In the literature there is at least 3 orders of magnitude dispersion in the values of Young's moduli determined by a number of experimental techniques. Optical tweezers give values around 1 GPa [2], whereas a previous AFM study gives 1 MPa [3]. Our goal is to find a reproducible method to enable comparison of healthy and "diseased" states.

This project stems from a method developed for measuring the elastic properties of individual carbon nanotubes by suspending them across pores in alumina filters [1]. The structural similarities between carbon nanotubes and microtubules are astonishing. Both nanotubes and microtubules are cylinders with diameters in the nanometer range and can be microns in length. Although Young's modulus of

CP544, *Electronic Properties of Novel Materials—Molecular Nanostructures*, edited by H. Kuzmany, et al.

microtubules may be 6 orders of magnitude lower, the flexural rigidity is high enough to allow measurements using AFM.

MICROTUBULES – STRUCTURE AND FUNCTION

Microtubules (MTs) are one of the protein polymers that make up the cytoskeleton. Besides fulfilling a role in the mechanical properties of the cell, they have many other functions. They polymerize outward from the nucleus of the cell directing the shape and polarity of the cell, particularly relevant for neural cells to make connections to neighbors. Other motor proteins travel along microtubules, carrying vesicles containing chemicals essential for the maintenance of the extremities of the cell. Microtubules are also central to the process of cell division, separating the divided chromosomes very rapidly through depolymerization.

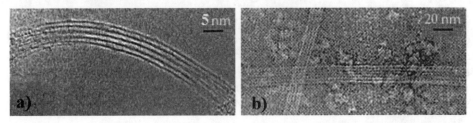

FIGURE 1. (a) High resolution TEM images of a single-walled carbon nanotube bundle and (b) of microtubules. Both consist of filament structures bundled together in a well-defined geometry.

In vivo, microtubules are continuously polymerizing and depolymerizing. This ability for dynamic instability seems to be crucial to their function although the exact mechanism of this instability is still under scrutiny. However, the possible depolymerization of microtubules presents challenging problems for measuring their mechanical properties. In order to obtain favorable experimental conditions, one needs to stabilize them using various methods.

The tubulin polymerizes into chains known as protofilaments, which then self-assemble in a parallel fashion into a cylindrical structure, usually containing 13 protofilaments, although populations can contain microtubules consisting of between 13 to 16 protofilaments [4]. The exact mechanism of self-assembly is still unclear.

The similar geometrical nature of the microtubules to carbon nanotubes, i.e. hollow cylinders with diameters in the nanometer range motivated our project. Figure. 1 shows TEM images of a carbon nanotube bundle and microtubules to illustrate their similarity. Although the Young's modulus of microtubules was expected to be orders of magnitude lower than carbon nanotubes, their flexural rigidity (Young's modulus multiplied by the second moment of area of beam), EI, turns out to be similar in magnitude (only one or two orders difference). This has been estimated using the highest E for microtubules quoted in the literature (see below). The similarity in flexural rigidity means that AFM cantilevers with a similar spring constant to the ones used to bend carbon nanotubes are also suitable for bending microtubules.

Several different groups have already made measurements of the Young's modulus (or more specifically in most cases, the flexural rigidity) by a number of different techniques. The range of Young's moduli quoted by these studies varies by at least 3 orders of magnitude from over 1 GPa [2] down to 1 MPa [3]. Some of this dispersion must arise from the different types of MTs studied, but undoubtedly the measurement technique plays an important role. Therefore, there is a need for greater understanding in how the measurement technique affects the result and to find a reliable method for measuring MTs under different conditions. The goal of our project was to find a suitable experimental configuration and valid analysis using AFM.

MECHANICAL MEASUREMENTS

Bending Over Pores

Before applying the method of Salvetat et al. [1], we had to overcome several problems specific to the microtubules. In order to obtain realistic results, and to hinder any damages to microtubules due to capillary forces during drying, one has to measure the mechanic properties of microtubules in liquid.

In liquid, one also needs to find a reliable method to fix the microtubules to a surface in order to enable reproducible AFM imaging in contact mode.

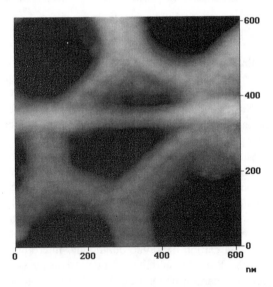

FIGURE 2. Contact mode image of a microtubule lying across the edge of a pore in the alumina filter. This MT was stable enough to make a two point force measurement and estimate the Young's modulus to be 74 MPa.

To date the method using glutaraldehyde to link MTs to the surface has proved to be the most satisfying one. A measurement was made of a microtubule suspended across a pore, shown in Figure. 2. The deflection δ of the microtubule is given by the clamped-beam formula:

$$\delta = \frac{FL^3}{192EI} \qquad (1)$$

where F is the applied load, L the suspended length, E Young's modulus and I the second moment of area of beam (microtubule).

In this case, the MT was lying over one edge of a pore and was stable enough for a Young's modulus of 74 MPa to be determined from a two force measurement. Future experiments will aim at reproducing this result with a higher degree of confidence.

Compression on a Flat Surface

Recent data shows that the MTs can be compressed on flat surfaces, such as AP-mica (See Fig. 3), under different loading forces (Figure 4) without any significant lateral movement. The Hertz approximation connects the loading force F to the indentation Δz:

$$F = \frac{4\sqrt{RE}}{3(1 - \mu^2)} \Delta z^{1.5} \qquad (2)$$

where E is the Young's modulus, μ is the Poisson ratio, and R is the radius of curvature of the AFM tip.

Using this approximation, a value of 22±6 MPa for the Young's modulus has been obtained. This value is comparable with the value of 6 MPa quoted in Vinckier et al. [3], obtained using a similar method of measurement and analysis. However, because of the similar curvatures of the microtubule and the AFM tip, we believe that a better analysis needs to be found to convert such measurements into meaningful numbers.

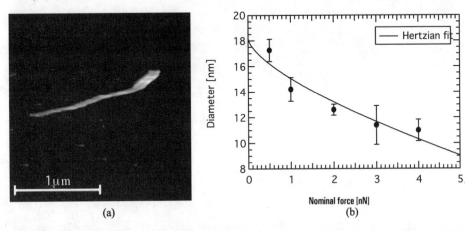

(a)   (b)

FIGURE 3. (a) Contact mode image of a microtubule bound to an AP-mica surface. (b) Plot of the diameter of the microtubule in dependence of the applied load. Line represents the fit of the measured values to the Hertz model (formula 2).

485

CONCLUSIONS

The principle of measuring the mechanical properties of microtubules using AFM and the modified version of the method of Salvetat et al. [1], originally conceived for measuring SWNTs, has been demonstrated. Furthermore, the range of measured values has been reduced from 3 orders of magnitude (1MPa-1GPa) quoted earlier in the literature, to within an order of magnitude (22-74 MPa).

There are, at least, three potential ways of measuring the elastic properties of MTs (two more than for the carbon nanotube study). Coupling this with the control of protofilament topology (i.e. sheet or tubes) and the suspended lengths of microtubules, should enable the investigation and separation of bending and shear behavior of MTs. Reliable comparisons between the properties of different types of MT can then be made, for example, between those from healthy and diseased tissue.

ACKNOWLEDGMENTS

The study of the mechanical properties of individual microtubules was supported by a UNIL-EPFL grant 1999. We thank prof. W. Benoît, prof. L. Zuppiroli and dr. J. M. Soletti for fruitful discussions. We also thank G. Beney for providing us with polished filters.

REFERENCES

1. Salvetat J.-P. et al. (1999) *Appl. Phys. A* **69** (3) 255-260.
2. Kurachi M., Hoshi M. and Tashiro H. (1995) *Cell Motility and the Cytoskeleton* **30**, 221-228.
3. Vinckier A. et al. (1996) *J. Vac. Sci. Technol. B.* **14** (2) 1427-1431.
4. Chrétien D. and R.H. Wade (1991) *Biol. Cell.* **71**, 161-174.

VIII. APPLICATIONS

Devices and Machines at the Molecular Level

M. Venturi, A. Credi, and V. Balzani

Dipartimento di Chimica "G. Ciamician", Università di Bologna, via Selmi 2, 40126 Bologna, Italy.

Abstract: The concept of (macroscopic) device can be extended to the molecular level. A *molecular–level device* can be defined as an assembly of a discrete number of molecular components designed to achieve a specific function. A *molecular–level machine* is a particular type of molecular–level device in which the component parts can display changes in their relative positions as a result of some external stimulus. The extension of the concept of a device to the molecular level is of interest not only for basic research, but also for the growth of nanoscience and the development of nanotechnology. A few examples of molecular–level devices and machines are illustrated.

INTRODUCTION

The concept of device can be extended to the molecular level [1]. A *molecular–level* device can be defined as an assembly of a discrete number of molecular components (that is, a *supramolecular* structure) designed to achieve a specific function. Each molecular component performs a single act, while the entire assembly performs a more complex function, which results from the cooperation of the various molecular components.

A *molecular–level machine* is a particular type of molecular device in which the component parts can display changes in their relative positions (i.e., motions) as a result of some external stimulus [2].

Most of the recently designed molecular–level machines are based on pseudorotaxanes, rotaxanes, and catenanes (Figure 1) [3]. Important features of these systems derive from noncovalent interactions between components that contain complementary recognition sites. Two types of interactions are mainly involved in these compounds: (i) charge–transfer (CT) interactions between components that contain electron–acceptor (e.g., 4,4'–bipyridinium) and electron–donor (e.g., dioxynaphthalene) units and/or (ii) hydrogen–bonding interactions between secondary ammonium functions and suitable crown ethers.

It is worth noting that molecular–level machines could play a role in information processing since their mechanical movements take place according to a binary logic. Systems of this type implementing the XOR [4] and XNOR [5] logic operations have already been reported.

Molecular–level devices, like macroscopic devices, are characterized by the following features.

CP544, *Electronic Properties of Novel Materials—Molecular Nanostructures*, edited by H. Kuzmany, et al.
© 2000 American Institute of Physics 1-56396-973-4/00/$17.00

(i) They need energy to work. Since molecular–level devices operate on chemical basis, such an energy is supplied by a chemical reaction, which can be caused by adding suitable reactants (protons, reductants, oxidants, etc.), or, most important, by light (photochemical reactions) or electricity (electrochemical reactions).

(ii) Their operation rely on motion. In molecular–level devices, motions correspond to electronic and/or nuclear rearrangements. To monitor such rearrangements, readable changes in some chemical or physical property of the system have to occur. In this regard, photochemical and electrochemical techniques are very useful since both photons and electrons (or holes) can play the dual role of "writing" (i.e., causing a change in the system) and "reading" (i.e., reporting the state of the system) [6].

(iii) The process must be cyclic. This means that in molecular–level devices the chemical reaction involved in the operation has to be reversible. From this viewpoint, photochemical and electrochemical processes are to be preferred since they do not lead to accumulation of waste products.

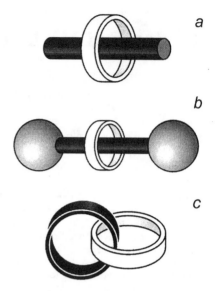

FIGURE 1. Schematic representation of pseudorotaxanes *(a)*, rotaxanes *(b)* and catenanes *(c)*.

(iv) The cyclic process occurs in a certain time scale. For molecular–level devices, the time scale can range from less than picoseconds to seconds, depending on the type of rearrangement (electronic or nuclear) and the nature of the components involved.

(v) A function has to be performed. Molecular–level devices performing various kinds of functions can be imagined, as it will be discussed below.

For space limitations, in this paper we will only mention a few examples of molecular–level devices studied in our laboratories.

MOLECULAR WIRES

An important function at the molecular level is photoinduced energy and electron transfer over long distances and/or along predetermined directions. This function can be obtained by linking donor and acceptor components by a rigid spacer, as illustrated in Figure 2. An example is given by the $[Ru(bpy)_3]^{2+}-(ph)_n-[Os(bpy)_3]^{2+}$ compounds (bpy=2,2'–bipyridine; ph=1,4–phenylene; n=3, 5, 7), in which excitation of the $[Ru(bpy)_3]^{2+}$ moiety is followed by electronic energy transfer from the excited $[Ru(bpy)_3]^{2+}$ unit to the $[Os(bpy)_3]^{2+}$ one, as shown by the sensitized emission of the latter [7]. For the compound with n=7, the rate constant for energy transfer over the 4.2

nm metal-to-metal distance is 1.3×10^6 s^{-1}. In the [Ru(bpy)$_3$]$^{2+}$–(ph)$_n$–[Os(bpy)$_3$]$^{3+}$ compounds, obtained by chemical oxidation of the Os–based moiety, photoexcitation of the Ru(bpy)$_3{}^{2+}$ unit causes the transfer of an electron to the Os–based one with a rate constant of 3.4×10^7 s^{-1} for compounds with n=7. Spacers whose energy or redox levels can be manipulated by an external stimulus can play the role of switches for the energy– or electron–transfer processes.

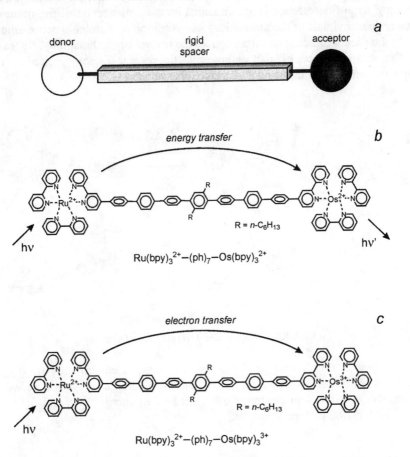

FIGURE 2. Schematic representation of a molecular–level wire *(a)*, and photoinduced energy *(b)* and electron *(c)* transfer processes in a wire based on metal complexes and polyphenylene rigid bridge.

PLUG/SOCKET AND RELATED SYSTEMS

Supramolecular systems have recently been designed that may be considered as molecular–level plug/socket devices [8]. Plug in/plug out is reversibly controlled by acid/base reactions, and the photoinduced flow of electronic energy (or electrons) takes place in the plug–in state. The plug–in function can be based on the threading of a (±)–binaphthocrown ether by a (9–anthracenyl)benzylammonium ion promoted, for instance, by protonation of the corresponding amine with an acid (Figure 3). The association process can be reversed quantitatively (plug out) by addition of a suitable base like tributylamine. In the plug–in (pseudorotaxane) state, the quenching of the binaphthyl–type fluorescence is accompanied by the sensitization of the fluorescence of the anthracenyl unit of the ammonium ion. Addition of a stoichiometric amount of base to the pseudorotaxane structure causes the revival of the binaphthyl fluorescence and the disappearance of the anthracenyl fluorescence upon excitation in the binaphthyl bands, demonstrating that plug out has happened.

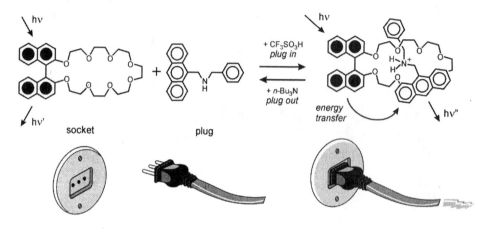

FIGURE 3. A molecular–level plug/socket system.

The plug/socket molecular–level concept can be extended straightforwardly to the construction of molecular–scale extensions and to the design of systems in which *(i)* light excitation induces an electron flow instead of an energy flow, and *(ii)* the plug in/plug out function is stereoselective.

LIGHT–DRIVEN PISTON/CYLINDER MOLECULAR SYSTEMS

Dethreading/rethreading of the wire and ring components of a pseudorotaxane reminds the movement of a piston in a cylinder. In suitably designed systems, the movement of such a rudimentary molecular machine can be driven by chemical energy or electrical energy and, most importantly, by light.

The "light–fueled" motor (i.e., a photosensitizer) can be incorporated in the wire (Figure 4a) [9] or in the macrocycle components of the pseudorotaxane (Figure 4b) [10]. In both cases, excitation of the photosensitiser with visible light in the presence of a sacrificial donor (Red) causes reduction of an electron–acceptor bipyridinium–type unit (the one contained in the wire, or one of those contained in the macrocycle), destroying the interaction responsible for self assembly. As a consequence, dethreading takes place. If oxygen is allowed to enter the solution (which had previously been deoxygenated), oxidation of the reduced bipyridinium unit restores the interaction and causes rethreading. The threading, dethreading, and rethreading processes can be easily monitored by absorption and fluorescence spectroscopy.

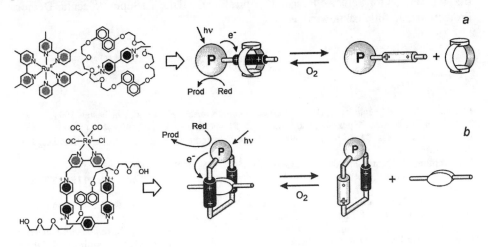

FIGURE 4. Light–driven dethreading of: *(a)* a pseudorotaxane incorporating a photosensitizer as a stopper in the wire–type component and *(b)* a pseudorotaxane incorporating a photosensitizer in the macrocyclic component. Red is a sacrificial reductant.

CONCLUSION

The miniaturization of components for the construction of useful devices, which is an essential feature of modern technology, is currently pursued by the large–downward (top–down) approach. This approach, however, which leads physicists and engineers to manipulate progressively smaller pieces of matter, has its intrinsic limitations. An alternative and promising strategy is offered by the small–upward (bottom–up) approach. Chemists, by the nature of their discipline, are already at the bottom, since they are able to manipulate molecules (i.e., the smallest entities with distinct shapes and properties) and are therefore in the ideal position to develop bottom–up strategies for the construction of nanoscale devices.

Apart from futuristic applications related, e.g., to the construction of a chemical computer [11], the design and realization of a *molecular–level mechanical and electronic set* (i.e., a set of molecular–level systems capable of playing functions that mimick those performed by macroscopic components in mechanical machines and electronic devices) is of great scientific interest since it introduces new concepts into the field of chemistry.

ACKNOWLEDGMENTS

Financial support from EU (TMR grant FMRX–CT96–0076), the University of Bologna (Funds for Selected Research Topics), and MURST (Supramolecular Devices Project) is gratefully acknowledged.

REFERENCES

1. Balzani, V., and Scandola, F., *Supramolecular Photochemistry*, Horwood, Chichester, 1991.
2. Balzani, V., Gómez–López, M., and Stoddart, J. F., *Acc. Chem. Res.* **31**, 405–414 (1998). Sauvage, J.-P., *Acc. Chem. Res.* **31**, 611–619 (1998). Balzani, V., Credi, A., Raymo, F. M., and Stoddart, J. F., *Angew. Chem. Int. Ed.*, in press.
3. Amabilino, D. B., and Stoddart, J. F. *Chem. Rev.* **95**, 2725–2828 (1995).
4. Credi, A., Balzani, V., Langford, S. J., and Stoddart, J. F., *J. Am. Chem. Soc.* **119**, 2679–2681 (1997).
5. Asakawa, M., Ashton, P. R., Balzani, V., Credi, A., Mattersteig, G., Matthews, O. A., Montalti, M., Spencer, N., Stoddart, J. F., and Venturi, M., *Chem. Eur. J.* **3**, 1992–1996 (1997).
6. Balzani, V., Credi, A., and Venturi, M., "Molecular-Level Devices", in *Supramolecular Science: Where It Is and Where It Is Going*, edited by R. Ungaro and E. Dalcanale, Kluwer, Dordrecht, 1999, pp. 1–22.
7. Schlicke, B., Belser, P., De Cola, L., Sabbioni, E., and Balzani, V., *J. Am. Chem. Soc.* **121**, 4207–4212 (1999), and unpublished results.
8. Ishow, E., Credi, A., Balzani, V., Spadola, F., and Mandolini, L., *Chem. Eur. J.* **5**, 984–989 (1999).
9. Ashton, P. R., Ballardini, R., Balzani, V., Constable, E. C., Credi, A., Kocian, O., Langford, S. J., Preece, J. A., Prodi, L., Schofield, E. R., Spencer, N., Stoddart, J. F., and Wenger, S., *Chem. Eur. J.* **4**, 2411–2422 (1998).
10. Ashton, P. R., Balzani, V., Kocian, O., Prodi, L., Spencer, N., and Stoddart, J. F., *J. Am. Chem. Soc.* **120**, 11190–11191 (1998).
11. Rouvray, D., *Chem. Br.* **34**, 26–29 (1998).

Field Emission from Nanostructured Carbon

J. Robertson

Engineering Department, Cambridge University, Cambridge CB2 1PZ, UK. jr@eng.cam.ac.uk

Abstract. Electron field emission from carbon nanotubes, nanostructured carbon, diamond and diamond-like carbon is compared. For many applications such as field emission displays, which require deposition at low substrate temperature, nanostructured carbon deposited at room temperature provides a good choice.

Many forms of carbon such as nanotubes, diamond and diamond-like carbon (DLC) are good electron field emitters. The emission properties of each type are different. Diamond is a semiconductor with a band gap of 5.5 eV. When its surface is terminated by hydrogen, it has a negative electron affinity (NEA), so that its conduction band edge lies above the vacuum level [1]. This means that any electrons in its conduction band could pass into the vacuum with no energy barrier [2]. However, in a field emission device, the electrons must flow round a complete circuit. In diamond, the large voltage barrier at the back contact and high resistivity of diamond means that, to date, the NEA property of diamond has not been used practically. The best electron emission occurs from micro-crystalline and nano-crystalline diamond, with emission varying inversely with grain size [3,4]. Diamond-like carbon (DLC) is amorphous carbon or hydrogenated amorphous carbon (a-C:H) containing a substantial sp^3 bonding. A hydrogen-free form of DLC with very high sp^3 content is called tetrahedral amorphous carbon (ta-C). It consists of very smooth films, usually grown by plasma deposition at room temperature. DLC is a reasonable emitter, particularly if doped with nitrogen [5-7]. However, the emission site density from DLCs is quite low, which limits its use in field emission displays. This is because emission occurs from local sp^2 regions created by conditioning [8].

Carbon nanotubes should have many of the best field emission qualities, such as a high currrent density, large emission site density and narrow electron energy distribution [9-12]. The emission stability may be controlled by adsorbed molecules [13]. However, the tradition fabrication process of laser ablation or arc followed by purification and attachment to the substrate is not ideal for commercial microlectronics. Direct plasma deposition onto prepared substrates would be better, but so far the low temperature limit is about 650C [10], despite optimistic reports.

There are various electronic devices, which can use efficient electron field emitters based on carbon. Field emission displays (FEDs) are flat panel displays in which the image is formed from a large array of pixels each addressed by field emission sources. The sources are 0.2 to 2 μm in diameter, so it is necessary that there is one emission site per source, equivalent to an emission site density (ESD) of 10^6 cm^{-2}. FEDs require emission current densities of about 0.1 mA.cm^{-2}. Low cost FEDS require the cathodes to be deposited on glass substrates, so deposition and processing temperatures must be kept below 500C and ideally below 250C.

CP544, *Electronic Properties of Novel Materials—Molecular Nanostructures*, edited by H. Kuzmany, et al.
© 2000 American Institute of Physics 1-56396-973-4/00/$17.00

Another electronic device is the terahertz power amplifier. The high frequency limit of silicon devices is set by the device dimensions and the relatively low limiting electron velocity in Si, 10^5 m.s^{-1}. Vacuum microelectronic devices can in principle operate at THz, as the limiting electron velocity in space is 3.10^8 m.s^{-1}. These devices require a large, stable emission current density, but place no limits on substrate or emission site density.

Carbon field emitters could be valuable in vacuum power switches. Solid state power switches have some difficulty to simultaneously withstand large breakdown voltages and have small forward voltage drops, as this places opposite constraints on the doping density. A vacuum emission device can separate these requirements – the vacuum withstands the off voltage, while the forward voltage drop can be low for a good field emitter.

Finally, carbon nanotubes make electron sources with extremely high brightness and narrow energy distributions for use as field emission tips for scanning electron microscopes.

	T_{dep}	Emission Field	Emission Site density	Stability
Diamond				*
nano-diamond		*	*	*
a-C:H	*	*		
ta-C	*	*		*
nanotubes		*	*	
ns-C	*	*	*	*

Table 1. Comparison of the field emission utility of various types of carbon.

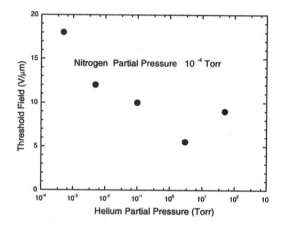

Figure 1. Variation of field emission threshold field with He partial pressure during deposition, for N$_2$ pressure of 10^{-4} Torr.

The emission properties of each form of carbon make it suitable for different devices, as summarised in Table 1. A fifth form of carbon, nanostructured carbon (ns-C) combines some of the good emission properties of nanotubes with the low deposition temperature of DLC [14].

Nanotubes are presently prepared by condensation from a carbon plume formed by laser ablation or by a graphite arc, thermalised in a He or Ar atmosphere, and aided by the presence of a Ni/Co/Y catalyst [15,16]. Based on such deposition conditions, disordered forms of carbon with onion- or nanotube-like nanostructure (ns-C) can be deposited at low temperature (20-250C) by running a cathodic arc of graphite in gaseous atmospheres such as He, hydrogen or nitrogen. Coll et al [14] deposited films with a strong external nanostructure in a He/N atmosphere. Amaratunga et al [17,18] used a N_2 atmosphere and a higher substrate temperature to provide relatively smooth films with an internal onion nanostructure.

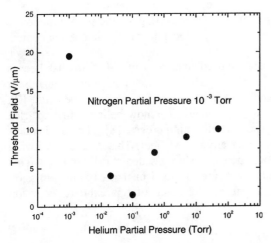

Figure 2. Variation of field emission threshold field with He partial pressure during deposition, for N_2 pressure of 10^{-3} Torr.

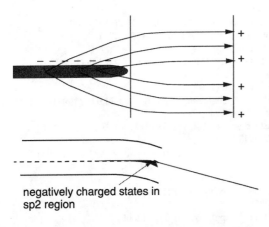

negatively charged states in sp2 region

Figure 3. Field lines focusing to negatively charged gap states in a sp2-rich channel in ta-C or a grain boundary in micro-crystalline diamond.

We describe here films prepared at room temperature using a cathodic arc with various background pressures of He and N_2 [19]. The rationale for using N is that N atoms in graphite sheets introduce 5-fold rings and curvature. Depositions were carried out at N_2 partial pressures of 10^{-4} and 10^{-3} Torr, the latter giving sizeable N incorporation. The substrates are Si. No catalyst is used to favour nanotube nucleation. The microstructure was studied by scanning electron microscopy and the bonding was studied by Raman. The field emission properties were measured in the parallel plate configuration at the pressure of 10^{-7} mbar using an ITO coated glass anode. The threshold emission field is defined as the field for an emission current density of $1 \mu A.cm^{-2}$.

Fig. 1 shows that the threshold field reaches a minimum of 5 V/μm at a He pressure of 2 Torr, for deposition at 10^{-4} Torr of N_2. Fig. 2 shows that the threshold field decreases to 1.5 V/μm at a He pressure of 0.1 Torr when the N_2 pressure is 10^{-3} Torr.

This is a very low threshold field. In addition, the emission site density is then found to be high, of order 10^4 to 10^5 cm^{-2}, for our measurement techniques. This material is very suitable for field emission displays.

Kelvin probe measurements have found that the work function of both the ta-C films and the nanostructured C films is 4.5 to 5 eV [20]. Thus the easy emission in these novel forms of carbon cannot be due to a small barrier to emission, but must be due to a large local field at the emission site. Explicit measurements of the width of the electron energy distribution have proved this for the cases of DLC, nanotubes and nano-crystalline diamond [4,12,21]. It has not so far been proved for nanostructured carbon. However, the predominantly sp^2 bonding of this material suggests the work function is also of order 5 eV, and that the emission occurs by a local geometric field enhancement creating a large local field.

The source of such large local fields and field enhancement in ta-C films has been a major theoretical question, as the films are nominally so smooth and homogeneous. It was first considered that local variations in the surface termination were the cause [7], but it was then realised that this effect was too small. It is now believed that sp^2-rich channels are formed below the ta-C by a soft-conditioning process [8]. The sp^2 channels are nm-sized and trap negative charge under an applied field. This causes downward band-bending and the necessary large local fields, as shown schematically in Fig 3. This mechanism of 'internal tips' is basically the same mechanism that explains emission from nano-crystalline diamond. Thus, a single mechanism of tips explains emission from all practical types of carbon.

REFERENCES

1. F J Himpsel, J S Knapp, J A VanVechten, D E Eastman, Phys Rev B **20** 624 (1979)
2. M W Geis, et al, IEEE Trans ED Let **12** 456 (1991)
3. W Zhu, et al, J Appl Phys **78** 2707 (1995); Science **282** 1471 (1998)
4. O. Gröning, O.M. Küttel, P. Gröning, L. Schlapbach, J. Vac. Sci. Technol. B **17**, 1064 (1999).
5. B S Satyanarayana, A Hart, W I Milne, J Robertson, App Phys Lett **71** 1430 (1997)
6. A Hart, B S Satyanarayana, J Robertson, W I Milne, App. Phys. Lett. **74** 1594 (1999).
7. J. Robertson, Mat Res Soc Symp Proc **417** 217 (1997); J. Vac. Sci. Technol. B **17**, 659 (1999)
8. A Ilie, T Yagi, A C Ferrari, J Robertson, Appl. Phys. Lett. **76** 2627 (2000)
9. W. A. de Heer, A. Chatelain, D. Ugarte, Science 270 (1995) 1179
10. J M Bonard, J P Salvetat, W A deHeer, L Forro, A Chaletain, App Phys Lett 73 918 (1998)
11. W. Zhu, G.P. Kochanski, and S. Jin, Science **282**, 1471 (1998)
12. M J Fransen, T L vanRooy, P Kruit, App Surface Science **146** 312 (1999)
13. K A Dean, B R Chalamala, App Phys Lett **75** 3017 (1999)
14. B.F. Coll, J.E. Jaskie, J.L. Markham, E.P. Menu, A.A. Talin, Mat. Res. Soc. Proc. **498**, 185 (1998).
15. A Thess et al, Nature **273** 483 (1996)
16. C Jourenet al, Nature **388** 756 (1997)
17. G.A.J. Amaratunga, New Diamond Frontier Carbon Technol **9**, 31 (1999).
18. G A J Amaratunga, et al, Nature **383** 321 (1996)
19. B.S. Satyanarayana, J. Robertson and W.I. Milne, J. Appl. Phys. **81** 3126 (2000).
20. A Ilie, J Robertson, submitted to J App Phys (2000)
21. O. Gröning, O.M. Küttel, P. Gröning, L. Schlapbach, App Phys. Lett. **71**, 2253 (1997)

Environmental Stability of Carbon Nanotube Field Emitters

L. Nilsson*, O. Gröning, P. Gröning, O. Küttel, L. Schlapbach

Institute of Physics, University of Fribourg, Chémin du Musée 3, Pérolles
CH-1700 FRIBOURG, SWITZERLAND
*) Lars-Ola.Nilsson@unifr.ch `

Abstract. Investigations of the current-time (I-t) characteristics of field emission (FE) from single-walled carbon nanotubes (SWNT), does not show any significant dependence on ambient partial pressures to 10e-5 mbar of hydrogen or water. It is however shown that oxygen causes substantial reduction of the FE current and this is believed to be related to reactive etching/ oxidation of the nanotubes. By heating the tubes up to 1100 Kelvin an increased instability in the FE current as well as on the field emission microscopy (FEM) screen can be observed. These instabilities become less and less pronounced after repeated annealing steps. FEM during short-time heating reveals dim five-fold as well as six-fold feinstructures. These feinstructure FEM patterns are not stable and have a short lifetime. Resonant tunnelling states as well as FE from nanotube cap states might explain these observations.

INTRODUCTION

World wide efforts to produce stable carbon nanotube (CNT) field emission sources are underway since the middle of the last decade [1-4] and at the IWEPNM 2000 in Kirchberg, Samsung presented a 9-inch full colour field emission display (FED) based on CNT emitters. CNT's have proven to be more robust than comparable unballasted metal field emitters [5], however the inherent short-term emission instability has not yet been understood. From field emission microscope (FEM) investigations it is recognised that emission does not occur homogeneously all over the CNT cap, but in a time-variable (~sec) lobed pattern, which reflects the electronic structure, local variation of the microscopic field and transmission probability. Explanations to these spatial pattern fluctuations, include FE from non-metallic cap states correlated to the presence of pentagonal defects [6], FE from open/closed tubes with dough-nut pattern [7] and FE through resonant tunnelling states caused by adsorbed gas species [8]. For reasons of long term stability and safe operation of e.g. FED's, a fundamental understanding of the underlying short-term emission instability mechanism is required.

Pressure and Temperature Dependence on FE (SWNT)

Experiments were carried out in a FEM with different phosphor-coated planar anodes at a typical base pressure of 5e-9 mbar. Commercial SWNT's were pasted to a

CP544, *Electronic Properties of Novel Materials—Molecular Nanostructures*, edited by H. Kuzmany, et al.
© 2000 American Institute of Physics 1-56396-973-4/00/$17.00

tungsten heating filament with conductive carbon cement. FE currents were recorded on the grounded anode as a function of temperature and partial gas pressures of high purify oxygen, hydrogen and water. Typical potentials of 2000 Volts were applied over a variable 2-cm anode-cathode separation and the filament temperature was determined with a pyrometer. FEM patterns were real time recorded using a charge coupled device camera.

As observed from I-t measurements, *initial* heating increases the noise (sub-sec fluctuations) of the FE current (Fig. 1a, point 1).

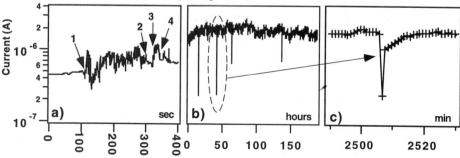

FIGURE 1. (a) FE current vs. time during heating to 900 Kelvin between point 1-2 and 3-4 causes increased noise level. Before point 1, between 2-3 and after point 4 the level of noise is decreased; i.e. at T=300 Kelvin. This behaviour is typical during *first* period heating. This sample was not annealed before. P~1.6e-7 mbar, 2600 Volts, 22 mm gap. **(b)** Current dependence during one week (one point/ minute) –stable FE at ~1.5 μA, 4 emitters, FEM pattern unchanged at 1900 Volts, 20 mm, T= 300 Kelvin and P~5e-9 mbar. Current decay at four occasions; second event resolved in **(c)** Possible spontaneous desorption (~1 sec) and current recovery during 15-16 minutes. Same current scale in (a)-(c). Different samples a and b.

Analogous to the increased current noise level, the FEM pattern rate of change is increased and the typical one-, two- and four-lobed symmetrical and asymmetrical patterns, fluctuate at a time scale of sub-seconds. Other, more exotic five/six-fold FEM patterns may occasionally be visible during heating, showing a dim structure, Fig. 2c, but do not seem to be long time stable. When the *initial* heating is turned off, point 2 and 4 in Fig. 1a, the level of noise is drastically reduced to the initial level, and a few minutes after heating the FEM pattern remain quite stable, Fig. 2e.

With repeated annealing steps, the increase of the noise level upon heating, becomes less and less pronounced. Repeated heating over 1000 Kelvin causes in general the overall current level to decrease, sometimes up to a factor of 10, and this current reduction is irreversible, depending mainly on the state of the screen (see below).

We further investigated the long-term stability (one week) of four emitters using a thermally outgassed phosphor screen, Fig. 1b, c. In contrast to unballasted metal field emitters no current runaway (associated with protrusion growth) is observed and the current level is maintained at approximately 1.5 μA with the exception of spontaneous

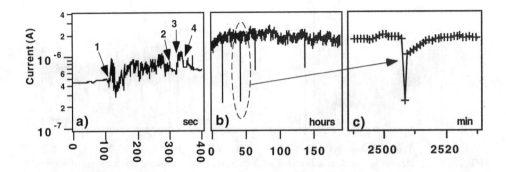

FIGURE 2. FEM time-resolved sequence of single SWNT cap at 2000 volts, d=21 mm, P~2e-7 mbar. Average diameter of spot ~1cm. **(a)** t=0: Long-term stable pattern at T= 300 Kelvin; possibly representing the surface electronic structure of an adsorbed two-atom configuration on the NT cap. **(b)** t= 39s: Short-term filament heating (3 sec) to T=1600 Kelvin. The stable atomic configuration in **a**, is disturbed and possibly a two-atom molecule is formed as the molecule is desorbed from the NT cap surface. **(c)** t=42s: Removing the lobed pattern reveals an intensity-weaker five-fold fine structure underneath. This form lasts only for some seconds at T= 300 Kelvin (Intensity magn. ✕3). **(d)** t=74s: Spatial charge fluctuations. The pattern in **c** is occasionally spotted. **(e)** after t=140s is the pattern stable again. T=300 Kelvin.

current reduction about on order of magnitude at four occasions (see Fig. 1b). It can be speculated that the rapid current decay is associated with the spontaneous desorption of chemisorbed molecules or atoms, and that the current recovery is due to readsorption from the gas phase. However we are not able to assign a certain relation between the current recovery time and the ambient gas pressure up to 10e-5 mbar. Current recovery times of normally ~8-20 minutes are observed independent of the ambient gas pressure. Preliminary investigations show stronger dependence on both current level and type of screen material used in experiment. This would indicate that the source of molecules/ions possibly being readsorbed after heating, is the screen. This dependence can be explained in terms of ion bombardment of the NT cap, where the ions are liberated from the screen due to the impinging electrons. When we used metal plate anodes or well thermally outgassed thin phosphor coatings, that had been electron scrubbed for a long time, we could *not* restore initial higher current levels after heating to 900, 1000 or 1100 Kelvin.

We further investigated the FE instability during high partial pressures (up to 10e-5 mbar) of hydrogen, water and oxygen, all of them to be expected in our chamber (or FED) background.

The samples were thermally annealed at 900-1500 Kelvin, before any gases were introduced into the ultrahigh vacuum system through a leak vent. Thermal annealing causes a reduction of both current level (up to a factor of 10) as well as of current noise.

Hydrogen or water exposure for periods of hours did not degrade or improve the emission, Fig. 3b,c. Nor could we secure a relation between the partial pressure and the level of noise in these cases. That is if impinging gas molecules or ions are responsible

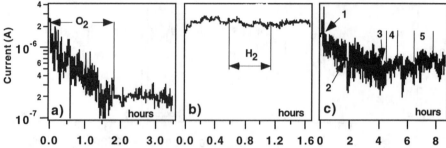

FIGURE 3. FE current vs. time as a function of high partial gas pressure, 10e-5 mbar, of oxygen, hydrogen and water at T=300 Kelvin. Current scale 1e-7 to 4e-6 Ampere. **(a)** Oxygen causes an irreversible damage by reactive ion etching. FE current is reduced by a factor 10. **(b)** Hydrogen does not seem to affect the FE current stability on a time scale of hours. **(c)** Point 1,2 and 3: heating of filament to 900, 1000 and 1100 Kelvin for 120 sec $\Rightarrow$ current reduction. H_2O at 10e-7 mbar (4) and 10e-6 mbar (5) during ~1 hour does not restore initial current of 2μA.

for chemisorbed states, we should be able to see a dependence on the ambient background pressure. Moreover we were never able to restore the initial current of 2.2 μA after heating to 900, 1000 and 1100 Kelvin, in Fig. 3c, by introducing water vapour. Repeated trials on several sample with shorter and longer (~hour) exposures of 10e-7, 10e-6 and 10e-5 mbar left us without success.

Operation of nanotubes in 10e-5 mbar of O_2 causes however a 10-fold irreversible degradation over a time of 1.5 hours. We propose that this decrease be due to a reactive ion etching effect [9], where C-O bonds are formed at the open NT cap.

FEM-Pattern During Short-Term Heating

Traditional one-, two- and four lobed patterns are commonly reported from NT's in the literature [8], and we observe them as well. We will however here not show any of these patterns but patterns of rather rare nature. They are obtained during short periods of filament heating and show five-fold fine structure, Fig. 2c, with lower intensity than before heating. In an adsorbate-tunnelling-state model they could be interpreted as an adsorbate-free NT cap, where the chemisorbed state is removed when given enough heat. We speculate that the remaining fine structure of the five-fold spatial charge distribution, is a result of pentagons introduced into a hexagonal C-bonding network of the NT cap. Such a five-fold fine structure has been theoretically calculated [10] from the NT cap.

In conclusion we have shown that the ambient gas phase pressure of both hydrogen and H_2O has a negligible effect on the long-term stability of the FE currents from SWNT's. Increased partial pressure of oxygen causes a degradation of the FE, and this effect is ascribed to reactive ion etching in the vicinity of the open NT cap.

The temperature dependence indicates that adsorbates are present on the as-introduced pristine NT cap and that these states can be removed during heating. Re-adsorption does not seem to come from the ambient gas phase but rather from the screen if not well thermally outgassed. The spatial charge fluctuations of the pristine

NT cap reflect the ion bombardment coming from the screen, which probably displace adsorbed species around the NT cap. This work was supported by NFP36 "Nanoscience" and TOP Nano 21 (ETH Board–KTI).

REFERENCES

1. Gulyaev Yu.V., Chernozatonskii L.A, Kosakovskaya Z.Ya., Sinitzyn N.I., Torgashov G.V., Zakharchenko Yu. F., "Field Emitter Arrays on Nanofilament Carbon Structure Films", Revue Le Vide, les Couches Minces" – Supplément au No 271-Mars-Avril 1994
2. Heer W.A, Châtelain A., Ugarte D., *Science*, vol 270, p. 1179 (1995)
3. Kuettel O., Groening O., Emmenegger C., Schlapbach L., *Appl. Phys. Lett.,* 73, 2113 (1998)
4. Nilsson L., Groening O., Emmenegger C., Schlapbach L., *Appl. Phys. Lett.,* 76, 2071 (2000)
5. Dean K.A., Chalamala B.R., *Appl. Phys. Letters*, vol 75, no 19, p. 3017-3019 (1999)
6. Bonard J-M., Salvetat J-P., Stöckli T., de Heer, W.A., Forro L., Châtelain, *Appl. Phys. Letters, vol* 73, no 7, p.918-920 (1998)
7. Saito Y., Hamaguchi K., Hata K., Tohji K., Kasuya A., Nishina Y., Uchida K., Tasaka Y., Ikazaki F., Yumura M., *Ultramicroscopy 73* (1998), p.1-6
8. Dean K , Chalamala B.R., *Jour. of Appl. Phys.,* vol 85, no 7 (1999), p.3832-3836
9. Mazzoni M.S.C., Chacham H.,*Phys. Rev. B*, vol 60, no 4 (1999), R2208
10. De Vita A., Charlier J-C, Blase X., Car R., *Appl. Phys. A* 68, 283-286 (1999)

Single-Wall Carbon Nanotube Devices Prepared By Chemical Vapor Deposition

P.R. Poulsen, J. Borggreen, J. Nygård, D.H. Cobden, M.M. Andreasen, and P.E. Lindelof

Ørsted Laboratory, Niels Bohr Institute, Universitetsparken 5, DK-2100 Copenhagen, Denmark

Abstract. We have fabricated single-wall carbon nanotubes by chemical vapor deposition from methane using a catalyst consisting of iron oxide with molybdenum supported by alumina particles. The nanotubes are grown both on loose catalyst powders, which allows for transmission electron microscopy studies, and on catalyst islands deposited on a substrate to enable direct growth on the substrate surface and subsequent evaporation of metal electrodes over the nanotubes for transport measurements. In the latter case, the nanotubes are characterized by scanning electron microscopy and atomic force microscopy. They are found to be single-walled and occur both individually and in small bundles.

INTRODUCTION

The standard methods for single-wall carbon nanotube (SWCNT) production rely on co-vaporization of graphite and transition metal catalysts in an inert gas atmosphere, either by an electrical arc [1,2] or by laser vaporization [3]. During the last three years, however, chemical vapor deposition (CVD) using hydrocarbons as feedstock has emerged as a promising alternative for SWCNT synthesis [4,5]. The advantages of CVD include lower preparation temperatures, simpler equipment, better prospectives for large-scale production, and the possibility to grow long and impurity-free carbon nanotubes in specified locations on a substrate for incorporation into electronic devices. In the present work we describe our success in CVD-growth following the technique introduced by Kong *et al.* [6,7]. The nanotubes are grown from methane using catalysts consisting of iron oxide with molybdenum supported by alumina particles. We confirm that this is indeed a practical and straightforward method for making SWCNT devices.

EXPERIMENTAL

Two types of catalysts were produced: a powder catalyst and a suspended catalyst. The powder catalyst was prepared in a similar way to that reported in ref. [5]. 1 g of alumina nanoparticles (Degussa, 'Aluminiumoxid C'), 228 mg $Fe(NO_3)_3 \cdot 9H_2O$, and 12 mg $MoO_2(acac)_2$ were added to 30 ml methanol. The suspension was stirred overnight before the solvent was evaporated in a rotary evaporator. The resulting powder was ground in a mortar and baked overnight at 180°C.

CP544, *Electronic Properties of Novel Materials—Molecular Nanostructures*, edited by H. Kuzmany, et al.
© 2000 American Institute of Physics 1-56396-973-4/00/$17.00

A suspended catalyst was made either by sonicating 1 mg of the powder catalyst in 1 ml methanol, or by a procedure similar to that described in ref. [6]. In the latter case, 15 mg alumina nanoparticles, 20 mg $Fe(NO_3)_3 \cdot 9H_2O$, and 5 mg $MoO_2(acac)_2$ were added to 15 ml methanol, sonicated for 1 h, stirred overnight, and sonicated for 1 h again. One drop of the suspension was placed on a substrate surface, which was then spun at 2000 r.p.m. for 60 seconds. The substrates were either bare Si wafers or thermally grown SiO_2 covered with a layer of polymethylmethacrylate (PMMA) in which windows had been patterned by electron beam lithography. In the latter case, the PMMA was removed in acetone after the catalyst deposition, leaving isolated islands of catalyst particles on the surface.

The CVD growth was performed with the powder catalyst or the substrates with spun-on catalyst particles in a tube furnace held at 900°C. A methane flow rate of 5000 bar·cm^3/min was maintained for typically 10 minutes at a pressure of ~1.1 bar.

After CVD, grains of the powder catalyst were placed on holey carbon films for high resolution transmission electron microscopy (HRTEM) characterization. The results with the catalysts spun on substrates were studied by atomic force microscopy (AFM) and scanning electron microscopy (SEM). SWCNTs grown on SiO_2 substrates were contacted electrically by a further lithography step and metal evaporation.

RESULTS

HRTEM images of the powder-grown material are presented in Fig. 1. Only SWCNTs were observed. When the nanotubes are very short, they are mostly individual and their closed ends can occasionally be resolved as for the 3 nm thick SWCNT in Fig. 1(a). Most nanotubes, however, grow to lengths of several micrometers. In that case, they often occur either as individual tubes or as thin bundles of a few tubes, which assemble to thicker bundles as illustrated in Fig. 1(b). Fig. 1(c) shows a part of a twisted bundle consisting of a few SWCNTs. Generally, the nanotubes appear defect-free with only occational traces of amorphous carbon deposits on their surface.

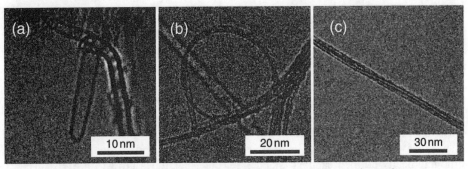

FIGURE 1. HRTEM images showing SWCNTs grown on the powder catalyst.

Fig. 2(a) and 2(b) are SEM micrographs of a large and a small powder catalyst grain, respectively, after CVD synthesis. A web-like network of nanotubes is covering the catalyst surface. As seen in Fig. 2(b), nanotubes are extending to, and lying on, the surface that supports the catalyst grains.

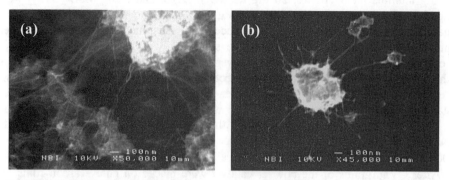

FIGURE 2. SEM images showing nanotubes grown on powder catalyst grains lying on a Si support.

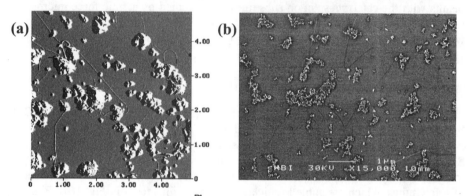

FIGURE 3. (a) AFM and (b) SEM images showing carbon nanotubes grown from suspended catalyst particles that have been deposited uniformly on a Si surface.

When suspended catalyst particles were deposited on a substrate, the CVD growth lead to nanotubes that lie on the substrate surface as illustrated in Fig. 3. Similar results were obtained with both types of catalyst suspensions.

The height of the nanotubes measured by AFM is in the range from 0.5 nm to ~4 nm, consistent with their being either individual SWCNTs or very thin bundles. Often nanotubes or thin bundles are seen to merge into a thicker bundle, or to change thickness abruptly. Also kinks are occationally observed.

In the SEM micrograph in Fig. 3(b), the nanotubes appear dark in regions where they touch the surface, and white in those parts that are supported on catalyst particles rather than on the silicon surface.

Fig. 4(a) is an example of nanotubes that have been grown on a substrate with lithographically defined catalyst islands. Selected nanotubes grown on SiO_2 surfaces were succesfully contacted by evaporating metal contacts over the nanotube ends. An example is shown in the SEM micrograph in Fig. 4(b) where the two contacts consist of 6 nm chromium under 43 nm of gold. The resistance of this particular nanotube at room temperature was 51 kΩ. With a metallic-doped silicon substrate used as gate, such devices can operate as field effect transistors [8].

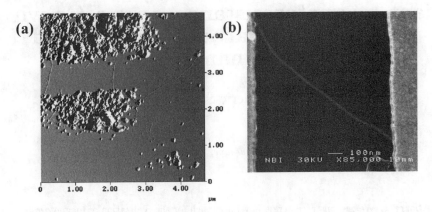

FIGURE 4. (a) AFM image of carbon nanotubes bridging catalyst islands on a Si surface. (b) SEM image showing two Cr/Au contacts evaporated on top of a nanotube which runs across a SiO_2 surface.

CONCLUSIONS

We have confirmed that CVD is a practical method for growth of SWCNTs directly on a substrate surface, allowing for integration of the nanotube growth into the fabrication of electronic devices based on carbon nanotubes.

ACKNOWLEDGMENTS

The authors are grateful to Jan-Olle Malm, National Center for HREM, Inorganic Chemistry 2, Lund University, Sweden for taking the HRTEM images and to Thomas Bjørnholm, Chemical Department, Copenhagen University for allowing us to his AFM facilities. We also acknowledge financial support from The Danish Natural Science Research Council, grants SNF9903274 and SNF9900525.

REFERENCES

1. Iijima, S. and Ichihashi, T., Nature 363, 603-605 (1993).
2. Bethune, D.S., Kiang, C.H., de Vries, M.S., Gorman, G., Savoy, R., Vazquez, J., and Beyers, R., Nature 363, 605-607 (1993).
3. Guo, T., Nikolaev, P., Thess, A., Colbert, D.T., and Smalley, R.E., Chem. Phys. Lett. 243, 49-54 (1995).
4. Kong, J., Cassell, A.M., and Dai, H., Chem. Phys. Lett. 292, 567-574 (1998).
5. Hafner, J.H., Bronikowski, M.J., Azamian, B.R., Nikolaev, P., Rinzler, A.G., Colbert, D.T., Smith, K.A., and Smalley, R.E., Chem. Phys. Lett. 296, 195-202 (1998).
6. Kong, J., Soh, H.T., Cassell, A.M., Quate, C.F., and Dai, H., Nature 395, 878-881 (1998).
7. Soh, H.T., Quate, C.F., Morpurgo, A.F., Marcus, C.M., Kong, J., and Dai, H., Appl. Phys. Lett. 75, 627-629 (1999).
8. Tans, S.J., Verschueren, A.R.M., and Dekker, C., Nature 393, 49-52 (1998).

Giant magnetoresistance from a Co wire incorporating carbon-encapsulated magnetic nanoparticles

Jean-Marc Bonard, Aymeric Sallin, Jean-Eric Wegrowe

Institut de Physique Expérimentale, Ecole Polytechnique Fédérale de Lausanne,
CH-1015 Lausanne, Switzerland[1]

Abstract. We demonstrate here a novel approach for the realization of magnetoresistive devices. Carbon-encapsulated cobalt nanoparticles are embedded in a Co wire that is electrodeposited in the pores of polycarbonate or alumina membranes. Preliminary results show giant magnetoresistive effects of 3 % at 5 K.

Since the discovery of giant magnetoresistive (GMR) effects in magnetic structures in 1988, the field has advanced at a rapid pace [1,2]. For example, wires with alternating stacks of magnetic layers and non-magnetic spacers allow one to reach magnetoresistivities of 50% in the current-perpendicular-to-plane (CPP) configuration (i.e., the direction of the current is perpendicular to the interfaces between the different materials). The realization of such structures is however quite demanding. One can produce GMR wires by electrodepositing alternatively magnetic and non-magnetic materials (e.g., Co and Cu) in the pores of a polycarbonate or alumina membrane [3]. However, the control of the deposition conditions is critical and small deviations in the composition of the electrolytic solution or in the deposition parameters can have drastic influences on the GMR properties. We present here an alternative method for producing GMR wires by incorporating carbon-encapsulated magnetic particles in a Co wire.

In a first step, we used a modified carbon arc discharge to produce carbon-encapsulated Co particles [4], as shown in Figure 1(a). The arc is struck between a 6.5 mm graphite cathode and an anode made of a graphite crucible of 25 mm diameter with a solid 15 mm diameter cylinder of pure Co in its middle. The spacing between anode and cathode during deposition is typically a fraction of a millimeter, with an applied voltage of 18 V and a current of 80 A. The deposition is performed under a static He pressure, p_0, that can be varied between 50 and 500 mbar. An additional He jet blowing perpendicularly from the side of the cathode-anode assembly is provided through a 0.5 mm diameter nozzle placed 5 cm away from the

---

[1] E-MAIL ADDRESS : jean-marc.bonard@epfl.ch, jean-eric.wegrowe@epfl.ch

CP544, Electronic Properties of Novel Materials—Molecular Nanostructures, edited by H. Kuzmany, et al.
© 2000 American Institute of Physics 1-56396-973-4/00/$17.00

cathode. The velocity of the jet is controlled by a calibrated microleak and can be varied between corresponding flowrates of $\nu_{jet} = 0$ and 400 mbar·l/s. Finally, a water-cooled heat shield encloses the deposition assembly. A significant proportion of as-produced particles was found to be only partly protected, and thus subject to oxidation. In order to remove these particles, we performed an acid treatment. The powder was kept in aqua regia (1:3 solution of HCl and HNO_3) for at least 2 h, with strong ultrasonication for the first 15 min. of the treatment. The powder was collected at the bottom of the beaker with a magnet, and 9/10 of the acid were replaced with distilled water with subsequent ultrasonic dispersion. This rinsing procedure was repeated once with water and twice with ethanol. Figure 1(b) shows a high resolution TEM micrograph of acid-treated particles. Only completely covered particles remain, along with the carbonaceous debris left over by the dissolution of unprotected particles. A significant advantage of this method is that the mean size of the particles can be tuned by changing the deposition conditions [4]. For the experiments described in this article we used $p_0 = 400$ mbar and $\nu_{jet} = 10$ mbar·l/s, for a resulting mean diameter of 13 nm.

In a second step, the purified particles were dispersed in distilled water and sonicated during at least 45 min. to obtain a homogenous suspension. The suspension was then drawn through a polycarbonate or an alumina microfiltration membrane with 200 nm diameter pores. Figure 2(a) shows a typical example of the cross section of an alumina membrane after this procedure. The tightly packed pores of the membrane are easily visible, and encapsulated particles are decorating the walls of the pores down to a few microns below the top of the membrane. After the decoration, a Au layer was deposited by sputtering on both sides of the membrane.

The third step consisted in growing the Co wire by electrodeposition in a $CoSO_4$

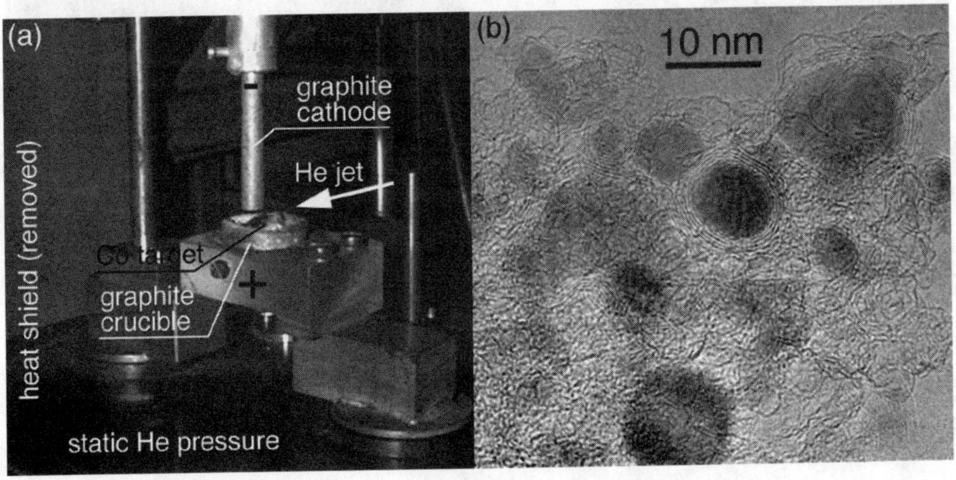

FIGURE 1. (a) Photograph of the modified arc discharge set-up; (b) TEM micrograph of carbon-encapsulated Co particles after purification.

+ H_2O and H_3BO_3 bath maintained at a pH of 3.2. The reference electrode was a Ag/AgCl electrode, and a platinum foil of 15×6 mm^2 was used as counterelectrode. The Au layer on one side of the membrane was taken as working electrode [3].

To measure reliably the magnetoresistance of the structures, it has to be ensured that the characterization is performed on one wire only. For this reason, the resistivity between the Au layers on the two sides of the membranes is monitored during the electrodeposition, and the deposition is stopped as soon as a metallic resistance is detected. Since the growth rate is slightly different from one pore to the next, this allows us to obtain samples where only one wire is contacted [5].

Figure 2(b) shows a TEM micrograph of the top of a wire after decoration and electrodeposition. The diameter of the wire is 160 nm, which corresponds approximately to the diameter of the pores, and is homogeneous over the whole length of the wire. At the extremities, carbon-encapsulating particles are embedded in the Co matrix.

Figure 3 displays the resistivity of the wire as a function of the magnetic field at 5 K. The field was applied in a direction perpendicular to the wire. The resistivity was normalized to the saturation value at high applied field. The maximal variation of resistivity amounted to 3 % at 5 K, and decreased with increasing temperature. This indicates clearly that the variation in resistivity is due to spin-diffusive effects, and not to the shape of the wire as is the case with a pure Co wire of identical

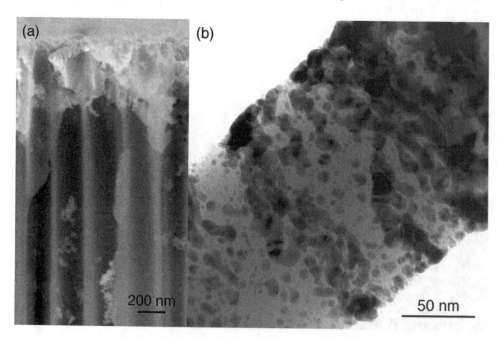

FIGURE 2. (a) SEM micrograph of the cross-section of an alumina membrane after decoration with the particles; (b) TEM micrograph of the wire after electrodeposition of the Co matrix.

diameter. The incorporation of Si nanoparticles of comparable diameter yielded only anisotropic magnetoresistive effects that were lower by a factor 4, showing that the nature of the incorporated particles influences strongly the magnetic properties of the wire.

In conclusion, we have successfully incorporated carbon-encapsulated cobalt nanoparticles in a Co wire of 160 nm diameter, and obtained GMR effects. The magnitude of the GMR is at present quite low (3 % at 5 K), which is due to two reasons. First, the GMR signal is obtained only at the extremities of the wire. Second, the particles are situated mostly on the outer regions of the wire. The filling is therefore far from ideal, and a more efficient decoration method should allow us to increase significantly the GMR ratio.

We acknowledge gratefully discussions and encouragement from André Chatelain and Jean-Philippe Ansermet. The electron microscopy was performed at the Centre Interdépartemental de Microscopie Electronique of EPFL.

REFERENCES

1. Baibich, M.N., Broto, J.M., Fert, A., Nguyen Van Dau, F., Petroff, P., Etienne, P., Creuzet, G., Friederich, A., Chazelas, J., *Phys. Rev. Lett.* **61**, 2472 (1988).
2. Binash, G., Grünberg, P., Saurenbach, F., Zinn, W., *Phys. Rev. B* **39**, 4828 (1989)
3. Blondel, A., J. P. Meier, J.P., Doudin, B., and Ansermet, J.-P., *Appl. Phys. Lett.* **65**, 3019 (1994).
4. Bonard, J.-M., Seraphin S., Wegrowe, J.-E., Jiao, J., and Chatelain, A., *Appl. Phys. Lett.*, submitted.
5. Wegrowe, J.-E., Kelly, D., Franck, A., Gilbert, S. E., and Ansermet, J. P., *Phys. Rev. Lett.* **82**, 3681 (1999).

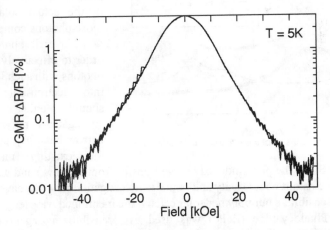

FIGURE 3. Resistivity of a single Co wire incorporating carbon-encapsulated Co particles at 5 K as a function of applied magnetic field.

Sensibilization of Polymer/Fullerene Photovoltaic Cells using Zinc Phtalocyanine Studied by Combinatorial Technique

D. Godovsky*, L. Chen, L. Petterson, and O. Inganäs

IFM, Linköping University, S-581 83, Linköping, Sweden
*EHF, Oldenburg University, D-26111, Oldenburg, Germany

Abstract. The influence of Zinc Phtalocyanine admixture to fullerene layers on top of PTOPT to the photovoltaic cells performance was studied. In order to investigate all the possible combinations of ZnPc and C_{60} the combinatorial technique was developed consisting in thermal co-evaporation of ZnPc and C_{60} from two different boats.

The significant increase in solar cells photocurrent was observed, coming from ZnPc absorbance bands, especially for the layers containing 1:1 molar ratio of the components.

INTRODUCTION

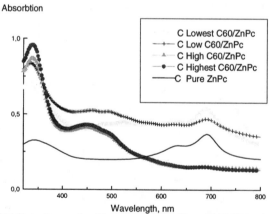

Fig.1 Absorption spectra of the structures with different ZnPc/C60 ratio

The harvesting of the photons with the wavelength in the range between 600 and 700 nm is an important task for photovoltaic applications since in the actual solar spectrum those photons comprise around ¼ of overall photon flux. In order to effectively convert the photons with such low energies into electron-hole pairs one should either use the conjugated polymers with narrow band gap or sensibilize the existing materials by organic dyes. Since the photoinduced charge transfer from the low band gap polymers to fullerene seems not occur easily and is under investigation now we have chosen the other way to sensibilize our polymer solar cells to the mentioned wavelength range.

Zink Phtalocyanine (ZnPc) was used as sensibilizing agent due to it's absorbance in the mentioned range of wavelengths and due to the fact, that the weak photoinduced charge transfer takes place between ZnPc and C_{60} [1,2].

CP544, *Electronic Properties of Novel Materials—Molecular Nanostructures*, edited by H. Kuzmany, et al.
© 2000 American Institute of Physics 1-56396-973-4/00/$17.00

EXPERIMENTAL

We co-evaporated ZnPc and C_{60} to form mixed layers on top of the polymer layers. Since we knew from the results of the collaborating groups [2] that the change in ratio between C_{60} and ZnPc significantly changes the photovoltaic properties of the devices we developed the combinatorial technique to study all the possible ratios between fullerene and ZnPc. The two substances were thermally evaporated from two boats, separated spatially and a number of ITO substrates was placed on the line of the maximum change of the flux ratio. Evaporation was carried out simultaneously from two sources with the fluxes being approximately equal to each other. Thus all the possible combinations of C_{60} and ZnPc on the 40 nm thick (poly (3- (4-octylphenyl) 2,2 – bithiophene "PTOPT" polymer layer were obtained during one evaporation cycle, allowing to study how the concentration of these components influences the photocurrent. The thickness of all evaporated layers was approx 20 nm. As a next step the aluminum or gold stripes were evaporated on top of the structure to serve as the cathode.

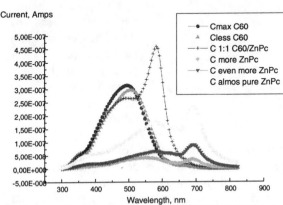

Current, Amps

- Cmax C60
- Cless C60
- C 1:1 C60/ZnPc
- C more ZnPc
- C even more ZnPc
- C almos pure ZnPc

Wavelength, nm

Fig.2 Photocurrent spectra for the structures with different ZnPc/C60 ratios

The absorbance spectra of the co-evaporated layers were taken by Perkin-Elmer Lambda 9 Spectrophotometer and can be seen from Fig.1. The photocurrent action spectra were measured by a set-up consisting of Keithley 485 picoammeter and the Oriel MS257 Monochromator Unit. As a light source the tungsten-halogen lamp (Oriel) was used. The monochromatic light intensity was measured by Pyroelectric Detector (Oriel). The I-V curves were taken using Keithley 2400 source meter.

RESULTS AND DISCUSSION

Analysing the absorbance spectra of the structures one can see, that the fullerene peak at 340 nm decreases in intensity with the increase of the ZnPc concentration in the mixed layer. At the same time the ZnPc absorbance peak at 700 nm becomes more and more pronounced.

The results of the photocurrent measurements on the same samples can be seen from Fig.2, which depicts the action spectra for different ZnPc/C_{60} ratios. It can be seen that the device with the ratio of ca 1:1 mole ZnPc:C_{60} (blue line) shows a significant peak of the photocurrent coming from 590 nm line (one of the two absorbance bands of ZnPc in the red part of VIS-spectrum).

Even though the improvement of IPCE values is quite moderate (ca.12% at peak middle), due to a large number of photons within the mentioned wavelength range in actual solar spectrum we should not treat the IPCE spectra as the reference of the real input from these bands to the photocurrent. The real picture is given if the photocurrent spectra are taken using illumination by solar light, with the maximum of the intensity coming to 600 nm (AM1.5). The corresponding current increase is of the order of 80-100%, and the photocurrent spectra taken under tungsten-halogen lamp illumination (Fig.2) give more adequate picture of the real input of ZnPc into the total photocurrent of the solar cell.

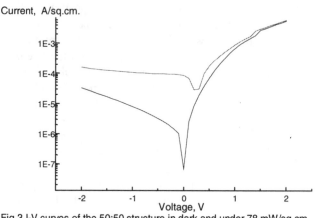

Fig.3 I-V curves of the 50:50 structure in dark and under 78 mW/sq.cm illumination

The I-V curve of the structure with 50:50 ZnPc to C_{60} ratio can be seen from Fig.3. It can be seen that the rectification is significant and the Open Circuit Voltage is ca. 0.3 V. The maximal values of the photocurrent were in the range 100-150 $\mu A/cm^2$. Unfortunately the evaporated layers of ZnPc/C_{60} were highly resistive, and the increase of the thickness of the layer resulted in the decrease of the devices photocurrent, even though our modelling calculations showed, that the maximal IPCE values should be expected from 60-80 nm thick dye sensibilized layers [4].

CONCLUSIONS

We used the combinatorial technique to investigate the influence of the Zinc Phtalocyanine admixture co-evaporated along with C_{60} on top of PTOPT on the performance of the solar cells.

We found out, that for the components ratios close to 1:1 molar the significant increase of the photocurrent comes from the range 590-600 nm associated with one of the ZnP absorbance bands. Even though the increase in IPCE values is not so dramatic, we should consider the fact, that the photocurrent under illumination by source with the solar spectrum must be taken, and at such conditions the additional input of ZnPc into the photocurrent equals to ca.80%.

ACKNOWLEDGMENTS

Authors would like to acknowledge the support in the frame of EC Joule 3 Project "Development of Molecular Plastic Solar Cells". The Göran Gustafsson Foundation is gratefully acknowledged for support, as well as TFR and SSF.

REFERENCES

1. C.Schlebusch, J.Morenzin, B.Kessler, and W.Eberhardt "Charge Transfer and Relaxation Dynamics of excited electronic states in organic photoreceptor materials with and without C_{60} ".Proceedings of IWEPNM-98, Kirchberg, Austria, p. 98.
2.Giampiero Ruani , Jean-Michel Nunzi , Carlo Tagliani , Paul Lane, Donal Bradley, Christine Videlot, Denis Fichou , Jörn Rostalski and Dieter Meissner "The Status of Molecular Organic Solar Cells as Investigated in the EUROSCI Project", Proceedings of ECOS 98, Cadarache, France, p.63.
3. Roman, L.S., Andersson, M.R., Yohannes, T. & Inganäs, O Photodiode performance and nanostructure of polythiophene/C60 blends. Adv. Mater. 9(15), 1164-1168 (1997).
4. Leif A. Petterson, Lucimara S. Roman and Olle Inganäs, Modelling photocurrent action spectra of photovoltaic devices based on organic thin films, Journal of Applied Physics, v.86 (1), pp. 487-496 (1999).

Field-Effect Mobility Measurements of Conjugated Polymer / Fullerene Photovoltaic Blends

W. Geens[*], S.E. Shaheen[§], C.J. Brabec[§], J. Poortmans[*], and N. Serdar Sariciftci[§]

[*]IMEC vzw, Kapeldreef 75, B-3001 Leuven, Belgium
[§]Christian Doppler Laboratory for Plastic Solar Cells, Physical Chemistry, Johannes Kepler University of Linz, A-4040 Linz, Austria

Abstract. Organic field-effect transistors (FETs) have been fabricated using an active region consisting of a conjugated polymer poly(2-methoxy-5-(3',7'-dimethyloctyloxy)-1,4-phenylene-vinylene) (MDMO-PPV) blended with a soluble derivative of C_{60} (6,6)-phenyl C_{61}-butyric acid methyl ester (PCBM). Such blends are promising for use in organic photovoltaic devices. Fabrication of the FET provides a technique for estimating both the hole and electron field-effect mobilities of the material in the active region by measuring the saturation current as a function of the gate voltage. The hole field-effect mobility for pristine MDMO-PPV was found to be 9×10^{-4} $cm^2/(Vs)$. Saturation of electron current could not be attained for pristine MDMO-PPV. Evidence for the saturation of electron current could be seen only at very high PCBM doping levels.

INTRODUCTION

Organic semiconductors exhibit attractive properties for applications in a variety of opto-electronic devices. These include organic light-emitting diodes, which are currently entering the first phase of commercialization[1]. Also, the fabrication of integrated circuits made entirely out of plastic has become a research field of major interest [2].

Photovoltaic cells based on a photoactive organic layer consisting of a conjugated polymer / fullerene blend sandwiched between two metal electrodes were first demonstrated by Sariciftci and Heeger [3, 4]. Blending the fullerene directly into the conjugated polymer matrix results in interpenetrating networks of the photoactive acceptors and donors, which act also as electron and hole conductors, respectively. After photoexcitation and electron transfer to the fullerene molecule [5], the separated charges are driven toward their respective contacts by a gradient in the electrochemical potential.

CP544, *Electronic Properties of Novel Materials—Molecular Nanostructures*, edited by H. Kuzmany, et al.
© 2000 American Institute of Physics 1-56396-973-4/00/$17.00

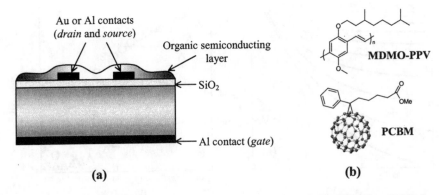

FIGURE 1. (a) FET device structure. (b) Chemical structures of MDMO-PPV and PCBM.

In this paper, the transport properties of both carriers within blends of the conjugated polymer MDMO-PPV and the soluble C_{60} derivative PCBM are investigated. This is accomplished by measuring the drain-source saturation current of an FET constructed with such blends as the semiconducting channel material.

EXPERIMENTAL

FETs (see Fig. 1(a)) were assembled on highly doped p^+ Si-substrates. An insulating thermal oxide (232 nm) was grown on one side of the substrate, and the backside was covered with an Al layer as the gate electrode. A structure of either Au or Al interdigitating fingers, forming the source and drain electrodes, was realized on top of the insulating SiO_2 layer with a combination of photolithography and a lift-off process. The following combinations of conduction channel width (W) and length (L) were produced: $W/L = 1075 \ \mu m/10 \ \mu m$, $W/L = 1035 \ \mu m/5 \ \mu m$, and $W/L = 550 \ \mu m/3$ μm. Finally, the organic semiconducting layer was spin-coated to fill the channel. Pristine MDMO-PPV (obtained from Covion) and blends of MDMO-PPV / PCBM were deposited from toluene solutions. The chemical structures of the materials are shown in Fig. 1(b). The measurement mode of the FET is determined by the gate voltage, which induces an accumulation layer of charges in the region of the conduction channel adjacent to the interface with the SiO_2. For p-channel (n-channel) operation, a negative (positive) gate voltage is applied to induce an accumulation layer of holes (electrons), allowing the measurement of the hole (electron) mobility, respectively. Au source and drain electrodes were used for p-channel mode measurements in order to facilitate the injection of holes into the highest occupied molecular orbital (HOMO) level of the channel material. Likewise, Al source and drain electrodes were used for n-channel model measurements in order to facilitate the injection of electrons into the lowest unoccupied molecular orbital (LUMO) of the channel material. FET characterization was performed using a HP4156A analyzer, with the source contact connected to ground. All measurements were performed in air.

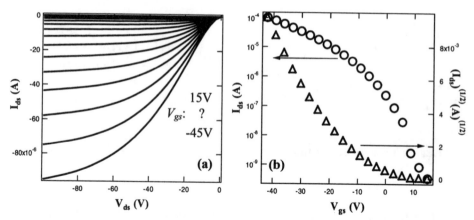

FIGURE 2 (a) I_{ds} versus V_{ds} characteristics of a pristine MDMO-PPV FET with Au contacts and $L = 3\mu m$. (b) (left axis) I_{ds} plotted as a function of V_{gs} for $V_{ds} = -99$ V on a logarithmic scale. (right axis) $(I_{ds})^{1/2}$ plotted as a function of V_{gs}. From the slope at high negative V_{gs}, the field-effect mobility for holes is calculated to be $\mu_{FE} = 9.05 \times 10^{-4}$ cm²/Vs.

RESULTS AND DISCUSSION

The field-effect mobilities μ_{FE} were calculated from the saturation regime of the drain-source current (I_{ds}) using the square-law theory [6]:

$$I_{ds_{sat}} = \frac{\mu_{FE}WC_{ox}}{2L}\left(V_{gs} - V_t\right)^2 \tag{1}$$

where W and L are respectively the conduction channel width and length, C_{ox} is the capacitance of the insulating SiO$_2$ layer, V_{gs} is the gate voltage, and V_t is the threshold voltage.

Fig. 2 shows the FET characteristics of a device in p-channel mode with pristine MDMO-PPV as the channel material. The hole current I_{ds} reaches saturation for negative applied V_{ds} and V_{gs} (Fig. 2(a)). The saturation point of $V_{ds} = -99$ V was used to plot the V_{gs} dependence in Fig. 2(b). From the logarithmic scale on the left axis, an on/off current ratio of 10^6 can be seen for the voltage range -45 V $< V_{gs} < 15$ V can be seen. From the slope of the linear fit at high negative V_{gs}, a field-effect hole mobility of $\mu_{FE} = 9.05 \times 10^{-4}$ cm²/Vs was calculated using Eq. 1.

FET characteristics of devices in p-channel mode using blends of MDMO-PPV / PCBM as the channel material are given in Figs. 3(a) – 3(c). For doping ratios of 1:3 and 1:4, one can see I_{ds} versus V_{ds} plots that do not show saturation behavior for small values of V_{gs}. This is a result of electron currents in the device arising from an accumulation layer of electrons that is induced near the drain contact. This behavior has been reported before by Dodabalapur et al. [7] for a bipolar multilayer FET structure. Here we demonstrate such bipolarity in an FET using a single layer of blended hole and electron transport materials.

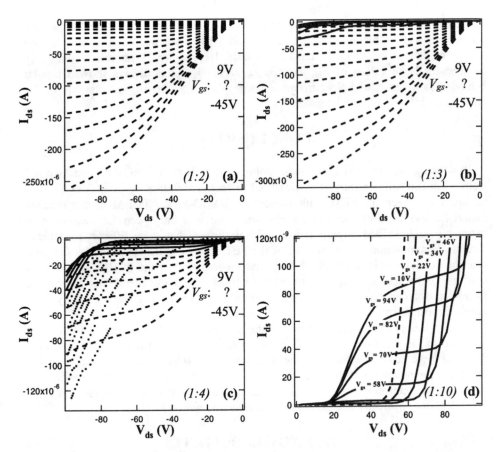

FIGURE 3 (dashed line = hole current; solid line = hole+electron current; dotted line = electron current) I_{ds} versus V_{ds} characteristics of FETs based on blends with the following MDMO-PPV : PCBM ratios: (a) *(1:2)* - Au contacts and $L = 3\mu m$; $\mu_{FEholes} = 9.64 \times 10^{-4}$ cm²/Vs. (b) *(1:3)* - Au contacts and $L = 3\mu m$; $\mu_{FEholes} = 1.68 \times 10^{-3}$ cm²/Vs. (c) *(1:4)* - Au contacts and $L = 3\mu m$; $\mu_{FEholes} = 1.26 \times 10^{-3}$ cm²/Vs. (d) *(1:10)* - Al contacts and $L = 3\mu m$; $\mu_{FEelectrons} = 9.73 \times 10^{-8}$ cm²/Vs.

The value of the hole mobility for the blends is seen to increase slightly as the PCBM concentration is increased, reaching a value of 1.26×10^{-3} cm²/Vs at a MDMO-PPV / PCBM ratio of 1:4. The origin of this is not clear. One would expect a decrease in the hole mobility of the blend as compared to the pristine polymer based on the argument that incorporation of the PCBM into the polymer dilutes the number and density of available hole carrier sites. A likely explanation for the apparent increase seen here is that the presence of electron currents produces deviations from the square-law theory given in Eq. 1. However, we do not rule out the possible case that incorporation of the PCBM into the polymer matrix affects the morphology in a way that increases the interchain interactions between neighboring polymer chains [8]. Further investigations on this issue are underway.

n-channel measurements were performed (using Al contacts) for the pristine polymer as well as blends of 1:2, 1:3, 1:4, and 1:10. Evidence of saturated electron current could only be seen for blends of 1:4 and 1:10 (Fig. 3(d)). Saturation could only be seen for large values of V_{gs}, and the I_{ds} versus V_{ds} plots always showed bipolar behavior as a result of hole currents at large V_{ds}. As a result, calculations of the electron mobility could not be made.

CONCLUSIONS

In conclusion, the field-effect hole mobility of pristine MDMO-PPV, spin-coated from a toluene solution, has been measured to be $\mu_{FE} = 9.05 \times 10^{-4}$ cm^2/Vs. Bipolar transport was seen for FETs with blends of MDMO-PPV / PCBM as the channel material. Evidence of electron current saturation was only seen for devices with extremely high PCBM doping ratios. This indicates that it is difficult to achieve electron accumulation in the channel of the FET. This can be explained by the poor electron conduction in the blend due to phase-segregation of the PCBM into isolated clusters, or by the high density of electron traps within the blend that inhibit the formation of an accumulation layer of electrons. Further investigations to study these possible mechanisms are underway.

Finally, it is of importance to note that the hole mobilities for the blends are not less than that of the pristine polymer. The incorporation of PCBM into the MDMO-PPV does not have adverse effects on the hole transport. This is an important point in respect to the operation of the solar cell based on such blends. This result implies that the hole and electron transport are independent processes that occur on electronically separated, but physically interpenetrating, conduction networks.

ACKNOWLEDGMENTS

Financial support from the Austrian FWF (Lise Meitner Stipendium, project no. M539-CHE for S. E. S.), the Magistrat Linz and the IWT (Flemish Institute for Promotion of Scientific Technological Research in the Industry for W. G.) is gratefully acknowledged.

REFERENCES

[1] Friend, R. H., Burroughes, J., and Shimoda, T., *Physics World* 12, 35-40 (June 1999).
[2] de Leeuw, D., *Physics World* 12, 31-34 (March 1999).
[3] Sariciftci, N. S., and Heeger, A. J., U.S. Patent 5,331,183.
[4] Yu, G., Gao, J., Hummelen, J. C., Wudl, F., and Heeger, A. J., *Science* 270 1789-1791 (1995).
[5] Sariciftci, N. S., Smilowitz, L., Heeger, A. J., and Wudl, F., *Science* 258, 1474-1476 (1992).
[6] Pierret, R.F., *Field Effect Devices* in Modular Series on Solid State Devices, **IV**, Addison-Wesley Publishing Company, Reading, Massachusetts, 1983, pp. 89-90.
[7] Dodabalapur, A., Katz, H. E., Torsi, L., and Haddon, R. C., *Appl. Phys. Lett.* **68**, 1108-1110 (1996).
[8] Nguyen, T.-Q., Martini, I. B., Liu, J., and Schwartz, B. J., *J. Phys. Chem. B* **104**, 237 – 255 (2000).

Reinforcement of an Epoxy Resin by Single Walled Nanotubes

L. Vaccarini[a], G. Désarmot[b], R. Almairac[a], S. Tahir[a], C. Goze[a], P. Bernier[a]

[a]GDPC, UMR 5581, University of Montpellier, 34095 Montpellier cedex 05, France.
[b]ONERA-DMSC, 92320 Châtillon, France.

Abstract. The anisotropic shape and the very high stiffness of carbon nanotubes make them good candidates for the reinforcement of various materials. For instance polymer-nanotubes blends could be used as a new material but also as a matrix for carbon fibre composites in order to improve their mechanical properties, in particular under compressive loading. In this study, we have prepared several SWNTs-epoxy blends containing various SWNTs weight percentages (from 0 to 35%). Blends are themselves composite materials, observed and characterized by SEM, TEM and X-ray diffraction. Two methods are used to evaluate their stiffness: Vickers microindentation and three points bending tests. We observe a linear increase of the Young's modulus with the weight percentage of SWNTs. For comparison SWNTs purified by a cross flow filtration process have also been used. The obtained results are discussed.

INTRODUCTION

The exceptional mechanical properties of carbon nanotubes open various fields of application. Nanotubes combine high axial stiffness, aspect ratio and flexibility [1,2]. They can be incorporated in a matrix to improve its mechanical properties. But they can also be used as super strong nanometric fibres, for example as a tip for AFM. Several studies have been reported on the properties of nanotube-polymer composites to deduce intrinsic mechanical properties of nanotubes [3] or to investigate the composite properties themselves [4,5]. In this contribution, we have processed and characterized single walled carbon nanotubes-epoxy resin composites and measured the induced reinforcement.

PREPARATION OF THE COMPOSITES

The composites were made of an epoxy resin (CIBA Araldite LY556 with HT972 hardener) and single walled carbon nanotubes (SWNTs). Most of the studies were done using raw SWNTs synthesized by the electric arc technique [6]: a composite rod (Ni:Y:C) is vaporized under helium atmosphere. The pristine material contains SWNTs ropes coexisting with metallic nanoparticles and other forms of carbon (graphite, amorphous nanoparticles, nano-horns,...). For comparison, one composite was also

CP544, *Electronic Properties of Novel Materials—Molecular Nanostructures*, edited by H. Kuzmany, et al.
© 2000 American Institute of Physics 1-56396-973-4/00/$17.00

prepared using purified nanotubes. The purification process [7] consists in a nitric acid treatment followed by a cross flow filtration and an annealing under nitrogen atmosphere at 1600°C.

The powder of SWNTs dispersed in CH_2Cl_2 is then sonicated with the resin. Then, the hardener is added and the solvent evaporated. The blend is introduced in a mould and heated at 140°C during 2 hours under a pressure of 70 bars. Depending on the choosen mechanical test, a specimen or a pellet is prepared. The surface of the moulded samples are then polished. Composites containing various weight percentages of raw SWNTs were processed (from 0 to 35 wt %).

CHARACTERIZATIONS, RESULTS AND DISCUSSION

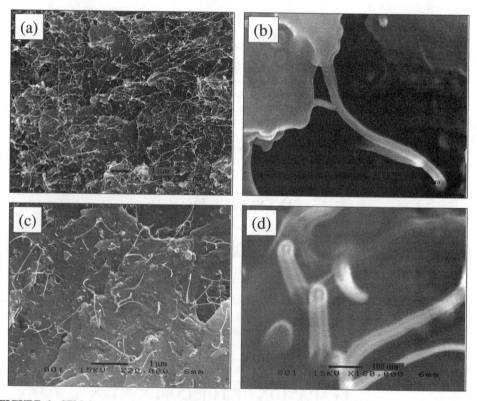

FIGURE 1. SEM images of several fractures of raw SWNTs-epoxy resin. (a) Large view of an homogeneous 9 wt % SWNTs composite; (b) high magnification of a rope impregnated into the resin; (c) Numerous cut bundles after the composite failure; (d) zoom of cut ropes embedded into the matrix.

Electron Microscopy characterizations were intensely carried out on the composites: Scanning Electron Microscopy (SEM) on fracture surfaces and Transmission Electron Microscopy (TEM) in cavities. These cavities were created during the preparation of the TEM samples with a microtome but the samples were not thin enough for the observation of the bulk.

The large views of the fractures (Fig. 1.a.) show a very good homogeneity of the material. The ropes of nanotubes are homogeneously impregnated by the resin (Fig. 1.b.). The artificial cavities observed by TEM consist in some thick pieces of composite completely separated or maintained by some stretched ropes of tubes (Fig. 2.). This suggests that some SWNTs-epoxy resin adhesion was formed. The nanotubes are partially extracted from the resin during the cavity opening, and the matrix seems to be sticky (see at the embedding) but the naked surface of ropes suggests a moderate SWNTs-resin interaction. Many cut ropes are also found at the border of the cavities (Fig. 2.b.c.): due to their large embedded length they cannot be pulled out, and break. SEM images suggest that bonding exists between the resin and SWNTs. Many ropes are also cut after the fracture process (Fig. 1.c.d.). The 2D organization of the tubes in bundles is still preserved in the composite (Fig. 2.c.). This shows that the interaction between SWNTs in a rope is rather high.

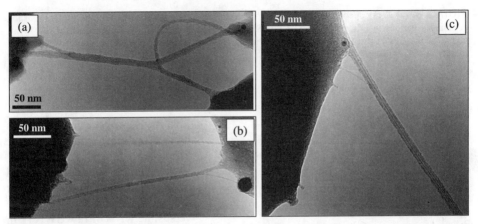

FIGURE 2. TEM pictures of artificial cavities of a 9 wt % raw SWNTs-epoxy composite. (a) (b) Stretched and cut ropes attached to the composite; (c) Fringes of the 2D lattice organization of the nanotubes in bundles.

X-ray diffraction experiments were also carried out to investigate the structural organization of the composite. Unfortunately, the resin has an uncomfortable diffraction component around 0.4 Å^{-1}, at almost the same position than the (1 0) peak of the SWNTs ropes. In the composites, the signal of SWNTs is superposed to the signal of the epoxy resin. Fig. 3.a. shows the X-ray spectra change with the weight percentage of raw SWNTs introduced in the matrix. To extract the data, a 3-steps treatment is applied to the spectra: subtraction of (i) the epoxy signal, (ii) a supplementary amorphous phase component centred around 1.6 Å^{-1}, and (iii) a power law component in Q to reveal the signal of the 2D lattice. The curves obtained (Fig. 3.b.) confirm the presence of bundles of nanotubes. The positions and the relative intensities of the peaks are not modified but a slight change of the shape of the peaks occurs. As they become asymmetric (see the (1 0)), it could reflect a modification of the parameter distribution in the bundles. A tentative explanation could be that the amorphous phase seen around 1.6 Å⁻¹ would intercalate only the ropes of high lattice parameters. Studies are under progress in order to verify these results.

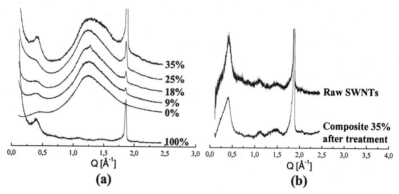

FIGURE 3. (a) X-ray spectra of the raw SWNTs (100%), the epoxy resin (0%), and the composites containing various weight percentage of raw SWNTs (from 0 to 35%). (b) Comparison of the X-ray signal of the SWNTs in the raw sample and in the composite containing 35 wt % of SWNTs after treatment of the spectra.

MECHANICAL PROPERTIES

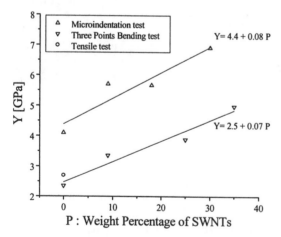

FIGURE 4. Evolution of the Young Modulus Y of the composite versus the weight percentage of raw SWNTs evaluated by microindentation and 3 points bending tests.

To measure the Young modulus of our isotropic composites, we used two mechanical tests: Vickers microindentation [8] and three points bending test. We get the stiffness of the composite versus the wt % of raw SWNTs in the composites. Both methods confirm the reinforcement effect due to the SWNTs ropes. The load is transferred from the matrix to the nanotubes. The observed linear behavior in the Young modulus by addition of nanotubes is presented in figure 4. There are several features that contribute to limit or improve the reinforcement of the composite: (i) the possible sliding of the SWNTs within the ropes when loading, (ii) the unknown SWNTs-resin interaction nature and amplitude, (iii) the bending of ropes limiting the effective load transfer length, (iv) the purity of the sample, (v) the exact wt % of SWNTs ropes in the raw material. The experiment performed with purified nanotubes did not give better results (Y=4,6 GPa for 10 wt % measured by microindentation). We think that the surface of the tubes was functionalized with undesirable chemical groups or surface layer during the purification treatments which weakened the SWNTs-epoxy resin interaction. Further experiments are under progress in order to optimize the composite performance by derivatization of the nanotubes surface in order to improve the resin bonding and by the alignment of the ropes in the matrix.

ACKNOWLEDGEMENTS

The authors would like to thank B. Giraud, J.P. Selzner and P. Azais for the electronic microscopy facilities. The three points bending tests were carried out in collaboration with P. Etienne in the University of Montpellier (LDV). This work was financially supported by the GDR Nanotubes 1752 (CNRS, France).

REFERENCES

1. Hernández, E., Goze, C., Bernier, P., and Rubio, A., *Phys. Rev. Lett.* **80**, 20, 4502-4505 (1998).
2. Krishnan, A., Dujardin, E., Ebbesen, T.W., Yianilos, P.N., and Treacy, M.M.J., *Phys. Rev. B* **58**, 20, 14013-14019 (1998).
3. Lourie, O., and Wagner, H.D., *J. Mater. Res.* **13**, 9, 2418-2422 (1998).
4. Schadler, L.S., Giannaris, S.C., and Ajayan, P.M., *Appl. Phys. Lett.* **73**, 26, 3842-3844 (1998).
5. Shaffer, M.S.P., and Windle, A.H., *Adv. Mat.* **11**, 11, 937-941 (1999).
6. Journet, C., Maser, W.K., Bernier, P., Loiseau, A., Lamy de le Chapelle, M., Lefrant, S., Deniard, P., Lee, R., and Fisher, J.E., *Nature* **388**, 756-758 (1997).
7. Vaccarini, L., Goze, C., Aznar, R., Micholet, V., Journet, C., and Bernier, P., *Synth. Met.* **103**, 2492-2493 (1999).
8. Loubet, J-L., *Etude des courbes d'indentation: cas du poinçon conique et de la pyramide Vickers*, Ph-D thesis; Ecole Centrale de Lyon, September 1983.

Sharpened Nanotubes, Nanobearings, and Nanosprings

A. Zettl and John Cumings

Department of Physics, University of California at Berkeley, and
Materials Sciences Division, Lawrence Berkeley National Laboratory,
Berkeley, CA 94720 U.S.A.

Abstract. We demonstrate a method whereby outer nanotube walls or shells can be successively removed near the end of multiwalled carbon nanotubes (MWNTs). This allows "sharpening" of the tubes, and has important implications for power dissipation in current-carrying nanotubes. We further exploit the technique to create low friction MWNT-based linear bearings and constant-force nanosprings. Our experiments are performed in-situ inside a high resolution transmission electron microscope, which allows simultaneous monitoring of atomic-scale mechanical deformation and wear of nanotube bearing surfaces.

Nanotubes are model systems for the study of fundamental physical properties of nanostructures. They also have far-reaching applications potential. Obvious applications include catalysts[1], biological cell electrodes[2], nanoscale electronics[3,4], scanned probe microscope and electron field emission tips[5,6], nanobearings, and nanosprings. For many such applications it would be desirable to control or shape nanotube geometry. For example, the "ideal" scanned probe, field emission, or biological electrode tip would be long, stiff and tapered for optimal mechanical response, and have an electrically conducting tip. Similarly, a constant-force nanospring might be formed from a configuration of concentric nanotubes where the van der Waals force provieds the extention-independent restoring force. Again using MWNTs, linear and rotational[7] bearings might might be achieved by using concentric shells of nanotubes as low-friction sliding surfaces.

We here describe the successful engineering of multiwalled carbon nanotubes into "ideal-geometry" tips for scanned probe microscopy, field emission, or biological insertion applications, as well as the construction of apparently wear-free linear nanobearings and constant-force nanosprings. All three constructions hinge on an initial "sharpening" of the tip of a MWNT, which results in a stepped or tapered nanotube diameter. At the end of the MWNT, the inner core nanotubes then freely protrude. The shaping process involves electrically-driven vaporization of successive layers (i.e. tube walls) of the MWNT. This decortication process can be repeatedly applied to the same MWNT until the very innermost small-diameter tube or tubes are exposed, often with a tip radius of curvature comparable to that of a single, single-walled nanotube.

CP544, *Electronic Properties of Novel Materials—Molecular Nanostructures*, edited by H. Kuzmany, et al.

Our MWNT "tip engineering" is performed *in-situ* inside a transmission electron microscope (TEM) configured with a home-built mechanical/piezo manipulation stage with electrical feedthroughs to the sample.   Fig. 1 shows high resolution TEM images of a conventional arc-grown MWNT at different stages in the tip engineering or peeling

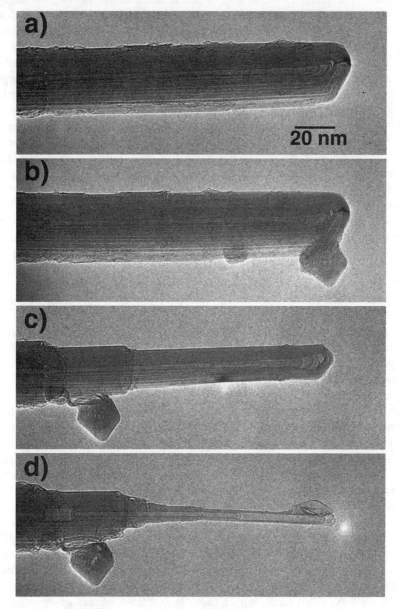

FIGURE 1. TEM images of a multiwall carbon nanotube being peeled and sharpened

process. The left end of the nanotube (beyond the figure border) is attached to a stationary zero-potential gold electrode. To the right (also not shown) is a second nanotube which serves as the "shaping electrode"; it is attached to a manipulator whose potential can be externally controlled. Fig. 1a shows the original MWNT before modification. For Fig. 1b, the shaping electrode has been momentarily brought into contact with the MWNT and a carbon onion has been inadvertently transferred from the shaping electrode to the MWNT, but no attempt has been made to shape the MWNT. For Fig. 1c, the shaping electrode has been brought into contact with the tip of the MWNT at 2.9V and 200mA; numerous layers of the MWNT have been peeled away near its end and the MWNT now has a stepped diameter and is significantly sharpened. The carbon onion has been displaced further down the tube. Most importantly, the newly exposed tip of the MWNT appears undamaged. For Fig. 1d, the peeling and sharpening process has been repeated, resulting in a MWNT with highly desirable characteristics for many nanotube applications. The dominant protruding segment now consists of a three-walled electrically conducting nanotube with a radius of just 2.5 nm. This sharpened MWNT constitutes, for example, a "near ideal" coducting AFM tip. The physical process by which only the outer MWNT shells are "blown away" by an applied electric current has important implications for charge conduction in MWNT's. It is possible that for the most part MWNTs conduct ballistically[8], but that dissipation does occur at defect sites (such as pentgons, etc). Such defects are most abundant at nanotube ends, which would lead to current-induced end-sharpening as described above.

The nested concentric shells of MWNTs interact primarily via the relatively weak van der Waals force. This suggests that MWNTs might form highly efficient linear and rotational bearings. A major difficulty in actually realizing such a construction is the commonly capped ends which seal in all inner core nanotube cylinders. Even if the MWNT ends are opened by methods such as acid etching, it is difficult to selectively contact just the core tubes and move them with respect to the outer shells. On the other hand, the peeling and sharpening process just described perfectly prepares MWNTs for bearing applications.

Fig. 2 shows schematically the configurations used inside the TEM for bearing and other mechanical experiments. The MWNT is first rigidly mounted (a), and the free end of the MWNT is then sharpened to expose the core tubes (b). In c) a nanomanipulator is brought into contact with the core tubes and, using electrical current, is spot-welded to the core. c) is the common starting point for numerous mechanical experiments. d), e), and f) show three different classes of such experiments. In d) the manipulator is moved right and left, thus telescoping the core out from, or reinserting it into, the outer housing of nanotube shells. The extration/reinsersion process can be repeated numerous times, all the while viewing the MWNT at high TEM resolution to test for atomic-scale nanotube surface wear and fatigue. In e) the manipulator first telescopes out the inner core, then fully disingages, allowing the core to be drawn back into the outer shells by the intertube van der Waals energy-lowering force. A real-time video recording of the core bundle dynamics gives information pertaining to van der Waals and frictional forces. In f), a partially-telescoped nanotube

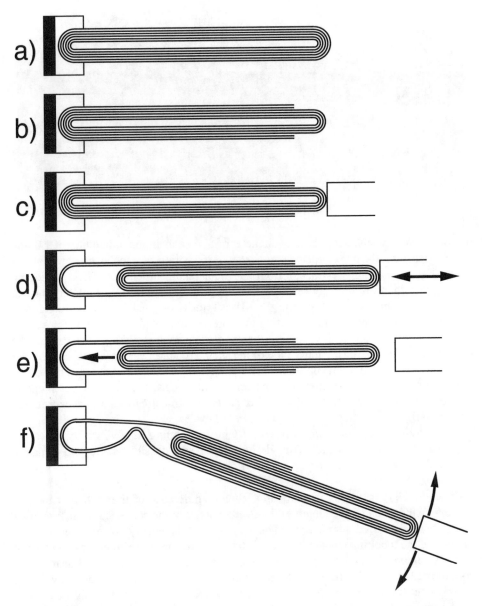

FIGURE 2. A schematic for MWNT processing and mechanical manipulations. A sliding nanobearing is illustrated in d).

is subjected to additional transverse displacements, and reversible mechanical failure modes such as buckling and complete collapse are induced.

Fig. 3 shows a TEM image of a MWNT in a fully telescoped position. Using higher resolution imaging than that used for Fig. 3, we determined that this MWNT

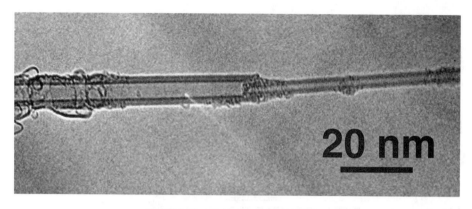

FIGURE 3. A fully telescoped MWNT.

originally had 9 walls, with an outer diameter of 8 nm and an inner diameter of 1.3 nm. After extension, a 4 nm diameter core segment (consiting of 4 concentric walls) has been almost competely extracted from the outer shell structure. The telescoping process was found to be fully reversible, in that the core could be completely pushed back into the outer shells, restoring the MWNT to its original "retracted" condition. The process of extending and retracting the core was repeated many times for several different MWNTs, and in all cases no apparent damage to the "sliding" surfaces, i.e. the outer tube of the core or the inner tube of the shell structure, was observed, even under the highest TEM resolution conditions (~2 Å). The apparent lack of induced defects or other structural changes in the nanotube contact surfaces at the atomic level suggests strongly that these near atomically-perfect nanotube structures may be wear-free and will not fatigue even after a very large number of cycles.

Several internal forces are associated with telescoping MWNTs. To first order these consist of the van der Waals-derived force and possible static and dynamic frictional forces. The van der Waals force is given by

$$F_{vdW} = -\nabla U(x) \qquad (1)$$

where $U(x) = 0.16Cx$ joules with C the circumference of the "active" nanotube bearing cylinders and x the length of the overlap between the core section and the outer walls, both measured in meters. The van der Waals energy lowering gained by increasing the tube-tube contact area tends to retract the extended core of a telescoped MWNT. Interestingly, since the active intertube contact area decreases linearly with core tube extension, this restoring force is independent of contact area, or equivalently, independent of core extension. Hence, a telescoped nanotube with only one active (sliding) surface pair is expected to act as a constant force spring.

To determine experimentally if F_{vdW} dominates nanotube linear bearing dynamics we have used the configuration described in Fig. 1e. The core tubes of a MWNT were first telescoped using the manipulator. Lateral deformations of the manipulator were used to fatigue and eventually break the spot weld, thus releasing the core segment. The resulting accelerated motion of the released core segment was recorded using a continuous video system tied to the TEM imaging electronics. The core tubes were observed to rapidly and fully retract back into the outer shells. From the dimensions of of the core tubes in one such experiment the FvdW retraction force for the nanotube in is calculated to be 9 nN. From the observation that the core spontaneously retracted, together with the experimentally determined upper bound for the retraction time, we determine that the static friction force is small, with $f_s < 2.3 \times 10^{-14}$ newtons per atom (6.6×10^{-15} newtons per Å^2), and conclude that the dynamic friction $f_k < 1.5 \times 10^{-14}$ newtons per atom (4.3×10^{-15} newtons per Å^2).

We briefly consider lateral deformations of partially telescoped nanotubes. It has been predicted and observed that nanotubes can, upon lateral deformation, form kinks[11] or even fully collapse[12]. MWNTs with large inner diameters and few concentric shells are particularly susceptible to kinking and collapse. Although kinked or fully collapsed nanotubes have been observed experimentally using static TEM[11,12] or atomic force microscopy methods[13], in-situ high-resolution controlled and reversible deformation studies have been difficult. As outlined in Fig. 1f), using partially telescoped nanotubes we can study the kinking and collapse of a controlled nanotube system. We can directly alter the inner diameter and aspect ratio of the hollow nanotube, apply lateral forces, and in real time observe resulting failure modes. As expected, we find that a MWNT will kink and collapse much more readily after the inner core has been removed. One particular nanotube had 60 original layers with an outer diameter of 43 nm, and upon telescoping a 40 layer core was pulled out up to a maximum extension of 150 nm, leaving an outer shell housing of just 20 layers with an inner diamter of 29 nm. The housing was supported at the base and the inner core section of the tube was still engaged in the housing for a length of 200 nm. When the manipulator was driven laterally to approximately ~5° angular displacement the housing shells developed a kink in the middle of large inner diameter section. At ~26° displacement the kink was severe and resembled the schematic in Figure 1f). At any displacement angle, the telescoped core section was still mobile, and could be moved back and forth inside the unkinked portion of the outer shell housing. At small kink angles less than ~10°, the core could be inserted past the kink position, forcing the kink to disappear and reinflating the outer shells to their original circular cross section. At more severe bending angles, in excess of ~20°, the kink blocked the inner core section from being fully inserted. Hence, suitable kinking of the outer shell housing provides an effective motion stop for nanotube core insertion.

ACKNOWLEDGEMENTS

This research was supported in part by the NSF Grants DMR-9801738 and CMR-9501156 and by the Director, Office of Energy Research, Office of Basic Energy Sciences, Materials Sciences Division of the U.S. Department of Energy under Contract No. DE-AC03-76SF00098.

REFERENCES

1.Freemantle, M. Filled carbon nanotubes could lead to improved catalysts and biosensors. Chemical & Engineering News 74, 62-66 (1996).

2.Britto, P. J., Santhanam, K. S. V. & Ajayan, P. M. Carbon nanotube electrode for oxidation of dopamine. Bioelectrochemistry and Bioenergetics 41, 121-125 (1996).

3.Collins, P. G., Zettl, A., Bando, H., Thess, A. & Smalley, R. E. Nanotube nanodevice. Science 278, 100-103 (1997).

4.Tans, S. J., Verschueren, R. M. & Dekker, C. Room temperature transistor based on a single carbon nanotube. Nature 393, 49-52 (1998).

5.Dai, H., Hafner, J. H., Rinzler, A. G., Colbert, D. T. & Smalley, R. E. Nanotubes as nanoprobes in scanning probe microscopy. Nature 384, 147-150 (1996).

6.de Heer, W. A., Chatelain, A. & Ugarte, D. A carbon nanotube field-emission electron source. Science 270, 1179-1180 (1995).

7.Kong, J., Soh, H. T., Cassell, A. M., Quate, C. F. & Dai, H. Synthesis of individual single-walled carbon nanotubes on patterned silicon wafers. Nature 395, 878-881 (1998).

8.Tsang, S. C., Chen, Y. K., Harris, P. J. F. & Green, M. L. H. A Simple Chemical Method of Opening and Filling Carbon Nanotubes. Nature 372, 159-162 (1994).

9.Frank, S., Poncharal, P., Wang, Z. L. & De Heer, W. A. Carbon nanotube quantum resistors. Science 280, 1744-1746 (1998).

10.Bachtold, A. et al. Aharonov-Bohm oscillations in carbon nanotubes. Nature 397, 673-675 (1999).

11.Iijima, S., Brabec, C., Maiti, A. & Bernholc, J. Structural flexibility of carbon nanotubes. Journal of Chemical Physics 104, 2089-2092 (1996).

12.Chopra, N. G. et al. Fully collapsed carbon nanotubes. Nature 377, 135-138 (1995).

13.Falvo, M. R. et al. Bending and buckling of carbon nanotubes under large strain. Nature 389, 582-584 (1997).

Storage of Energy in Supercapacitors from Nanotubes

Elzbieta Frackowiak*, Krzysztof Jurewicz*, Sandrine Delpeux,
Valérie Bertagna, Sylvie Bonnamy and François Béguin

*ICTE, Poznan University of Technology, ul. Piotrowo 3, 60-965 Poznan, Poland
CRMD, CNRS-Universite, 1B rue de la Férollerie, 45071 Orléans, France

Abstract. Different types of nanotubular carbon materials have been used for storage of energy in supercapacitors. Application of various multi-walled and single walled nanotubes as electrodes for capacitors proved the high ability of these materials for the accumulation of charges in electrode/electrolyte interface due to their unique mesoporosity. Apart from electrostatic attraction, pseudocapacitance reactions have been observed because of functionalization of nanotubes, deposition of conducting polypyrrole or presence of metallic particles. The values of capacitance obtained from modified nanotube electrodes varied from 20 to 170 F/g.

INTRODUCTION

Electrochemical capacitors based on carbon electrodes represent attractive energy storage devices with a high power density and a long durability. Storage of energy in supercapacitor combines the pure electrostatic attraction of ions in electrical double layer and pseudo-capacitance faradaic reactions. Enhancement of capacitance values by pseudo-capacitance effects can be especially realized through introducing of electroactive metallic particles and electroconducting polymers [1-3]. Application of nanotubes as capacitor electrodes [4-6] proved the high ability of this material for the accumulation of charges in the electrode/electrolyte interface due to their special morphology.

EXPERIMENTAL

Different types of multi-walled carbon nanotubes (MWNTs) have been used as electrodes for electrochemical capacitors. Decomposition of acetylene at 700^0C and 900^0C, using cobalt catalyst supported on silica, supplied two different types of nanotubes. MWNTs with an open central canal and partly amorphous carbon on the outer layers were obtained at 700^0C (MWNT/700Co) whereas nanofilaments of fishbone morphology with hardly defined central canal were elaborated at 900^0C (MWNT/900Co). Nanotubular material was purified by hydrofluoric and nitric acids that allowed to remove the silica and residual Co catalyst.

CP544, *Electronic Properties of Novel Materials—Molecular Nanostructures*, edited by H. Kuzmany, et al.
© 2000 American Institute of Physics 1-56396-973-4/00/$17.00

Commercially available nanotubes called Hyperion Graphite Fibrils[TM] have been also analysed as capacitor material without any purification. Electroconducting polypyrrole (PPy) was deposited on this material by chemical polymerization of pyrrole in acidic solution.

All the MWNT material was characterized by Transmission Electron Microscopy and nitrogen adsorption at 77K. For comparison a bucky paper from single-walled nanotubes (RICE) has been also used for assembly a capacitor. Two electrode swagelok type cells from teflon were applied using 6M KOH and 1M H_2SO_4 as electrolytic solution. Electrodes were prepared in the form of pellet with 85 wt% content of MWNTs, 5 wt% of acetylene black and 10 wt% of binding substance (PVDF- Kynar flex Atochem).

Capacitor performance has been investigated by galvanostatic charge/discharge and voltammetry techniques (VMP-Biologic, France) and impedance spectroscopy (Solartron 1260 Schlumberger) in the range of frequencies from 100 kHz to 1 mHz.

RESULTS AND DISCUSSION

Accessible network of entangled nanotubes, their highly mesoporous character confirmed by IV type nitrogen adsorption/desorption isotherm and surface functionality are at the origin of capacitance and pseudocapacitance properties. Presence of mesopores due to the open central canal and mutual entanglement facilitate transport of ions from solution to the charged interface.

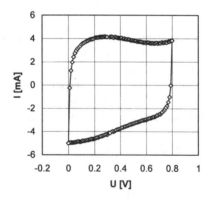

Figure 1. Voltammetry characteristics (10 mV/s) of capacitor from MWNTs obtained at 700°C and treated by nitric acid (80 °C, 1h); m=8.5 mg; electrolyte 6MKOH.

MWNT/700Co pointed out the regular rectangular shape of voltammograms with capacitance values of 80 F/g. After treatment by hot nitric acid (80°C, 1h) even if the specific surface area of the sample remains the same order of 410 m²/g, the values of capacitance increased to 136 F/g due to the presence of rich surface functionality well remarkable in Fig. 1 at around 0.2 V. Galvanostatic charge/discharge characteristics of such capacitor are shown in Fig. 2. The values of capacitance (per nanotube material) were calculated from discharge curve and have some tendency to decrease with cycling (typically for pseudocapacitance).

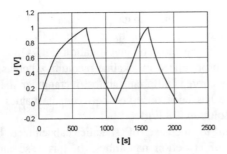

Figure 2. Galvanostatic charge/discharge characteristics of capacitor from MWNT700/Co treated by HNO$_3$; I=1mA; m=8.5 mg.

The values of capacitance for MWNT900/Co with a comparable surface area (396 m^2/g) but hardly accessible central canal were smaller, i.e. 62 F/g. It seems to confirm the role of central canal in charging the electrical double layer.

For Hyperion sample a striking behaviuor has been observed in voltammetry (Fig. 3) and galvanostatic charge/discharge characteristics (Fig. 4) due to the presence of remaining catalyst confirmed by elemental analysis (1.2 wt% Fe). This sample has BET surface area of 260 m^2/g, hence a low capacitance was obtained in 6 M KOH (20 F/g).

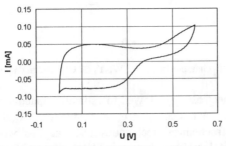

Figure 3. Voltammetry curves (2mV/s) of capacitor built from Hyperion nanotubes; m=4mg; 6M KOH

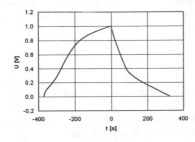

Figure 4. Galvanostatic charge/discharge of capacitor built from Hyperion nanotubes. I=0.1 mA; m=4 mg; 6M KOH

Two step charging/discharging of this capacitor is connected with electrochemical activity of iron in the range of potential from 0 to 0.3V. The same two step characteristics are remarkable in galvanostatic investigation (Fig. 4).

Effect of iron was especially pronounced in alkaline medium, whereas in acidic medium (1M H_2SO_4) the voltammetry curve is different (almost rectangular shape) because of Fe solubility. A great difference has been observed between the values of capacitance in alkaline (20 F/g) and acidic solution (78 F/g).

The trials were undertaken to increase of specific capacitance by deposition of polypyrrole on the surface of Hyperion nanotubes. In this case modified nanotubes reached values of 172 F/g.

For comparison of multi and single-walled nanotubes, the specific capacitance of SWNTs (RICE) was evaluated and the value of 40 F/g has been found.

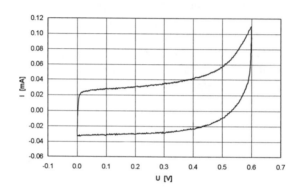

Figure 5. Voltammetry characteristics of SWNTs (RICE) at 2 mV/s; m=1mg; 6M KOH.

Results of impedance spectroscopy allowed to estimate capacitance values, resistance of capacitors, time constant and dependence on frequency. Good correlation has been found for all the electrochemical techniques used at this work. The values from impedance, calculated at 1 mHz, are comparable to these obtained from galvanostatic discharge. An oxidative treatment of nanotubes and deposition of PPy definitively increase the capacitance, however with limited durability. Performed investigations proved that nanotubes represent a quite attractive material as electrodes for capacitors even if they have very moderate surface area from 200 to 410 m^2/g. Further enhancement of capacitance through stable pseudoeffects is possible.

REFERENCES

[1] Conway, B.E., *Electrochemical Supercapacitors*, Kluwer Academic/Plenum Publishers, New York, 1999.
[2] Jurewicz, K., and Frackowiak, E., *Mol. Phys. Reports*, **27**, 36-43 (2000).
[3] Frackowiak, E., and F. Beguin, *Carbon*, 2000 submitted.
[4] Niu, C., Sichel, E.K., Hoch, R., Moy, D., and Tennet, H., *Appl. Phys. Lett.*, **70**, 1480-1482 (1997).
[5] Frackowiak, E., Metenier, K., Kyotani, T., Bonnamy, S., and Beguin, F., *Ext. Abstr. 24th Bien. Conf. on Carbon*, Charleston (USA), American Carbon Society, 1999, II, pp. 544-545.
[6] Frackowiak, E., Metenier, K., Pellenq, R., Bonnamy, S. and Beguin, F., *Electronic Properties of Novel Materials*, ed. H. Kuzmany et al, American Institute of Physics, 1999, pp. 429-432.

Report on the evening discussion: "Hydrogen storage in carbon materials"

Andrea Quintel

Max Planck Institute for Solid State Research, Heisenbergstr.1, 70569 Stuttgart, Germany

Abstract. Hydrogen may be the most important energy carrier of the future as soon as the problem of hydrogen storage is solved. Storing of hydrogen under high pressure or as liquid costs much energy. Furthermore, a high pressure or liquid hydrogen tank in a fuel cell driven vehicle would be much larger and heavier compared to a typical gasoline tank. In metal hydride tanks the stored hydrogen density is higher, but the tank would be much too heavy (for a comparison see Fig. 1). Since the first promising results of Heben et al. in 1997 on hydrogen storage in single walled carbon nanotubes and the spectacularly large storage capacities in carbon nanofibers from the Baker and Rodriguez group in 1998, considerable research activity has been started all over the world to investigate hydrogen storage in carbon materials. Especially, car industry is very interested and is waiting for a material with a reversible hydrogen storage capacity above 6.5 wt%. In this report, the evening discussion on "Hydrogen storage in carbon materials" is summarised.

CP544, *Electronic Properties of Novel Materials—Molecular Nanostructures*, edited by H. Kuzmany, et al.

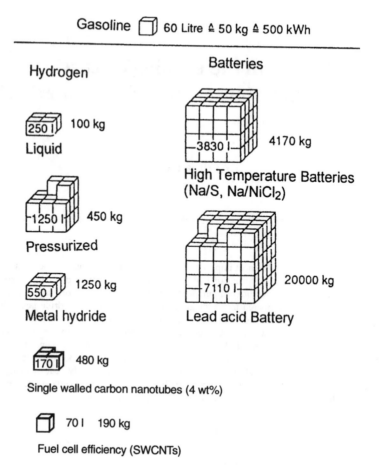

FIGURE 1. Comparison of energy storage tanks for vehicles to reach 1000 km[1-4]. The modern Li-ion batteries for laptops are not considered.

JACK FISCHER
(UNIVERSITY OF PENNSYLVANIA, USA)

Jack Fischer opened as chairman the discussion. He listed some important points, which have to be investigated on the way to a powerful hydrogen storage device: Differences between adsorption and desorption concerning the amount of hydrogen, loading and release times will influence the reversibility and cyclibility of the process. Structural modifications like alkali doping in graphite could act as promoters for the storage. What are the binding sites of hydrogen in carbon? In the case of in single walled carbon nanotubes: Is hydrogen stored inside the tubes or between them in the bundles? In real materials, defects and impurities might play important roles for the mechanism of hydrogen storage.

ANDREA QUINTEL
(MAX PLANCK INSTITUTE FOR SOLID STATE RESEARCH, GERMANY)

As second speaker, A. Quintel reviewed the publications, which appeared up to IWEPNM 2000 in this field.

Single walled carbon nanotubes (SWCNTs)

Dillon et al. [3] was the first group who could store hydrogen in SWCNTs. From thermal desorption measurements by mass spectroscopy on samples of a very low SWCNT content, they extrapolated a hydrogen storage capacity of 5-10 wt% in pure SWCNT material. The storage conditions were: 10 min under 0.5 bar hydrogen at 273K. More details were given by M. Heben himself, who was invited to tell about their newest results (see below). The second publication was in 1999 from Ye et al. [5]. At high pressure (110 bar) and low temperature (80K) they succeeded in storing 8 wt% hydrogen in purified SWCNTs, which corresponds to a ratio of ~1.0 H/C. In order to cut the tubes, disrupt the rope structure and perhaps open the tube caps, the purified SWCNT material was sonicated in dimethylformamide, until the sample was completely suspended in the solvent. A change in pressure dependence of the storage capacity was explained by a phase transition around 40 bar where the individual tubes start to separate from each other and hydrogen is physisorbed on their exposed surfaces. The hydrogen adsorption and desorption isotherms were measured by volumetry (Sieverts apparatus). In the same year Liu et al. [6] stored up to 4,2 wt% hydrogen at 100 bar and 300K in arc discharge material of an estimated SWCNT purity of 50 wt%. The highest storage capacity they measured by volumetry in samples which were pre-treated by soaking in conc. HCl and vacuum annealing. They claimed that the enlarged tube diameter of 1,85 nm favours the hydrogen storage. 3.3 wt% hydrogen could be released at 1 bar and room temperature, whereas to desorb the remaining 0.9 wt%, the sample had to be heated up to 473K.

Carbon nanofibers (CNF)

In 1998 Chambers et al. [7] published extremely large values of up to 67 wt% hydrogen storage capacity in herringbone carbon nanofibers. However, these values have never been confirmed by others, e.g. [8]. Fan et al. [9] found 1999 much lower (but still fairly large) hydrogen storage capacities of 10-13 wt% in CNFs. As in their work on SWCNTs [6] the CNFs were pre-treated by soaking in conc. HCl and vacuum annealing. Again, a volumetric setup was used to measure the hydrogen adsorption and desorption. Meanwhile, they published a review article on their work in this field [10]. There the storage capacities in SWCNTs are the same as in their Science paper but their values of hydrogen storage in CNFs decreased significantly by factors of 2-3. A group in Singapore, Chen et al. [11], measured in alkali-doped herringbone CNFs and even in alkali-doped graphite hydrogen storage capacities up to 20 wt% (in Li-doped herringbone CNF, storage conditions: 1 bar, 673K). All values are summarised in Tab. 1. For doping a solid state reaction between carbon and alkali salts was used. The storage/release cycles were measured as weight changes (thermogravimetry). Keeping the sample for a long time under hydrogen flow near the maximum of the weight hysteresis, the high values in Tab. 1 were achieved. Recent efforts to reproduce their results using the same commercially available graphite material are presented in the contribution of M. Hirscher.

TABLE 1:   Comparison of H_2 storage properties of various carbon materials

Material	Max wt% H_2	T [K]	p [MPa]	Reference
SWCNTs (low purity)	5 - 10	273	0,05	[3]
SWCNTs (high purity)	8,25	80	11,0	[5]
SWCNTs (~50 wt%)	4,2	300	10 - 12	[6]
SWCNTs (98 wt%)	6,5 - 7,5	300	0,05	M. Heben, IWEPNM 2000
CNFs (tubular)	11,26	298	11,35	[7]
CNFs (herringbone)	67,55	298	11,35	[7]
CNFs (platelet)	53,68	298	11,35	[7]
Graphite	4,52	298	11,35	[7]
CNFs (herringbone)	10 - 13	300	8 - 11	[9]
CNFs (herringbone)	0,4	298 - 773	0,101	[11]
Li-CNFs	20,0	473 ~ 673	0,101	[11]
Li-Graphite	14,0	473 ~ 673	0,101	[11]
K-CNFs	14,0	< 313	0,101	[11]
K-Graphite	5,0	< 313	0,101	[11]

MICHAEL HEBEN
(NATIONAL RENEWABLE ENERGY LABORATORY, USA)

M. Heben started his talk by reviewing their first results [3]: For comparison, arc discharge material and activated carbon were loaded for 10 min with 0,5 bar hydrogen at 273K. After cooling to 133K, the sample chamber was evacuated, which lowered the temperature to 90K. Then the sample was heated with a heating rate of 1K/sec and the desorbing hydrogen was detected by mass spectroscopy (thermal desorption spectroscopy). At 288K, a small and broad desorption maximum occurred only in the SWCNT material after heating in vacuum to 970K. The corresponding hydrogen storage capacity was 0,01 wt%. With an estimated SWCNT content of 0,1 - 0,2 wt%, they extrapolated a hydrogen storage capacity of 5 - 10 wt% in pure SWCNTs. After this very promising result, they searched for a material with a higher SWCNT content. For this purpose, they produced a new SWCNT material by a special laser ablation technique [12] and developed a 3 step purification method [13]: i) oxidation in 3M nitric acid to remove the catalyst (16h at 120°C), ii) oxidation in air (30min at 550°C) to remove the amorphous carbon and iii) short vacuum annealing at 1500°C to reassemble the SWCNT bundles. By this purification method the SWCNT content increased from ~ 20 - 30 wt% in their raw material to 98 wt% in the purified samples [1]. Sonication of this 98 wt% pure SWCNT material in conc. nitric acid (ultrasound power: $100W/cm^2$) disordered the bundle structure in a way, that the NT bundles are shortened and the tubes seemed to be not densely packed any more. If the NTs are opened by this cutting procedure, could not be decided due to the low resolution of their TEM pictures. Directly after a short vacuum annealing of this pure and cut SWCNTs (to 1000K with 1K/sec), 0.5 bar hydrogen was exposed at room temperature for 2 min. After cooling to LN_2 temperatures, the sample was heated with a constant heating rate of 1K/sec and the hydrogen desorption was detected by mass spectroscopy (setup see [2]). Two desorption maxima were observed: the first around 400K with a desorbing amount of 2,5 wt% hydrogen and the second around 600K with a desorbing amount of 4 - 5 wt% hydrogen. Cycling up to 10 times showed that totally a reversible hydrogen storage capacity of 6,5 to 7,5 wt% (with a volumetric hydrogen density of ~ 35 kgH_2/m^3) can be achieved in

---

[1] Applying this 3 step purification method to the commercial arc discharge product CarboLex AP grade led to a SWCNT purity below 5 wt% and a metal content greater than 25 wt%. They argued, that their purification method only works for their laser ablation material, where the catalytic particles are not encapsulated in amorphous carbon. Only then the catalyst could be removed completely by the first oxidation step. Remaining catalytic particles could act as hot spots in the heating steps and could therefore reduce the SWCNT content. The SWCNT contents were determined by transmission electron microscopy and thermogravimetry.

[2] setup: the sample is wrapped in a Platinum foil, which is placed at the end of a copper rod. The Platinum can be heated up to 1200°C by resistivity heating. The other end of the copper rod can be cooled to liquid nitrogen temperature outside the high vacuum chamber. The chamber can be filled with 1 bar of different gases (H_2O, H_2, O_2). A mass spectrometer is connected to measure gas desorption as function of time and temperature, respectively.

98 wt% pure and cut SWCNTs. The first physisorbed 2,5 wt% will desorb, if the sample is evacuated at room temperature. Cutting and hydrogen loading of as-prepared laser ablation material led to a total hydrogen storage capacity of ~ 0.2 wt% (estimated from the shown desorption curve). Without cutting, no hydrogen adsorption occurred at all. The activation by vacuum annealing to 1000K just before hydrogen storage (at high vacuum: $<10^{-7}$ bar in the same setup!) is needed in order to remove CO_2, which blocks the adsorption of hydrogen. While H_2O is competing with H_2 for probably the same sites, the CO_2 shows a different desorbing behaviour. Exposing a hydrogen loaded sample to air for a long time (months) and measuring the desorption after evacuation shows, that first around 400K the CO_2 desorbs. The desorption of H_2 shifts to higher temperatures (around 650K), which means, that CO_2 blocks also the desorption of hydrogen. Raman investigations of the C-C stretching mode showed that there might be a charge transfer between C and H_2 in the high temperature site. M. Heben assumed, that the high temperature site is something between physisorption and chemisorption. In order to learn more about the adsorption/desorption kinetics they start with low pressure Sieverts volumetry. To investigate the nature of the binding sites, in situ resistivity, IR and Raman spectroscopy measurements are planned.

MICHAEL HIRSCHER
(MAX PLANCK INSTITUTE FOR METAL RESEARCH, GERMANY)

In a German hydrogen storage project, financed by the German ministry for education and research (BMBF), the work was first concentrated on reproducing the experiments of Chen et al.. R. Ströbel and L. Jörissen at the centre for solar energy and hydrogen research in Ulm succeeded in verifying the weight change behaviour for Li-doped graphite. Because of the same particle size of < 50μm in the graphite from Merck, one can assume, that the experiments were made with the same material as Chen et al. The reaction steps of the solid state reaction for doping were elucidated by X-ray diffraction and mass spectroscopy. After heating the sample to 800°C for some hours, only a mixture of Li_2O and graphite remains in the product (and no Li-intercalated graphite!). From literature, it is well known, that Li_2O can take up H_2O to form $Li_2O*(H_2O)$, which is stable just around 450°C. This is the temperature where the highest weight uptake occurred in the thermogravimetric experiments from Chen et al. and R. Ströbel et coworkers. Therefore most of the weight changes can be explained by adsorption and desorption of H_2O and not of H_2. Our complete investigations will be published elsewhere. Just before this winterschool, Yang et al. [14] published his studies on reproduction of the thermogravimetric results from Chen et al.. The same large weight changes as claimed by Chen et al., he found only for wet hydrogen flow, whereas for dry hydrogen flow, only a small hydrogen storage capacity below 2 wt% remained in the case of alkali-doped herringbone CNFs.

Parallel to these reproduction experiments hydrogen storage in SWCNT material from the group of P. Bernier in Montpellier was investigated. As-produced and mechanically cut SWCNT samples were exposed to a constant hydrogen flow at 523K for 1h. After cooling to room temperature and evacuation, the sample was heated with a constant heating rate of 3.8 K/min and the hydrogen desorption was detected by mass spectroscopy (thermal desorption spectroscopy, setup comparable to Heben's). In the case of cut SWCNTs, 0.05 wt% hydrogen released around 500K and 0.15 wt% around 700K. For as-produced SWCNTs, the hydrogen desorption curve showed no significant maxima and the totally released amount was one order of magnitude smaller. Compared to Heben's results, a total hydrogen storage capacity of 0.2 wt% seems to be very small, but one has to take into account, that no purification and no activation was done before storage. Furthermore, the efficiency of the cutting process in unknown (see contribution of P. Bernier below). The BMBF project is now focussing on the optimisation of the hydrogen storage capacity in SWCNTs.

TSUYOSHI SUGIMOTO
(TOYOTA, JAPAN)

Toyota is working on hydrogen storage in metal alloys and activated carbon. In the first part of his talk, Sugimoto focused on the question: Is the hydrogen stored as H_2 molecule or is it dissociated to H atoms? They investigate this problem by the help of the gas mixture equilibrium $H_2 + D_2 \leftrightarrow 2HD$. Charging the samples by a mixture of $H_2 + D_2$ and measuring the H_2, D_2 and HD desorption by mass spectroscopy, they can distinguish between atomic H storage as metal hydride in $LaNi_5$ (strong HD formation) and molecular H_2 storage in activated carbon (weak HD formation). In the second part of his talk, Sugimoto presented grand canonical Monte Carlo simulations of hydrogen storage in graphite. The variation of the layer distance led to a minimum layer distance of 5,2 Å and an optimum interval of 6 - 8 Å layer distance for hydrogen storage in graphite (273K, 10 bar). They conclude, that pure graphite will not adsorb H_2 and therefore intercalated graphite is of interest. In activated carbon a maximum H_2 storage capacity of 4 wt% at -200 °C and 40 bar was found experimentally as well as by simulation. Furthermore, they simulated hydrogen storage in a bundle of 4 SWCNTs as function of the wall to wall distance. From the simulated values, they conclude, that SWCNT based materials could be very promising for hydrogen storage.

PATRICK BERNIER
(UNIVERSITY OF MONTPELLIER, FRANCE)

He contributed to the discussion in order to explain their new cutting procedure, used for the samples, which M. Hirscher presented as promising candidates for hydrogen storage. By now all investigations in this field indicate, that cutting of the tubes plays a key role in activating a SWCNT material for hydrogen storage. From literature, chemical cutting by a combination of acid treatment and sonication or thermal treatment and oxidation is known. In Montpellier, a mechanical cutting technique was developed. The idea is, that nanodiamonds act as kind of scissors to cut the tubes. The nanodiamonds are fixed as thin layer on an abrasive paper or suspended in an aqueous solution. Rubbing between two sheets of nanodiamond paper or sonication in nanodiamond suspension cut an unknown fraction of tubes in the arc discharge material, produced in Montpellier. In the presented HRTEM pictures, partially broken bundles of SWCNTs were clearly visible. In one HRTEM picture, no caps on the free ends of some broken tubes could be resolved. This was interpreted as a strong evidence, that these tubes are open.

OTTO ZHOU
(UNIVERSITY OF NORTH CAROLINA, USA)

During their investigations of electrochemical intercalation of Li into SWCNT material, they studied ball milling to open the tubes. After ball milling (1 - 20 min) a lot of individual nanotubes with damaged ends occurred in TEM pictures. The TEM resolution was not high enough to decide, if the tubes were open or not, but in ball milled samples the Li charging capacity was improved.

O. Zhou claimed, that nuclear magnetic resonance (NMR) would be a powerful method to study hydrogen storage in carbon materials. In the case of SWCNTs, he emphasised, that magnetic catalysts will influence the NMR signal. In preliminary experiments, they loaded a raw laser ablation SWCNT powder with 0,5 bar hydrogen at room temperature and measured in situ the NMR response. Besides the hydrogen gas peak, they measured a broad peak of hydrogen in carbon, which increased in time. From that, they calculated a maximum value of 0,3 wt% hydrogen storage capacity after 9 h loading.

GOTTHARD SEIFERT
(UNIVERSITY OF PADERBORN, GERMANY)

From a theoretical point of view he asked the question: Can hydrogen pass through the walls of SWCNTs? With a tight binding model, he found two stable hydrogenated SWCNTs, an armchair form, where all H are bound from the outside to the SWCNT and a zigzag form, where the C-H bonds are standing alternating to the outside and inside of the tube. In the armchair form, he observed a "flip in" process with a low barrier of ~ 1 eV, where the H-C bond flips in from the outside to the inside of the tube. In this situation, an additional H from the surrounding could bind from the outside to the same C by kicking the H inside away from the C. This free H could recombine with another free H in the inside. At the end the tube will be filled with H_2 molecules, which came in by a three step process: "flip in", "kick in" and recombination. He calculated a total hydrogen storage capacity of 14 wt% for this loading mechanism through the wall.

WERNER ZITTEL
(L-B-SYSTEMTECHNIK, GERMANY)

As representative form industry, he emphasised, that storage is the only missing link for mobile application of hydrogen as secondary energy carrier. Liquid or high pressure tanks would need complicated infrastructures for filling the tank. As example of the